BIOREMEDIATION AND NATURAL ATTENUATION

ENVIRONMENTAL SCIENCE AND TECHNOLOGY

A Wiley-Interscience Series of Texts and Monographs

Edited by JERALD L. SCHNOOR, *University of Iowa*
ALEXANDER ZEHNDER, *Swiss Federal Institute for Water Resources and Water Pollution Control*

A complete list of the titles in this series appears at the end of this volume.

BIOREMEDIATION AND NATURAL ATTENUATION

Process Fundamentals and Mathematical Models

Pedro J. J. Alvarez
Rice University

Walter A. Illman
University of Iowa

A JOHN WILEY & SONS, INC., PUBLICATION

For general information on our other products and services please contact our Customer Care Department
within the U.S. at 877-762-2974, outside the U.S. at 317-572-3993 or fax 317-572-4002.

Wiley also publishes its books in a variety of electronic formats. Some content that appears in print,
may not be available in electronic format. For more information about Wiley products, visit our web site
at www.wiley.com

Library of Congress Cataloging-in-Publication Data:

Alvarez, Pedro J. J.
  Bioremediation and natural attenuation : process fundamentals and mathematical models /
Pedro J. J. Alvarez, Walter A. Illman.
    p. cm. – (Environmental science and technology)
  Includes index.
  ISBN-13 978-0-471-65043-0 (cloth)
  ISBN-10 0-471-65043-9 (cloth)
  1. Groundwater–Purification. 2. Groundwater flow–Mathematical models. 3.
In situ bioremediation. 4. Hazardous wastes–Natural attenuation. I. Illman,
Walter A. (Walter Arthur) II. Title. III. Series.
  TD426.A483 2006
  628.5–dc22   2005005175

10 9 8 7 6 5 4 3 2 1

CONTENTS

Preface   ix

1. Introduction to Bioremediation   1

2. Geochemical Attenuation Mechanisms   25

3. Biodegradation Principles   49

4. Fundamentals of Groundwater Flow and Contaminant
   Transport Processes   115

5. Fate and Transport Equations and Analytical Models for
   Natural Attenuation   169

6. Numerical Modeling of Contaminant Transport, Transformation,
   and Degradation Processes   201

7. Field and Laboratory Methods to Determine Parameters for
   Modeling Contaminant Fate and Transport in Groundwater   283

8. Bioremediation Technologies   351

9. Performance Assessment and Demonstration of Bioremediation
   and Natural Attenuation   457

Appendix A. Chemical Properties of Various Organic Compounds   527

Appendix B. Free Energy and Thermodynamic Feasibility
            of Chemical and Biochemical Reactions   535

Appendix C. Commonly Used Numerical Groundwater Flow and
            Solute Transport Codes   543

Appendix D. Nonparametric Statistical Tests for Determining the Effectiveness of Natural Attenuation   **551**

Appendix E. Critical Values of the Student *t*-Distribution   **557**

Glossary   **559**

Index   **577**

PREFACE

Groundwater represents about 98% of the available fresh water of the planet. Thus, protecting and restoring groundwater quality is of global strategic importance. One common threat to groundwater resources is soil and aquifer contamination by hazardous wastes. This widespread problem represents a significant technical and economical challenge because underground contamination is difficult to locate and remove by traditional extraction and excavation methods. Consequently, there is a need for wider application of cost-effective, in situ remediation approaches that take advantage of natural phenomena, such as bioremediation and natural attenuation.

This book aims to provide fundamental biological, chemical, mathematical, and physical principles related to the fate and transport of hazardous wastes in aquifer systems, and their connection to natural attenuation and bioremediation engineering. The book is based on the authors' extensive experience as educators, researchers, and consultants in environmental biotechnology and hydrogeology and is meant to serve as a textbook for advanced undergraduate or graduate students in environmental engineering and related sciences who are interested in the selection, design, and operation of groundwater treatment systems. This work is also intended to serve as a reference book for practitioners, regulators, and researchers dealing with contaminant hydrogeology and corrective action. Thus, recent advances and new concepts in bioremediation are emphasized throughout the book.

It is difficult to write a book about emerging technologies that have not reached pedagogical maturity, as is the case with bioremediation and natural attenuation. The implementation of these technologies has experienced a relatively rapid growth in the past ten years, from negligible levels to more than 15% of all hazardous waste site remediation approaches. Furthermore, monitored natural attenuation has been selected for managing more than 50% of all sites contaminated by petroleum product releases from leaking underground storage tanks. This recent increase in demand suggests the need for a textbook that organizes, synthesizes, and analyzes scientific and technological information to rapidly discern the applicability, merits, and limitations of these innovative technologies. Such a text is useful not only for process selection and performance evaluation, but also to enhance the acceptance of bioremediation and natural attenuation as bona fide mainstream technologies. Many decision makers have historically considered the application of biological processes to treat natural systems as unpredictable and unreliable. Such erroneous perception dates back to the origins of bioremediation, when failures to meet the expectations raised by technology salespeople occurred sporadically due to inadequate understanding of process fundamentals and inappropriate process design and implementation. We hope that this book will address such deficiencies and stimulate the optimization and innovation of environmental biotechnologies.

We are very grateful for the encouragement and stimulating discussion and reviews provided by our friends and colleagues, Richard Heathcote, Joe Hughes, John McCray, Larry Nies, Gene Parkin, Michelle Scherer, Jerry Schnoor, Richard Valentine, Timothy Vogel, Herb Ward, and You-Kuan Zhang. We thank Andy Craig for making many of the illustrations in the book. We also would like to thank our students in environmental engineering and earth sciences for their editorial contributions and intellectual sparring. As engineering educators, we strive to build bridges for students to link theory and sustainable practice and hope that some of our students will maintain these bridges and build better ones for future generations.

PEDRO J. J. ALVAREZ
WALTER A. ILLMAN

Rice University
University of Iowa

1

INTRODUCTION TO BIOREMEDIATION

"Water, water everywhere, nor any drop to drink."
   —*The Rime of the Ancient Mariner*, Samuel Taylor Coleridge
"All the water that will ever be is, right now."
   —*National Geographic*, October 1993

1.1. ENVIRONMENTAL CONTAMINATION BY HAZARDOUS SUBSTANCES: MAGNITUDE OF THE CONTAMINATION PROBLEM

Throughout human history, many societies have developed and thrived at the expense of inefficient and unsustainable exploitation of the environment. In the twentieth century, the tension that existed between civilization and nature grew disproportionately. For example, about 39–50% of the land surface of our planet was modified due to human activities such as agriculture and urbanization, and the atmospheric CO_2 concentration increased by 40% over the past 140 years (mainly due to hydrocarbon combustion and deforestation), which raises serious concerns about global warming (Intergovernmental Panel on Climate Change 2001). Biodiversity has also been significantly impacted, and more than 20% of bird species have become extinct in the last 200 years (Wilson 2002).

Many natural resources show some degree of anthropogenic impact, including the widespread contamination of groundwater aquifers by hazardous wastes. This is particularly significant because groundwater represents about 98% of the available fresh water of the planet (Figure 1.1). Table 1.1 provides some statistics on the magnitude of the environmental contamination problem in the United States, and Table 1.2 summarizes the main sources of groundwater contamination. The fact that we are already using about 50% of readily available fresh water makes groundwater protection and cleanup of paramount importance.

Bioremediation and Natural Attenuation: Process Fundamentals and Mathematical Models
By Pedro J. J. Alvarez and Walter A. Illman   Copyright © 2006 John Wiley & Sons, Inc.

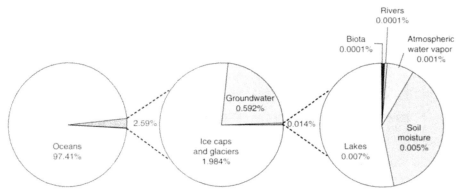

Figure 1.1. Distribution of the world's water. Approximately 70% of the Earth is covered by water, and fresh water accounts for less than 3% of the total amount of water available (including ice caps). Less than 1% of the world's fresh water (or 0.01% of all water) is usable in a renewable fashion. If ice caps and glaciers are ignored, groundwater represents about 98% of the available fresh water. (From Speidel and Agnew 1988.)

Table 1.1. Magnitude of the Hazardous Waste Contamination Problem in the United States

- United States generates 100 million tons of hazardous wastes per year.
- In 1987, 4 million tons of toxic chemicals was released to streams, 0.87 million tons was discharged to wastewater treatment plants, 1.2 million tons was disposed in landfills, 1.5 million tons was injected into deep wells for disposal.
- There are 300,000–400,000 sites in the United States that are highly contaminated by toxic chemicals and require remedial action.
- As of 1997, cleanup activities had not began at 217,000 of these sites.
- Forty million U.S. citizens live within 4 miles of a Superfund site (there are thousands of Superfund sites).
- About 440,000 out of 2 million underground tanks storing gasoline in the United States have leaked (11 million gallons of gasoline leaked every year in the U.S., and BTEX account for 60% of soluble components).
- Fifty percent of U.S. population drinks groundwater, and 1–2% of readily available groundwater is contaminated (1.2 trillion gal of contaminated groundwater infiltrates every day).
- According to a 1982 EPA survey, 20% of drinking water wells showed contamination by synthetic organic chemicals. Approximately 30% of 48,000 public drinking water systems serving populations in excess of 10,000 were contaminated, and 3% of these systems had groundwater contamination in excess of EPA standards.
- The estimated cost of environmental cleanup and management for the United States is on the order of $1 trillion.

Sources: LaGrega et al. (1994) and NRC (1994, 1997).

Table 1.2. Principal Sources of Groundwater Contamination in the United States

- Leaking underground storage tanks (450,000)
- Municipal solids and hazardous waste landfills (93,000)
- Hazardous waste management sites (18,000)
- Unlined pits, ponds, and lagoons (180,000)
- Household septic systems (20 million)
- Pesticide application areas (75×10^6) tons/yr in Iowa)
- Abandoned petroleum wells
- Saltwater intrusion along the U.S. coastline
- Surface spills

Sources: LaGrega et al. (1994) and NRC (1994, 1997).

Remediation costs for sites contaminated with hazardous wastes in Europe are expected to exceed $1.5 trillion in the near future (ENTEC 1993). In the United States, the Office of Technological Assessment (OTA) of the U.S. Congress estimates that the cost of cleaning up more than 300,000 highly contaminated sites will exceed $500 billion (National Research Council (NRC) 1994). This does not include costs associated with about 440,000 sites impacted by gasoline releases from leaking underground storage tanks (U.S. Environmental Protection Agency (USEPA) 2003) or about 19,000 landfill sites used for disposal of municipal and industrial wastes (USEPA 1989). Thus, there is an urgent need for cost-effective treatment approaches.

Bioremediation—which will be broadly defined here as *a managed or spontaneous process in which biological catalysis acts on pollutants, thereby remedying or eliminating environmental contamination present in water, wastewater, sludge, soil, aquifer material, or gas streams*—holds great potential as a practical and cost-effective approach to solve a wide variety of contamination problems. Therefore, it is expected that bioremediation will play an increasingly important role in the cleanup of soils, sediments, and groundwater contaminated with hazardous organic chemicals.

1.1.1. Common Groundwater Pollutants

Groundwater contamination by hazardous substances is commonly the result of accidental spills that occur during production, storage, or transportation activities. Table 1.3 lists the top 25 hazardous groundwater contaminants in North America and Europe. Appendix A includes a more exhaustive list of organic pollutants that contaminate groundwater aquifers, including their physicochemical properties. As will be discussed in the following chapters, many of these "traditional" hazardous wastes can be degraded by a wide variety of chemical (Chapter 2) and biological mechanisms (Chapter 3), and bioremediation can be utilized to enhance their removal to different degrees of efficacy.

Table 1.3. The 25 Most Frequently Detected Priority Pollutants at Hazardous Waste Sites in North America and Europe

1. Trichloroethylene (TCE)	14. Cadmium (Cd)
2. Lead (Pb)	15. Magnesium (Mg)
3. Tetrachloroethylene (TCE)	16. Copper (Cu)
4. Benzene	17. 1,1-Dichloroethane (1,1-DCA)
5. Toluene	18. Vinyl chloride
6. Chromium (Cr)	19. Barium (Ba)
7. Dichloromethane (DCM)	20. 1,2-Dichloroethane (1,2-DCE)
8. Zinc (Zn)	21. Ethylbenzene (EB)
9. 1,1,1,-Trichloroethane (TCE)	22. Nickel (Ni)
10. Arsenic (As)	23. Di(ethylhexyl)phthalate
11. Chloroform (CF)	24. Xylenes
12. 1,1-Dichloroethene (1,1-DCE)	25. Phenol
13. 1,2-Dichloroethene (1,2-DCE)	

Source: NRC (1994).

The most common classes of organic groundwater pollutants include aromatic hydrocarbons (Figure 1.2), chlorinated solvents (Figure 1.3) and pesticides (Figure 1.4). Common inorganic groundwater pollutants include nitrate (NO_3^-), arsenic (As), selenium (Se), and toxic heavy metals such as lead (Pb), cadmium (Cd), and chromium (Cr^{6+}). Such metals and radionuclides (e.g., uranium, technetium, and strontium) cannot be destroyed by microorganisms. Thus, their removal does not commonly rely on bioremediation. However, as discussed in Chapter 3, many microorganisms can mediate oxidation–reduction reactions that decrease the toxicity and/or mobility of these inorganic species, resulting in enhanced risk reduction. The following discussion will focus on common priority pollutants that are treated using bioremediation.

Petroleum Hydrocarbons

The extensive use of petroleum hydrocarbons as fuels and industrial stock has resulted in widespread soil and groundwater contamination by this group of contaminants. Common sources of contamination include leaking underground storage tanks, pipelines, oil exploration activities, holding pits near production oil wells, and refinery wastes.

Petroleum hydrocarbons comprise a diverse group of compounds, including alkanes, alkenes, and heterocyclic and aromatic constituents. Jet fuel, for example, typically contains more than 300 different hydrocarbons. Gasoline, which is a very common groundwater pollutant that is amenable to bioremediation, is also a complex mixture. The most abundant gasoline constituents are generally isopentane, *p*-xylene, *n*-propylbenzene, 2,3-dimethylbutane, *n*-butane, *n*-pentane, and toluene, which together make up over 50% of the mixture. However, the most important

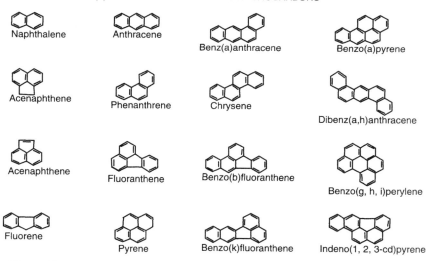

Figure 1.2. Common aromatic hydrocarbons that contaminate soils and groundwater aquifers.

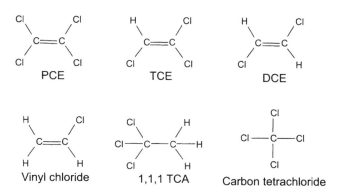

Figure 1.3. Common chlorinated solvents found in contaminated groundwater aquifers.

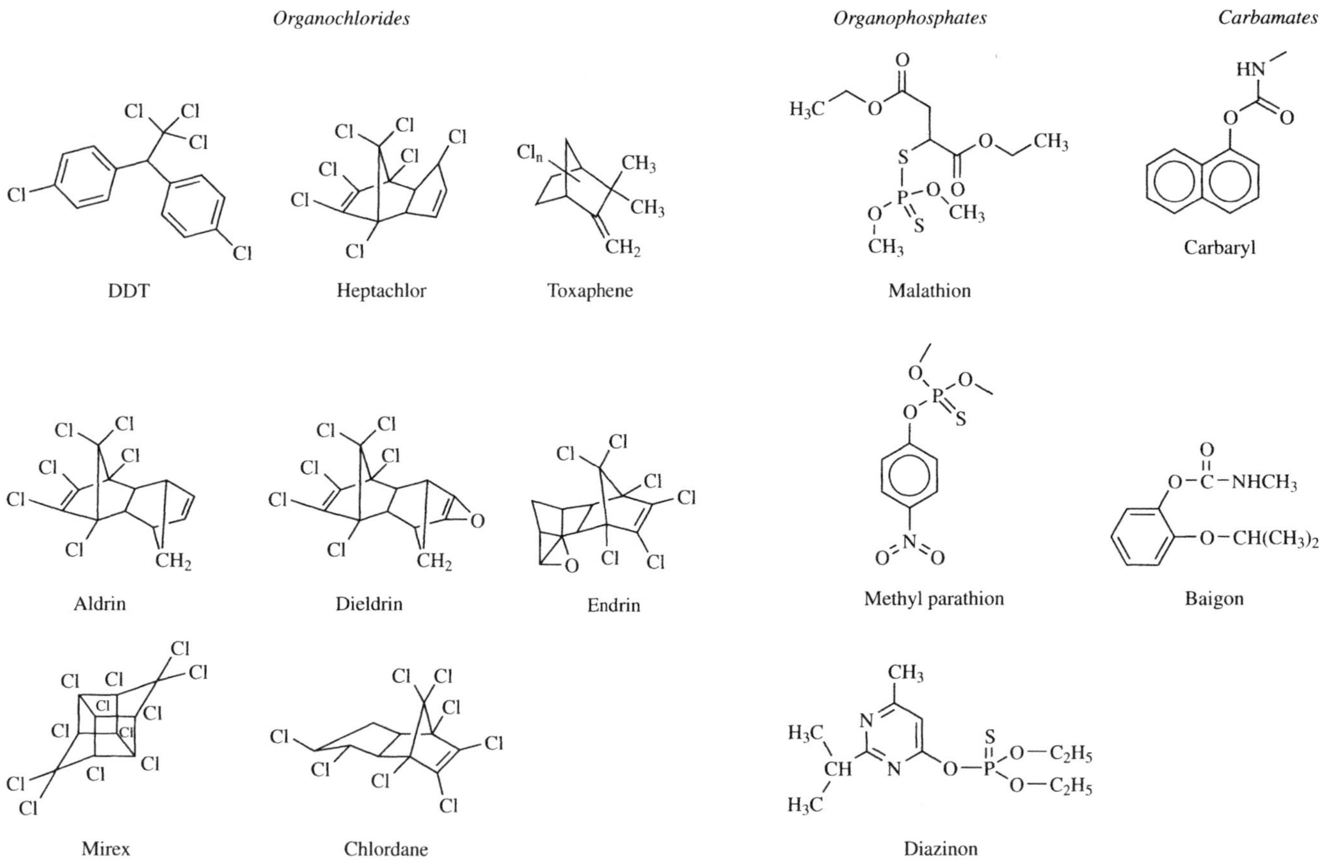

Figure 1.4. Selected pesticides found in contaminated groundwater aquifers.

constituent from a public health risk perspective is probably benzene, which is highly soluble (and thus highly mobile in aquifers) and is a known human carcinogen. Indeed, the drinking water standard for benzene (5 µg/L) is much more stringent than that of other monoaromatic hydrocarbons such as toluene (1000 µg/L) and xylenes (10,000 µg/L), and the presence of benzene is often the driving force for the bioremediation of gasoline-contaminated sites.

Another important group of pollutants are the polynuclear aromatic hydrocarbons (PAHs), which are commonly found near coal conversion facilities and petroleum plants. These hydrophobic pollutants are of concern to both public and environmental health because of their tendency to concentrate in food chains (McElroy et al. 1989) and acute toxicity (Heitkamp and Cerniglia 1988), and some PAHs (e.g., benzo[a]pyrene) are recognized mutagens and carcinogens (Mortelmans et al. 1986). Consequently, the U.S. Environmental Protection Agency (EPA) has listed 16 PAH compounds as priority pollutants (Keith and Telliard 1979) (Figure 1.2b). PAHs are the principal constituents of creosote, which is a complex mixture of about 200 compounds also containing phenolic and heterocyclic pollutants.

Hydrocarbons are generally lighter than water and tend to float on top of the water table if present in a separate organic phase—the so-called light nonaqueous-phase liquid (LNAPL) (Figure 1.5). However, different hydrocarbons in the mixture exhibit different physicochemical properties that affect their transport, fate, and principal removal mechanism. For example, short-chain alkanes tend to be volatile and are readily stripped from groundwater whereas monoaromatic hydrocarbons such as benzene, toluene, ethylbenzene, and xylenes (which are collectively known as BTEX) (Figure 1.2a) tend to be relatively soluble and are transported over longer distances by groundwater. In fact, BTEX typically represent about 60% of the water-soluble hydrocarbons in gasoline. Furthermore, dissimilar hydrocarbons exhibit different levels of resistance to biodegradation. Thus, biodegradation and abiotic weathering processes (e.g., volatilization, sorption and dilution) result in differential removal of specific hydrocarbons, which changes the relative composition of the hydrocarbon mixture over time.

Chlorinated Compounds

Chlorinated aliphatic and aromatic compounds make up an important group of organic pollutants that are both ubiquitous and relatively persistent in aquifers. Chlorinated ethenes fall into a class of chemically stable compounds commonly known as "safety solvents." Because they are resistant to combustion and explosion, these compounds were widely used as industrial solvents and degreasers for most of the twentieth century. The combination of extensive use, volatility, and chemical stability has led to widespread contamination of groundwater and soil by such ubiquitous and recalcitrant pollutants. Common volatile organic compounds (VOCs) in the chlorinated solvents group include tetrachloroethylene (PCE, $CCl_2=CCl_2$), trichloroethylene (TCE, $CCl_2=CHCl$), dichloroethylene (DCE, $CHCl=CHCl$), and vinyl chloride or chloroethylene (VC, $CH_2=CHCl$)

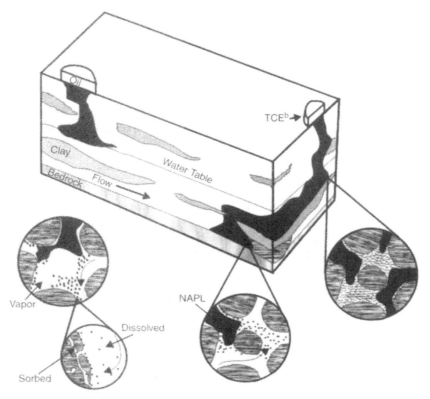

Figure 1.5. Schematic diagram showing two types of nonaqueous-phase liquid (NAPL) spills in an alluvial aquifer. Oil has a lower specific gravity than water and floats on the water table, forming a light nonaqueous-phase liquid (LNAPL). Trichloroethylene (TCE), on the other hand, is heavier than water and sinks, forming a dense nonaqueous-phase liquid (DNAPL). The dissolved phase travels with flowing groundwater. (Adapted from Pankow and Cherry 1996.)

(Figure 1.3). All of these VOCs are potential carcinogens. Groundwater contamination by 1,1,1-trichloroethane (TCA) and chlorinated methanes, such as carbon tetrachloride (CCl_4) and chloroform ($CHCl_3$), is also common.

Chlorinated solvents generally have a higher specific gravity than water and tend to sink to the bottom of the aquifer if present in a separate organic phase—the so-called dense nonaqueous-phase liquid (DNAPL) (Figure 1.5). These DNAPLs represent a major challenge to site remediation due to their persistence and relative inaccessibility.

Unlike the chlorinated solvents, chlorinated aromatic compounds such as hexachlorobenzene and pentachlorophenol (which are common fungicides used as wood preservers) or polychlorinated biphenyls (PCBs, which were common dielectric fluids in transformer oil) are similar to PAHs in terms of their potential carcinogenicity and lipophilic nature (i.e., high affinity for fatty tissue), which is conducive to bioaccumulation. These compounds also have a strong tendency to sorb to soil and aquifer sediments, and their dispersal is often due to cotransport with sorbents such as colloidal matter or eroded sediments.

Pesticides

Agricultural applications of pesticide represent an important source of soil and groundwater pollution. Pesticides that contaminate aquifers are often insecticides and herbicides, although some fungicides and rodenticides are also found. Pesticides are problematic because they are designed to be persistent (for long-lasting action), and many are lipophilic—often accumulating in animal's fatty tissue through food webs.

According to their chemical structure, pesticides can be classified as organochlorides, organophosphates, and carbamates (Figure 1.4). Organochlorides such as DDT, Aldrin, Chlordane, Endrin, Dieldrin, Heptachlor, Mirex, and Toxaphene represent an early generation of pesticides that are characterized by their environmental persistence and high toxicity. Modern pesticides tend to be organophosphates and carbamates. Organophosphates such as Malathion, Methyl parathion, and Diazinon are not as persistent in the environment but are more toxic to humans and can be absorbed through the skin, lungs, and intestines. Carbamates such as Carbaryl and Baigon have also some side effects at acute exposure.

Pesticides account for eight of the top-twelve list of persistent organic pollutants identified by the United Nations Environment Programme (i.e., the UNEP "dirty dozen") (Table 1.4), which are compounds that were recently banned worldwide for production due to their ubiquitous distribution by atmospheric deposition, high bioaccumulation potential, and propensity to affect reproduction by disrupting the endocrine system (Figure 1.6). However, no pesticides are included in the list of top 25 priority pollutants frequently found at hazardous waste sites in North

Table 1.4. The United Nations Environment Programme (UNEP) List of Top Twelve Persistent Organic Pollutants (POPs) Recently Banned from Production Worldwide

1. Dioxins[a]	7. Toxaphene
2. Furans[a]	8. Dieldrin
3. PCBs	9. Aldrin
4. HCB	10. Endrin
5. DDT	11. Heptachlor
6. Chlordane	12. Mirex

[a]*Note*: Dioxins and furans represent a family of over 200 compounds that are not produced for industrial purposes: these are toxic by-products that may be formed during the chlorine bleaching process at pulp and paper mills and are also released into the air in emissions from municipal solid waste and industrial incinerators.

Source: UNEP (1999).

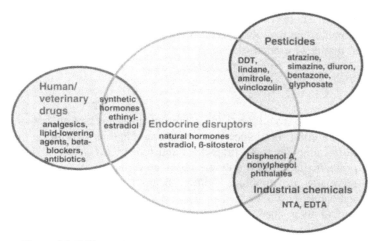

Figure 1.6. Different classes of emerging pollutants that affect reproduction.

America and Europe (Table 1.3), which are predominantly contaminated with industrial chemicals.

1.1.2. Emerging Pollutants

During the past decade, a wide variety of environmental contaminants have become recognized as potentially important in the fields of environmental science and engineering. These include endocrine disrupting compounds produced by the pharmaceutical and personal care products industries, oxidized energetic and rocket propellants, and small ethers (Figure 1.7). In addition, there is mounting evidence that pharmaceuticals, hormones, and other organic wastewater contaminants are being detected in water resources throughout the United States (Kolpin et al. 2002). Little is known about the potential interactive effects of these compounds on the biosphere.

Endocrine disrupting compounds (EDCs) are of concern to environmental health because of their potential to bioaccumulate through food chains and affect reproduction (Sedlak and Alvarez-Cohen 2003). Although groundwater contamination by EDCs is not yet a priority issue, it is important to recognize that surface water contamination by these compounds is widespread (possibly due to atmospheric transport and deposition) and that surface and groundwater resources are interconnected. Therefore, a brief review of this emerging class of pollutants is appropriate.

Perfluorinated octanes (PFOs) are an important class of EDCs that constitute the building blocks of 3M's perfluorinated surfactants, such as Scotchgard®. The persistence of PFOs is partly due to the fact that the carbon–fluorine bond is one of the strongest bonds in nature (110 kcal/mol). Potential adverse effects of PFOs include bioaccumulation, interference with mitochondrial bioenergetics and biogenesis,

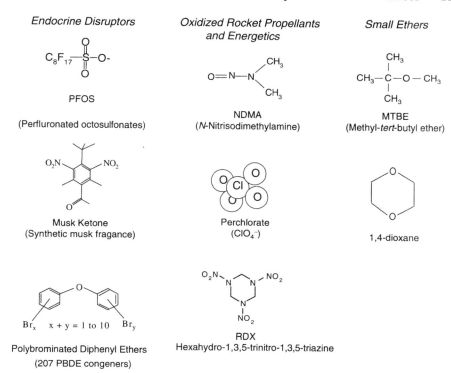

Figure 1.7. Selected emerging pollutants.

increase in liver size, neuroendocrine disruption, and acute aquatic toxicity. Due to these concerns, the 3M Company recently decided to remove this product line from market production.

Synthetic musk fragrances are another important class of EDCs. These are semi-volatile compounds used for perfumes, cosmetics, soaps, shampoos, and detergents and are commonly released by down-the-drain disposal of consumer products. Similar to PFOs, these EDCs are subject to atmospheric distribution and deposition, which results in widespread and worldwide contamination at trace levels.

Polybrominated diphenyl ethers (PBDEs) are also an important class of EDCs that are often referred to as the new PCBs. These compounds are used as fire retardants in electronics, plastics, car upholstery, furniture, and cables. PBDEs are also very persistent and tend to bioaccumulate in wildlife and human tissue. An exponential increase in the concentration of PBDEs has been observed in fish tissue in the Great Lakes and in marine wildlife, and levels on the order of pg/m^3 have been detected in human blood and milk (Snyder et al. 2003).

Oxidized rocket propellants and energetics represent another important class of emerging pollutants and include the rocket propellants NDMA and perchlorate and the military explosive RDX (Figure 1.7). These compounds are increasingly being found in groundwater aquifers, which is of concern due to their potential carcinogenicity. NDMA, which can also be a by-product of wastewater chlorination, is an extremely potent carcinogen. For example, the USEPA has proposed an action level of 0.0007 μg/L (for 10^{-6} risk), which is orders of magnitude more stringent than the drinking water standard for common carcinogens such as benzene or trichloroethylene (5 μg/L).

Groundwater contamination by small ethers is also a widespread problem. The most common pollutants in this category are 1,4-dioxane, which is used as a chlorinated solvent stabilizer, and MTBE, which is a common gasoline oxygenate that is added to minimize air pollution by hydrocarbon and carbon monoxide emissions during combustion. These compounds are suspected carcinogens and tend to impact large volumes of groundwater due to their high solubility, low tendency to volatilize, and low tendency to be retarded by sorption. The recent finding that 1,4-dioxane is commonly found as a trace contaminant in chlorinated solvent plumes has earned it the nickname of "the MTBE of the chlorinated solvent sites."

The remediation of sites contaminated with some emerging pollutants can be a challenge, not only because of the persistent nature of the contaminants but also because of logistic issues. Specifically, the chemical structure of many new pharmaceuticals and cosmetics are intellectual property, and standards for their analysis may be difficult to obtain. Thus, it is very difficult to manage and control contamination that cannot be accurately measured. Furthermore, there are multiple uptake pathways for compounds such as EDCs that are probably more important than drinking contaminated groundwater. For example, we ingest pharmaceuticals, spray insecticides in the air, apply fragrances, and sit on couches that have both brominated flame retardant and perfluorinated surfactant protectants. Therefore, groundwater contamination by some of these emerging pollutants is viewed by many decision makers as relatively low risk in relation to the groundwater pollutants listed in Table 1.4. Consequently, this book will focus on the bioremediation of common groundwater pollutants such as hydrocarbons and chlorinated solvents.

1.2. HISTORICAL DEVELOPMET OF ENVIRONMENTAL BIOTECHNOLOGY AND BIOREMEDIATION

The term "biotechnology" is analogous to "applied biology" and refers to the application of biological knowledge and techniques to develop products or to provide beneficial services. This term was coined in 1919 by the Hungarian engineer Karl Ereky, although the use of biological processes for the production of food products dates back to the beginning of human history (Table 1.5). Common biotechnology products that date to ancient times include yogurt, cheese, wine, beer, and spirits. Microorganisms are also used currently to produce large quantities of industrial products such as organic acids, solvents, polymers, enzymes, antibiotics, steroids, hormones, and fine chemicals for the pharmaceutical and cosmetic industries. The use of microorganisms to produce such compounds is motivated mainly

Table 1.5. Selected Biotechnology Landmarks

Year	Event
6000 B.C.	Yeast was used to make beer by Sumerians and Babylonians.
4000 B.C.	The Egyptians discovered how to bake leavened bread using yeast. Other fermentation processes were established in the ancient world, notably in China. The preservation of milk by lactic acid bacteria resulted in yogurt. Molds were used to produce cheese, and vinegar and wine were manufactured by fermentation.
400 B.C.	Hippocrates (460–377 B.C.) determined that the male contribution to a child's heredity is carried in the semen. By analogy, he guessed there is a similar fluid in women, since children clearly receive traits from each in approximately equal proportion.
320 B.C.	Aristotle (384–322 B.C.), choosing to reject the theories of Hippocrates, told his students that all inheritance comes from the father. The male semen, he asserted, determines the baby's form, while the mother merely provides the material from which the baby is made. He suggested that female babies are caused by "interference" from the mother's blood.
A.D. 100	Romans speculated that mares can be fertilized by the wind.
1100–1700	Spontaneous generation is the dominant explanation that organisms arise from nonliving matter. Maggots, for example, were supposed to arise from horse hair.
1300	The Aztecs in Mexico harvested algae from lakes as a food source.
1400	While distillation of a variety of spirits from fermented grain was widespread, Egypt and Persia largely gave up brewing as a result of the influence of Islam. Fermented breads and cereals still maintained their hold in the African diet.
1668	Francesco Redi used an experiment to compare two competing ideas that seek to explain why maggots arise on rotting meat. He observed that meat covered to exclude flies did not develop maggots, while similar uncovered meat did. This is regarded as the first disproof of spontaneous generation and was among the first uses of a controlled experiment.
1673	Anton van Leeuwenhoek, a Dutch merchant and civic administrator who ground glass lenses as a hobby, used his microscopes to make discoveries in microbiology. He was the first scientist to describe protozoa and bacteria and to recognize that such microorganisms might play a role in fermentation.
1724	Cross-fertilization in corn was discovered.
1856	In contrast to the ideas of Justis Liebig, Louis Pasteur asserted that microbes are responsible for fermentation. His experiments in the ensuing years proved that fermentation is the result of activity of yeasts and bacteria and conclusively disproved the theory of spontaneous generation.
1859	Charles Darwin hypothesized that animal populations adapt their forms over time to best exploit the environment, a process he referred to as "natural selection." He theorized that only the creatures best suited to their environment survive to reproduce.
1863	Louis Pasteur invented the process of pasteurization, heating wine sufficiently to inactivate microbes (that would otherwise turn the "vin" to "vin aigre" or "sour wine") while at the same time not ruining the flavor of the wine.

Table 1.5. (*Continued*)

Year	Event
1865	Gregor Mendel, an Augustinian monk, presented his laws of heredity to the Natural Science Society in Brunn, Austria. Mendel proposed that invisible internal units of information account for observable traits, and that these "factors"—which later became known as genes—are passed from one generation to the next.
1884	Robert Koch stated his "postulates" for testing whether a microbe is the causal agent of a disease (i.e., (1) the suspected pathogenic organism should be present in all cases of the disease and absent from healthy animals; (2) the suspected organism should be grown in pure culture away from the animal body; (3) such a culture, when inoculated into susceptible animals, should cause disease; and (4) the organism should be reisolated and shown to be the same as the original). Pasteur also concurrently developed a rabies vaccine.
1900	It was first shown that key industrial chemicals (glycerol, acetone, and butanol) could be generated using bacteria.
1909	Wilhelm Johannsen coined the terms "gene" to describe the carrier of heredity, "genotype" to describe the genetic constitution of an organism, and "phenotype" to describe the actual organism, which results from a combination of the genotype and the various environmental factors.
1914	Ardern and Lockett invent the activated sludge process to treat municipal wastewater, and the process is first used in Salford, England.
1940–1953	Large-scale production of penicillin and cortisone was achieved.
1952	Joshua Lederberg introduces the term plasmid to describe the bacterial structures he has discovered that contain extrachromosomal genetic material.
1953	James Watson and Francis Crick proposed the double-stranded, helical, complementary, antiparallel model for DNA. William Hayes discovered that plasmids can be used to transfer introduced genetic markers from one bacterium to another.
1972	Engineered bioremediation is used for the first time to clean up a Sun Oil pipeline spill in Ambler, Pennsylvania.
1974	Dick Raymond patents a bioremediation approach based on recirculation of nutrient-amended groundwater.

Sources: Alleman and Prakasam (1983), Brock (1961, 1990), Bunch and Hellemans (1993), and Hellemans & Bunch, (1988).

by economic factors, since biotechnology is often less costly and environmentally friendlier than chemical synthesis. Furthermore, biological processes can be used to produce specific stereochemicals or complex organic molecules such as vitamin B_{12} and riboflavin, which could not be feasibly produced by a chemical synthesis process.

Similar to many industrial processes, the main motivation to use biotechnology for environmental cleanup is also economics. For example, bioremediation is generally more cost-effective than conventional physical and chemical processes to treat excavated soil (Table 1.6). In addition, bioremediation can be a more practical

Table 1.6.   Costs of Alternative Methods to Treat Soil Contaminated with
Hazardous Wastes

Treatment Method	Average Cost (U.S. $/m^3$)	Typical Range (U.S. $/m^3$)
Incineration	975	350–1600
Landfilling	350	100–600
Thermal desorption	125	50–200
Soil washing	237	125–350
Above-ground bioremediation	95	40–150

Adapted from *Bioremediation Report* (1993).

approach to clean up contaminated aquifers than traditional technologies such as pump and treat or incineration of contaminated soil. Unlike many physicochemical treatment processes that mainly transfer the pollutants from one phase (or location) to another, bioremediation offers a terminal solution. Indeed, bioremediation often destroys organic pollutants, thereby eliminating future liability costs. Therefore, it is not surprising that bioremediation has recently gained an important place among alternatives to clean up sites contaminated with a wide variety of hazardous wastes.

The use of biological processes for waste treatment dates back to the Roman Empire, although the systematic design and application of microorganisms to treat wastewater only began about one century ago. The need to treat water and wastewater was traditionally driven by public health protection. Prior to the Asian cholera epidemic of London (1848), deaths due to environmental contamination received little attention. In 1849, Dr. John Snow, the father of modern epidemiology, demonstrated the connection between fecal contamination of a drinking water source (i.e., the Broad Street water pump in London) and deaths due to cholera (Stainer et al. 1986). Around that time, the discoveries of Louis Pasteur and the etiological postulates of Robert Koch (Table 1.5) were also beginning to transform microbiology from an observational hobby to a hypothesis-driven experimental science. This provided the basis for the rational development of environmental biotechnology and bioremediation.

In 1860, Pasteur demonstrated the connection between chemical changes and microbial activities. Pasteur also proved that fermentation was a microbial process, and in related experiments, he also conclusively disproved the theory of spontaneous generation (advocated by Aristotle), which held the notion that life (e.g., maggots) could originate from inanimate matter (e.g., horse hair).

In 1882, Dr. Angus Smith demonstrated that municipal wastewater could be "stabilized" by aeration, and Ardern and Lockett advanced this concept to develop the activated sludge treatment process, which was first used in Salford, England, in 1914 (Alleman and Prakasam 1983). This process was named activated sludge because it relied on an activated mass of microorganisms to treat the wastewater. The development of biological treatment processes in the United States was initiated at the Lawrence Experimental Station in Massachusetts (established in 1886), which was a unique facility aimed at experimental verification of different possible wastewater treatment procedures. In fact, one of the greatest impulses for the invention of the activated sludge process in England was the trip of Dr. Gilbert Fowler to the Lawrence Experimental Station in 1912. Dr. Fowler, from the University of Manchester, undertook this journey as a consultant chemist to the Manchester Corporation, and upon his return to England, he encouraged Ardern and Lockett to repeat the experiments with wastewater aeration he observed at the Lawrence Experimental Station. The activated sludge process soon found application outside the United Kingdom. The first experimental activated sludge plant in the United States was built in Milwaukee in 1915 with the help of Dr. Fowler as a consultant. Imhoff performed the first tests with the activated sludge process in 1924 and the first full-scale plant was built in 1926 in Essen-Rellinghausen, Germany. Other key engineers and scientists that contributed to the development of the activated sludge process include Allen Hazen, chemical engineer at the Lawrence station; George C. Whipple, civil engineer at the Massachusetts Institute of Technology (MIT); Harrison P. Eddy, consultant from Boston; and William R. Nichols, chemistry professor at MIT.

The history of in situ bioremediation is considerably shorter, and it reflects many upturns and downturns as a result of political and economic forces (Table 1.7). Interest in the use of microorganisms to degrade specific hazardous organic chemicals probably dates back to Gayle (1952), who proposed the microbial infallibility hypothesis. Gayle postulated that for any conceivable organic compound, there exists a microorganism that can degrade it under the right conditions. If not, evolution and adaptation would produce such a strain (Alexander 1965). This hypothesis cannot be proved wrong, because failure to degrade a contaminant can be attributed to the researcher's failure to use the right strain under the right conditions. In other words, using a popular adage, *"the absence of evidence is not in itself evidence of absence."*

In the 1970s, environmental statutes of unprecedented scope passed, such as the Occupational Safety and Health Act (OSHA) of 1970, the Clean Air Act (CAA) of 1970, the Clean Water Act (CWA) of 1972, the Safe Drinking Water Act (SWA) of 1974, and the Toxic Substance Control Act (TSCA) of 1976. This regulatory pressure stimulated interest in site remediation technologies, including bioremediation. However, bioremediation failed to meet the expectations raised by many technology salespeople who, for example, commonly advocated the addition of specialized bacteria to contaminated sites (i.e., bioaugmentation). Early proponents of this approach generally did not recognize that indigenous bacteria already present at a contaminated site were probably better predisposed physiologically and genetically to mediate the degradation of the target pollutants, but were not accomplishing this task because of a number of limiting factors discussed in Chapter 3. These include a lack of essential nutrients (including oxygen for aerobic processes), insufficient access of the microorganisms to sorbed contaminants (i.e., lack of bioavailability), or unfeasible thermodynamics (e.g., attempts to oxidize highly chlorinated pollutants that resist further oxidation and require reducing (anaerobic) conditions to undergo biotransformation). The failure to meet high performance expectations

Table 1.7. Historical Perspective of the Development of Bioremediation

Decade	Events
1900	Development of biological processes to treat municipal wastewater and sludge.
1950	Development of industrial wastewater treatment processes. Microbial infallibility hypothesis proposed by Gayle (1952), borne out of aerobic lab studies.
1960	Research on the biodegradation of synthetic organic chemicals present as "micropollutants" in wastewaters.
1970	Environmental statutes of unprecedented scope pass, and regulatory pressure (1976 RCRA and TSCA, 1980 CERCLA-Superfund) stimulates development of remediation technologies. Bioremediation successes are reported for the cleanup of aquifers contaminated by gasoline releases, and the first bioremediation patent is granted to Richard Raymond (1974). Nevertheless, adding acclimated microorganisms to contaminated sites becomes common practice. Earlier proponents of bioaugmentation often fail to recognize that indigenous bacteria already present might be better suited genetically and physiologically to degrade the pollutants, but biodegradation might be limited by contaminant bioavailability or unfavorable redox conditions rather than by a lack of catabolic potential. Failure to meet expectations creates a bandwagon effect and prompts a major downturn in bioremediation.
1980	It becomes clear that fundamental hydrogeologic and biogeochemical processes inherent to bioremediation need to be understood before a successful technology can be designed. This realization, along with the fear of liability and federal funding (e.g., Superfund), stimulates the blending of science and engineering to tackle environmental problems and improve bioremediation practice.
1990	Many successful bioremediation technologies (mainly ex situ) and hybrid (microbial/chemical) approaches are developed (often reflecting that adaptation is part of innovation). However, many decision makers continue to regard bioremediation as a risky technology and continue to select (relatively ineffective) pump-and-treat (P&T) technologies for remediation purposes. A 1994 NRC study reports that conventional P&T technologies restored contaminated groundwater to regulatory standards at only 8 of 77 sites. Superfund is depleted. Poor cleanup record and resource allocation problems stimulate paradigm shift toward natural attenuation and risk-based corrective action (RBCA).
2000	In situ bioremediation and monitored natural attenuation are widely accepted as cost-effective cleanup alternatives for sites contaminated with a wide variety of organic pollutants, and the interest in bioaugmentation grows for enhancing the removal of recalcitrant compounds.

created a bandwagon effect and prompted a major downturn in bioremediation applications.

In the 1980s, it became clear that a fundamental understanding of microbiology, microbial ecology, hydrogeology, and geochemistry was needed to successfully design and implement bioremediation systems. Bioremediation research and the blending of science and engineering to tackle environmental problems boomed in the 1980s, partly stimulated by federal funding through initiatives such as the Comprehensive Environmental Response, Compensation, and Liability Act of 1980 (CERCLA) and its Superfund program. Numerous bioremediation trials were successful—primarily those involving the clean up of petroleum product releases— and several hybrid technologies were developed. However, many decision makers insisted on pump-and-treat technologies that were relatively expensive and largely ineffective due to the hydrophobic nature of many pollutants that sorbed to the aquifer material. This led to a poor cleanup record, which stimulated a recent paradigm shift toward monitored natural attenuation (Chapter 8) and risk-based corrective action (RBCA) (NRC 1997, 2000).

RBCA is a decision-making process for the assessment of and response to subsurface contamination, based on the protection of human health and environmental resources. The objectives of implementing risk-based corrective action are to (1) reduce the risk of adverse human or environmental impacts to appropriate levels (e.g., the maximum contaminant level or MCL) at the point where the receptor is or may be potentially located, (2) ensure that site assessment activities are focused on collecting only information that is necessary to make risk-based corrective action decisions, (3) ensure that limited resources are focused toward those contaminated sites that pose the greatest risk to human health and environmental resources at any time, (4) ensure that the preferred remedial option is the most economically favorable one that has a high probability of achieving the negotiated degree of exposure and risk reduction, and (5) evaluate the compliance relative to site-specific standards (American Society for Testing and Materials (ASTM) 1994).

RBCA utilizes a three-tiered approach in assessing petroleum-contaminated sites. In Tier 1 (which involves screening the contamination data with highly conservative corrective action goals), sites are classified by the urgency of need for initial corrective action. Tier 1 assumes, for example, that the receptor will drink the groundwater with the highest contaminant concentrations within the site. If Tier 1 analysis concludes that a site needs cleanup, the responsible party has two options: meet Tier 1 (most stringent) cleanup goals or conduct a Tier 2 evaluation. At the Tier 2 level (which considers site-specific conditions), the user is provided with an option for determining site-specific cleanup levels and appropriate points of compliance (e.g., the receptor may be located offsite). The Tier 2 approach generally involves the use of analytical models (Chapter 5) to estimate the risk at a point of exposure. This analysis may conclude that a site does not pose a significant risk to existing receptors, and that monitoring rather than aggressive cleanup is all that is needed. However, similar to the previous step, if this analysis concludes that a site needs cleanup, the responsible party has two options: meet Tier 2 goals or conduct a Tier 3 evaluation. At Tier 3 (which also considers site-specific conditions), the assessment is similar to Tier 2 except that less conservative but more rigorous (and often more costly) numerical modeling (Chapter 6) and analysis is conducted to assess risk (ASTM 1994).

Overall, RBCA is intended to be a solution to a resource allocation problem, but its misuse or abuse could lead to an excuse to do nothing. RBCA contains the mandates that allow for passive bioremediation (natural attenuation), which involves thorough monitoring and modeling of the contamination. However, there is no treatment required if the contamination occurs at a sufficient rate that the plume does not spread. Natural attenuation, by no means, is a do- nothing option. One has to monitor the site and prove that natural attenuation is actually happening. Soil samples have to be taken and analyzed to determine the existence of microbes at the site and the disappearance of contaminants. Such analyses and monitoring can be quite expensive and time consuming. Nevertheless, these RBCA provisions are not as stringent as those mandated by CERCLA and the Superfund Amendment and Reauthorization Act of 1986 (SARA), where responsible parties are required to clean up everything to meet the compliance requirements (Bedient et al. 1994). Thus, for RBCA at the Tier 2 and Tier 3 levels of analysis, it is necessary to have models to predict the behavior of contaminants at a site (Chapters 5 and 6) as well as evidence that the implemented remediation strategies are working (Chapter 9).

1.3. MERITS AND LIMITATIONS OF BIOREMEDIATION

Bioremediation is an emerging technology that holds great promise for the cost-effective removal of a wide variety of environmental pollutants. Successful applications of bioremediation have been well documented for many sites contaminated with three major classes of hazardous wastes that are amenable to bioremediation: petroleum hydrocarbons (33% of all applications), creosotes (22%), and chlorinated solvents (9%). Bioremediation has also been applied, to a lesser degree, to cleanup sites contaminated with pesticides, munitions wastes, and other chemical mixtures. Note that these percentages include applications involving aboveground bioreactors to treat contaminated soil.

Bioremediation offers several advantages and limitations compared to traditional site remediation approaches such as pump-and-treat or soil excavation followed by incineration (Table 1.8). The principal advantages generally include lower cost and the ability to eliminate pollutants in situ, often transforming them into innocuous by-products such as CO_2 and water. This eliminates potential liability costs associated with hazardous waste transportation and storage. However, bioremediation is not universally applicable and it may be marginally effective for recalcitrant pollutants or toxic heavy metals if the necessary catabolic capacity is not present or expressed. For example, adverse environmental conditions such as extreme pH, temperature, or the presence of heavy metals at toxic concentrations may hinder specific microbial activities (Chapter 3).

Many hazardous substances are persistent in the environment when environmental conditions are not conducive to the proliferation and activity of specific microorganisms that can metabolize them. Thus, successful bioremediation requires an understanding of site-specific factors that limit desirable biotransformations or

Table 1.8. Advantages and Disadvantages of Bioremediation Relative to Traditional Remediation Approaches Such as Pump-and-Treat Contaminated Groundwater and Incineration of Source Soils

Advantages

- Cleanup occurs in situ, which eliminates hazardous waste transportation and liability costs.
- Organic hazardous wastes can be destroyed (e.g., converted to H_2O, CO_2, and mineral salts) rather than transferred from one phase to another, thus eliminating long-term liability.
- Relies on natural biodegradation processes that can be faster and cheaper (at least $10\times$ less expensive than removal and incineration, or pump and treat).
- Minimum land and environmental disturbance.
- Can attack hard-to-withdraw hydrophobic pollutants.
- Environmentally sound with public acceptance.
- Does not dewater the aquifer due to pumping.
- Can be used in conjunction with (or as a follow up to) other treatment technologies.

Disadvantages

- Certain wastes, such as heavy metals, are not eliminated by biological processes (although many metals can be bioreduced or biooxidized to less toxic and less mobile forms).
- It may require extensive monitoring.
- Requirements for success and removal efficiency may vary considerably from one site to another.
- Some contaminants can be present at high concentrations that inhibit microorganisms.
- Can be a scientifically intensive technique.
- There is a risk for accumulation of toxic biodegradation products.

that result in unintended consequences such as the production of toxic metabolites (Chapter 3). However, many site-specific limitations to natural biodegradation processes can be overcome through engineered manipulations of the environment (Chapter 8).

The requirements for bioremediation are depicted in pyramidal fashion in Figure 1.8. In order of importance, first, we need the presence of microorganisms with the capacity to synthesize enzymes that can degrade the target pollutants. These enzymes catalyze metabolic reactions that often produce cellular energy and building blocks for the synthesis of new cell material. Many contaminant degradation processes involve oxidation–reduction reactions that are discussed in Chapters 2 and 3. Briefly, a contaminant or a substrate may serve as an energy source (i.e., cellular fuel), and when it is oxidized, the substrate-derived electrons are transferred to an electron acceptor such as oxygen. Therefore, the second level of the pyramid shows that appropriate energy sources (i.e., electron donors) and electron acceptors must be present. The third level shows the need for sufficient moisture and acceptable pH, and the fourth level reminds us of the importance of avoiding extreme temperatures and ensuring the availability of inorganic

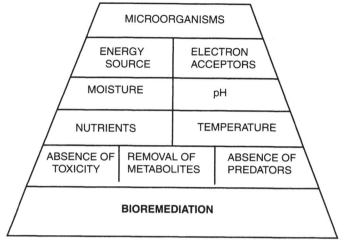

Figure 1.8. Requirements for bioremediation. (Adapted from Cookson 1995.)

nutrients such as nitrogen, phosphorus, and trace metals. Finally, at the base of the pyramid, we have three environmental requirements that are important for the sustainability of bioremediation: the absence of high concentrations of substances that are toxic to the microorganisms, the removal of metabolites that may inhibit specific microbial activities (perhaps by other members of the microbial community), and the absence of high concentrations of protozoa that act as predators on the bacteria responsible for contaminant degradation.

In summary, bioremediation engineers and scientists need to satisfy the physiological and nutritional requirements of specific degraders and ensure that a competitive advantage is provided to desirable (but not undesirable) biotransformation pathways. Achieving this can be a very complex task, especially when mass transfer limitations and difficulties in distributing stimulatory materials are considered. Therefore, bioremediation is often implemented as an art rather than as an empirical science. However, recent advances in our fundamental understanding of geochemical (Chapter 2), microbial (Chapter 3), and transport processes (Chapter 4) have improved our ability to control and improve the efficacy of bioremediation.

Despite many significant technological and scientific advances in the past ten years, bioremediation is currently an underutilized technology. According to the Organization for Economic Cooperation & Development (OECD), the global market potential for environmental biotechnology doubled during the 1990s to about $75 billion in the year 2000. In the United States, the cost associated with the cleanup of about 400,000 highly contaminated sites has been estimated to be from $500 billion to $1 trillion (NRC 1994). However, the current bioremediation market is only about $0.5 billion (Glass 2000). This relatively small market share probably reflects an often erroneous perception that bioremediation is a risky technology with uncertain results, cleanup time, and remediation costs. Whereas this may have been the case for some of the earlier empirical applications, the incorporation of sound scientific and engineering principles into the design of modern bioremediation systems has significantly improved the reliability and robustness of the process. As discussed in Chapter 8, many different bioremediation strategies exist that can be tailored to specific contamination scenarios. Therefore, as the bandwagon effect caused by earlier shortcomings fades, bioremediation is likely to achieve a more prominent role in hazardous waste treatment and site cleanup. We hope that this book will contribute to a broader acceptance of bioremediation and stimulate further optimization and innovation efforts.

REFERENCES

American Society for Testing and Materials (1994). *1994 Annual Book of ASTM Standards: Emergency Standard Guide for Risk-Based Corrective Action Applied at Petroleum Release Sites* (Designation: ES 38-94). ASTM, West Conshohocken, PA.

Alexander, M. (1965). Biodegradation: problems of molecular recalcitrance and microbial fallibility. *Adv. Appl. Microbiol.* **7**:35–80.

Alexander, M. (1999). *Biodegradation and Bioremediation*, 2nd ed. Academic Press, San Diego, CA.

Alleman J. and T.B.S. Prakasam (1983). On seven decades of activated sludge history. *Journal WPCF* **55**(5):436–443.

Bedient, P.B., H.S. Rifai, and C.J. Newell (1994). *Groundwater Contamination: Transport and Remediation*, PTR Prentice-Hall, Inc., Englewood Cliffs, NJ.

Brock, T.D. (1961). *Milestones in Microbiology*. Science Tech Publishers, Madison, WI.

Brock, T.D. (1990). *The Emergence of Bacterial Genetics*. Cold Spring Harbor Laboratory Press, Cold Spring Harbor, NY.

Bunch, B. and A. Hellemans (1993). *The Timetables of Technology*. Simon & Schuster, New York.

Hellemans, A. and B. Bunch (1988). *The Timetables of Science*. Simon & Schuster, New York.

Bioremediation Report (1993). King Publishing Group, Washington, DC.

Cookson, J.T. (1995). *Bioremediation Engineering. Design and Application.* McGraw Hill, New York.

ENTEC (1993). *Directory of Environmental Technology.* Earthscan Publications and Lewis Publishers/CRC Press, Ann Arbor, MI.

Gayle, E.F. (1952). *The Chemical Activities of Bacteria.* Academic Press, London.

Glass, D.J. (2000). The United States Remediation Market. Report from D. Glass Associates, Inc. (http://www.channel1.com/dglassassoc/BIO/usrem.htm).

Heitkamp, M.A. and C.E. Cerniglia (1988). Mineralization of polycyclic aromatic hydrocarbons by a bacterium isolated from sediments below an oil field. *Appl. Environ. Microbiol.* **54**:1612–1614.

Intergovernmental Panel on Climate Change (2001). The Third Assessment of Climate Change. United Nations.

Keith, L.H. and W.A. Telliard (1979). Priority pollutants I—a perspective view. *Environ. Sci. Technol.* **13**:416–423.

Kolpin, D.W., E.T. Furlong, M.T. Meyer, E.M. Thurman, S.D. Zaugg, L.B. Barber, and H.T. Buxton (2002). Pharmaceuticals, hormones, and other organic wastewater contaminants in U.S. streams, 1999–2000: a national reconnaissance. *Environ. Sci. Technol.* **36**:1202–1211.

La Grega, M.D., P.L. Buckingham, and J.C. Evans (1994). *Hazardous Waste Management.* McGraw Hill, New York.

McElroy, A.E., J.W. Farrington, and J.M. Teal (1989). Bioavailability of polycyclic aromatic hydrocarbons in the aquatic environment. In *Metabolism of Polycyclic Aromatic Hydrocarbons in the Aquatic Environment*, U. Varanasi (Ed.). CRC Press, Boca Raton, FL.

Mortelmans K., S. Harworth, T. Lawlor, W. Speck, B. Tainerand, and E. Zeiger (1986). *Salmonella* mutagenicity tests. II. Results from the testing of 270 chemicals. *Environ. Mutagen.* **8**(Suppl. 7):1–119.

National Research Council (1994). *Alternatives for Ground Water Cleanup.* National Academy Press, Washington, DC.

National Research Council (1997). *Innovations in Ground Water and Soil Cleanup: From Concept to Commercialization.* National Academy Press, Washington, DC.

National Research Council (2000). *Natural Attenuation for Ground Water Remediation.* National Academy Press, Washington, DC.

Pankow, J.F. and J.A. Cherry (1996). *Dense Chlorinated Solvents and Other DNAPLs in Groundwater: History, Behavior, and Remediation.* Waterloo Press, Guelph, Ontario.

Raymond, R.L. (1974). Reclamation of hydrocarbon contaminated groundwater. U.S. Patent 3,846,290.

Sedlak, D. and L. Alvarez-Cohen (2003). Emerging contaminants in water. *Environ. Eng. Sci.* **20**(5):387–388.

Snyder, S.A., P. Westerhoff, Y. Yoon, and D.L. Sedlak (2003). Pharmaceuticals, personal care products, and endocrine disruptors in water: implications for the water industry. *Environ. Eng. Sci.* **20**(5):449–470.

Speidel, D.H. and A.F. Agnew (1988). The Wold budger. In *Perspectives in Water Uses and Abuses,* D.H. Speidel, L.C. Ruedisili, and A.F. Agnew (Eds.). Oxford University Press, New York.

Stainer, R.Y., J.L. Ingraham, M.L. Wheelis, and P.R. Painter (1986). *The Microbial World,* 5th ed. Prentice Hall, Englewood Cliffs, NJ.

United Nations Environmental Protection Programme (1999). *Inventory of Information Sources on Chemicals: Persistent Organic Pollutants.* UNEP Chemicals, Geneva, Switzerland.

U.S. Environmental Protection Agency (1989). Toxic Release Inventory. Magnetic Tape Number P.B. 89–186-118, NTIS or TRI Data Base, National Library of Medicine, Bethesda, MD.

U.S. Environmental Protection Agency (2003). Washington, DC. (www.epa.gov/swerust1/pubs/ustfacts.pdf.)

Wilson, E.O. (2002). *The Future of Life.* Little Brown & Company, Warner Books, Lebanon, IN.

2

GEOCHEMICAL ATTENUATION MECHANISMS

Ulises: "One touch of nature makes the whole world kin."
—Troilus and Cressida, Shakespeare

"In time and with water, everything changes."
—Leonardo da Vinci

"Speak to the earth, and it shall teach thee."
—Bible, Job 12:8

All processes that result in structural or phase changes of environmental pollutants can be considered to have a chemical basis, including those that are mediated directly or indirectly by microorganisms. Under certain circumstances, the main role of microorganisms is to provide the ideal conditions for chemical transformations to proceed, as opposed to directly degrading the contaminants (Heijman et al. 1993, 1995). For organizational purposes, however, it is convenient to consider nonbiological (abiotic) and biological attenuation processes separately.

This chapter will discuss several abiotic processes that can contribute to (or inhibit) the natural attenuation of groundwater pollutants. These will be grouped into (1) processes that *transform* contaminants to less harmful compounds and (2) processes that *immobilize* some contaminants within the aquifer matrix. Abiotic transformation processes that could be important in groundwater systems include hydrolysis, oxidation–reduction reactions at mineral interfaces, elimination reactions where molecules undergo spontaneous rearrangements, and radioactive decay. Photolysis, which involves chemical reactions initiated by the absorption of photons, are negligible in aquifers due to the lack of light penetration and will not be discussed in this book. Immobilization processes that will be considered include sorption, humification, ion exchange, and precipitation reactions. Emphasis will be placed on the geochemical principles responsible for such abiotic natural attenuation mechanisms.

Bioremediation and Natural Attenuation: Process Fundamentals and Mathematical Models
By Pedro J. J. Alvarez and Walter A. Illman   Copyright © 2006 John Wiley & Sons, Inc.

2.1. CHEMICAL TRANSFORMATION PROCESSES

2.1.1. Hydrolysis

Hydrolysis can be defined as the addition of the hydrogen and hydroxyl ions of water to a molecule, with its consequent splitting into two or more simpler molecules that are commonly easier to biodegrade. This is illustrated below for a generic molecule RX, which reacts with water to form a new R—O bond with the oxygen atom from water, and displaces the electron-withdrawing X group (e.g., an attached halogen, sulfur, phosphorus, or nitrogen) with OH. Therefore, hydrolysis is the result of a nucleophilic substitution in which water or a hydroxide (a nucleophile) attacks electrophilic carbon or phosphorus atoms:

$$RX + H_2O \rightarrow ROH + HX \qquad [2.1]$$

Hydrolytic degradation provides a baseline loss rate for organic pollutants in aqueous environments and has been shown to be an important attenuation mechanism for some common groundwater pollutants that can be hydrolyzed within the one- to two-decade time span of general interest to site remediation (e.g., organophosphate pesticides and 1,1,1-trichloroethane) (NRC 2000).

Some classes of pollutants are generally resistant to hydrolysis, such as alkanes, aromatic hydrocarbons, alcohols, ketones, glycols, phenols, ethers, carboxylic acids, sulfonic acids, and polycyclic and heterocyclic hydrocarbons (Neely 1985). The classes of pollutants that are generally susceptible to hydrolysis include some alkylhalides, amides, amines, carbamates, carboxylic acid esters, epoxides, lactones, nitriles, phosphonic acid esters, phosphoric acid esters, sulfonic acid esters, and sulfuric acid esters (Mabey and Mill 1978; Harris 1982). Nevertheless, the hydrolytic rates of such compounds can be highly variable. For example, triesters of phosphoric acid hydrolyze in pH-neutral water at ambient temperatures with half-lives ranging from several days to several years (Wolfe 1980), and the chlorinated alkanes pentachloroethane, carbon tetrachloride, and hexachloroethane exhibit hydrolytic half-lives (pH 7, 25 °C) of about 2 hours, 50 years, and 1000 years, respectively (Mabey and Mill 1978; Jeffers et al. 1989). The rate of hydrolysis also depends on the pollutant concentration, the pH, and the groundwater temperature. For example, the half-life of 1,1,1-trichloroethane is about 12 years at 12 °C and decreases to about 2.5 years at 20 °C (Rittman et al. 1994). Sorption can also affect the rate of hydrolysis, which generally decreases with increasing "protection" by sorption (Burkhard and Guth 1981). There are, however, exceptions to this trend, and some organophosphate insecticides have been observed to hydrolyze faster when sorbed, presumably due to the catalytic role of some mineral surfaces (Konrad and Chesters 1969). Examples of environmentally important hydrolytic reactions are provided below.

Alkyl Halides

Halogenated alkanes with a single halogen substituent are hydrolyzed to alcohols, which are easy to biodegrade. This is illustrated below for the conversion of methyl bromide (CH_3Br) to methanol (CH_3OH):

$$CH_3Br + H_2O \rightarrow CH_3OH + H^+ + Br^- \qquad [2.2]$$

The hydrolysis of polyhalogenated alkanes can be relatively complex, involving both elimination and substitution reactions and yielding different products. This is illustrated below for 1,1,1-trichloroethane (1,1,1-TCA), 80% of which is converted by sequential hydrolysis via 1,1-dichloroethanol to acetic acid (vinegar), and the remaining 20% undergoes dehydration of the alcohol to form 1,1-dichloroethylene (1,1-DCE).

$$CH_3CCl_3 + H_2O \rightarrow CH_3CCl_2OH + H^+ + Cl^-$$
(1,1,1-TCA)   (1,1-dichloroethanol)

$$CH_3CCl_2OH + H_2O \rightarrow CH_3COOH + 2H^+ + 2Cl^-$$
(acetic acid)

$$CH_3CCl_2OH \rightarrow CH_2{=}Cl_2 + H_2O$$
(1,1-DCE) $[2.3]$

Another common hydrolytic reaction for polyhalogenated alkanes is the 1,2-elimination reaction, where vicinal hydrogen and halogen atoms leave the molecule with concomitant formation of a double carbon–carbon bond. This reaction usually proceeds faster under alkaline conditions (i.e., pH > 7) because it involves a nucleophilic attack by a hydroxyl group, as illustrated for the transformation of 1,2-dichloroethane to chloroethene:

$$\underset{\underset{Cl}{|}}{\overset{\overset{Cl}{|}}{H_2C}}{-}CH_2 + OH^- \longrightarrow H_2C{=}CHCl + Cl^- + H_2O \qquad [2.4]$$

Carboxylic Acid Esters

These compounds are hydrolyzed to carboxylic acids and alcohols. The reaction mechanism involves nucleophilic attack by a hydroxide group at the carbonyl group. Consequently, these compounds hydrolyze faster at higher pH values (i.e., "base-catalyzed" hydrolysis).

$$\underset{ester}{R_1{-}\overset{\overset{O}{\|}}{C}{-}O{-}R_2} + H^- \longrightarrow \underset{carboxylic\ acid}{R_1{-}\overset{\overset{O}{\|}}{C}{-}O^-} + \underset{alcohol}{HOR_2} \qquad [2.5]$$

Amides

Although amides are less susceptible to hydrolysis than esters, measurable rates can be observed under highly acidic or basic conditions. In such cases, carboxylic acids and amines are produced:

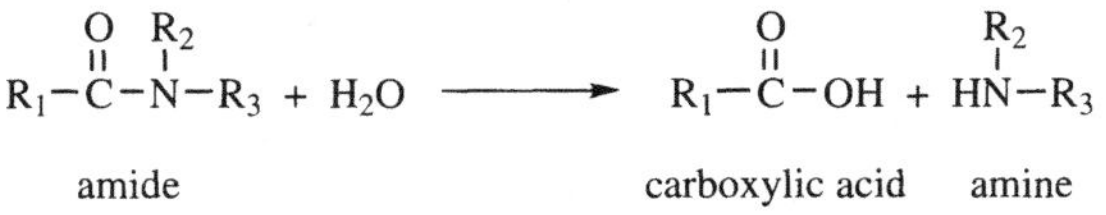

$$R_1-\overset{\overset{\text{O}}{\|}}{C}-\overset{\overset{\text{R}_2}{|}}{N}-R_3 + H_2O \longrightarrow R_1-\overset{\overset{\text{O}}{\|}}{C}-OH + H\overset{\overset{\text{R}_2}{|}}{N}-R_3 \qquad [2.6]$$

amide   carboxylic acid   amine

Nitriles

Nitriles are organic compounds characterized by a triple bond between carbon and nitrogen atoms, such as cyanide (HC≡N). These compounds can be hydrolyzed to amides first, and then to carboxylic acids and amines (or ammonia) as illustrated above:

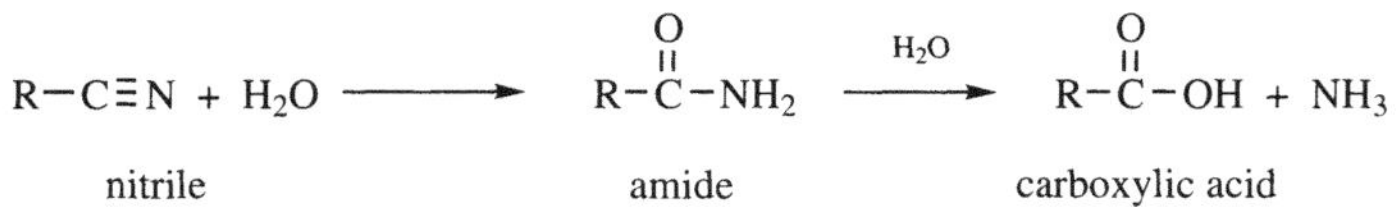

$$R-C\equiv N + H_2O \longrightarrow R-\overset{\overset{\text{O}}{\|}}{C}-NH_2 \xrightarrow{\text{H}_2\text{O}} R-\overset{\overset{\text{O}}{\|}}{C}-OH + NH_3 \qquad [2.7]$$

nitrile   amide   carboxylic acid

Epoxides

Epoxides are functional groups characterized by a triangular arrangement between two carbon atoms and one oxygen atom. Some epoxides can be hydrolyzed even at neutral pH, yielding the corresponding diol and sometimes rearranged products also:

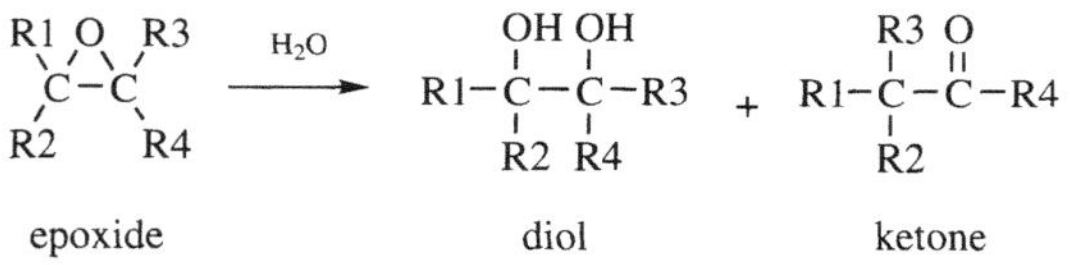

$$\overset{\overset{\text{R1 O R3}}{\diagdown\diagup}}{\underset{\text{R2   R4}}{\diagup\diagdown}}{C-C} \xrightarrow{\text{H}_2\text{O}} R1-\overset{\overset{\text{OH}}{|}}{\underset{\underset{\text{R2}}{|}}{C}}-\overset{\overset{\text{OH}}{|}}{\underset{\underset{\text{R4}}{|}}{C}}-R3 + R1-\overset{\overset{\text{R3}}{|}}{\underset{\underset{\text{R2}}{|}}{C}}-\overset{\overset{\text{O}}{\|}}{C}-R4 \qquad [2.8]$$

epoxide   diol   ketone

Carbamates

This group of pesticides could be rapidly hydrolyzed to amines and alcohols, depending on the substituents on the N atom. For example, the insecticide carbaryl (Ar = napthyl) undergoes rapid alkaline hydrolysis at room temperature, even at pH 7, with half-lives on the order of a few hours (Wolfe et al. 1978):

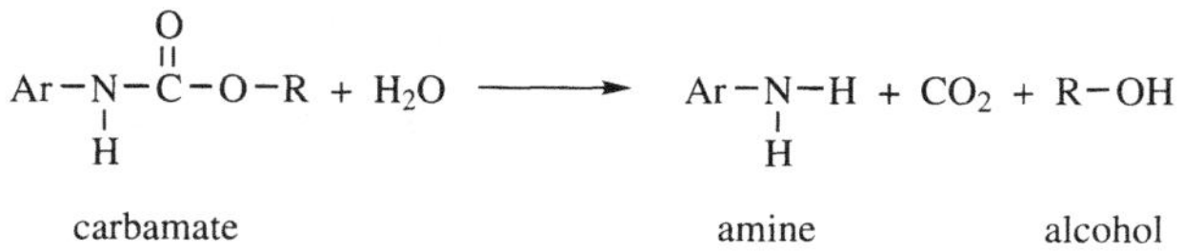

$$Ar-\underset{\underset{\text{H}}{|}}{N}-\overset{\overset{\text{O}}{\|}}{C}-O-R + H_2O \longrightarrow Ar-\underset{\underset{\text{H}}{|}}{N}-H + CO_2 + R-OH \qquad [2.9]$$

carbamate   amine   alcohol

Hydrolysis is much slower if the substituent on the N atom is an alkyl group. For example, the half-life of chlorpropham (Ar = chlorophenyl) is six to seven orders of magnitude slower under the same conditions:

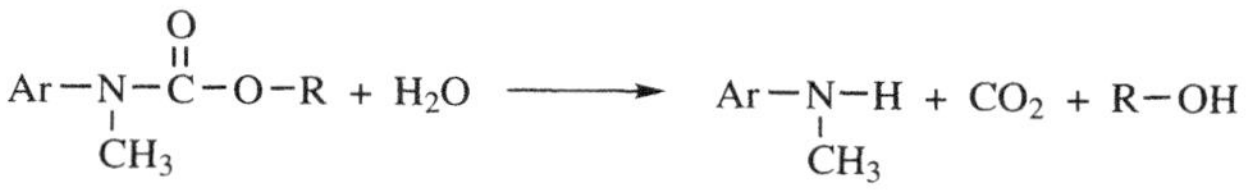

$$\underset{\underset{\overset{|}{CH_3}}{}}{Ar-N-\overset{\overset{O}{\parallel}}{C}-O-R} + H_2O \longrightarrow \underset{\underset{CH_3}{|}}{Ar-N-H} + CO_2 + R-OH \qquad [2.10]$$

chlorpropham   amine   alcohol

The difference in rates is related to differences in reaction mechanisms. Carbaryl hydrolyzes by an elimination process in which the proton acidity of the nitrogen atom determines reactivity. Chlorpropham, on the other hand, hydrolyzes similarly to carboxylic acid esters, and electron-withdrawing substituents accelerate the reaction. Conversely, the reaction is much slower when electron-donating substituents are present, as is the case with the methyl group in chlorpropham above.

Sulfonylureas

These pesticides undergo rapid acid-catalyzed hydrolysis, with cleavage occurring typically at the sulfonylurea bridge:

$$R-SO_2-\overset{\overset{O}{\parallel}}{C}-\overset{\overset{H}{|}}{N} + H_2O \longrightarrow R-SO_2-H + HO-\overset{\overset{O}{\parallel}}{C}-\overset{\overset{H}{|}}{N}-R \qquad [2.11]$$

sulfonylurea   amide

The resulting amide could be hydrolyzed further to the corresponding carboxylic acid and amine, as illustrated above (Eq. [2.6]).

Organophosphate Ester

Many pesticides belong to this group of chemicals, which includes organophosphates, organophosphorothioates, organophosphorothionates, and organophosphorodithionates:

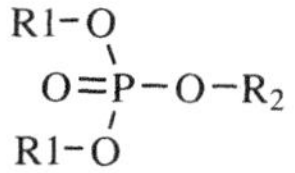

$$\underset{R1-O}{\overset{R1-O}{\diagdown\diagup}}O=P-O-R_2 \qquad\qquad \underset{R1-O}{\overset{R1-O}{\diagdown\diagup}}S=P-O-R2$$

organophosphate 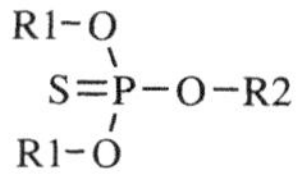   organophosphorothioate

$$[2.12]$$

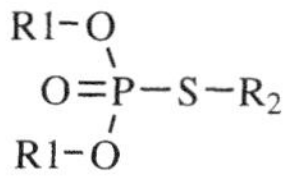

$$\underset{R1-O}{\overset{R1-O}{\diagdown\diagup}}O=P-S-R_2 \qquad\qquad \underset{R1-O}{\overset{R1-O}{\diagdown\diagup}}S=P-S-R2$$

organophosphorothionate 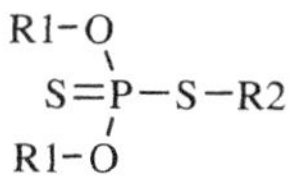   organophosphorodithioate

Hydrolysis usually occurs by direct nucleophilic attack by OH^- at the P atom:

$$\underset{R1-O}{\overset{R1-O}{\diagdown\diagup}}O=P-O-CH_2-R2 + OH^- \longrightarrow \underset{R1-O}{\overset{R1-O}{\diagdown\diagup}}O=P-OH + HOCH_2R_2 \qquad [2.13]$$

Nucleophilic attack with water can also occur at one of the methyl or ethyl susbtituents (Lacorte et al. 1995):

$$
\begin{array}{ccc}
\text{R1-O} & & \text{R1-O} \\
\diagdown & & \diagdown \\
\text{O=P-O-CH}_2\text{-R2 + H}_2\text{O} & \longrightarrow & \text{O=P-OCH}_2\text{R2 + R1-OH} \\
\diagup & & \diagup \\
\text{R1-O} & & \text{H-O}
\end{array}
\qquad [2.14]
$$

2.1.2. Radioactive Decay

Radioactive decay can be an important attenuation mechanism for elements with 83 or more protons, which are unstable or radioactive and whose atomic nuclei spontaneously change form into so-called daughter products. These by-products are elementally different and can therefore behave differently in the environment. Common radioactive environmental problems include radon, which is a naturally occurring gas that leaks into houses and poses an inhalation hazard; plutonium, which is a military waste product; uranium and decay products; radioactive heavy metals excavated in mines; and radioactive elements used as tracers in medical applications, such as cesium and iodine. Figure 2.1 illustrates the relationship between common radioactive materials that are likely to pose a threat to residential housing.

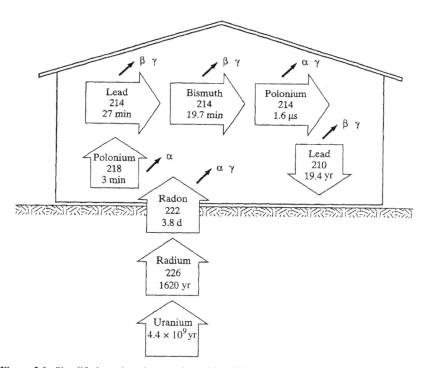

Figure 2.1. Simplified uranium decay series, with half-lives and emissions. (From Masters 1997.)

Table 2.1. Several Units of Radiation

Unit	Symbol	Description
Curie	Ci	Used to measure emission at source; 1 Ci = disintegration of 3.7×10^{10} atoms/s (radium)
Becquerel	Bq	Used to measure emission at source; 1 Ci = 3.7×10^{10} Bq
Roentgen	R	Gives ionizations produced in a given amount of air by x- or γ-rays
Rad	rad	(radiation absorbed dose) used for exposure. Corresponds to absorption of 100 ergs of energy
Rem	rem	(roentgen equivalent man) takes into account differences in effect of radiation type; for example, if 10 rads of α has same rem as 1 rad of β, you get the same rem

The main concern with the radioactive decay process is the emission of radiation (Table 2.1). Three types of radiation could be emitted:

Alpha Radiation. This consists of massive particles (i.e., $^4_2$He) that have difficulty penetrating the skin, although such particles are of concern if one breathes them into the lungs.

Beta Radiation. These are electrons that can penetrate the skin a few centimeters and pose a greater health risk than alpha particles.

Gamma Radiation. This consists of very damaging electromagnetic energy of short wavelength that has virtually no mass. This radiation causes ionization, making biological molecules unstable.

Depending on the intensity and duration of exposure, radioactivity can be very damaging to organisms. Detrimental effects range from somatic effects such as cancer, sterility, and cataracts to genetic effects such as mutation of chromosomes. Nevertheless, subsurface solids absorb radioactive emissions and radionuclides that remain in the subsurface pose little or no risk provided that exposure to contaminated groundwater does not occur (NRC 2000).

The half-life $(t_{1/2})$ of a radioactive element is an important parameter related to its decay rate and refers to the time required for half of the atoms to decay to other elements. Radioactive decay rates are independent of temperature and follow first-order kinetics. Thus, the decay rate (dC/dt) is proportional to the radioactive material concentration (C):

$$\frac{dC}{dt} = -kC \qquad [2.15]$$

Integrating and substituting initial conditions,

$$C = C_0 e^{-kt} \qquad [2.16]$$

This is the expression that describes the residual concentration of a radioactive element that decays exponentially as a function of time, as illustrated in Figure 2.2.

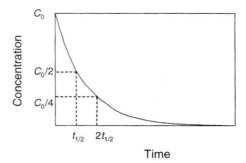

Figure 2.2. Characteristics of exponential decay and half-life.

Equation [2.16] can be rearranged to derive an expression for the half-life, as follows:

$$t = \frac{\ln(C_0/C)}{k} \qquad [2.17]$$

Also, by definition, $C = C_0/2$ when $t = t_{1/2}$.
Thus,

$$t_{1/2} = \frac{\ln(2)}{k} = \frac{0.693}{k} \qquad [2.18]$$

Half-lives of radioactive elements can vary over many orders of magnitude (Table 2.2). Natural attenuation by radioactive decay may be possible for some elements with relatively short half-lives, such as tritium —^{3}H (12.5 yr), cesium —^{137}Cs (30 yr), strontium —^{90}Sr (29 yr), iodine —^{131}I (8 yr), and radon —^{222}Rn (3.8 days). However, some radionuclides have very long half-lives and are very persistent, such as uranium —^{238}U (4.4×10^9 yr) and plutonium —^{239}Pu (24,390 yr).

Example 2.1   Radioactive iodine (^{131}I) was accidentally spilled from a hospital into the sewer system. If the average wastewater flow is 0.1 ft/s and the wastewater treatment plant is located 10 miles away, what percent of the initial radioactivity will reach the plant?

Assuming leaks from the sewer system are negligible, the average travel time from the hospital to the wastewater treatment plant would be 52,800 ft ÷ 0.1 ft/s = 528,000 s = 6.1 days.

The decay rate coefficient for ^{131}I can be determined from Eq. [2.18], using $t_{1/2} = 8$ yr:

$$k = 0.693/8 \text{ yr} = 0.0866 \text{ yr}^{-1} = 0.00024 \text{ day}^{-1}$$

Table 2.2. Half-Lives of Common Radioactive Isotopes

Nuclides[a]	$t_{1/2}$ (years)	Nuclides[a]	$t_{1/2}$ (years)
^{3}H	12.3	^{231}Pa	3×10^4
^{14}C	5×10^3	^{241}Am	432
^{36}Cl	3.1×10^3	^{243}Am	7×10^3
^{63}Ni	100	^{79}Se	6.5×10^4
^{90}Sr	29	^{93}Mo	3.5×10^3
^{93}Zr	1.5×10^6	^{99}Tc	2×10^5
^{94}Nb	2×10^4	^{99}Tc	2×10^5
^{107}Pd	7×10^6	^{126}Sn	1×10^5
^{129}I	2×10^7	^{151}Sm	90
^{135}Cs	3×10^6	^{147}Sm	$1.3 \times 10^{1-}$
11			
^{137}Cs	30	^{106}Ru	1.0
^{154}Eu	8.2	^{235}U	7×10^8
^{210}Pb	22	^{238}U	4.5×10^9
^{226}Ra	1.6×10^3	^{237}Np	2×10^6
^{227}Ac	22		
^{230}Th	8×10^4		
^{232}Th	1.4×10^{10}		

[a]Superscript denotes the atomic mass.

Source: Domenico and Schwartz (1998).

Using Eq. [2.16],

$$C/C_0 = \exp(-0.00024 \text{ day}^{-1} \times 6.1 \text{ days}) = 0.9986$$

Thus, only 0.14% of the initial radioactivity would be removed during the 6.1 days that it flows through the sewer system.

A common observation with some natural decay series is that of *secular equilibrium*. This is a form of radioactive equilibrium where all nuclides in the chain decay at the same rate and their ratios are the same as the ratios of their decay constants. This can be expressed in terms of "activity" (i.e., amount of radiation), calculated as the product of the abundance and the decay constant of a nuclide (IUPAC 1997). Two conditions are necessary for secular equilibrium. First, the parent radionuclide must have a half-life much longer than that of any other radionuclide in the series. Second, a sufficiently long period of time must have elapsed, for example, ten half-lives of the decay product having the longest half-life, to allow for ingrowth of the decay products (Figure 2.3). The relevance of secular equilibrium is that the activity of the parent radionuclide undergoes no significant changes during many half-lives of its decay products. At this time (and for a long time) all members of the series disintegrate the same number of atoms per unit time and the activity of the daughter nuclide equals the activity of the parent.

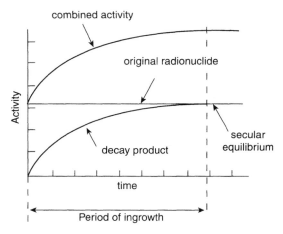

Figure 2.3. The concept of secular equilibrium. When the half-life of the original radionuclide is much longer than the half-life of the decay product, the decay product generates radiation more quickly. Within about 7–10 half-lives of the decay product, their activities are equal, and the amount of radiation (activity) is doubled. Beyond this point, the decay product decays at the same rate it is produced—a state called "secular equilibrium." (From http://www.epa.gov/radiation/understand/equilibrium.htm#transient.)

An important undesirable consequence of secular equilibrium is that the hazard associated with one type of radioactive element can greatly increase due to the accumulation of its by-products that emit more harmful forms of radiation. For example, radium decay leads to a whole host of radioactive elements that produce gamma and beta radiation (see Figure 2.1).

Secular equilibrium is never exactly attained but is essentially established in nature for the decay series of uranium-238, thorium-232, and uranium-235. One commonly observed example of secular equilibrium is the decay of radium to ^{210}Pb (see Figure 2.1). The half-life of radium is about 1620 yr, of radon, approximately 3.82 days, and of each of the subsequent members, a few minutes. After about a month, essentially the equilibrium amount of radon is present; and then all members of the series disintegrate at the same rate. Other examples of secular equilibrium include the generator system ^{113}Sn/^{113m}In, where the half-lives of the parent nucleus (^{113}Sn) and the daughter (^{113m}In) are 118 and 1.7 days, respectively, and the ^{68}Ge/^{68}Ga generator with parent–daughter half-lives of 270 days and 68 min, respectively.

2.1.3. Chemical Reduction and Oxidation (Redox) Reactions

Reduction and oxidation (redox) reactions are very important for natural attenuation. Such reactions are typically irreversible and transform contaminants into different chemicals by means of electron transfer. The contaminant is reduced if it gains electrons, and the electron donor becomes oxidized. On the other hand, the contaminant is oxidized if it gives up electrons, and the electron acceptor becomes reduced.

Table 2.3. Half-Reactions of Potential Electron Acceptors

$$\frac{1}{2}O_2 + 2H^+ + 2e^- \rightarrow H_2O$$

$$2NO_3^- + 12H^- + 10e^- \rightarrow N_2 + 6H_2O$$

$$NO_3^- + 10H^- + 3e^- \rightarrow NH_4^+ + 3H_2O$$

$$2NO_2^- + 8H^+ + 6e^- \rightarrow N_2 + 4H_2O$$

$$NO_2^- + 8H^+ + 6e^- \rightarrow NH_4^+ + H_2O$$

$$Fe(OH)_3 + 3H^+ + e^- \rightarrow Fe^{2+} + 3H_2O$$

$$FeOOH + 3H^+ + e^- \rightarrow Fe^{2+} + 2H_2O$$

$$\frac{1}{2}Fe_2O_3 + 3H^+ + e^- \rightarrow Fe^{2+} + \frac{3}{2}H_2O$$

$$MnO_2 + 4H^+ + 2e^- \rightarrow Mn^{2+} + 2H_2O$$

$$SO_4^{2-} + 9H^+ + 8e^- \rightarrow HS^- + 4H_2O$$

$$CO_2 + 8H^+ + 8e^- \rightarrow CH_4 + 2H_2O$$

$$C_6H_4O_2 + 2H^+ + 2e^- \rightarrow C_6H_6O_2$$

$$CH_2O + 4H^+ + 4e^- \rightarrow CH_4 + H_2O$$

Redox reactions are often mediated by microorganisms and thus will be discussed in greater detail in Chapter 3. Nevertheless, some interactions between contaminants and aquifer solids or redox-sensitive constituents in groundwater can result in abiotic redox transformations that contribute to natural attenuation (NRC 2000; Scherer et al. 2000). This occurs to an appreciable extent only when electron transfer is both thermodynamically feasible and kinetically facile (see Appendix A).

The oxidation capacity of an aquifer (OXC) is defined as its ability to restrict the development of reducing conditions due to the availability of oxidized species. To estimate this value, we need to consider both the electron-accepting capacity and concentration of potential electron acceptors. The electron-accepting capacity of commonly occurring compounds is given in Table 2.3. For example, considering the half-reaction for the oxygen–water couple, we can see that 1 mole of O_2 can accept 4 moles of electrons. Similarly, 1 mole of NO_3^- accepts 5 moles of electrons when it is reduced to N_2, and organic matter (represented by CH_2O, see the last half-reaction in Table 2.3) can accept up to 4 electrons per molecule. Thus, based on Table 2.3:

$$OXC = 4[O_2] + 3[NO_2^-] + 5[NO_3^-] + [Fe^{3+}] + 2[Mn^{4+}]$$
$$+ 8[SO_4^{2-}] + 8[HCO_3^-] + 4[TOC] \quad \text{(equivalents per volume)} \quad [2.19]$$

Note that these electron acceptors may require the presence of appropriate catalysts (e.g., microbial enzymes) to oxidize organic pollutants at measurable rates. It should also be emphasized that OXC is often grossly underestimated due to lack of information of solid Fe(III)- and Mn(VI)-bearing minerals.

We can similarly define the total reduction capacity (TRC) of an aquifer as the ability of reduced species in an environment to become oxidized. Thus, TRC is the sum of all electron-donating species (mainly precipitated and reduced electron acceptors) multiplied by their electron-donating capacity.

High TRC is desirable for reductive treatment processes (e.g., reductive dechlorination) and undesirable for oxidative bioremediation, where TRC constituents could compete with the target pollutants (e.g., hydrocarbons) for the readily available electron acceptors (e.g., O_2).

Organic contributions to the TRC include hydroquinones ($C_6H_6O_2 \rightarrow C_6H_4O_2 + 2H^+ + 2e^-$) and mineralization of organic matter ($CH_2O + H_2O \rightarrow CO_2 + 4H^+ + 4e^-$). Nevertheless, similar to OXC, the TRC is mainly associated with the solid phase, and it is often underestimated.

The most important abiotic reductant is generally ferrous iron (Fe(II)). The electron transfer between the Fe(II)–Fe(III) couple exhibits a wide range of reduction potentials, ranging from $+1.1$ to -0.5V (Klausen et al. 1995). Therefore, Fe(II) species can often serve as electron donors or electron transfer mediators. For example, dissolved Fe(II) can reduce hexavalent chromium to insoluble and less toxic $Cr(OH)_3$ (Buerge and Hug 1997). Furthermore, when Fe(II) is associated with a mineral surface, such as iron-bearing minerals, it can reduce a wide variety of organic pollutants and metals that are not reducible by dissolved Fe(II). These include chlorinated aliphatics (Kriegman-King and Reinhard 1991, 1992, 1994; Sivavec and Horney 1997; Butler and Hayes 1998; Haderlein and Pecher 1998), substituted nitrobenzenes (Klausen et al. 1995), and uranium (Hansen et al. 1996; Peterson et al. 1997). Several investigators have shown that Fe(II) adsorbed onto iron oxides (e.g., FeS (ferrous sulfide), FeS_2 (pyrite), and $Fe(CO_2)_2$ (iron carbonate) and mixed-valence iron oxides (e.g., magnetite, goethite, and lepidocrocite) becomes a strong reductant (Klausen et al. 1995; Hofstetter et al. 1999; Strathmann and Stone 2002). These studies have also shown that the rate of reaction increases with the extent of Fe(II) adsorbed, which increases with pH. Apparently, at higher pH, deprotonation of surface hydroxyl groups on the mineral surface creates a negative charge that facilitates adsorption of Fe(II), which results in faster reduction rates. The ability of iron oxides to serve as reductants depends both on the nature of the oxide or mineral present and the manner in which Fe(II) is incorporated into the mineral structure (Klausen et al. 1995; Pecher et al. 1997; Sivavec and Horney 1997).

Recent work has shown that dissimilatory iron-reducing bacteria can reductively dissolve iron oxides and produce Fe(II), which can be adsorbed to the mineral surface and enhance its reactivity toward redox-sensitive pollutants (Heijman et al. 1993, 1995; Hofstetter et al. 1999; Gerlach et al. 2000; Gregory et al. 2001; McCormick et al. 2002). Some Fe(II) could also be present as structural Fe(II) ions within the oxide lattice or within a freshly precipitated oxide coating (e.g., mixed-valent iron minerals such as magnetite and highly reactive green rust) (Stucki 1988). Both adsorbed Fe(II) sites and lattice Fe(II) sites (Figure 2.4) are relatively strong reductants that contribute to natural attenuation of many classes of priority pollutants.

Sulfides, which are produced by sulfate-reducing bacteria or are present as structural components of some minerals such as pyrite, can also contribute to the

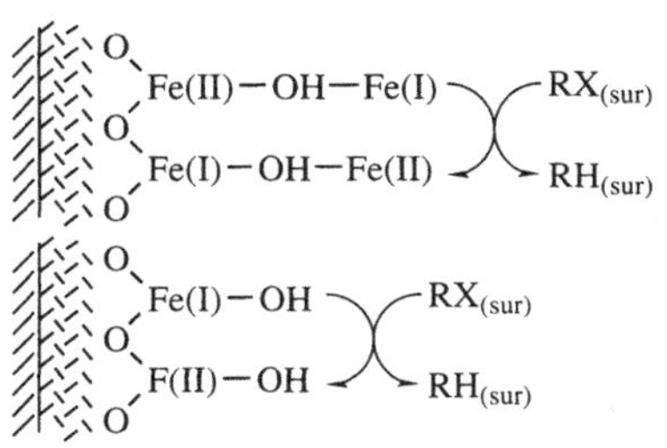

Figure 2.4. Conceptual models of electron transfer from adsorbed or lattice Fe(II) surface sites. (Adapted from Scherer et al. 2000.)

reduction of some priority pollutants. Dissolved sulfides can participate in nucleophilic substitution reactions with halogenated methanes (Roberts et al. 1992). Similar to Fe(II), surface-bound sulfur species can also act as reductants to dechlorinate some chlorinated solvent (Kriegman-King and Reinhard 1994) and scavenge some heavy metals from water (e.g., Cu, Zn, Ni, Cd, and As) serving as immobilizing agents that reduce their aqueous concentrations. Nevertheless, further research is needed to better understand how groundwater chemistry and microbial activities affect the reactivity of such mineral surfaces.

Natural organic matter that originates from plant decay and/or microbial activities can also participate in oxidation–reduction reactions with some priority pollutants. It has been postulated that humic substances with quinone groups can be reduced by some bulk reductant (e.g., Fe(II) or HS^-) to semiquinones or hydroquinones (Dunnivant et al. 1992). These reduced quinines can then act as reductants of Mn(IV), Fe(III) oxides, U(VI), Cr(VI), and a variety of organic contaminants (NRC 2000).

2.2. IMMOBILIZATION AND PHASE CHANGE PROCESSES

Some contaminants can be immobilized by sorption to the aquifer matrix or precipitation from the dissolved phase (Scherer et al. 2000). Sorption is an abiotic reaction where the contaminant is attracted to the surface via hydrophobic interactions, electrostatic attraction, and/or surface complexation (Westall 1987; Schindler 1990). The most common mechanism for organic compounds (particularly nonpolar organic compounds) is sorption due to hydrophobic expulsion from water. Sorption of chlorinated organics onto activated carbon is an important example of a removal process controlled by hydrophobic expulsion (Perrich 1981). Metals, on the other hand, tend to sorb via an electrostatic attraction or surface complexation reaction. In addition, metals can be immobilized by manipulating the solubility of the metal by raising the pH or adding excess ions to form an insoluble mineral. Precipitation of metals by reduction to a less soluble form is a combination of a transformation process followed by an immobilization process and is reviewed in the sections on chemical reaction barriers and biological barriers (Chapter 8).

Both sorption and precipitation processes are generally reversible and may therefore require removal of the reactive materials and accumulated products, depending on the stability of the immobilized compound and the geochemistry of the groundwater.

2.2.1. Sorption

Sorption processes remove contaminants from a groundwater plume via partitioning from the dissolved phase to a solid medium. Thus, sorption affects a wide variety of fate processes, including bioavailability, biodegradability, and volatilization. Sorption is a general term that is used to describe several mechanisms by which a contaminant may partition to a surface or onto a solid matrix. Sorption mechanisms are often classified as adsorption, absorption, and precipitation reactions (Brown 1990). Precipitation reactions involve the formation of a solid of different composition than the solid medium. The precipitate phase is defined by a three-dimensional structure that is not influenced by the solid medium. Adsorption, on the other hand, is defined as accumulation at the solid–water interface without the formation of a three-dimensional structure. Absorption implies that the contaminant has diffused or partitioned within the bulk of the solid medium.

Adsorption is a combination of three possible mechanisms: (1) hydrophobic expulsion (dislike of water), (2) electrostatic attraction (opposite charges attract), and/or (3) surface coordination reactions (hydrolysis, metal complexation, ligand exchange, or hydrogen bonding) (Westall 1987; Schindler 1990). Table 2.4 outlines various terms used to describe sorption mechanisms and provides examples of contaminants that are strongly influenced by the individual mechanisms. Hydrophobic expulsion is the dominant mechanism for most nonpolar (uncharged) organic compounds (e.g., most chlorinated solvents). Partitioning into organic matter present in soils, however, should be viewed as an absorption process, whereby the organic contaminant dissolves into the soil organic matter (Stumm 1992). Hence, the degree of partitioning is strongly correlated to both the compound's octanol–water coefficient, K_{ow}, and the fraction of organic matter

Table 2.4. Major Adsorption Mechanisms

Mechanism	Other Terminology	Examples
Hydrophobic expulsion	Partitioning	Nonpolar organics (e.g., PCBs, PAHs)
Electrostatic attraction	Outer-sphere nonspecific physisorption physical ion exchange	Some anions (e.g., NO_3^-) Alkali and alkaline earth metals (Ba^{2+}, Ca^{2+})
Complexation reaction	Inner-sphere specific chemisorption Chemical ligand exchange	Transition metals (e.g., Cu^{2+}, Pb^{2+}, CrO_4^{2-})

Source: Scherer et al. (2000).

present in the solid material, f_{oc} (Schwarzenbach and Westall 1981; Karickhoff 1984).

Polar compounds such as metals and inorganic nutrients, on the other hand, tend to sorb via electrostatic attraction and surface complexation reactions. Electrostatic attraction is the weaker of the two forces and involves nonspecific adsorption of a charged compound at an oppositely charged surface. A layer of coordinated water molecules separates the compound from the surface, providing only an indirect attachment to the surface, known as outer sphere adsorption. The replacement of one weak (or outer-sphere) ion with another is also known as ion exchange. Unlike electrostatic interactions, surface coordination reactions, such as metal complexation and ligand exchange, involve direct contact between the surface and the contaminant (Westall 1986; Dzombak and Morel 1990; Sposito 1995). The surface reaction is not influenced by a layer of intervening coordinated water molecules and is thus considered inner-sphere or specific adsorption. Surface complexation is the dominant mechanism of sorption for most transition metals (with the exception of some of the alkali and alkaline earth metals, group I and group IIA of the periodic table) (Schindler 1990).

Distribution coefficients generally describe the partitioning of a substance between two immiscible phases:

$$K_D = \frac{C_{X/A}}{C_{X/B}} \qquad [2.20]$$

where K_D = distribution (or partitioning) coefficient
  $C_{X/A}$ = concentration of solute X in phase A (g/m^3)
  $C_{X/B}$ = concentration of solute X in phase B (g/m^3)

Sorption coefficients similarly describe the extent to which an organic chemical is distributed (at equilibrium) between an environmental solid (e.g., aquifer material) and the aqueous phase. For aquifer systems, the partitioning coefficient K_p is often used to characterize the equilibrium distribution of a contaminant between aquifer solids (phase A) and groundwater (phase B). For simplicity, linear partitioning is assumed. K_p is compound specific and increases with contaminant hydrophobicity and natural organic matter content of the aquifer material. K_p can be estimated using empirical correlations, such as that proposed by Karickhoff (1984):

$$K_p = 0.63(K_{ow})(f_{oc})\left(\frac{n}{\rho_b}\right) \qquad [2.21]$$

where K_{ow} = octanol–water partition coefficient (mass in water)/(mass in octanol)
  f_{oc} = mass fraction of organic carbon in the solids
  n = total porosity (water volume)/(total volume)
  ρ_b = bulk density (mass solids)/(total volume)

Other correlations have been proposed for families of structurally similar compounds. For example, the following correlation was proposed by Schwarzenbach et al. (1993) to describe the partitioning characteristics of aromatic hydrocarbons in soil–water systems.

$$K_p = 0.14 f_{oc} K_{ow}^{0.82} \qquad [2.22]$$

Natural Sorbents

Humic materials, particularly peat and activated carbon, are effective sorbents (Perrich 1981; Couillard 1994). Humic materials are complex organic molecules with molecular weights ranging from 500 to 20,000 (Leventhal 1980). Humic materials contain a wide variety of functional groups, including phenolic and carboxylic groups, which provide exchange sites for the sorption of cations, such as heavy metals (e.g., Pb^{2+} and Cd^{2+}) and anions (e.g., CrO_4^{2-} and MoO_4^{2-}). In addition, the high carbon content of these materials may make humic materials a suitable sorbent for nonpolar organic compounds. Some natural organic matter could also contribute to natural attenuation by serving as a sink for pollutant transformation products. For example, aniline products resulting from the reduction of nitroaromatic compounds are strong nucleophiles that can undergo nucleophilic addition to humic and fulvic acids (Thorn et al. 1996; Weber 1996). Further research is needed to determine if such "sequestration" reactions would lead to an acceptable "treatment end point."

Clays and zeolites have a high capacity for ion exchange, particularly cation exchange. A net negative charge is created on clays and zeolites by substitution of lower-valent cations (e.g., Al^{3+}) for higher-valent cations (e.g., Si^{4+}) within the mineral structure (Bohn et al. 1985). The negatively charged surface, in combination with the high surface area of these materials, creates a strong affinity for positively charged contaminants, such as transition metal cations (e.g., Pb^{2+} and Cd^{2+}), but little affinity for anions and nonpolar organic compounds.

The sorption and exchange of anions and cations to naturally occurring oxides and clays has been recognized for well over a century (Thomas 1977). Sorption onto these natural materials has been used for the treatment of industrial wastewaters (Aoki and Munemori 1982) and may provide a promising alternative for containment of metals (Morrison and Spangler 1993; Morrison et al. 1995). Natural oxides that have been investigated for use in a reactive barrier include amorphous ferric oxide ($Fe(OH)_3$), goethite (α-FeOOH), hematite (α-Fe_2O_3), magnetite (Fe_3O_4), and hydrous titanium oxide ($Ti(OH)_4$). In addition, the sorption of metals onto a variety of oxides and clays has been studied extensively to determine the fate and transport of metals in subsurface environments (Zachara et al. 1987; Ainsworth et al. 1989; Ford et al. 1997; Raven et al. 1998; Robertson and Leckie 1998).

The sorption of an ion at a charged oxide surface can be a combination of both electrostatic attraction and surface complexation. In general, metal ions tend to sorb via a complexation reaction (Buffle and Altmann 1987). Both mechanisms are strongly influenced by solution pH and it is experimentally difficult to distinguish between the two without direct spectroscopic measurements (Wersin et al. 1984; Charlet and Manceau 1992). Example reactions describing the surface complexation of U(VI), Mo(II), and Cr(VI) are represented by Eqs. [2.23]–[2.25] (Zachara et al. 1987; Morrison and Spangler 1992).

$$SOH + UO_2^{2+} + 3H_2O \rightarrow SOH - UO_2(OH)_3 + 2\ H^+ \qquad [2.23]$$

$$2\ SOH + MoO_4^{2-} + 2H^+ \rightarrow (SOH_2)^{2-}MoO_4 \qquad [2.24]$$

$$SOH + CrO_4^{2-} + H^+ \rightarrow (SOH_2^+ - CrO_4^{2-})^- \qquad [2.25]$$

where S = oxide surface site.

Anions, such as molybdenate and chromate, tend to sorb more at lower pH values because of increased surface protonation, which creates a more positively charged surface. Cations, on the other hand, tend to sorb more at higher pH because of deprotonation of the surface, which creates a negatively charged surface. The sorption of uranium (UO_2^{2+}), for example, increases significantly at pH values greater than 4. Batch studies conducted by Morrison and Spangler found that a pH value of 6 was effective for the sorption of both molybdenate and uranium (Morrison and Spangler 1992). Note that contaminant removal by sorption can be hindered by the presence of some sorption anions (such as sulfate and carbonate) that compete for surface sites (Zachara et al. 1987; Morrison and Spangler 1993, 1995).

Spodic soil materials (B horizon soils) are comprised of various aluminum and iron oxides that have the potential for surface complexation and ion exchange processes. Lindberg and co-workers (1997) have shown that spodic material can effectively remove arsenic from groundwater. In natural waters, arsenic exists in two oxidation states, As(III) and As(V). As(III), or arsenite, exists primarily as an uncharged species (H_3AsO_3), whereas the dominant forms of As(V) are anions ($H_2AsO_4^-$ and $HAsO_4^{2-}$). As found with other sorbent materials (e.g., humic materials), the effect of competing or complexing ions, such as sulfate and carbonate, has to be considered.

The sensitivity to geochemical changes, such as pH, and presence of competing ions will require careful management of field applications that use oxides as a sorbent material. It may be necessary to incorporate additional materials to maintain optimal pH conditions. The optimal pH range will depend on both groundwater composition and contaminant speciation.

2.2.2. Precipitation

Immobilization of contaminants via precipitation occurs when the solubility limit is exceeded in solution. The precipitate phase has a three-dimensional structure that is not influenced by the surface of the solid medium (Brown 1990).

Precipitation is typically due to increases in pH, such as when groundwater flows through a limestone formation. Contaminants that can be attenuated by this mechanism include uranium, which precipitates as calcium urinate, and phosphorus, which precipitates as calcium phosphate.

Another condition that is conducive to precipitation is when an ion that coprecipitates with the contaminant is present in excess. For example, high concentrations of phosphate in the presence of Pb^{2+} will cause the precipitation of insoluble lead phosphates. The abundance of phosphate minerals, such as apatite $(Ca_{10}(PO_4)_6(OH)_2)$, leads to the precipitation of low-solubility metal phosphates (Ma et al. 1993, 1994a,b, 1995; Chen et al. 1997). The major mechanism for Pb^{2+} removal appears to be the formation of hydroxypyromorphite $(Pb_{10}(PO_4)_6 (OH)_2)$. Other removal mechanisms, however, such as ion exchange and surface adsorption, may be important in the phosphate-based removal of Cd and Zn (Chen et al. 1997). Aqueous aluminum, cadmium, copper, ferrous iron, nickel, and tin can hinder the removal of lead by precipitation of phosphate minerals. The presence of anions, such as nitrate, chloride, and sulfate, does not significantly affect lead removal; however, high carbonate concentrations result in lower lead removal efficiency due to lower hydroxyapatite solubility at increased pH values.

Immobilization by precipitation has the advantage that precipitation reactions tend to be less dependent on groundwater conditions than adsorption reactions, which can be highly dependent on pH. The most significant uncertainty for both immobilization techniques (i.e., adsorption and precipitation) as natural attenuation mechanisms is assessing the stability of the contaminant and the risk of future remobilization by desorption or dissolution.

2.2.3. Volatilization

Volatilization is the conversion of liquid or solid compounds with a high vapor pressure into gaseous form. Volatile organic compounds (VOCs) such as monoaromatic hydrocarbons and chlorinated solvents can escape to the atmosphere when they come into contact with a gas phase in the unsaturated zone of an aquifer, or during sampling and analysis. Field studies have shown that up to 10% of monoaromatic hydrocarbons can be removed by volatilization (Chiang et al. 1989). Volatilization is an important process to quantity (1) to account for contaminant "loss" from groundwater and evaluate hazards if toxic or explosive vapors accumulate in basements, (2) to preclude sampling and analytical errors, (3) to understand pollutant fate and transport, including the potential formation and migration of vapor-phase plumes, and (4) as a basis for contaminant detection and soil monitoring. Figure 2.5 illustrates volatilization from a liquid or solid surface.

In addition to the vapor pressure of the compound, which represents its tendency to evaporate, volatilization also depends on pH, temperature, soil organic matter content, humidity, sorption, and air flow. Volatilization generally decreases with increasing compound solubility and molecular weight.

Henry's law is commonly used to describe volatilization of organic compounds from water:

$$P_{org} = K_H C_w \tag{2.26}$$

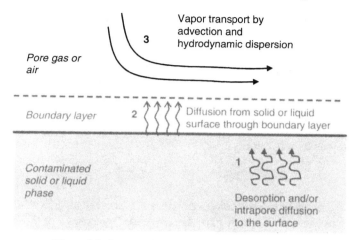

Figure 2.5. Processes involved in volatilization of VOCs.

where P_{org} is the partial pressure of the organic compound in the vapor phase (atm), C_w is its concentration in water (mol/m^3 or mol/L), and K_{H} is Henry's constant (atm-m^3/mol). K_{H} is approximately equal to the vapor pressure of the pure compound divided by its solubility in water. Therefore, K_{H} values are an excellent indication of the tendency of a compound to volatilize.

Compounds with K_{H} values lower than 3×10^{-7} atm-m^3/mol are generally considered nonvolatile (e.g., DDT, phenol, and pentachlorophenol), whereas those with K_{H} values higher than 10^{-3} atm-m^3/mol are considered volatile (e.g., BTEX, TCE, and carbon tetrachloride) and can be effectively removed from water by air stripping.

On occasion, Henry's law is expressed as

$$C_{\text{air}} = K'_{\text{H}} C_w \qquad [2.27]$$

where C_{air} is the organic concentration in air and K'_{H} is the dimensionless Henry's constant. The two different Henry's constants are related to each other according to

$$K'_{\text{H}} = K_{\text{H}}/RT \qquad [2.28]$$

where R is the universal gas constant (8.21×10^{-5} atm-m^3/mol-K) and T is the absolute temperature in kelvin units (K). Recall that zero in the Celsius scale equals 273.15 K. Appendix C lists Henry's constant and other physicochemical parameters for organic compounds commonly found in the environment.

2.2.4. Complexation

Complexation reactions can facilitate the solubilization and transport of toxic heavy metals such as cadmium, chromium, uranium, and plutonium. Complexes are ions formed by combining simple cations or anions with other molecules that act as *ligamentum*, or binding agents (i.e., ligands). The cation is often a heavy metal that serves as a central atom. The anion, which serves as the ligand, can be a common inorganic species such as Cl^-, F^-, Br^-, SO_4^{2-}, PO_4^{3-}, and CO_3^{2-}. For example, uranium often forms uranyl complexes with phosphate $(UO_2(HPO_4)_2^{2-})$, carbonate $(UO_2(CO_3)_2^{2-})$, and fluoride $(UO_2F_2^0)$, which can account for the majority of the uranium found dissolved in groundwater (Langmuir 1978). Organic ligands that commonly bind and cotransport heavy metals include amino acids and humic acids.

Complexation reactions can be simple, as when a metal and a ligand associate due to electrostatic attraction:

$$Mn^{2+} + Cl^- \rightarrow MnCl^+ \qquad [2.29]$$

Some heavy metals can also combine with water-derived hydroxyl ions to form hydroxo complexes, as illustrated for the sequential hydrolysis of Cr^{3+}:

$$
\begin{aligned}
Cr^{3+} + OH^- &\rightarrow Cr(OH)^{2+} \\
Cr(OH)^{2+} + OH^- &\rightarrow Cr(OH)_2^+ \\
Cr(OH)^{2-} + OH^- &\rightarrow Cr(OH)_3^0 \\
Cr(OH)_3^0 + OH^- &\rightarrow Cr(OH)_4^-
\end{aligned}
\qquad [2.30]
$$

Such reactions can result in negatively charged complexes that are less prone to sorption onto negatively charged soil surfaces (e.g., silica minerals), which enhances their mobility in groundwater.

REFERENCES

Ainsworth, C.C., D.C. Girvin, J.M. Zachara, and S.C. Smith (1989). Chromate adsorption on goethite: effects of aluminum substitution. *Soil Sci. Soc. Am. J.* **53**:411–418.

Aoki, T. and M. Munemori (1982). Recovery of chromium(VI) from wastewater with iron(III) hydroxide—I. *Water Res.* **16**:793–796.

Bohn, H.L., B.L. McNeal, and G.A. O'Connor (1985). *Soil Chemistry*. John Wiley & Sons, Hoboken, NJ.

Brown, G.E. (1990). Spectroscopic studies of chemisorption reaction mechanisms at oxide–water interfaces. In *Mineral–Water Interface Geochemistry*, M.F. Hochella, Jr. and A.F. White (Eds.). Mineralogical Society of America, Washington, DC, Vol. 23, pp. 309–363.

Buerge, I. and S. Hug (1997). Kinetics and pH dependence of Cr(VI) reduction by Fe(II). *Environ. Sci. Technol.* **31**:142–143.

Buffle, J. and R.S. Altmann (1987). Interpretation of metal complexation by heterogeneous complexants. In *Aquatic Surface Chemistry*, W. Stumm (Ed.). John Wiley & Sons, Hoboken, NJ, pp. 351–383.

Burkhard, N., and J.A. Guth (1981). Chemical hydrolysis of 2-chloro-4,6-bis(alkylamino)-1,3,5-triazine herbicides and their breakdown in soil under the influence of adsorption. *Pestic. Sci.* **12**:45–52.

Butler, E.C. and K.F. Hayes (1998). Effects of solution composition and pH on the reductive dechlorination of hexachloroethane by iron sulfide. *Environ. Sci. Technol.* **32**(9):1276.

Charlet, L. and A.A. Manceau (1992). X-ray adsorption spectroscopic study of the sorption of Cr(III) at the oxide–water interface: II. Adsorption, coprecipitation, and surface precipitation on hydrous ferric oxide. *J. Colloid Interface Sci.* **148**(2):443–458.

Chen, X., J.V. Wright, J.L. Conca, and L.M. Peurrung (1997). Effects of pH on heavy metal sorption on mineral apatite. *Environ. Sci. Technol.* **31**(3):624–630.

Chiang, C.Y., J.P. Salanitro, E.Y. Chai, J.D. Colthart, and C.L. Klein (1989). Aerobic biodegradation of benzene, toluene, and xylene in a sandy aquifer—data analysis and computer modeling. *Ground Water* **27**:823–834.

Couillard, D. (1994). The use of peat in wastewater treatment—review. *Water Res.* **28**(6):1261–1274.

Dunnivant, F.M., R.P. Schwarzenbach, and D.L. Macalady (1992). Reduction of substituted nitrobenzenes in aqueous solutions containing natural organic matter. *Environ. Sci. Technol.* **26**:2133–2141.

Dzombak, D.A. and F.M.M. Morel (1990). *Surface Complexation Modeling: Hydrous Ferric Oxide*. John Wiley & Sons, Hoboken, NJ.

Ford, R.G., P.M. Bertsch, and K.J. Farley (1997). Changes in transition and heavy metal partitioning during hydrous iron oxide aging. *Environ. Sci. Technol.* **31**(7):2028–2033.

Gerlach, R., A. Cunningham, and F. Caccavo (2000). Dissimilatory iron-reducing bacteria can influence the reduction of CCl_4 by iron metal. *Environ. Sci. Technol.* **34**:2461–2464.

Gregory, K., M. Von Arb, P.J.J. Alvarez, M. Scherer, and G.F. Parkin (2001). Biogeochemical removal of RDX using iron oxide and *Geobacter metallireducens* GS-15. In *In Situ and On-Site Bioremediation: The Sixth International Symposium*, San Diego, CA, June 4–7, 2001. Battelle Press, Columbus, OH, Vol. 6, No. 3, pp. 1–8.

Haderlein, S.B. and K. Pecher (1998). Pollutant reduction in heterogeneous Fe(II)/Fe(III) systems. In *Mineral–Water Interfacial Reactions: Kinetics and Mechanisms*, T. Grundl and D. Sparks (Eds.) ACS Symposium Series 715, American Chemical Society, Washington, DC.

Hansen, H.C.B., C.B. Koch, H. Nancke-Krogh, O.K. Borggaard, and J. Sorensen (1996). Abiotic nitrate reduction to ammonium: key role of green rust. *Environ. Sci. Technol.* **30**(6):2053–2056.

Harris, J.C. (1982). Rate of hydrolysis. In *Handbook of Chemical Property Estimation Methods: Environmental Behavior of Organic Compounds*, W.J. Lyman, W.F. Reehl, and D.H. Rosenblatt (Eds.). McGraw Hill, New York.

Heijman, C.G., C. Holliger, M.A. Glaus, R.P. Schwarzenbach, and J. Zeyer (1993). Abiotic reduction of 4-chloronitrobenzene to 4-chloronitroaniline in a dissimilatory iron-reducing enrichment culture. *Appl. Environ. Microbiol.* **59**:4350–4353.

Heijman, C.G., E. Greider, C. Holliger, and R.P. Schwarzenbach (1995). Reduction of nitroaromatic compounds coupled to iron reduction in laboratory aquifer columns. *Environ. Sci Technol.* **29**:775–783.

Hofstetter, T.B., C.G. Heijman, S.B. Haderlein, C. Holliger, and R.P. Schwarzenbach (1999). Complete reduction of TNT and other (poly)nitroaromatic compounds under iron-reducing subsurface conditions. *Environ. Sci. Technol.* **33**:1479–1487.

IUPAC (1997). *Compendium of Chemical Terminology*, 2nd ed.

Jafvert, C.T. and N.L. Wolfe (1987). Degradation of selected halogenated ethanes in anoxic sediment–water systems. *Environ. Tox. Chem.* **6**:827–837.

Jeffers, P.M., L. Ward, L. Woytowitch, and N.L. Wolfe (1989). Homogeneous hydrolysis rate constants for selected methanes, ethanes, ethenes, and propanes. *Environ. Sci. Technol.* **23**:965–969.

Karickhoff, S.W. (1984). Organic pollutant sorption in aquatic systems. *J. Hydraul. Eng. ASCE* **10**(6):707–735.

Klausen, J., S.P. Trober, S.B. Haderlein, and R.P. Schwarzenbach (1995). Reduction of substituted nitrobenzenes by Fe(II) in aqueous mineral suspensions. *Environ. Sci. Technol.* **29**:2396–2404.

Konrad, J.G. and G. Chesters (1969). Degradation in soils of Ciodrin, an organophosphate insecticide. *J. Agric. Food Chem.* **17**:226–230.

Kriegman-King, M.R. and M. Reinhard (1991). Reduction of hexachloroethane and carbon tetrachloride at surfaces of biotite, vermiculite, pyrite, and marcasite. In *Organic Substances and Sediments in Water*, R. Baker (Ed.). Lewis Publishers, Chelsea, MI, Vol. 2, pp. 349–364.

Kriegman-King, M.R. and M. Reinhard (1992). Transformation of carbon tetrachloride in the presence of sulfide, biotite, and vermiculite. *Environ. Sci. Technol.* **26**(11):2198–2206.

Kriegman-King, M.R. and M. Reinhard (1994). Transformation of carbon tetrachloride by pyrite in aqueous solution. *Environ. Sci. Technol.* **28**(4):692–700.

Lacorte, S., S.B. Lartiges, P. Garrigner, and D. Barcelo (1995). Degradation of organophosphorous pesticides and their transformation products in estuarine waters. *Environ. Sci. Technol.* **29**:431–438.

Langmuir, D. (1978). Uranium solution–mineral equilibria at low temperatures with applications to sedimentary ore deposits. *Geochim. Cosmochim. Acta* **42**:547–569.

Larson, R.A. and E.J. Weber (1994). *Reaction Mechanisms in Environmental Organic Chemisty.* Lewis Publishers, Boca Raton, FL.

Leventhal, J.S. (1980). *Geology and Mineral Technology of the Grants Uranium Region 1979.* Memoir 38, New Mexico Bureau of Mines & Mineral Resources, pp. 75–85.

Lindberg, J., J. Sterneland, P.-O. Johansson, and J.P. Gustafsson (1997). Spodic material for *in situ* treatment of arsenic in ground water. *Ground Water Monitoring and Remediation* **Fall**:125–130.

Ma, Q.Y., S.J. Traina, and T.J. Logan (1993). *In situ* lead immobilization by apatite. *Environ. Sci. Technol.* **27**:1803–1810.

Ma, Q.Y., S.J. Traina, T.J. Logan, and J.A. Ryan (1994a). Effects of aqueous Al, Cd, Cu, Fe(II), Ni, and Zn on Pb immobilization by hydroxyapatite. *Environ. Sci. Technol.* **28**:1219–1228.

Ma, Q.Y., S.J. Traina, T.J. Logan, and J.A. Ryan (1994b). Effects of NO_3^-, Cl^-, F^-, SO_4^{2-}, and CO_3^{2-} on Pb^{2+} immobilization by hydroxyapatite. *Environ. Sci. Technol.* **28**:408–418.

Ma, Q.Y., T.J. Logan, and S.J. Traina (1995). Lead immobilization from aqueous solutions and contaminated soils using phosphate rocks. *Environ. Sci. Technol.* **29**(4):1118–1126.

Mabey, W. and T. Mill (1978). Critical review of hydrolysis of organic compounds in water under environmental condition. *J. Phys. Chem. Ref. Data* **7**:383–415.

Macalady, D.L., P.G. Tratnyek, and T.J. Grundl (1986). Abiotic reduction reactions of anthropogenic organic chemical in anaerobic systems: a critical review. *J. Contam. Hydrol.* **1**:1–28.

Masters, G.M. (1997). *Introduction to Environmental Engineering and Science*, 2nd ed. Prentice Hall, Upper Saddle River, NJ.

McCormick, M.L., E.J. Bouwer, and P. Adriaens (2002). Carbon tetrachloride transformation in a model iron-reducing culture: relative kinetics of biotic and abiotic reactions. *Environ. Sci. Technol.* **36**(3):403–410.

Morrison, S.J. and R.R. Spangler (1992). Extraction of uranium and molybdenum from aqueous solution: a survey of industrial materials for use in chemical barriers for uranium mill tailing remediation. *Environ. Sci. Technol.* **26**(10):1922–1931.

Morrison, S.J. and R.R. Spangler (1993). Chemical barriers for controlling groundwater contamination. *Environ. Prog.* **12**(3):175–181.

Morrison, S.J., R.R. Spangler, and V.S. Tripathi (1995). Adsorption of uranium(VI) on amorphous ferric oxyhydroxide at high concentrations of dissolved carbon(IV) and sulfur(VI). *J. Contam. Hydrol.* **17**:333–346.

NRC (National Research Council) (1994). Committee on Groundwater Cleanup Alternatives. In *Alternatives for Groundwater Cleanup*. National Academy Press, Washington, DC.

NRC (National Research Council) (2000). *Natural Attenuation for Groundwater Remediation*. National Academy Press, Washington, DC.

Neely, W.N. (1985). Hydrolysis. In *Environmental Exposure from Chemicals*, W.B. Neely and G.E. Blau (Eds.). CRC Press, Boca Raton, FL.

Pecher, K., S.B. Haderlein, and R.P. Schwarzenbach (1997). Transformation of polyhalogenated alkanes in suspensions of ferrous iron oxides. 213th National Meeting, San Francisco, CA, American Chemical Society.

Perrich, J.P. (1981). *Activated Carbon Adsorption for Wastewater Treatment*. CRC Press, Boca Raton, FL.

Peterson, M.L., A.F. White, G.E. Brown, and G.A. Parks (1997). Surface passivation of magnetite by reaction with aqueous Cr(VI): XAFS and TEM results. *Environ. Sci. Technol.* **31**(5):1573–1576.

Raven, K.P., A. Jain, and R.H. Loeppert (1998). Arsenite and arsenate adsorption on ferrihydrite: kinetics, equilibrium, and adsorption envelopes. *Environ. Sci. Technol.* **32**:344–349.

Rittman, B.E., E. Seagren, B.A. Wrenn, A.J. Valocchi, C. Ray, and L. Raskin (1994). *In Situ Bioremediation*. Noyes Publications, Park Ridge, NJ.

Roberts, A.L., P.N. Sanborn, and P.M. Gschwend (1992). Nucleophilic substitution reactions of dihalomethanes with hydrogen sulfide species. *Environ. Sci. Technol.* **26**:2236–2274.

Robertson, A.P. and J.O. Leckie (1998). Acid/base, copper binding, and Cu^{2+}/H^+ exchange properties of goethite, an experimental modeling study. *Environ. Sci. Technol.* **32**:2519–2530.

Scherer, M.M., S. Richter, R.L. Valentine, and P.J.J. Alvarez (2000). Chemistry and microbiology of permeable reactive barriers for in situ groundwater cleanup. *Crit. Rev. Environ. Sci. Technol.* **30**:363–411.

Schindler, P.W. (1990). Co-adsorption of metal ions and organic ligands: formation of ternary surface complexes. In *Mineral–Water Interface Geochemistry*, M. Hochella Jr. and A.F. White (Eds.). Mineralogical Society of America, Washington, DC, Vol. 23, pp. 281–307.

Schwarzenbach, R.P. and J. Westall (1981). Transport of nonpolar organic compounds from surface water to groundwater. Laboratory sorption studies. *Environ. Sci. Technol.* **15**:1360–1367.

Schwarzenbach, R.P., P.M. Gschwend, and D.M. Imboden (1993). *Environmental Organic Chemistry*. John Wiley & Sons, Hoboken, NJ.

Schwarzenbach, R.P., R. Stierli, K. Lanz, and J. Zeyer (1990). Quinone and iron porphyrin mediated reduction of nitroaromatic compounds in homogeneous aqueous solution. *Environ. Sci. Technol.* **24**:1566–1574.

Simon, R., D. Colón, C. Stevens, and E.J. Weber (2000). Effect of redox zonation on the reductive transformation of *p*-cyanonitrobenzene in a laboratory sediment column. *Environ. Sci. Technol.* **34**:3617–3622.

Sivavec, T.M. and D.P. Horney (1997). Reduction of chlorinated solvents by Fe(II) minerals. 213th National Meeting, San Francisco, CA, American Chemical Society.

Sposito, G. (1995). Adsorption as a problem in coordination chemistry. The concept of the surface complex. In *Aquatic Chemistry: Interfacial and Interspecies Processes*, C.P. Huang, C.R. O'Melia, and J.J. Morgan (Eds.). Advances in Chemistry Series No. 244, American Chemical Society, Washington, DC.

Strathmann, T. and A. Stone (2002). Reduction of the pesticides oxamyl and methomyl by Fe(II): effect of pH and inorganic ligands. *Environ. Sci. Technol.* **36**:653–661.

Stucki, J.W. (1988). Structural iron in smectites. *Nat. Adv. Sci. Inst. Ser., Ser. C* **217**:625–675.

Stumm, W. (1992). *Chemistry of the Solid–Water Interface: Processes at the Mineral–Water and Particle–Water Interface of Natural Systems.* John Wiley & Sons, Hoboken, NJ, p. 428.

Thomas, G.W. (1977). Historical developments in soil chemistry: ion exchange. *Soil Sci. Soc. Am. J.* **41**:230–238.

Thorn, K.A., P.J. Pettigre, and W.S. Goldenberg (1996). Covalent binding of aniline to humic substances.2. ^{15}N NMR studies of nucleophilic addition reactions. *Environ. Sci. Technol.* **30**:2784–2775.

Weber, E.J. (1996). Iron-mediated reductive transformations: investigation of reaction mechanism. *Environ. Sci. Technol.* **30**(2):716–719.

Werner, P. (1985). A new way for the decontamination of polluted aquifers by biodegradation. *Water Supply* **3**:41–47.

Wersin, P., M.F. Hochella, P. Persson, and G. Redden (1984). Interaction between aqueous uranium and sulfide minerals: spectroscopic evidence for sorption and reduction. *Geochim. Cosmochim. Acta* **58**(13):2829–2845.

Wolfe, N.L. (1980). Organophosphate and organophosphorothioate esters: application of linear free energy relationships to estimate hydrolysis rate constants for use in environmental fate assessment. *Chemosphere* **9**:571–579.

Wolfe, N.L., R.G. Zepp, J.A. Gordon, G.L. Baughman, and D.M. Cline (1977). Kinetics of chemical degradation of malathion in water. *Environ. Sci. Technol.* **11**:88–93.

Wolfe, N.L., R.G. Zeppand, and D.F. Paris (1978). Carbaryl, propham, and chloropropham: a comparison of the rates of hydrolysis and photolysis with the rate of biolysis. *Water Res.* **12**:565–571.

Westall, J. (1987). Adsorption mechanisms in aquatic chemistry. In *Aquatic Surface Chemistry*, W. Stumm (Ed.). John Wiley & Sons, Hoboken, NJ, pp. 3–32.

Westall, J.C. (1986). Reactions at the oxide–solution interface: chemical and electrostatic models. In *Geochemical Processes at Mineral Surfaces*, J.A. Davis and K.F. Hayes (Eds.). American Chemical Society, Washington, DC, Vol. 323, pp. 54–78.

Zachara, J.M., D.C. Girvin, R.L. Schmidt, and C.T. Resch (1987). Chromate adsorption on amorphous iron oxyhydroxide in the presence of major groundwater ions. *Environ. Sci. Technol.* **21**(6):589–594.

3

BIODEGRADATION PRINCIPLES

"The role of the infinitely small in nature is infinitely large."

Louis Pasteur

The ability of microorganisms to degrade and recycle environmental pollutants has been known for many centuries, and biodegradation is often the most important mechanism that attenuates the migration of dissolved organic contaminants in groundwater (National Research Council 1993). Indigenous microorganisms often exploit the biodegradation of many different types of organic compounds as a metabolic niche to obtain energy and building blocks for the synthesis of new cellular material, although the rate and extent of biodegradation can be highly variable, depending on the type of microbial community structure and the prevailing environmental conditions. Sometimes these natural degradation processes proceed in contaminated environments without the need for human intervention, whereas other times some degree of intervention is necessary to stimulate biodegradation.

The astonishing diversity of microbial capabilities to degrade organic compounds and the variability of site-specific conditions make it difficult to predict the rate and extent at which different pollutants may be degraded in different environments. This difficulty often invalidates the extrapolation of experience from one site to another. Nevertheless, some generalizations can be made regarding the hydrogeological, chemical, and biological factors that affect the success of biodegradation as a natural attenuation mechanism. These factors will be discussed below, following a review of microbial physiology and genetics and a definition of terms commonly used within the context of biological treatment processes. The effect of chemical structure on the propensity of organic pollutants to be biodegraded will also be addressed along with common biodegradation mechanisms. Microbial interactions will also be reviewed within the context of biodegradation. This chapter will then conclude with a discussion of biodegradation kinetics.

Bioremediation and Natural Attenuation: Process Fundamentals and Mathematical Models
By Pedro J. J. Alvarez and Walter A. Illman   Copyright © 2006 John Wiley & Sons, Inc.

3.1. THE BACTERIAL ENGINE

Bacteria are unicellular microorganisms that belong to the prokaryotes phylum. Thus, they lack a true nucleus (i.e., chromosomes surrounded by a nuclear membrane). Rather, bacterial DNA is generally stored in a single chromosome that is circular and supercoiled; although smaller extrachromosomal genetic elements known as plasmids may also be present.

Whereas fungi and other eukaryotic organisms can contribute to bioremediation, bacteria are often the prime workers responsible for biodegradation of organics. Bacteria are also the prime agents of biogeochemical cycles. Thus, it is important to review their physiologic and metabolic characteristics as well as the genetic basis for the expression of specific biotransformation capabilities.

3.1.1. Bacterial Physiology

The primary constituents of the bacterial cell are depicted in Figure 3.1. Considering these constituents from the inside to the outside of the cell, their primary functions are:

1. *Nuclear Region.* The bacterial nucleoid is a double-stranded, helical polymer of deoxy-ribonucleic acid (DNA) that contains the genetic information necessary for the reproduction of all cell components and for performing specific traits (e.g., biodegradation of pollutants).
2. *Ribosomes.* These are structures of ribonucleic acid (RNA) and protein that catalyze the synthesis of enzymes and other proteins.
3. *Flagellum.* The flagella are helical, rigid, locomotive organelles that rotate as a propeller. Some bacterial strains have multiple flagella (polytrichous) while others may have none (atrichous).

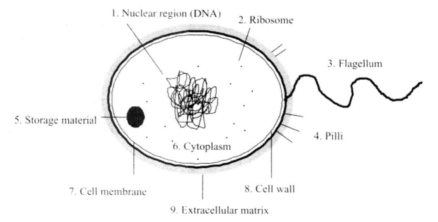

Figure 3.1. Principal components of the bacterial cell.

4. *Pilli*. These filaments are similar to flagella that are used for attachment and sexual conjugation (i.e., transfer of genetic material from one cell to another) rather than for locomotion.

5. *Storage Material*. Cells can store excess carbonaceous material as poly-β-hydroxybutyrate (PHB) as energy reserves. Some bacteria store phosphorus in polyphosphate granules (volutin) and others accumulate sulfur in similar granules.

6. *Cytoplasm*: This is a fibrillar material inside the cell that contains the enzyme systems required for the degradation of soluble food and the synthesis of new cell material.

7. *Cell Membrane*. The membrane is a semipermeable phospholipid bilayer that allows soluble food and nutrients to pass into the cell and waste products to pass out; it contains exocellular enzymes that break down substrates into smaller subunits outside the cell to facilitate uptake, as well as the enzyme transport system and the enzymes for respiration and energy transduction (i.e., ATP synthesis).

8. *Cell Wall*. The cell wall is a rigid container made of peptidoglycan that provides shape and protection against osmotic stress. Bacterial shapes include spherical (coccus), rodlike (bacillus), and comma-shaped (vibrio).

9. *Extracellular Matrix*. This slime layer (a.k.a. glycocalyx) is an extracellular, natural polyelectrolyte (long-chain polysaccharide with some protein) that helps organisms to flocculate (stick together) and attach to surfaces.

Bacterial cells are typically very small, on the order of 1–3 μm in size. However, giant bacteria that are visible to the naked eye have recently been discovered (e.g., the green sulfur bacterium *Thiomargarita namibiensis*, which is about 0.75 mm long). The relatively small size of bacteria endows them with a relatively large surface area in relation to their volume. For example, for spherical cells with radius r, the surface area (A) to volume (V) ratio is

$$\frac{A}{V} = \frac{4\pi r^2}{\frac{4}{3}\pi r^3} = \frac{3}{r}$$

Thus, the smaller the radius, the larger the surface area to volume ratio. A large A/V ratio is advantageous for coming into contact with nutrients and substrates, including environmental pollutants.

3.1.2. Bacterial Metabolism

Similar to other living beings, bacteria assimilate nutrients and substrates, convert them into cell components, and excrete waste products (Figure 3.2). The synthesis of new cellular materials is known as *anabolism*, which is often an endergonic process (i.e., it requires energy). The energy required for anabolism and to perform work and drive other important cell functions (e.g., locomotion and signal

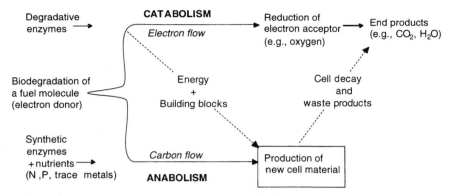

Figure 3.2. Simplified view of bacterial heterotrophic metabolism.

amplification) is obtained by the oxidation of fuel molecules. This energy-yielding biodegradation process, which also produces organic building blocks for cell synthesis, is known as *catabolism*. The term *metabolism* thus refers to all biochemical reactions occurring in a cell, both anabolic and catabolic.

Most energy-yielding biochemical reactions are oxidative in nature. That is, they involve the transfer of electrons from electron donors (e.g., the target pollutant to be oxidized) to electron acceptors (e.g., O_2) (Figure 3.3). In some anaerobic environments, oxidized pollutants such as highly chlorinated solvents can serve as electron acceptors, with hydrogen serving as the electron donor.

The energy harvested in these redox reactions is then stored in the two high-energy phosphoanhydride bonds of adenosine triphosphate (ATP) (Figure 3.4), which is the main energy currency in all biochemical systems. This energy can be released upon ATP hydrolysis:

$$ATP + H_2O \rightarrow ADP + H_3PO_4 + Energy(7.5\,kcal/mol)$$

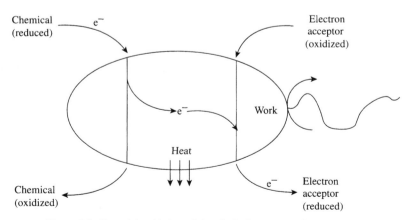

Figure 3.3. Bacterial oxidation of chemicals that serve as fuel molecules.

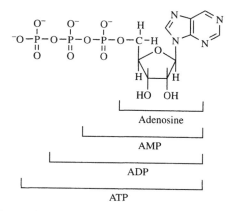

Figure 3.4. Structure of adenosine triphosphate (ATP), adenosine diphosphate (ADP) and adenosine monophosphate (AMP). ATP hydrolysis produces ADP, which can subsequently hydrolyze to AMP.

Figures 3.2 and 3.3 show that bacterial metabolism requires an energy source, a carbon source, an electron donor, and an electron acceptor. These four requirements are the basis for the metabolic diversity encountered in nature. Table 3.1 summarizes bacterial classifications according to these metabolic requirements and traits.

3.1.3. Gene Expression

Bacterial cells are highly structured and nonrandom entities that are capable of replication, growth, and biodegradation of many substances of natural or anthropogenic origin. The ability of cells to perform specific functions is strain-specific, and it is coded in a complex and highly regulated genetic system that serves as the "master plan" or "blueprint" of the cell. Thus, many microbiological phenomena

Table 3.1. Bacterial Classification According to Metabolic Traits

1. Energy source	(a) Phototrophs (use light)
	(b) Chemotrophs (use reduced chemical compound)
2. Carbon source	(a) Heterotrophs (use organic carbon)
	(b) Autotrophs (use inorganic carbon such as CO_2 and bicarbonate)
3. Electron donor	(a) Organotrophs (use organic compounds as fuel molecules)
	(b) Lithotrophs (use inorganic compounds such as NH_4^+, H_2S, and H_2)
4. Electron acceptor[a]	(a) Aerobes (use molecular oxygen and reduce it to water; O_2/H_2O)
	(b) Anaerobes (use combined inorganic oxygen). For example,

Denitrifiers	NO_3^-/N_2
Sulfate reducers	SO_4^{2-}/H_2S
Iron reducers	$Fe(III)/Fe(II)$
Methanogens	CO_2/CH_4

[a]*Note*: Fermenters do not use an external electron acceptor. Fermentation involves the use of different parts of an organic substrate as electron donor and electron acceptor simultaneously.

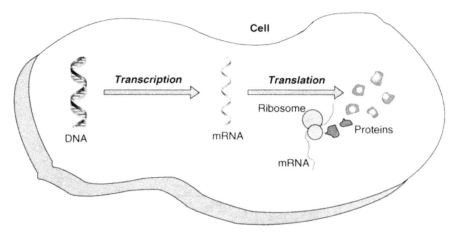

Figure 3.5. Transcription of the genetic code and translation into proteins: roles of DNA, mRNA, and ribosomes.

important for environmental pollution control are easily unmasked with only a limited knowledge of how bacteria express their genetic code.

The enzymes that mediate biodegradation and other metabolic reactions, as well as other proteins that serve as structural components, are coded in genes. Most genes encode proteins and most proteins are enzymes, which are polymers of amino acids. Two processes are involved in protein and enzyme synthesis: transcription and translation (Figure 3.5).

During transcription, the genetic code is transferred to the messenger RNA (mRNA). Recall that the genetic code is imparted onto the nitrogen base sequence of DNA. Specifically, the code is a sequence of four bases: adenine (A), thymine (T), guanine (G), and cytosine (C). During transcription, a complementary based sequence is synthesized (i.e., mRNA) (Figure 3.6). This is a sequence of the

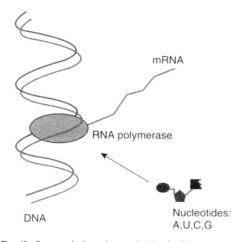

Figure 3.6. Detail of transcription: the synthesis of mRNA by RNA polymerase.

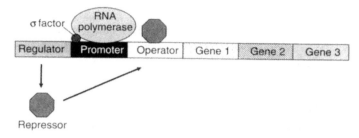

Figure 3.7. Example of an operon whose expression is repressed by a regulatory protein (repressor).

same bases, except that uracil (U) replaces thymine. The resulting mRNA is a single-stranded nucleic acid polymer that carries the code to the ribosome, the site of protein synthesis. The synthesis of mRNA is catalyzed by RNA polymerase. This enzyme system is composed of a core enzyme, which is responsible for mRNA elongation by catalyzing the polymerization of individual nucleotides, and the sigma factor (σ), which is a small protein involved in recognition of the start site. RNA polymerase binds to DNA with the help of the sigma factor at the promoter region, which signals the beginning of a gene (Figure 3.7). The binding of RNA polymerase creates a localized separation of the two DNA strands, exposing the base pairs to be transcribed. RNA polymerase then zips down a DNA strand, adding ribose nucleotides to a growing mRNA strand that is complementary to the nitrogen base sequence of the DNA (i.e., A on DNA produces U on mRNA, T yields A, G yields C, and C yields G).

During translation, the nucleotide sequence in mRNA directs polymerization of a specific sequence of amino acids forming a specific enzyme or protein (i.e., the "message" is read and decoded). This occurs in the ribosome, which is an assemblage of ribosomal RNA (rRNA) and proteins that acts as the catalyst for translation. This process also requires transfer RNA molecules (tRNA), which transport amino acids to the ribosome. Different tRNA molecules exist in the cell, each of which can bind only one specific amino acid.

Considerable energy and building blocks are required to synthesize enzymes. Unnecessary enzymes and proteins act as a drain on energy reserves that could be used for life-sustaining processes. Thus, enzyme synthesis is a highly regulated process. Some enzymes are always present (i.e., *constitutive* enzymes), such as those that participate in central metabolic functions. However, the concentration of constitutive enzymes can vary in response to the cell needs. In contrast, *inducible* enzymes are produced only under specific conditions, such as when some environmental pollutants are present and the cell exploits this opportunity to use them as carbon or energy sources. Therefore, not all genes need to be expressed all of the time. Regulation of gene expression occurs through several different processes—most of which occur at the level of transcription.

Different DNA segments are important for the regulation of gene expression (Figure 3.7). First is the *promoter*, which is a specific nucleotide sequence on the DNA located upstream of the gene to be transcribed. Promoter-specific sigma factors recognize and bind to the promoter, and the core enzyme of RNA polymerase subsequently attaches to the sigma factor–DNA complex to initiate transcription. Thus, the cell can control transcription by regulating the abundance of sigma factors available. Second, there is the *operator*, which is a DNA segment that can bind a repressor protein that precludes transcription. Specifically, the repressor blocks the binding of RNA polymerase to the promoter. The repressor protein is encoded in another DNA segment called the *regulator*. Note that multiple enzymes that participate in a given metabolic pathway or other related functions are often coded in gene clusters that are under the influence of a common promoter. Such gene clusters that are similarly regulated are called *operons*.

Figure 3.7 is an example of genes subject to negative control regulation. In this case, gene *induction* occurs only after repression is relieved. Derepression triggers the synthesis of catabolic enzymes when these are needed to degrade a specific substrate or pollutant. This requires the presence of an inducer (e.g., the target pollutant) that binds to and inactivates the repressor protein, which is then unable to bind to the operator to block transcription. An example of this regulatory mechanism is the induction of the *lac* operon (Figure 3.8). This gene cluster codes for enzymes needed for the uptake and metabolism of the sugar lactose. The *lac* genes are repressed by lacI repressor binding to the operator. When lactose (also considered an inducer) is present, it binds the lacI repressor, causing a conformation change in the repressor that decreases its binding affinity for the operator. With the repressor no longer binding to the operator, RNA polymerase binds to the promoter and initiates transcription of the structural genes *lacZ*, *lacY*, and *lacA*, eventually resulting in the synthesis of the encoded enzymes.

A similar negative control mechanism involving repressor proteins is gene *repression*. This mechanism prevents the synthesis of enzymes that are not needed,

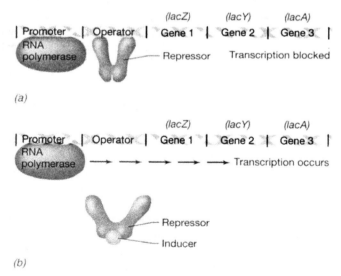

(a)

(b)

Figure 3.8. Induction of the *lac* operon by inactivation of the repressor. (From Madigan et al. 2003.)

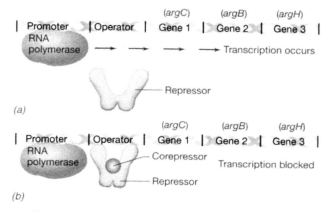

Figure 3.9. Enzyme repression. (From Madigan et al. 2003.)

such as when the end product of an enzyme is already present in excess. Similar to gene induction, a repressor protein binds to the operator and blocks transcription by RNA polymerase. Unlike induction, however, the operator is typically derepressed (i.e., free of the repressor, which cannot bind by itself). Repression occurs only when the substrate or product that is present in excess (i.e., the corepressor) binds to the repressor and alters its conformation, increasing its affinity for the operator (Figure 3.9). A well studied example of a corepressor is the amino acid arginine (Madigan et al. 2003).

Some gene regulation systems do not require a repressor protein and are subject to positive control mechanisms. In this case, transcription requires the presence of an *activator* protein that binds to RNA polymerase and facilitates its binding to the promoter (Figure 3.10). Such genes cannot be transcribed until the activator is activated by an inducer (in a mechanism comparable to how a repressor is inactivated), leading to gene induction. An example of this mechanism is the induction of the *tol* operon, which codes for enzymes involved in the degradation of toluene.

Another important regulatory mechanism under positive control is *catabolite repression*. In this case, binding of RNA polymerase to the promoter also requires a catabolite activator protein (CAP). This activator, however, is effective only when it binds another molecule, cyclic AMP (cAMP). The cAMP–CAP complex binds to DNA and enhances the binding of RNA polymerase, thus promoting transcription. When the cell is rich in energy (ATP), cAMP levels are low and the scarcity of cAMP–CAP complex limits transcription. Cyclic AMP is an alarmone, a chemical alarm signal that the cell uses to respond to environmental or nutritional stress.

Similar to enzyme repression, catabolite repression saves cellular energy by preventing the synthesis of enzymes that are not needed. Catabolite repression, however, prevents the synthesis of catabolic enzymes even when an inducer is present (e.g., if other substrates are being metabolized by constitutive enzymes). A classical example of catabolite repression is the control of β-galactosidase, an inducible enzyme used in lactose catabolism that is coded in the *lac* operon. When glucose and lactose are fed concurrently to *Escherichia coli* in a batch system, a preferential

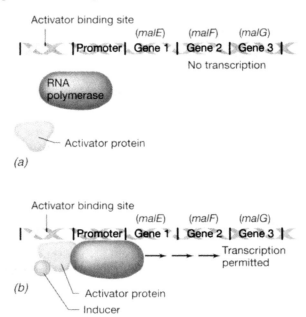

Figure 3.10. Positive control of induction involving an activator protein. (From Madigan et al. 2003.)

substrate utilization pattern called *diauxy* is observed. *Escherichia coli* prefers glucose (which is metabolized by constitutive enzymes through central metabolic pathways such as glycolysis and the Krebs cycle). During growth on glucose, *E. coli* does not expend energy to synthesize β-galactosidase. If glucose is available, the cell maintains low cAMP levels (glucose inhibits cAMP synthesis), and many inducible enzymes like β-galactosidase are not produced. Once glucose is depleted, cAMP levels increase (Figure 3.11). This promotes the binding of RNA polymerase to initiate transcription of the *lac* operon. This example implies that the presence of labile substrates can exert catabolite repression and hinder the degradation of target pollutants.

3.1.4. Definition of Biodegradation Terms

There are many terms that have similar meanings but are often used in different contexts, and defining such terms is important to avoid confusion. Recognizing that there may be alternative definitions, environmental engineers and scientists often use the term *biotransformation* broadly to signify the alteration of the molecular structure or oxidation state of a chemical by microbiological, usually enzymatic, catalysis. Such alterations generally decrease the number or complexity of intramolecular bonds, but the term biotransformation also includes reactions that increase the size of the molecule (e.g., complexation during metabolism).

Biodegradation will be used in this book to describe biotransformations that simplify an organic compound's structure by breaking molecular bonds. The sim-

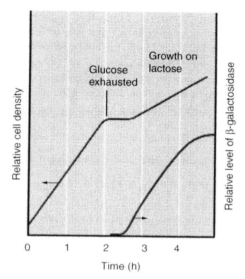

Figure 3.11. Catabolite repression resulting in diauxic growth of *E. coli* on glucose and lactose. (From Madigan et al. 2003.)

plification may be subtle, involving merely a substituent functional group, or severe, resulting in complete breakdown.

Mineralization is a form of biodegradation that results in conversion of an organic molecule into its inorganic constituents (e.g., CO_2, CH_4, H_2O, SO_4^{2-}, and PO_4^{3-}) or mineral salts. In general, mineralization occurs when central catabolic pathways transform an organic compound. The responsible organism(s) typically benefits from mineralization reactions. Thus, microbial growth is expected, and a portion of the carbon on the original organic molecule is usually incorporated into biomass. Examples of common pollutants that can be mineralized include hydrocarbons (e.g., saturated, unsaturated, monoaromatic, and polycyclic hydrocarbons and related carboxylic acids, aldehydes, alcohols, ketones, phenols, and biphenyls), lightly chlorinated solvents (e.g., methylene chloride, vinyl chloride, and 1,2-dichloroethane), lightly chlorinated aromatics (e.g., chlorobenzene, dichlorobenzenes, chlorophenols, and chlorinated biphenyls), fuel additives (e.g., ethanol, methanol, and MTBE), and some nitroaromatics (e.g., nitrobenzene, nitrotoluene, 2,4-dinitrotoluene, and 2,6-dinitrotoluene).

Biodegradation of most priority pollutants generally results in products of lesser toxicity. Such transformation of a compound to a less toxic form is known as *detoxification*. Nevertheless, there are some noteworthy exceptions where a pollutant is biotransformed to a product of greater toxicity. Examples include the reductive dechlorination of the solvent trichloroethylene (TCE) to vinyl chloride under anaerobic conditions, and the epoxidation of the pesticide aldrin to dieldrin under aerobic conditions. Such transformations that increase the toxic nature of the compound are known as *activation* and are important to consider when assessing risks associated with natural attenuation.

Cometabolism is a poorly constrained term that is used to describe the transformation of a compound by a microorganism that is unable to use the molecule as a source of energy or as a building block for new cell material. This is illustrated below. The compound being cometabolized (e.g., A′) serves no metabolic purpose and is not mineralized by the microorganism that transforms it. The cometabolic attack is generally mediated by a given enzyme with relaxed specificity (e.g., *a*), which routinely acts on another (primary substrate) molecule that supports cell growth (e.g., A) (Alexander 1999).

Metabolism: $\quad$ A $\xrightarrow{a}$ B $\xrightarrow{b}$ C $\xrightarrow{c}$ → → → CO_2 + Energy + Cell C

Cometabolism: $\quad$ A′ $\xrightarrow{a}$ B′ $\nrightarrow$

When pure cultures are used, cometabolism of the target pollutant results only in partial modification and no cell growth occurs. In this case, transformation by-products accumulate because the enzyme required for subsequent metabolism (e.g., *b*) has higher specificity and does not recognize the by-product (e.g., B′) as a usable substrate. In the environment, however, mixed cultures prevail, and the products of cometabolic transformations can often be metabolized by other species in the microbial community, resulting in some cases in complete mineralization. Examples of such a concerted (sequential) attack, which is known as cometabolic commensalism, are given at the end of this chapter.

Chlorinated pesticides, polychlorinated biphenyls (PCBs), and alkylbenzene sulfonates (ABS) are compounds that are often transformed through cometabolism. Table 3.2 provides examples of microorganisms and enzymes reported to exhibit cometabolism and the target compounds that are degraded. Note that exploiting cometabolism for site cleanup requires the selection of appropriate substrates that induce growth of specific organisms and their cometabolic enzymes and minimization of potential negative effects: in some cases, the reaction is harmful to the cell through product toxicity or loss of energy (Hughes et al. 2002).

3.2. REQUIREMENTS FOR BIODEGRADATION

Biodegradation of environmental pollutants is a "self-purification" process that occurs when conditions in a polluted medium (e.g., wastewater in a treatment plant, soil or groundwater contaminated by a leaking underground storage tank) are conducive to the growth of specific degraders and the functioning of their enzymes. These conditions are discussed below.

1. Existence of Organism(s) with Required Degradation Potential

Organic compounds will be degraded to a measurable extent only if the organism has enzymes that catalyze its conversion to a product that can be funneled into an existing metabolic pathway. The greater the differences in chemical structure between the target pollutant and the constituents of living organisms,

Table 3.2. Some Microbial Genera and Enzymes that Co-oxidize Organic Pollutants

Genera	Enzyme	Primary Substrates	Some Compounds Cometabolized	References
Pseudomonas *Rhodococcus*	Toluene dioxygenase	Toluene, phenol, others	TCE, many other aliphatic olefins, chlorobenzene, nitrobenzene, nitrotoluene isomers, alkylsulfides	Li and Wackett 1992; Nelson et al. 1988; Wackett and Hershberger 2001
Burhkolderia *Pseudomonas* *Acinetobacter*	Toluene monooxygenase	Toluene, phenol, others	TCE, chlorobenzene, nitrobenzene, nitrotoluene isomers, alkenes.	Kim and Hao 1999; Shields et al. 1991
Methylosynus *Methylococcus* *Methylomicrobium*	Methane monooxygenase	Methane, formate, methanol	TCE, all DCE isomers, chloroform, dichloromethane, cyclohexane, *n*-alkanes	Oldenhuis et al. 1989; Imai et al. 1986; Haber et al. 1983; Little et al. 1988
Nitrosomonas	Ammonia monooxygenase	Ammonia	TCE, 1,1-DCE, dimethylether, thioethers, methyl fluoride	Hyman et al. 1994; Juliette et al. 1994; Rasche et al. 1990, 1991
Pseudomonas	Naphthalene dioxygenase	Naphthalene	Toluene, ethylbenzene, aryl alkyl sulfides, indan, phenanthrene	Lee and Gibson 1996; Lee et al. 1995
Burkholderia *Pseudomonas* *Sphingomonas* *Comanmonas*	Biphenyl dioxygenase	Biphenyl	Lightly chlorinated PCB congeners, dibenzofurans and dioxins, carbazole	Habe et al. 1991; Seeger et al. 2001
Mycobacterium *Pseudomonas* *Acinetobacter* *Bacillus* *Rhodocossus*	Propane monooxygenase	Propane	TCE, DCE isomers, vinyl chloride, chloroalkanes, benzene, MTBE	Wackett et al. 1989

the less likelihood that the contaminant will be recognized as a substrate. This also holds for cases of a pollutant that occurs less commonly as building blocks in living matter. In such cases, extensive degradation is unlikely to occur.

Only a few central metabolic pathways exist among the many microbial species. Accordingly, some structural features in organic compounds that are not common in nature, called "xenophores" (e.g., substitutions with Cl, NO_2, CN, and SO_3 groups), make the molecule difficult to be recognized by these pathways (Alexander 1999). Thus, contaminants that contain such xenophores tend to be recalcitrant to microbial degradation.

There is a notion known as the "microbial infallibility hypothesis," which states that for virtually any organic compound available, there may be an organism that will break it down under suitable conditions: if not, evolution and adaptation will produce such a strain in a relatively short period of time (Gale 1952; Focht 1988). This hypothesis cannot be proved wrong because the absence of evidence is not in itself evidence of absence. That is, failure to demonstrate biodegradation could be ascribed to failure of the experimenter to provide the appropriate strain under the appropriate conditions. Nevertheless, this hypothesis is supported by numerous observations of "new biodegradation" capabilities for compounds previously thought to be recalcitrant (e.g., MTBE, 1,4-dioxane, and atrazine). The extent to which these observations are due to new enzymes that have evolved to handle new compounds, or to pre-existing enzymes that have proliferated due to selective pressure posed by the pollutant is still unclear. Another confounding factor is that researchers are getting better at finding enzymes in the environment.

2. Presence of Specific Degrader(s) in the Contaminated Zone

The assumption of microbial infallibility holds true for many common organic pollutants, such as petroleum hydrocarbons, alcohols, and ketones, and, as a result, this requirement is easily overlooked. For example, hydrocarbons have a natural pyrolytic origin and have been in contact with microorganisms throughout eons of geologic time. Thus, it is not surprising that many bacteria have evolved and acquired the ability to utilize hydrocarbons as food. The ability of microorganisms to degrade benzene, toluene, ethylbenzene, and xylenes (BTEX) has been known since 1908, when Stormer isolated the bacterium *Bacillus hexabovorum* by virtue of its ability to grow on toluene and xylene aerobically (Gibson and Subramaniam 1984). In an early review, Zobell (1946) identified over 100 microbial species from 30 genera that could degrade hydrocarbons. The ubiquity of soil bacteria capable of degrading BTEX was first demonstrated in 1928 by Gray and Thornton, who reported that 146 out of 245 uncontaminated soil samples contained bacteria capable of metabolizing hydrocarbons (Gibson and Subramaniam 1984). Therefore, this requirement is easily met for BTEX compounds. Nevertheless, differences in the relative abundance of dissimilar specific degraders (i.e., phenotypes) may lead to apparent discrepancies in biodegradability of a given BTEX compound at different sites. This is illustrated in the following example.

Example 3.1   Differences in the distribution of dissimilar phenotypes may lead to apparent discrepancies in the biodegradability of a given BTEX compound at different sites.

Species	Degradation Rate	Occurrence of Species	Compound Supporting Growth					
			B	T	E	m-X	p-X	o-X
1	Fast	Rare	+	+	−	−	−	−
2	Fast	Widespread	−	+	+	−	−	−
3	Slow	Widespread	−	+	−	+	−	−
4	Slow	Rare	−	−	+	+	+	−

In this hypothetical example, B would be degraded rapidly in some sites, T would always be degraded rapidly, E would be degraded rapidly, slowly, or not at all, m-X would always be degraded slowly, p-X would be degraded slowly, but only in some environments, and o-X would not be degraded. Consider what would happen at a nearby site if only species 3 and 4 were present.

Some "relatively new" environmental pollutants (i.e., not naturally occurring) such as MTBE, 1,4-dioxane, and TCE tend to be persistent because the microorganisms that could use them as growth substrates are relatively scarce in nature. It is often found that common, indigenous microorganisms cannot mineralize such compounds even under optimum conditions. Because of this, *bioaugmentation*, which refers to the introduction of specialized strains that degrade recalcitrant contaminants, is receiving increasing attention. The introduced microorganism augments the indigenous population's degradation capacity, hence the term bioaugmentation. This approach has been successfully used to enhance in situ bioremediation of sites contaminated with MTBE (Salanitro et al. 2000) and TCE (Ellis et al. 2000). In such cases, survival of the added bacteria and their adequate distribution throughout the contaminated zone are common limiting factors (Walter 1999).

3. Accessibility of Target Pollutants to the Microorganisms

A common limitation of natural degradative processes is the lack of adequate contact between pollutants and microorganisms (i.e., *bioavailability*). The target pollutants must be accessible in various aspects. Regarding physical state, microorganisms generally assimilate pollutants from the liquid phase and can not effectively degrade a pollutant until it desorbs from aquifer solids, diffuses out of nanopores, or dissolves from nonaqueous phase liquids (NAPLs) into the bulk solution. In such cases, the rate of biodegradation can be controlled by the diffusion, desorption, or dissolution rates (Alexander 1999). Bioavailability also implies that bonds requiring cleavage must be exposed and not be blocked by large atoms such as chlorine (i.e., steric hindrance), and that the target pollutant must be able to pass through the cellular membrane. Many common priority pollutants such as BTEX and TCE are relatively soluble and bioavailability rarely limits their biodegradation. The recalcitrance of more hydrophobic compounds such as polycyclic aromatic hydrocarbons (PAHs) and PCBs is often the result of their poor bioavailability. A mathematical analysis of bioavailability is included at the end of this chapter.

4. Induction of Appropriate Degradative Enzymes

Specific regions of the bacterial genome must be activated for degradative enzymes to be produced. Some enzymes, such as those participating in central metabolic pathways, are always present (at some level) regardless of environmental conditions. These are known as *constitutive enzymes*. The enzymes that initiate the biodegradation of many priority pollutants, however, are generally inducible. When an inducer is present (often, but not always, the target substrate), it initiates a cascade of biochemical reactions that result in the transcription of genes that code for the synthesis of the necessary degradative enzymes. Many catabolic enzymes require induction, and the inducing substrate must be present at a higher concentration than a strain-specific threshold level (Linkfield et al. 1989). In general, this threshold is relatively low (e.g., about 50 µg/L for toluene) (Robertson and Button 1987) and enzyme induction is rarely a limiting factor for BTEX bioremediation. Nevertheless, the presence of easily degradable substrates or co-occurring contaminants (e.g., ethanol in gasoline) could repress the synthesis of BTEX-degrading enzymes (Lovanh and Alvarez 2004). In such cases, a lag period may be observed during which the easily degradable substrate is metabolized before any significant BTEX degradation occurs.

5. Availability of Appropriate Electron Acceptors and/or Electron Donors

Microorganisms obtain energy to drive life functions by a complex sequence of oxidation and reduction reactions known as respiration, where electrons are transferred from an electron donor (i.e., the "fuel" molecule) to an electron acceptor though a series of biological electron carriers. This produces chemical energy, which is stored in adenosine triphosphate (ATP) molecules (Figure 3.4). In this context, oxidation refers to the removal of electrons from an atom or a molecule, and reduction is the addition of electrons. The type of electron acceptor used determines the metabolism mode (e.g., aerobic versus anaerobic) and determines to a great extent the specific reactions that can or cannot occur.

Perhaps the most important attribute of electron acceptors is their oxidation–reduction potential (ORP), or ability to accept electrons. Aerobic metabolism is the most energetically favored electron-accepting regime that microorganisms use to oxidize organic compounds. Therefore, molecular oxygen (O_2) is preferentially utilized over anaerobic electron acceptors because this yields more energy to the microbial community and results in faster contaminant oxidation rates. For example, for hydrocarbon-contaminated sites, the rate-limiting attenuation mechanism is frequently the influx of oxygen, which in turn limits aerobic hydrocarbon degradation kinetics (National Research Council 1993). In the absence of molecular oxygen, anaerobic microorganisms use other forms of combined oxygen. For example, denitrifying bacteria use nitrate (NO_3^-), nitrite (NO_2^-), or nitrous oxide (N_2O); dissimilatory metal-reducing bacteria use manganese or ferric iron oxides (e.g., MnO_2, $Fe(OH)_3$, or $FeOO^-$); sulfate-reducing bacteria use sulfate (SO_4^{2-}); and methanogens use carbon dioxide (CO_2) or bicarbonate (HCO_3^-) as electron acceptors.

Although aerobic biodegradation reactions tend to be very feasible thermodynamically, the biochemical oxygen demand exerted by organic contaminants often depletes the available dissolved oxygen and exceeds the restorative diffusion rate: consequently, anaerobic conditions develop. In such cases, electron acceptors are generally used up in a set sequence determined by the appropriate redox potentials of the oxidation reactions under consideration. Classical thermodynamic concepts imply the following sequence of electron acceptor utilization: $O_2 \rightarrow NO_3^- \rightarrow Mn^{4+} \rightarrow Fe^{3+} \rightarrow SO_4^{2-} \rightarrow HCO_3^-$. The thermodynamic hierarchy of these electron-accepting regimes is summarized in Figure 3.12.

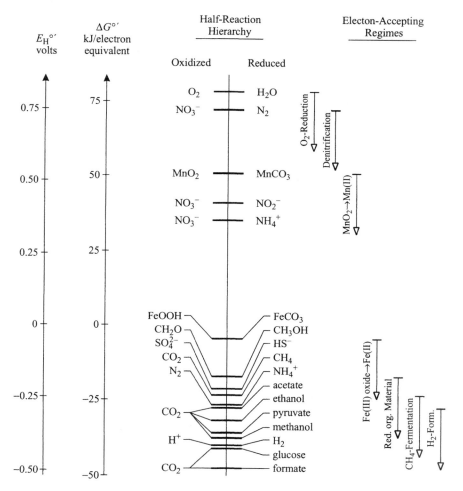

Figure 3.12. Free-energy diagram for common electron acceptors and donors. The hierarchy of electron acceptors provides a simple means to integrate thermodynamics, microbiology, and physiology of oxidation–reduction reactions. $E^{\circ\prime}$ is the equilibrium redox potential and $\Delta G^{\circ\prime}$ is the half-reaction free-energy. These values are for unit activities of oxidant and reductant in water with a pH of 7.0 (Adapted from Zehnder and Stumm 1988.)

The implication of this thermodynamic analysis is that when the electron acceptor demand is relatively high (e.g., near the source zone), microbial degradation would sequentially deplete the available oxygen, then nitrate, manganese, ferric iron, and sulfate before methanogenesis becomes predominant. Thermodynamic considerations also imply that heterotrophic microorganisms capable of deriving the maximum amount of energy per unit of carbon oxidized would have a competitive advantage over other species, and their respiration mode would become dominant until their specific electron acceptor is used up. This principle is invoked to explain the geochemical transitions in gasoline-contaminated sites, where strongly anaerobic conditions (e.g., methanogenesis) are observed near the source where the hydrocarbon concentration (and the electron acceptor demand) is highest (Figure 3.13). Away from the source, the hydrocarbon concentration decreases and electron acceptors are replenished from the surrounding groundwater, and electron-

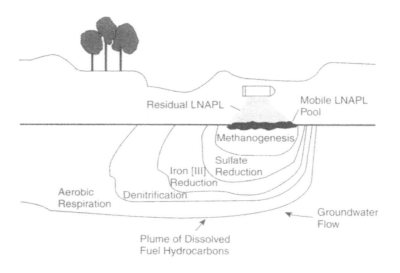

Toluene degradation reactions occurring in different electron accepting zones	
Aerobic:	$C_7H_8 + 9\,O_2 \rightarrow 7\,CO_2 + 4\,H_2O$
Denitrifying:	$C_7H_8 + 7.2\,NO_3^- + 7.2\,H^+ \rightarrow 7\,CO_2 + 7.6\,H_2O + 3.6\,N_2$
Iron-Reducing:	$C_7H_8 + 36\,Fe(OH)_3 + 72\,H^+ \rightarrow 7\,CO_2 + 36\,Fe^{2+} + 94\,H_2O$
Sulfate-Reducing:	$C_7H_8 + 4.5\,SO_4^{2-} + 3\,H_2O \rightarrow 7\,HCO_3^- + 2.25\,H_2S + 2.25\,HS^- + 0.25\,H^+$
Methanogenic:	$C_7H_8 + 5\,H_2O \rightarrow 2.5\,CO_2 + 4.5\,CH_4$

Figure 3.13. Electron-accepting regimes commonly observed in fuel-contaminated sites.

accepting conditions transition through sulfate-reducing, metal-reducing, and denitrifying zones before aerobic conditions are reached near the edge of the hydrocarbon plume.

The stoichiometric reactions occurring in these different zones are depicted in Figure 3.13, using toluene (C_7H_8, molecular weight = 92 g/mol) mineralization as an example. Based on these relationships, the complete mineralization of 1 mg/L of toluene would require 3.13 mg/L O_2 (i.e., 9 mol-O_2 per mol-toluene × (32 g-O_2/mol)/(92 g-toluene/mol) = 3.13), 4.85 mg/L NO_3^-, 21.85 mg/L Fe^{3+}, or 4.70 mg/L SO_4^{2-}. If the electron acceptor pool of the aquifer is known, these relationships can be used to estimate the amount of toluene (or other similar BTEX compounds) that could be mineralized, which provides an estimate of the assimilative capacity of the aquifer at that point in time. Such estimations, however, could be confounded by the fact that there may be other organic compounds present (including soil organic matter) that contribute to the electron acceptor demand, and that soluble electron acceptors could be replenished by mixing with uncontaminated groundwater and infiltrating rainwater in the case of oxygen.

It should be pointed out that hydrogeological heterogeneities could facilitate the development of spatial gradients with respect to electron donors and acceptors, and such spatial heterogeneities as well as temporal discontinuities might result in nonideal zonation of electron-accepting conditions. For example, it is possible to observe anaerobic microniches within aerobic environments.

Whether an organic pollutant serves as electron donor (i.e., becomes oxidized) or electron acceptor (i.e., becomes reduced) depends on the potential energy associated with the electron transfer reaction. A convenient way to depict the thermodynamic feasibility of electron transfer during biodegradation is to imagine a vertical tower (Figure 3.14). The "electron tower" represents the range of standard reduction potentials ($E^{\circ\prime}$) for different molecules, arranged from the most negative (i.e., the strongest reductant—or "electron abundant") at the top to the most positive (i.e., the strongest oxidant—or "electron hungry") at the bottom. Note that the E_h axis of Figure 3.14 is inverted with respect to Figure 3.13. As electrons from the electron donor at the top of the tower fall, they can be "caught" by electron acceptors at various levels. The difference in potential between the two molecules ($\Delta E^{\circ\prime}$) is directly proportional to the free energy available ($\Delta G^{\circ\prime}$), according to Nernst's equation:

$$\Delta G^{\circ\prime} = -nF\,\Delta E^{\circ\prime} \qquad\qquad [3.1]$$

where n is the number of electrons transferred per mole, and F is Faraday's constant (96.63 kJ/V). The use of Eq. [3.1] is illustrated in Figure 3.14.

The chemical potential for a molecule to transfer electrons to another molecule in the electron tower is analogous to the gravitational potential energy of an object that is dropped to the floor from a given height. In the gravitational case, the potential energy associated with the object equals the product of its weight times the height. The greater the height through which the object falls, the greater the kinetic energy it acquires once it hits the floor (notwithstanding free-falling objects that

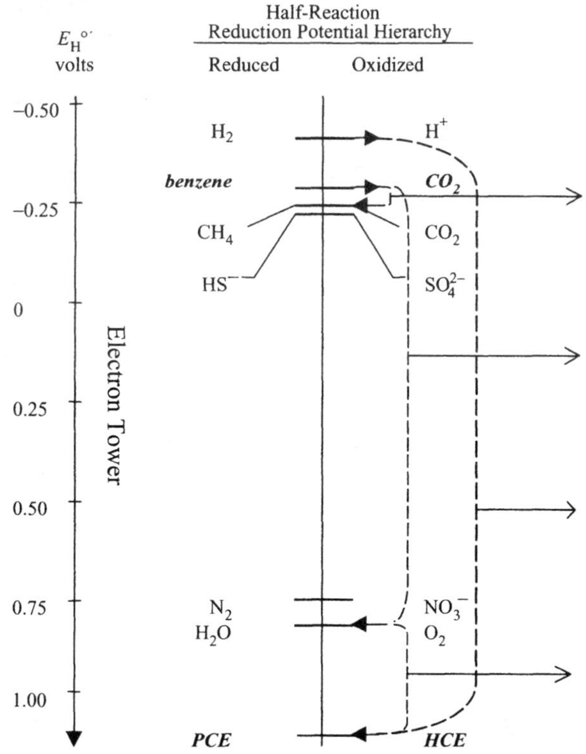

$E_H^{\circ\prime}$ volts
Electron Tower
Half-Reaction Reduction Potential Hierarchy
Reduced
Oxidized
−0.50
−0.25
0
0.25
0.50
0.75
1.00
H_2
H^+
benzene
CO_2
CH_4
CO_2
HS^-
SO_4^{2-}
N_2
NO_3^-
H_2O
O_2
PCE
HCE

Benzene degradation to CO_2 and CH_4 under methanogenic conditions
$C_6H_6 + 4.5\ H_2O \rightarrow 2.25\ CO_2 + 3.75\ CH_4$
$\Delta G^{\circ\prime}$ = -(30 e-/mol) (96.63 kJ/V) (-0.24 -(-0.29) V)
$\Delta G^{\circ\prime}$ = -133 kJ/mol of benzene, or - **4.5 kJ/e⁻ equiv transferred**
(barely feasible)

Benzene degradation to CO_2 under aerobic conditions
$C_6H_6 + 7.5\ O_2 \rightarrow 6\ CO_2 + 3\ H_2O$
$\Delta G^{\circ\prime}$ = -(30 e-/mol) (96.63 kJ/V) (+0.82 -(-0.29) V)
$\Delta G^{\circ\prime}$ = -3,200 kJ/mol of benzene, or - **107 kJ/e⁻ equiv transferred**
(highly feasible)

HCE dechlorination to PCE with H_2 as e^- donor under anaerobic conditions
$C_2Cl_6 + H_2 \rightarrow C_2Cl_4 + 2\ Cl^- + 2\ H^+$
$\Delta G^{\circ\prime}$ = -(2 e-/mol) (96.63 kJ/V) (+1.14 -(-0.42) V)
$\Delta G^{\circ\prime}$ = -302 kJ/mol of HCE, or - **151 kJ/e⁻ equiv transferred**
(highly feasible)

HCE dechlorination to PCE with H_2O as e^- donor under aerobic conditions
$C_2Cl_6 + H_2O \rightarrow C_2Cl_4 + 0.5\ O_2 + 2\ Cl^- + 2\ H^+$
$\Delta G^{\circ\prime}$ = -(2 e-/mol) (96.63 kJ/V) (+0.84 -1.14 V)
$\Delta G^{\circ\prime}$ = +60 kJ/mol of HCE, or + **30 kJ/e⁻ equiv transferred**
(Not feasible since $\Delta G^{\circ\prime}$ is positive, and does not occur)

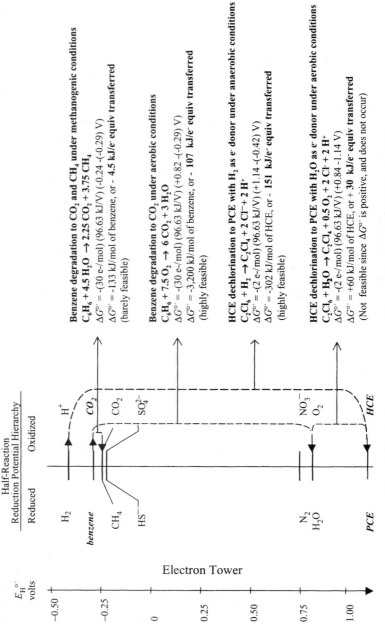

Figure 3.14. The electron tower concept. Free energies are calculated using Eq. [3.1].

achieve a constant terminal velocity due to the drag force). Similarly, the greater the $E^{o\prime}$ drop of the electron from the donor to the acceptor molecule, the greater the amount of chemical energy that can be harvested. For example, the half-reaction reduction potential for the benzene–CO_2 couple is negative (i.e., -0.28 V), which means that benzene is a good electron donor and its oxidation under aerobic conditions (using oxygen as the electron acceptor) is much more feasible than its oxidation under methanogenic conditions (using CO_2 as the electron acceptor). On the other hand, the half-reaction reduction potential for the perchloroethylene (PCE) and hexachlorethane (HCE) couple (i.e., $+1.14$ V) is more positive than that of the O_2–H_2O couple (i.e., $+0.84$ V), and its oxidation is not thermodynamically feasible. Yet, the reduction by HCE by H_2 as electron donor (-0.42 V) is highly feasible.

This example illustrates that, whereas many organic pollutants can serve as fuel molecules that donate electrons to an electron acceptor, the oxidation of highly halogenated (i.e., chlorinated, brominated, or fluorinated) or nitrated compounds (e.g., trinitrotoluene (TNT)) is not as feasible thermodynamically (if at all). Carbon atoms in such molecules are in a relatively high oxidation state and resist further oxidation. Nevertheless, these compounds can be reduced by many anaerobic microorganisms, provided that appropriate electron donors are present. Hydrogen (H_2), which is a by-product of the anaerobic degradation of organic compounds, is generally a very effective electron donor for such reductive biotransformations. Other organic substrates that are frequently added to stimulate reductive degradation mechanisms include acetate, formate, methanol, and ethanol. Oxidative and reductive degradation mechanisms are discussed later in this chapter.

Therefore, a number of common organic contaminants are biodegraded via reduction by serving as an acceptor of electrons. In some cases, the contaminant could serve as a terminal—or respiratory—electron acceptor in the same way that O_2 is reduced during aerobic metabolism. While only a small number of organic contaminants undergo this form of metabolism (e.g., perchloroethene, trichloroethene, cis-dichloroethene, vinyl chloride, 4-chlorobenzoate, carbon tetrachloride, and perchlorate), it is often important in both natural and engineered bioremediation systems, as it does occur with some of the most common contaminants of groundwater such as chlorinated ethenes (Alleman and Leeson 1993 2(1), 1995 3(4), 1999 5(2); Hinchee et al. 1994; Wickramanayake and Hinchee 1998 C-1 and C-2; Wickramanayake et al. 2000 C2-1 and C2-2). The reductive dechlorination of such compounds is discussed in greater detail in this chapter, and engineered systems that exploit such reactions are discussed in Chapter 8.

6. Availability of Nutrients

Microorganisms need macronutrients to synthesize cellular components. Examples of macronutrients are nitrogen for amino acids and enzymes, phosphorus for ATP and DNA, sulfur for some coenzymes, calcium for stabilizing the cell wall, and magnesium for stabilizing ribosomes. In general, microbial growth in subsoils is not limited by nitrogen and phosphorus as long as the contaminant concentrations are in the sub part per million (mg/L) range (Tiedje 1993). A C:N:P ratio of 30:5:1 is generally sufficient to ensure unrestricted growth in aquifers (Paul and Clark 1989). Microbes also need micronutrients to perform certain metabolic functions. For example, trace metals such as Fe, Ni, Co, Mo, and Zn are needed for some enzymatic activities. In general, aquifer minerals contain sufficient micronutrients to support microbial activity. Nevertheless, geochemical analyses or laboratory biodegradation assays should be performed to verify that the availability of inorganic nutrients is sufficient for biodegradation to proceed. If biodegradation is a cometabolic process, the presence of "recognizable" substrate(s) that can serve as energy and carbon source(s) will be a limiting factor. Last but not least, water must be present for biodegradation to occur. Moisture levels of at least 40% of soil field capacity (or 7% H_2O on a weight basis) should be provided for biodegradation of pollutants trapped in the unsaturated soil zone, with optimum results often obtained at greater than 80% field capacity (English and Loehr 1991).

7. Adequate pH and Buffering Capacity

Most microorganisms grow best in a relatively narrow range of pH around neutrality (pH 6–8). Enzymes are polymers of amino acids, and their activity requires the proper degree of amino acid protonation. This is controlled by pH, with the optimum value usually near neutral (pH 7). Most aquifer microorganisms can, however, perform well between pH values of 5 and 9. This range generally reflects the buffering capacity of the carbonate or silicate minerals present in aquifers (King et al. 1992; Chapelle 2001). Groundwater is typically well buffered within this range, so microbial pH requirements are generally met in aquifers (Chapelle 2001). Nevertheless, aquifers contaminated by municipal landfill leachates may contain elevated concentrations of volatile fatty acids (e.g., acetic acid) resulting in pH values as low as 3. In these cases, acidity may suppress microbial activity.

8. Adequate Temperature

Temperature is one of the most important environmental factors influencing the activity and survival of microorganisms. Microbial metabolism accelerates with increasing temperatures up to an optimum value at which growth is maximal. Most of the bacteria present in subsurface environments operate most effectively at 20–40 °C, which is a little higher than typical groundwater temperatures in the United States (Chapelle 2001). Low temperatures reduce the fluidity and permeability of the cellular membrane, which hinders nutrient (and contaminant) uptake. For example, a recent study of natural attenuation in subarctic climates suggests that biodegradation does not likely represent an important contribution to contaminant removal in cold climates (Richmond et al. 2001). Higher temperatures are associated with higher enzymatic activity and faster biodegradation rates, up to an optimum value that is species specific. In this range, degradation rates can double or triple due to a temperature increase of 10 °C (Corseuil and Weber 1994). If the temperature rises much beyond the optimum value, proteins, enzymes, and nucleic acids become denatured and inactive (Figure 3.15). The temperature of the upper

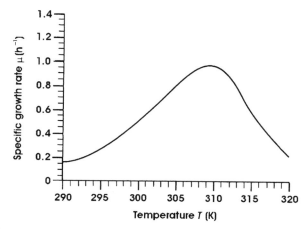

Figure 3.15. Growth rate dependence on temperature. Above example corresponds to *Klebsiella pneumoniae* and used $K_1 = 6.74 \times 10^{14}$ h^{-1}, $K_2 = 1.6 \times 10^{48}$ h^{-1}, $\Delta H_1 = 86.8$ kJ/mol, and $\Delta H_2 = 287.6$ kJ/mol (Bellgardt 1991).

10 m of the subsurface may vary seasonally; however, that between 10 and 100 m approximates the mean annual air temperature of a particular region (Lee et al. 1988).

The effect of temperature can be described mathematically by the superposition of activation and inactivation effects (Bellgardt 1991). As a result of both, an optimum temperature exists. The maximum reaction rate due to activation can be expressed by the Arrhenius relationship:

$$r_{max}(T) = K_1 C_{EA} \, e^{-\Delta H_1/RT} \qquad [3.2]$$

where K_1 is a constant, C_{EA} is the relative concentration of the active form of a key enzyme, ΔH_1 is the activation enthalpy, R is the universal gas constant, and T is the absolute temperature (K).

For fast inactivation, the equilibrium between active (C_{EA}) and inactive (C_{EI}) enzymes is given by

$$C_{EI} = K_2 C_{EA} \, e^{-\Delta H_2/RT} \qquad [3.3]$$

where K_2 is a constant, ΔH_2 is the enthalpy change of the inactivation reaction, and all other terms are as defined previously.

Normalizing the active and inactive forms of the enzyme to total enzyme concentration gives

$$C_{EI} + C_{EA} = 1$$

and substituting for C_{EI} in the previous equation implies that

$$1 - C_{EA} = K_2 C_{EA} \, e^{-\Delta H_2/RT} \qquad [3.4]$$

Solving for C_{EA},

$$C_{EA} = \frac{1}{1 + K_2\, e^{-\Delta H_2/RT}} \qquad [3.5]$$

Substituting for C_{EA} into the Arrhenius equation gives the maximum rate that can be attained at a given temperature:

$$r_{max}(T) = \frac{K_1\, e^{-\Delta H_1/RT}}{1 + K_2\, e^{-\Delta H_2/RT}} \qquad [3.6]$$

9. Absence of Toxic or Inhibitory Substances

Some contaminants can be present in aquifers at sufficiently high concentrations that inhibit microbial activity. In such cases, biodegradation may not proceed until hydrodynamic processes disperse and dilute the pollutant to noninhibitory levels. Sometimes, corollary pollutants can cause inhibition. For example, it is not uncommon for aquifer microorganisms to encounter potentially toxic heavy metals such as Pb, Hg, Cd, and Cr. While heavy metals are required in trace quantities for nutritional purposes, they can be bactericidal if present in soluble form at concentrations greater than about 1 mg/L. Some by-products of the degradation could also be inhibitory if they accumulate at relatively high concentrations. When sulfate is used as the terminal electron acceptor during anaerobic bioremediation, sulfate is reduced to hydrogen sulfide (H_2S), which can be toxic to microorganisms at about 200 mg/L (Cunnigham et al. 2001). Finally, the presence of easily degradable substrates can inhibit the degradation of target pollutants due to preferential substrate utilization. This concept can be extended to inorganic nutrients, since some compounds can be used as nitrogen or phosphorus sources and the absence of such nutrients in the medium exerts selective pressure for the biodegradation for relatively recalcitrant compounds that contain these elements. This has been observed for the explosive RDX (hexahydro-1,3,5-trinitro-1,3,5-triazine) (Sheremata and Hawari 2000), which can be used as a nitrogen source, and for the pesticide methylphosphonate, which can be used as a phosphorus source in the absence of inorganic sources (Alexander 1999).

3.2.1. Implications of Recalcitrance

Recalcitrance is a term used for organic compounds that are relatively resistant to biodegradation. This resistance is generally due to physiological limitations of the bacteria present and/or environmental properties. There are several levels of recalcitrance. Some compounds are very recalcitrant and are generally not degraded to an appreciable extent (e.g., some synthetic organic polymers) or degrade very slowly (e.g., structural plant polymer, lignin). Other compounds can be degraded rapidly in the lab under some conditions, but their rapid degradation in situ is rarely observed (e.g., TCE). Finally, there are pollutants that can serve as suitable substrates for certain populations and habitats but occasionally are quite persistent (e.g., benzene under anaerobic conditions). Many organic compounds are designed to be persistent, such as the pesticides DDT, dieldrin, and lindane, which have been reported to last for several decades in soil (Alexander 1999).

The following conditions are generally conducive to recalcitrance and should be kept in mind when considering natural attenuation as a risk management strategy:

1. Organisms with the required biodegradation capacity are absent, or present, at very low concentrations. This is likely to occur when "new" synthetic organic chemicals are introduced into the environment faster than biodegradation pathways evolve. This can result in temporary recalcitrance, in which case a lag phase would be observed before the microbial community adapts and degrades the pollutant. This lag often reflects the time required to achieve a "critical" concentration of competent strains capable of exerting measurable degradation rates and can range from as short as a few hours to a couple of years or longer.

2. The specificity of the enzyme(s) is too restrictive. Only a few central metabolic pathways exist. Some structural features in organic chemicals make them too different from the normal substrates of these pathways and they are not "recognized" by degradative enzymes.

3. Lack of primary substrate to drive cometabolism. Cometabolism might proceed relatively rapidly in the lab, where one can easily add primary substrates and attain high microbial concentrations. This is more difficult to achieve in the field.

4. Lack of an essential nutrient or electron acceptor. This common limitation can often be fixed through engineered manipulations discussed earlier.

5. Unfavorable environmental conditions. Low permeability can hinder the flow of nutrients and the replenishment of electron acceptors from surrounding groundwater. Low pH, high concentration of protonated fatty acids, extreme temperatures, high salinity, and predation of bacteria by protozoa can also hinder microbial activities.

6. Toxic concentrations of target compound or its degradation products. High concentrations of some chemicals can cause gross physical disruption to bacteria (e.g., membrane dissolution by solvents) or competitive binding of a single enzyme. In such cases, physical processes such as dilution, sorption, precipitation, and volatilization become important mechanisms to decrease the concentration of the contaminants to subinhibitory levels.

7. Inhibition or inactivation of enzymes. Some cometabolic transformations, such as the aerobic co-oxidation of TCE by methane-degrading bacteria produce toxic (free radical) intermediates that inactivate the enzyme. Also, clays, colloidal mater, and humic substances have been reported to bind and hinder the activity of some extracellular (protein-degrading) enzymes.

8. Failure of the pollutant to enter the cell. Most degradative reactions occur intracellularly, and uptake can be a problem for large polymers and long alkanes whose molecular weight exceeds 500. In such cases, biodegradation

may have to be initiated by exocellular enzymes excreted by fungi and other microorganisms.

9. The concentration of the target compound is too low to sustain a viable microbial population. In this case, auxiliary substrates may have to be added so that the target compound is degraded as a "secondary" substrate. Note also that an enzyme inducer (which is not necessarily the target substrate) may have to be present at a sufficiently high concentration to be effective.

10. Lack of bioavailability. The presence of a compound in nonaqueous phase liquid (NAPL) such as the fuel phase, or its sorption to soil organic matter and diffusion into nanopores might result in subthreshold concentrations in the aqueous phase that are insufficient to trigger enzyme induction and/or sustain a viable microbial population. Lack of bioavailability can sometimes be overcome by adding nonionic surfactants that promote the solubilization of hydrophobic compounds.

3.3. ACCLIMATION

When microorganisms are exposed to a new chemical stress (e.g., a recently spilled xenobiotic), there is a characteristic lag time before they adapt and begin to degrade the compound and grow and reproduce actively. This phenomenon is known as adaptation, and it refers to adaptive changes that occur at the individual, bacterial population and microbial community levels that increase the rate of biodegradation as a result of exposure. Subsequent exposure to the same xenobiotic to an acclimated microbial community results in a much shorter lag, if any (Figure 3.16).

The lag time is highly variable (Table 3.3), depending on the type of pollutant, exposure and environmental conditions, and microbial community structure. Long lags are of concern due to higher potential for dispersal of the pollutant and prolonged exposure.

Adaptation can be the result of several mechanisms. At the individual cell level, relatively short lag times (on the order of hours) may be the result of enzyme induction. As discussed above, this process involves activation of specific regions of the bacterial genome to produce specific enzymes that are synthesized only when the substrate (or a structurally related chemical or metabolite) is present. The lag can also reflect the time needed for individual cells to overcome catabolite repression,

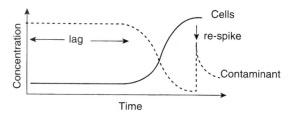

Figure 3.16. Acclimation of microorganisms to a xenobiotic. The lag reflects the length of time needed for microorganisms to adapt and exert measurable degradation rates.

Table 3.3. Reported Lengths of Acclimation Phases for Several Organic Pollutants

Chemical	Environment	Length of Acclimation Period Prior to Onset of Biodegradation
Several aromatic compounds	Soil	10–30 h
Dodecyltrimethyl-ammonium chloride	Fresh water	24 h
4-Nitrophenol	Water sediment	40–80 h
Amitrole	Soil	7 days
Chlorinated benzenes	Biofilm	10 days to 5 months
Dinitro-ortho-cresol (DNOC)	Soil	16 days
Pentachlorophenol (PCP)	Stream water	21–35 days
Mecoprop	Enrichments	30–37 days
Nitriloacetic acid (NTA)	Estuary	50 days
Halobenzoates	Anaerobic sediments	3 weeks to 6 months
2,4,5-Triphenoxyacetic acid (2,4,5-T)	Soil	4–10 weeks

Adapted from Alexander (1999).

which is a mechanism that saves cellular energy associated with the synthesis of inducible enzymes that are not needed when the cell is feeding on easily degradable substrates. Thus, significant degradation of the target pollutant occurs after the depletion of more labile substrate such as sugars or ethanol, a common oxygenate added to gasoline (Da Silva and Alvarez 2002).

Lag periods can also reflect temporarily adverse environmental conditions, such as inhibitory (high) concentrations of organic pollutants or co-occurring heavy metals. In such cases, the lag may reflect the time necessary for natural weathering processes (e.g., volatilization, dilution, sorption, and/or precipitation) to decrease concentrations of inhibitory compounds below inhibitory levels.

At the microbial population and community levels, adaptation may involve inter- and intraspecific genetic exchange. Catabolic genes can be exchanged among cells of the same genus (intra) or different genera (inter) (e.g., Pseudomonas to Alcaligenes), although the likelihood of genetic exchange and the compatibility of the transferred genes is often genetically predetermined. Many plasmids that contain genes coding for catabolic enzymes are transmissible (Table 3.4), and such genetic exchanges endow recipient microbes with broader catabolic capacity while enhancing molecular breeding.

Genetic rearrangements and mutations can also be important (long-term) adaptation mechanisms that contribute to the evolution of novel catabolic pathways. This is particularly important to adapt to thousands of newly synthesized emerging pollutants and commonly involves two potential mechanisms. First, there are genetic rearrangements resulting from the addition or deletion of DNA segments via transposable sequences (transposons). Such "accidental" rearrangements usually decrease metabolic efficiency but may fortuitously result in altered enzymes that can degrade new substrates. Second, there are point mutations where a single nucleotide in the DNA sequence is replaced. This often changes the enzymes' substrate specificity, occasionally broadening it, which results in an enzyme capable of

Table 3.4. Common Plasmids that Code for Catabolic Enzymes

Plasmid	Compounds	Transmissibility
CAM	Camphor	+
OCT	*n*-Octane	−
SAL	Salicylate	+
NAH	Naphthalene	+
TOL	Toluene/*m*- *p*-xylene	+
XYL-K	Xylene/toluene	+
2-HP	2-Hydroxypyridine	?
NIC	Nicotine/nicotinate	+
pJP1	2,4-Dichlorophenoxy acetic acid	+
pAC8	Xylene/toluene	+
pAC21	*p*-Chlorobiphenyl	+
pAC25	3-Chlorobenzoate	+

Source: Chakrabarty (1982).

degrading more different compounds. Note that such mutations can be spontaneous or induced; the former can occur as a result of natural radiation or exposure to toxic chemicals whereas the latter can be the result of exposure to specific chemicals in the laboratory. In both cases, according to the Darwinian evolution theory, if the mutation is detrimental, the microorganism dies and the mutation is not inherited, but if it is beneficial, the mutant acquires an inheritable competitive advantage that enhances its natural selection.

Lag periods can also reflect microbial population shifts; that is, the time required to achieve "critical" concentration of competent strains (or catabolic plasmids) capable of exerting significant biodegradation rates. This implies that contaminant biodegradation might be occurring during the lag period, but we may not detect it until a critical biomass concentration (on the order of 10^6 cells/g-soil) is reached (Alexander 1999).

In addition to exposure history, several environmental factors can affect the length of the acclimation period. In general, faster acclimation occurs at lower xenobiotic concentrations, due to potential toxicity of the target compound at high concentrations. As discussed earlier, the chemical structure of the xenobiotic is also an important factor in the duration of the acclimation phase, with faster acclimation corresponding to xenobiotics made up of building blocks that are similar to those of substrates normally utilized by the microorganisms. Finally, the presence of protozoa may lengthen the acclimation period due to predation on the bacteria that initiate biodegradation of the xenobiotic.

3.4. COMMON BIOTRANSFORMATION MECHANISMS

Numerous mechanisms and pathways have been elucidated for the biodegradation of a wide variety of organic compounds. For example, the University of Minnesota's

Biocatalysis/Biodegradation Database (http://umbbd.ahc.umn.edu/) lists 143 degradation pathways and 934 reactions for 877 compounds, including 109 reactions catalyzed by toluene dioxygenase and 74 by naphthalene 1,2-dioxygenase. The reader is referred to this comprehensive database for details on biodegradation pathways for a large number of pollutants. This section will discuss general mechanisms and unifying principles that apply to the biodegradation of common environmental pollutants.

The general strategy used by microorganisms to feed on organic pollutants is to transform them into potential substrates that could be funneled into central metabolic pathways such as beta-oxidation, glycolysis, and the Krebs cycle. For example, large compounds are usually broken into smaller molecules that can easily be cleaved to yield intermediate metabolites.

All metabolic reactions are mediated by enzymes, which are proteins that have specific catalytic properties. The term "enzyme" was coined in 1876 by the German researcher Wilhem Friedrich Kühne, who was working at the time on fermentation using yeast. Its Greek root simply means "in yeast." Kühne sided with those who correctly believed that many enzymes can exert catalytic activity independently of cells. The opposing view was led at the time by none other than Louis Pasteur (Gutfreund 1976).

Six major divisions of enzymes exist. These are listed below in order of importance to bioremediation.

1. *Oxidoreductases.* As the name implies, this class of enzymes specializes in catalyzing oxidation and reduction reactions, which are the most common reactions in biodegradation. This is the largest class of enzymes, reflecting that mineralization of organic compounds often involves oxidation, and biosynthesis requires adjusting the oxidation state of carbon. Common oxidoreductase enzymes include dehydrogenases, which remove hydrogen atoms, and oxygenases, which activate hydrocarbons to facilitate their further metabolism by adding molecular oxygen as hydroxyl (—OH) functional groups. Many of the oxygenase enzymes that attack aromatic hydrocarbons have a remarkably wide degradation capacity due to their relaxed substrate specificity. For example, toluene dioxygenase is capable of degrading more than 100 different compounds, including TCE, nitrobenzene, and chlorobenzene.

2. *Hydrolases.* This is the second largest class, and it specializes in performing hydrolysis of C—O, C—N, and some C—C bonds, including amides, esters, phosphate esters, monohaloalkanes, and epoxides. Examples include esterases, which break down ester bonds by the addition of water; depolymerases, which hydrolyze polymers; and dehalogenases, which remove halogen atoms such as chlorine and replace them with —OH groups.

3. *Lyases.* Enzymes in this group catalyze either the nonhydrolytic removal of a functional group from a substrate with the resulting (sometimes transient) formation of a double bond, or the reverse reaction. One common biodegradation step mediated by this enzyme class is the removal of CO_2 groups (i.e., decarboxylation), which is catalyzed by decarboxylases. Another

common reaction is the addition of —OH groups (derived from water) to double bonds (e.g., hydratases add water to alkenes converting them into secondary alcohols).

4. *Transferases.* These enzymes transfer functional groups to organic compounds, mainly for cell synthesis purposes. Some of these reactions can be beneficial for detoxification. For example, glutathione *S*-transferase transfers the thiol group to chlorinated compounds with concomitant dechlorination. Some of these reactions, however, can be detrimental, such as the methylation of heavy metals (e.g., mercury, arsenic, and strontium), which increases their toxicity and bioaccumulation potential.

5. *Isomerases.* The function of these enzymes is to catalyze intramolecular rearrangements and transform organic molecules into isomers that are more amenable for subsequent oxidation. For example, racemases catalyze L- and D-amino acid interconversions.

6. *Ligases.* These are used to catalyze covalent bond formation but, similarly to isomerases, the ligases can also catalyze reactions that facilitate subsequent metabolism. One example is *CoA*-ligase, which adds —S-CoA to fatty acids during beta-oxidation.

Hundreds of different enzymes have been isolated and studied. Whereas there is considerable variability regarding their mechanisms, the general types of biodegradation reactions that enzymes catalyze are not numerous; namely, oxidative, reductive, hydrolytic, and synthetic processes. These reactions are illustrated below using common priority pollutants as examples.

3.4.1. Oxidative Transformations

Hydroxylation

This transformation involves the addition of —OH groups and is often the first step for the aerobic biodegradation of hydrocarbons. Hydroxylation can be catalyzed by hydroxylase, monooxygenase, or mixed-function oxidase enzymes. A hydrogen donor (e.g., NADH or NADPH coenzymes) and molecular oxygen (O_2) are required for the functioning of these enzymes. In general, hydroxylation is a detoxification reaction that increases the solubility and subsequent biodegradability of hydrocarbons and some nitroaromatic and (lightly) chlorinated aromatic compounds.

Example 3.2. Aerobic Biodegradation of Alkanes. A simple example of hydroxylation is the oxidation of alkanes in sites contaminated with petroleum products. This aerobic reaction is mediated by monooxygenase enzymes that attack the end of the chain and transform the alkane into a primary alcohol.

$$CH_3-(CH_2)_n-CH_3 \xrightarrow[\text{monooxygenase}]{O_2} CH_3-(CH_2)_n-CH_2OH \qquad [3.7]$$

Subsequent reactions normally involve further oxidation of the alcohol to an aldehyde and then to a fatty acid:

$$CH_3-(CH_2)_n-CH_2OH \xrightarrow[\text{dehydrogenation}]{2H} CH_3-(CH_2)_n-CHO \xrightarrow[\text{oxidation}]{O_2} CH_3-(CH_2)_n-COOH$$

[3.8]

The fatty acid can then be metabolized via beta-oxidation, which is a central metabolic pathway that cleaves straight-chain fatty acids, two carbon fragments at a time (i.e., acetyl-CoA) (Figure 3.17). This pathway does not require aerobic conditions to operate and is inhibited when the chain is branched or if some of the carbon atoms have substituents that are uncommon in biological tissue, such as Cl or NO_2.

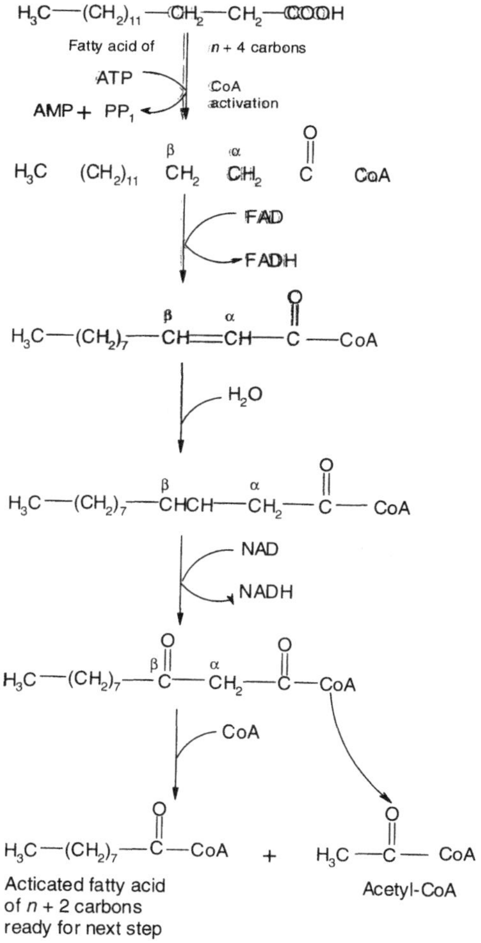

Figure 3.17. Mechanism of beta-oxidation of fatty aicd, which leads to successive formation of two-carbon fragments of acetyl-CoA.

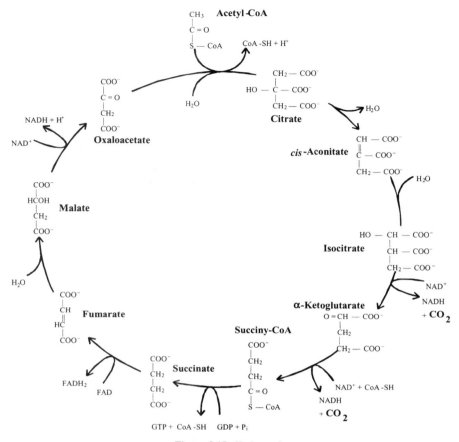

Figure 3.18. Krebs cycle.

The resulting acetyl-CoA fragments can subsequently enter the Krebs cycle, which completes the mineralization process (i.e., CO_2 is generated). Alternatively, Krebs cycle intermediates can be used as building blocks for the synthesis of new cell material (Figure 3.18).

Hydrocarbon hydroxylation can also occur anaerobically. In this case, molecular oxygen is absent and the oxygen atom is derived from water. Specifically, hydratase enzymes catalyze a nucleophilic attack of an electrophilic carbon atom using H_2O, as illustrated below for the hydration of a double bond in an alkene molecule, which becomes a single bond.

$$[3.9]$$

Dealkylation

Dealkylation is the oxidative cleavage of alkyl chains attached to nitrogen or oxygen atoms. This reaction is catalyzed by mixed-function oxidases and is one of the most important detoxification mechanisms for pesticides. Examples include the cleavage of ethyl and isopropyl groups attached to N atoms in atrazine, and the cleavage of ether bonds of chloroneb or 2-4-diphenoxyacetic acid (2,4-D).

N-Dealkylation of Atrazine

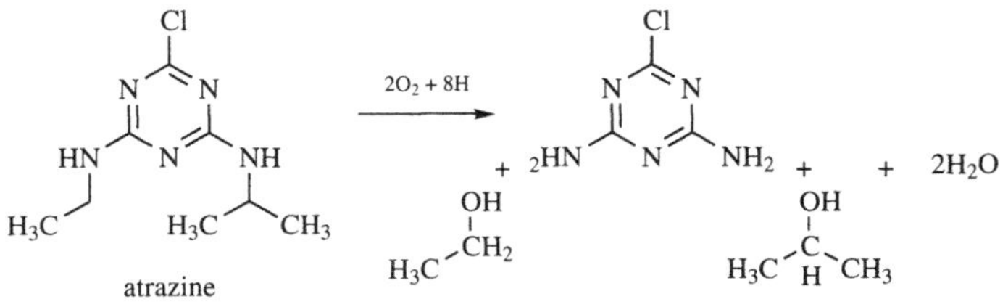

[3.10]

atrazine

O-Dealkylation of the Ether Groups of Chloroneb

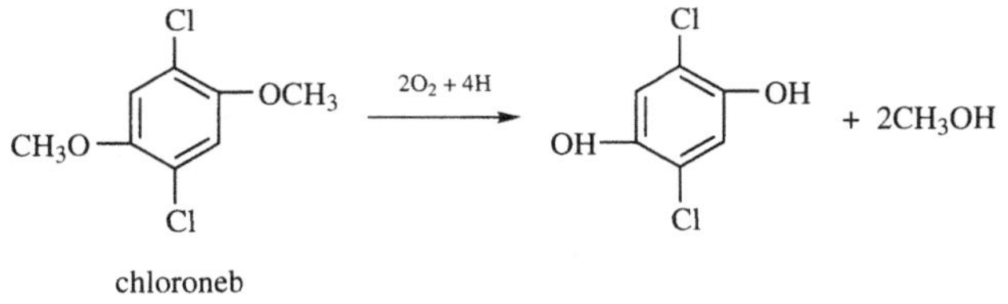

[3.11]

chloroneb

Epoxidation

Some oxygenase enzymes can catalyze the insertion of an oxygen atom between double-bonded carbon atoms, forming a triangular arrangement (i.e., an epoxide) as illustrated below for the oxidation of TCE by methane monooxygenase (Vogel et al. 1987).

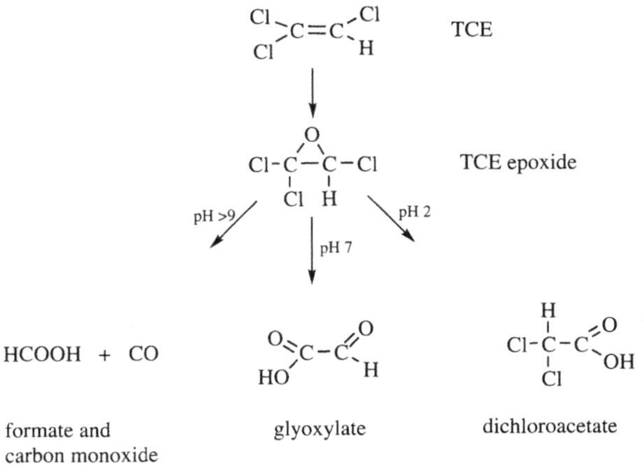

TCE

TCE epoxide

[3.12]

HCOOH + CO

formate and
carbon monoxide

glyoxylate

dichloroacetate

The epoxidation of TCE is a harmful reaction to the microorganisms that mediate it due to the formation of toxic free-radical intermediates that inactivate the enzyme. Some epoxides can be more persistent and more toxic than the parent compound, as is the case for the epoxidation of the pesticide aldrin to dieldrin.

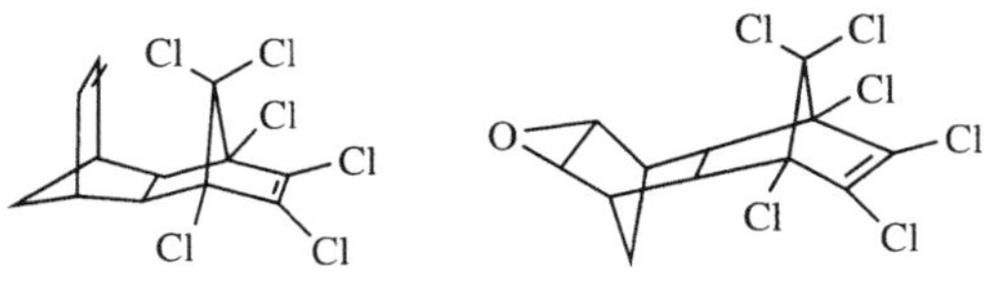

aldrin   dieldrin

[3.13]

Oxidative Ring Cleavage

Aromatic (benzene) rings are common constituents of organic pollutants, and their aerobic degradation is mediated by oxygenase enzymes (i.e., enzymes that "activate" O_2 and add it to carbon atoms in the aromatic molecule). These enzymes need O_2 as a cosubstrate for two critical steps. First, the ring is dihydroxylated by either monooxygenase enzymes that add one oxygen atom from O_2 at a time, or by dioxygenase enzymes that add both oxygen atoms simultaneously. Dihydroxylation is a prerequisite for ring cleavage. The second step is the oxidative cleavage of the resulting catechol, which is catalyzed by dioxygenase enzymes. The ring can be cleaved either at the *ortho* position (i.e., between the two —OH groups) or more commonly at the *meta* positions (i.e., adjacent to one of the —OH groups) as shown below. The enzymes associated with the *meta* cleavage pathway are generally associated with relatively broad specificity and can cometabolize many other contaminants, while the *ortho* pathway results in faster growth since these open-ring products can be more easily metabolized through the Krebs cycle as illustrated below for benzene (Figure 3.19).

There is considerable diversity among the initial steps that dissimilar bacteria use to initiate the degradation of alkylbenzenes, which is illustrated in Figure 3.20 using toluene as an example. Starting at the top, we have the TOL pathway, which was discovered in a plasmid harbored by *Pseudomonas putida* mt-2. This strain initiates catabolism by oxidizing toluene at the methyl group (Williams and Murray 1974). The TOD pathway was identified in *P. putida* F1, which uses toluene dioxygenase to add two oxygen atoms to the ring (Gibson et al. 1968). The TOM pathway was found in *Burkholderia cepacia* G4 (formerly known as *P. cepacia* G4), which uses toluene *ortho*-monooxygenase in the initial attack to form *o*-cresol (Shields et al. 1989). The TBU pathway occurs in *B. pickettii* PK01 (formerly known as *P. pickettii* PK01) which uses toluene *meta*-monooxygenase to form *m*-cresol (Kukor and Olsen 1991). The T4MO pathway was identified in *P. mendocina* KR1, which uses toluene *para*-monooxygenase to form *p*-cresol (Whited and Gibson 1991). The expression of these pathways is not exclusive to the five archetypes mentioned above.

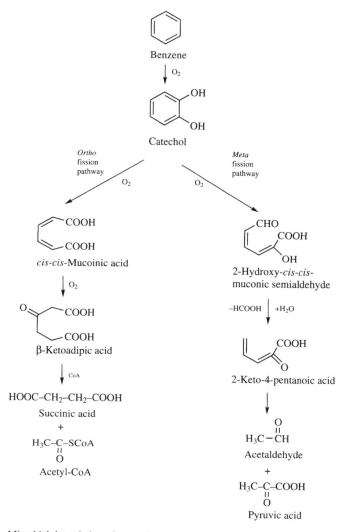

Figure 3.19. Microbial degradation of aromatic compounds by the *ortho* and *meta* fission pathways, as illustrated for benzene.

When aromatic rings are substituted with a xenophore such as NO_2, Cl, and SO_3, some oxygenase enzymes can hydroxylate the substituted carbon atom with concomitant elimination of the xenophore (Figure 3.21). Such aerobic biotransformations detoxify and increase the subsequent biodegradability of some nitroaromatic and (lightly) chlorinated aromatic compounds (Rieger and Knackmuss 1995).

Aromatic hydrocarbons can also be degraded under anaerobic conditions, although these reactions are usually much slower. In this case, benzenoid aromatic rings are generally carboxylated and transformed to CoA-esters as illustrated below

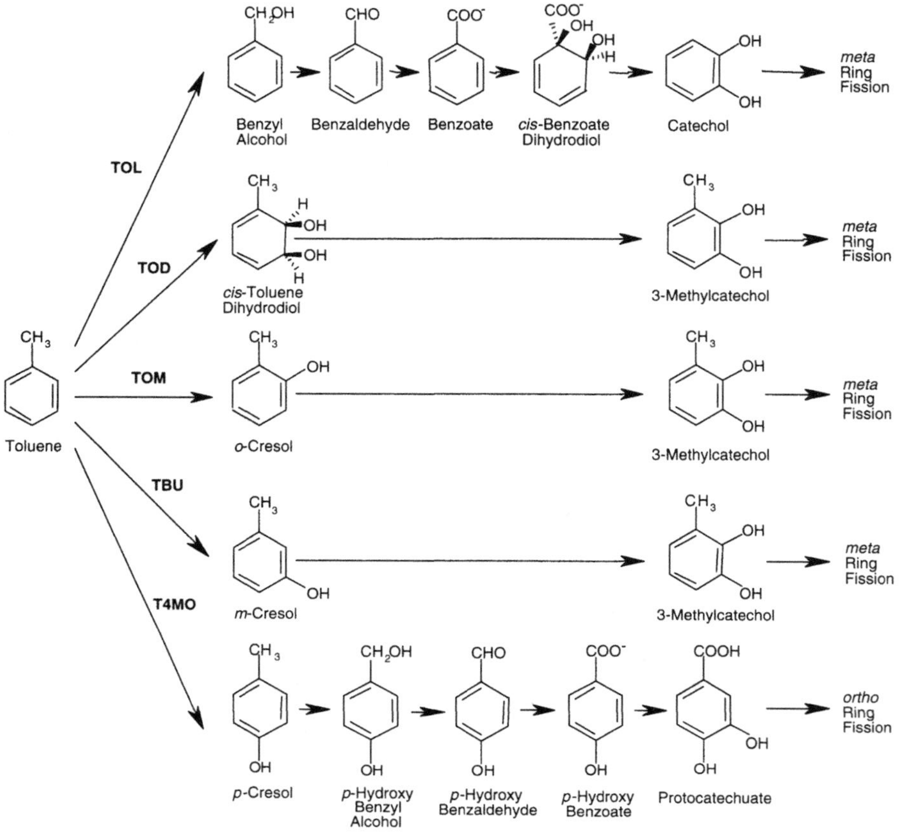

Figure 3.20. Diversity of aerobic toluene degradation pathways.

for toluene. The enzyme benzylsuccinate synthase (E) initiates catabolism of toluene under a wide variety of anaerobic electron acceptor conditions by adding fumarate to activate the molecule, via a free-radical intermediate (Beller and Spormann 1998).

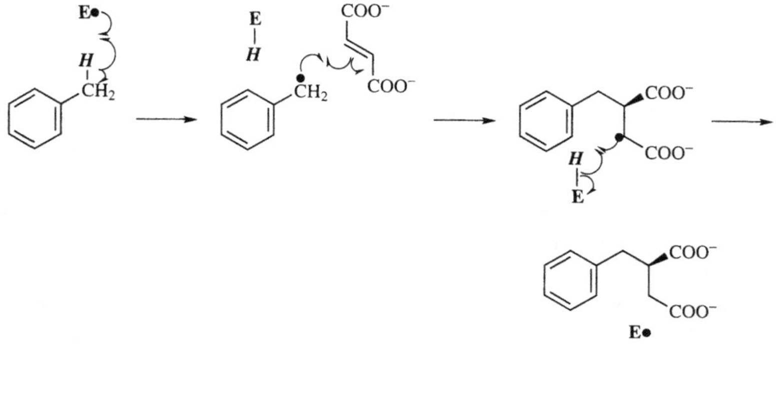

[3.14]

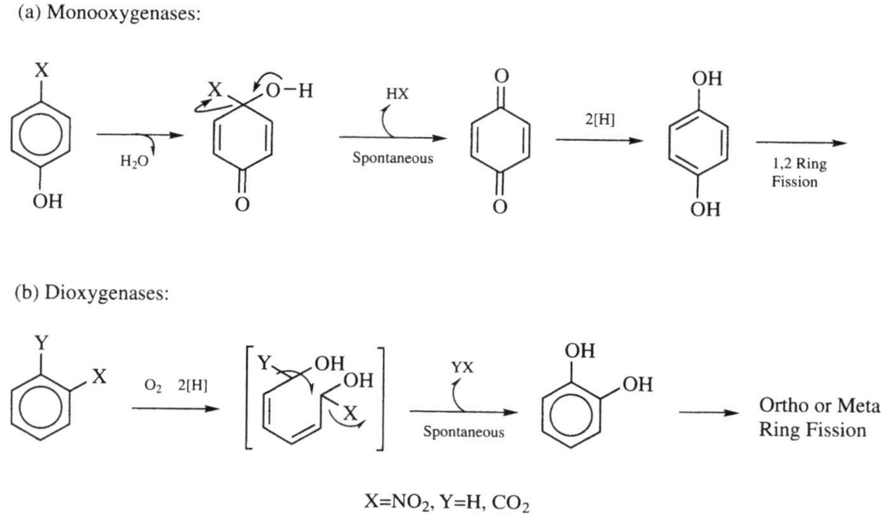

(a) Monooxygenases:

(b) Dioxygenases:

$X=NO_2, Y=H, CO_2$

Figure 3.21. Initial aerobic biotransformation of substituted aromatic compounds. The known anionic elimination of substituents like $X = Cl$, SO_3, O-Alkyl, or O-Aryl through (a) monooxygenases or (b) dioxygenases gives rise to the release of nitrite in the case of nitroaromatics ($X = NO_2$). Hydrogen and COO^- in the *ortho* position were eliminated as cations or CO_2.

The resulting benzylsuccinate is transformed via beta-oxidation of the fatty acid chain to benzoyl-CoA, which is a pivotal intermediate that is reductively cleaved. Benzoyl-CoA can be hydrolyzed to form benzoate, and the ring is reduced, hydroxylated with water, and eventually cleaved by hydrolysis. Other monoaromatic hydrocarbons are similarly transformed to benzoyl-CoA prior to ring cleavage (Figure 3.22).

3.4.2. Reductive Biotransformations

These transformations typically occur under anaerobic conditions and involve using the target pollutant as a sink for electrons transferred from a reduced molecule such as H_2 or internal electron donors such as reduced coenzymes (e.g., NADH or NADPH).

Reductive Dehalogenation

Halogenated compounds such as chlorinated solvents and pesticides are very common environmental pollutants that pose a significant risk due to their toxicity and persistence. The removal of halogen atoms such as Cl, Br, and F is a very important reaction for the natural attenuation of such compounds. This mechanism is often observed in anaerobic aquifers, and it involves the addition of electrons to the target molecule with concomitant release of halogen ions.

There are two reductive dehalogenation mechanisms. The first is hydrogenolysis (also known as hydrodehalogenation), which involves replacing a halogen atom by

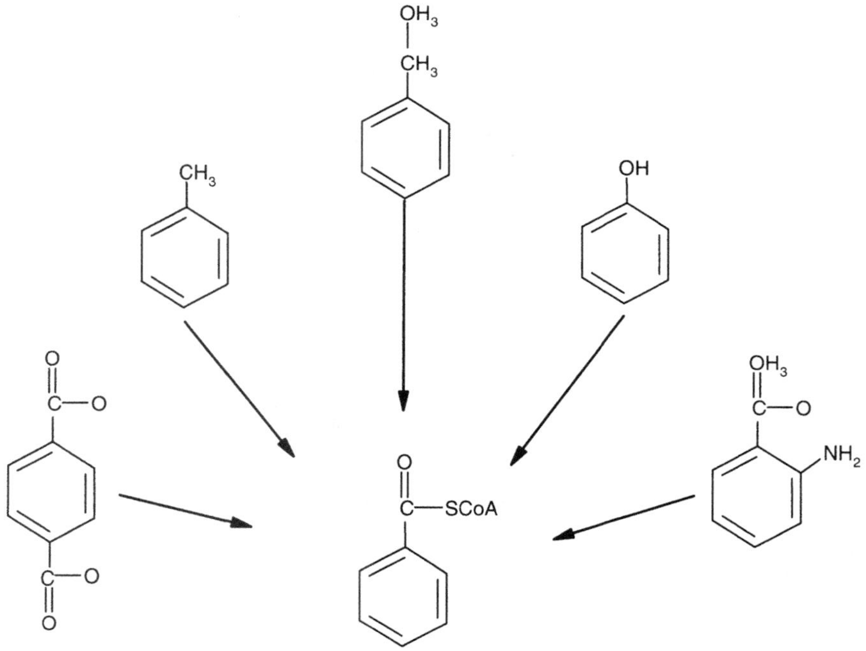
OH3
CH3
CH3
OH
O
C—O
C—O
O
O
C—SCoA
OH3
C—O
NH2

Figure 3.22. Anaerobic metabolism of many substituted aromatic compounds proceeds through the intermediacy of benzoyl-CoA.

a hydrogen atom. This is illustrated in Figure 3.23 for the stepwise reduction of TCE via dichloroethylene to vinyl chloride, and ultimately to ethene.

Microorganisms almost exclusively dechlorinate TCE through the left pathway (through *cis*-DCE), and the presence of *cis*-DCE is often taken as a signature for a microbial mediated process. TCE can also be dechlorinated abiotically at the surface of reduced minerals such as pyrite, and this usually (but not always) proceeds via *trans*-DCE.

The other reductive dehalogenation mechanism is dihaloelimination, which involves the simultaneous removal of two halogen atoms after two electrons are transferred. This results in the formation of a double bond, as illustrated for the conversion of hexachloroethane (HCA) to perchloroethylene (PCE).

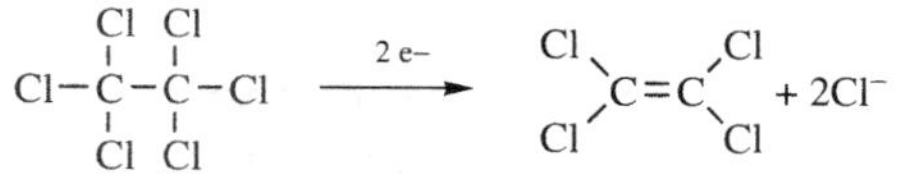
Cl Cl
Cl−C−C−Cl
Cl Cl
2 e−
Cl Cl
C=C
Cl Cl
+ 2Cl⁻

[3.15]

hexachloroethane (HCA) perchloroethylene (PCE)

Reductive dechlorination generally decreases the toxicity and enhances the solubility (and bioavailability) of the pollutant, but there are exceptions where the toxicity can be accentuated (e.g., TCE reduction to vinyl chloride). Reductive dechlorination is often a cometabolic reaction since the microorganisms that cata-

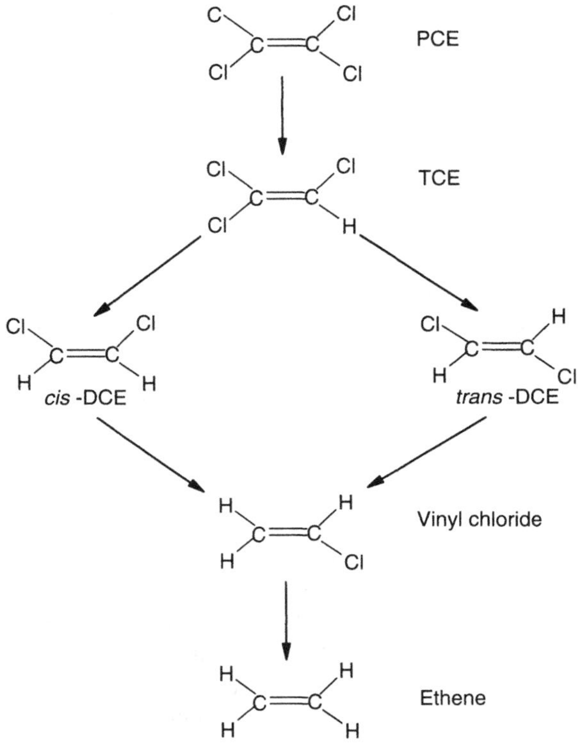

Figure 3.23. Reductive dechlorination of chlorinated ethenes via hydrogenolysis.

lyze it cannot harvest the energy released by the redox process. Recently, however, many strains have been found that can utilize PCE and TCE as a terminal electron acceptor during respiration, using H_2 and even formate, acetate, and pyruvate as electron donor. This process is known as halorespiration, and it can be mediated by species such as *Desulfomonile tiedjei, Dehalobacter restrictus, Desulfitobacter-ium* sp., and *Dehalococcoide ethenogenes* (Mohn and Tiedje 1990; Holliger et al. 1993; Gerritse et al. 1996; Maymogatell et al. 1997).

Nitro-Group Reduction

Nitro groups are common substituents in energetic compounds (e.g., RDX, HMX, and TNT) and some pesticides (e.g., 4,6-dinitro-*o*-cresol, EPN, and parathion) and can easily be reduced to amino groups under anaerobic conditions, as illustrated below for parathion:

parathion   aminoparathion   [3.16]

Such reductive biotransformations usually detoxify the organic molecule. One notable exception is the reduction of nitrobenzene to the highly toxic aniline:

$$\text{NO}_2 \qquad \text{NH}_2$$

nitrobenzene   aniline

[3.17]

Anaerobic reduction of multiple nitro groups attached to an aromatic ring can result in a product that is easily polymerized and/or covalently bound to humic substances. This strategy holds promise for treating TNT-contaminated soil. In this strategy, TNT is not removed, but it is detoxified and immobilized, becoming part of the naturally occurring organic matter. Polymerization involves TNT reduction to triaminotoluene (TAT) and subsequent formation of ring–N–ring bonds (Figure 3.24).

Other Reductions

Other reductive biotransformations that are commonly observed under anaerobic conditions include the reduction of sulfoxides ($-SO_2$) to sulfides ($-SH$), aldehydes ($-CHO$) to alcohols ($-CH_2OH$), and double bonds to single bonds. Anaerobic microorganisms can also reduce many organic pollutants and heavy metals indirectly, by producing relatively strong reductants such as H_2S from SO_4^{2-} and surface-bound Fe(II) from Fe(III). Biogenic formation of reactive mineral surfaces that can reduce pollutants may also be an important natural attenuation mechanism at some sites.

3.4.3. Hydrolytic and Other Biotransformations Not Involving Redox Processes

Hydrolysis

Hydrolysis is a cleavage mechanism with the simultaneous addition of water. This process can be abiotic, or mediated by enzymes such as esterases, amidases, phosphatases, and nitrilases, and can proceed under aerobic or anaerobic conditions. Examples of reactions mediated by such enzymes are provided below.

1. Esterases are enzymes that hydrolyze ester bonds (e.g., fats), yielding a carboxylic acid and a primary alcohol:

$$\underset{\text{R1}-\overset{\text{O}}{\overset{\|}{\text{C}}}-\text{O}-\text{R2}}{} \xrightarrow{\text{H}_2\text{O}} \underset{\text{R1}-\overset{\text{O}}{\overset{\|}{\text{C}}}-\text{OH}}{} + \text{R2}-\text{OH}$$

[3.18]

2. Phosphatases hydrolyze phosphoester bonds in naturally occurring compounds such as ATP and DNA as well as synthetic organic chemicals

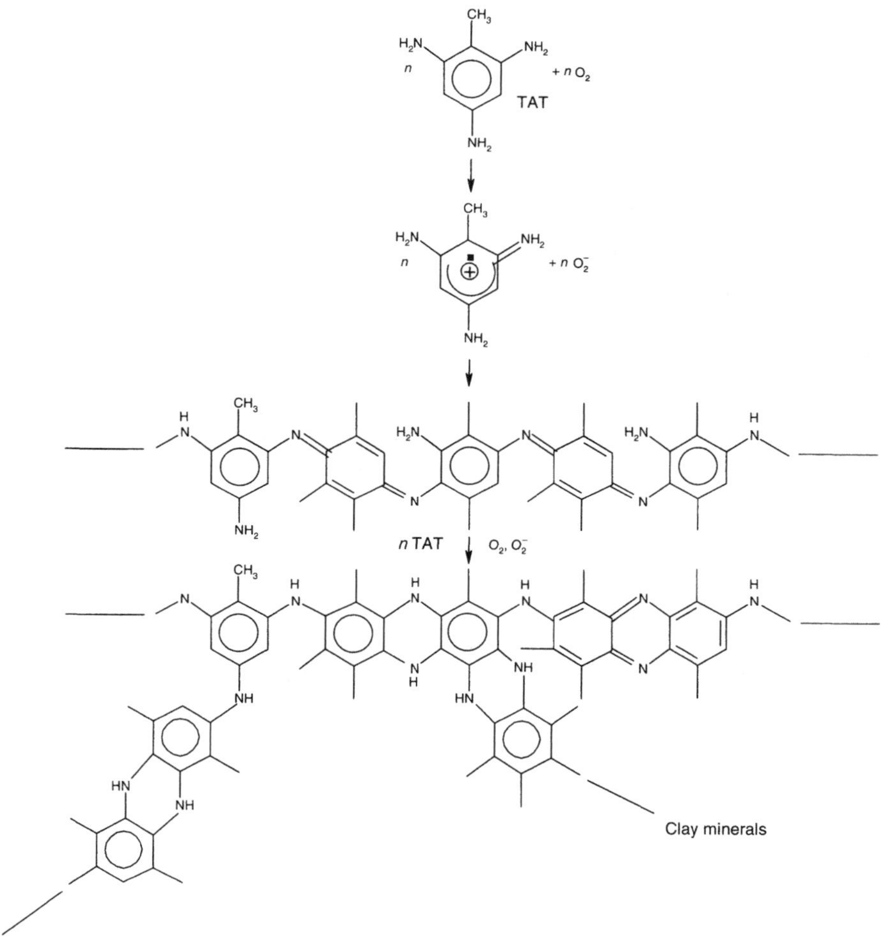

Figure 3.24. Potential mechanism of oxidative polymer formation from triaminotoluene.

such as pesticides:

$$R1-O-\overset{\overset{O}{\|}}{\underset{\underset{R3}{\overset{|}{O}}}{P}}-O-R2 \xrightarrow{3H_2O} R1-OH \;+\; R2-OH \;+\; R3-OH \;+\; H_3PO_4 \qquad [3.19]$$

3. Amidases hydrolyze amide bonds (e.g., proteins) yielding a carboxylic acid and an amine:

$$R1-\overset{\overset{O}{\|}}{C}-NH-R2 \xrightarrow{H_2O} R1-\overset{\overset{O}{\|}}{C}-OH \;+\; R2-NH_2 \qquad [3.20]$$

4. Nitrilases add water to carbon–nitrogen triple bonds (e.g., cyanide), forming amide bonds that can subsequently be cleaved by hydrolysis by amidase enzymes as illustrated above.

$$R1-C\equiv N \xrightarrow{H_2O} R1-\overset{\displaystyle O}{\overset{\|}{C}}-NH_2 \qquad [3.21]$$

Chemically facile hydrolysis reactions generally have metabolic priority and occur early in the biodegradation process. For example, many biodegradable compounds that contain repeating structures (e.g., polysaccharides and proteins) can be hydrolyzed into these substructures (e.g., sugars and amino acids, respectively). Hydrolysis is also a very important mechanism for detoxifying several pesticides (e.g., carbamates and organophosphates such as malathion). However, hydrolysis can also activate some herbicides, where the actual phytotoxin is the organic acid (e.g., flamprop-methyl, and dichlorfop-methyl).

Hydrolytic Dehalogenation

This reaction involves a nucleophilic substitution where a halogen atom is replaced by an —OH group and is mediated by dehalogenase enzymes. This substitution removes a xenophore (Cl, Br, or F) and facilitates subsequent metabolism.

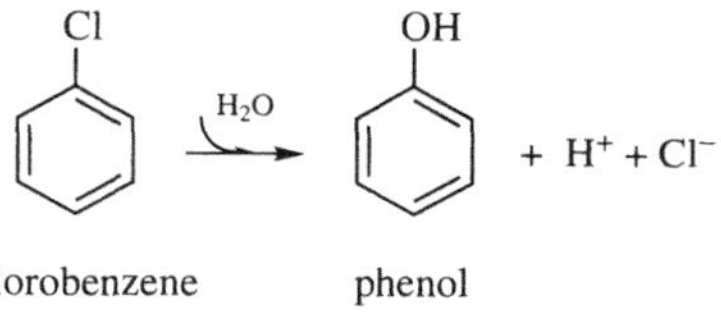

$$\text{chlorobenzene} \xrightarrow{H_2O} \text{phenol} \; + H^+ + Cl^- \qquad [3.22]$$

Dehydrohalogenation

This is an elimination reaction where vicinal hydrogen and halogen atoms are removed as ions with the concurrent formation of a double bond, as illustrated in Eq. [3.23]. Unlike reductive dehalogenation, no input of electrons is required for this elimination process. Dehydrohalogenation can occur spontaneously in water without the need of biological catalysts. Examples of dehydrohalogenation include the conversions of DDT to DDE, and lindane to 2,3,4,5,6-pentachlorocyclohexene (Eq. [3.24]).

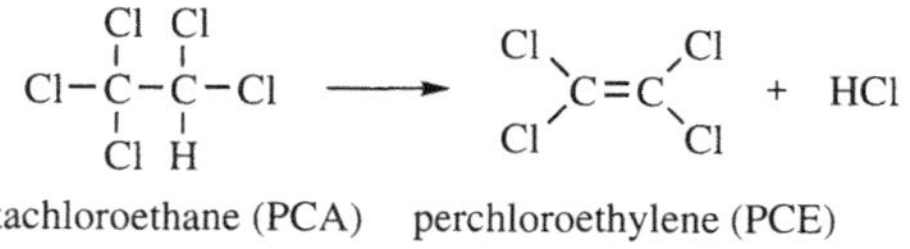

$$[3.23]$$

pentachloroethane (PCA)   perchloroethylene (PCE)

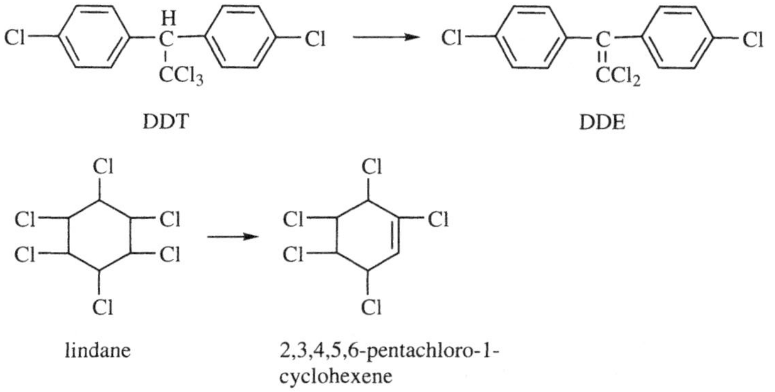

DDT → DDE

[3.24]

lindane → 2,3,4,5,6-pentachloro-1-cyclohexene

3.4.4. Synthetic Reactions

Synthetic reactions refer to the addition of molecules to target pollutants. Such reactions are rarer in biodegradation than in anabolism. Nevertheless, some xenobiotics can be detoxified by conjugation with intermediates in metabolic pathways (e.g., glutathione, sugars, and amino acids) or linked with themselves (e.g., dimerization of radicals produced by reductive dechlorination), carboxylated (—COOH addition), methylated (—CH$_3$ addition), and humified (e.g., polymerization of catechols or triaminotoluene).

The methylation of the oxygen atom of some pesticides (i.e., ROH → ROCH$_3$) can reduce their toxicity. This has been reported for the fungicide pentachlorophenol and for the broad-spectrum poison 2,4-dinitrophenol. However, the importance of such reactions as a natural attenuation mechanism has not been established. Note that some synthetic reactions can also have a detrimental effect. For example, the methylation of some heavy metals, as illustrated below, increases their toxicity and bioaccumulation potential.

$$Hg \rightarrow CH_3Hg^+ \rightarrow CH_3HgCH_3$$
$$As \rightarrow CH_3AsH_2 \rightarrow (CH_3)_2AsH \rightarrow (CH_3)_3As \qquad [3.25]$$
$$Sn \rightarrow (CH_3)_3Sn$$

3.4.5. Effect of Organic Contaminant Structure on Biodegradability

Researchers have attempted to correlate biodegradation rates with contaminant structure or physicochemical properties such as bond strength, solubility, octanol–water partitioning coefficient (K_{ow}), molecular weight, molar refractivity, and molecular topology (size, shape, van der Waals radius, molecular connectivity). However, such structure–activity correlations have been much more successful at predicting chemical rather than biological degradation rates. In the environment, compounds undergo enzyme-specific as well as fortuitous redox reactions, and the rate-limiting step may vary in different environments. Variations in microbial communities contribute to wide variations in biodegradation mechanisms, pathways, and related rates. This makes it difficult to predict in situ biodegradation

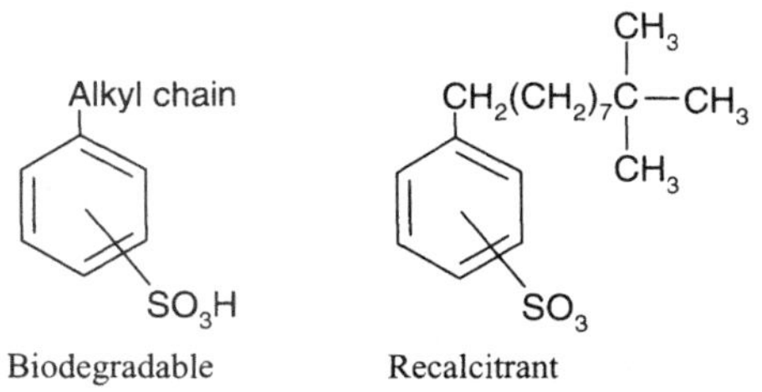

Figure 3.25. Structure of alkylbenzene sulfonates with linear and branched chains.

rates ab initio. Nevertheless, some heuristic generalizations can be made about how the structure of an organic compound affects its relative biodegradability.

1. Highly branched compounds are more resistant to biodegradation than straight chains. This is probably due to the fact that branched (tertiary and quaternary) carbon atoms impede beta-oxidation. For example, alkylbenzene sulfonates (ABSs) are detergents that were relatively persistent and caused foaming problems in wastewater treatment plants and rivers. A simple redesign of the molecule—the replacement of the branched hydrocarbon chain by a straight chain—dramatically increased the biodegradability of ABSs and solved the foaming problem (Figure 3.25).

2. Very long chains (MW > 500) become increasingly resistant to biodegradation due to the difficulty of bacteria to transport relatively large molecules through the cell membrane and cell wall.

3. Highly oxidized materials, such as halogenated or oxygen-rich materials, may resist further oxidation under aerobic conditions (in the presence of oxygen) but may be more rapidly degraded under anaerobic conditions—via reductive processes.

4. With halogenated alkanes containing three or more halogen substituents, the halogen is removed by reductive dechlorination.

5. With halogenated alkanes containing two or fewer halogen substituents, the halogen is removed by reductive hydrolytic or oxygenative reactions.

6. Benzene rings with four or more halogen substituents are recalcitrant under aerobic conditions and typically undergo reductive dechlorination first under anaerobic conditions.

7. More highly polar (and more soluble) compounds tend to be more biodegradable than less polar (and less bioavailable) compounds.

8. Unsaturated aliphatic compounds (i.e., with one or more carbon–carbon double bonds) are more readily biodegraded than saturated aliphatic compounds because of water's role as a nucleophile in attacking double bonds.

9. Increased substitution on aliphatic compounds impedes biodegradation.

10. Polycyclic aromatic hydrocarbons with more than four rings tend to be recalcitrant due to poor bioavailability and relative inability to induce the enzymes that can degrade them.

11. Functional groups or atoms that are unusual in living matter (i.e., *xenophores*) contribute to recalcitrance. For example, adding Cl, Br, NO_2, SO_3H, CN, or CF_3 to simple aromatic molecules, fatty acids, or other readily utilizable molecules increases their resistance to biodegradation. On the other hand, —OH, —COOH, or amide, ester, or anhydride functional groups often enhance biodegradation.

3.5. COOPERATION BETWEEN DIFFERENT MICROBIAL SPECIES FOR ENHANCED BIODEGRADATION

The mineralization of most organic hazardous compounds in the environment is rarely accomplished by single bacterial strains. Bioremediation is often the result of an intricate web of microbial consortia and activities, and the probability of successful bioremediation generally increases with greater microbial diversity. Often, different species collaborate in the stepwise or sequential degradation of complex organic chemicals, or exchange genetic materials (e.g., plasmids) that endow the recipient with enhanced biodegradation abilities. Some bacteria also excrete substances that provide nutrients or growth factors to other microorganisms; other bacteria remove and neutralize compounds that inhibit the activity of specific degraders. Examples of such beneficial microbial interactions are provided below.

3.5.1. Commensalism

Commensalism is a beneficial microbial interaction involving "consortial metabolism," where one population benefits from the "table scraps" of another. This is a unidirectional interaction since the population producing the "table scrap" is not affected by subsequent reactions. Mineralization of recalcitrant compounds such as PCBs, chlorinated solvents, and some pesticides is often based on *cometabolic commensalism*. As discussed earlier in this chapter, cometabolism results in dead-end metabolites that cannot be metabolized further by the specific degrader. However, other microorganisms with detoxifying capabilities can further degrade and mineralize these by-products. This is illustrated in Figure 3.26 for the cometabolism of cyclohexane by *Mycobacterium vaccae*, which uses the enzyme propane monooxygenase to convert cyclohexane to cyclohexanol. This compound is a "table scrap" that is mineralized by a *Pseudomonas* (Beam and Perry 1974).

The sequential anaerobic/aerobic biodegradation of polychlorinated compounds is another commensalistic reaction that is very important to bioremediation. Reductive dechlorination of such compounds is highly feasible under anaerobic conditions and proceeds readily when appropriate electron donors are present. This results in the accumulation of mono- and nonchlorinated by-products that are difficult to degrade further anaerobically. Such "table scraps," however, can often be readily degraded under aerobic conditions because their oxidation is thermodynamically more feasible than their further reduction, as illustrated for chlorinated methane, ethanes, and ethene homologues (Figure 3.27). This implies that the

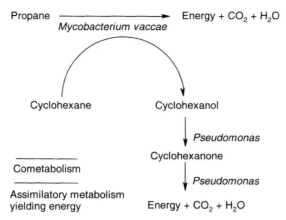

Figure 3.26. An example of commensalism based on cometabolism. Cyclohexane is cometabolized in the presence of propane by *Mycobacterium*, allowing for commensal growth of *Pseudomonas* on cyclohexane. (From Beam and Perry 1974.)

overall mineralization rate can be increased by a sequential anaerobic/aerobic attack. Such a sequential degradation occurs when either oxygen is introduced after the target compounds are dechlorinated in an anerobic system (i.e., a temporal sequence) or if the dechlorinated by-products flow out of the anaerobic zone into an aerobic environment (i.e., a spatial sequence).

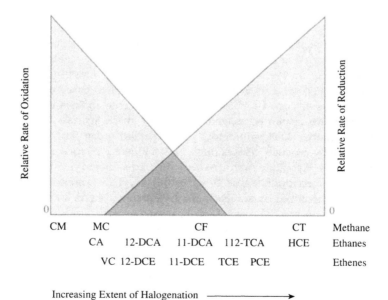

Figure 3.27. Relative rates of reduction and oxidation as a function of the extent of halogenation. Rates of reduction are highest for the more chlorinated homologues and decrease with degree of chlorination up to a point when oxidation rates become faster.

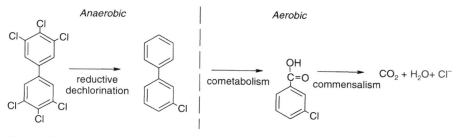

Figure 3.28. Sequential anaerobic/aerobic degradation of a highly chlorinated PCB. Hexachlorobiphenyl is reductively dechlorinated to monochlorobiphenyl under anaerobic conditions, which is then cometabolized to chlorobenzoate under aerobic conditions. Chlorobenzoate is then mineralized by other aerobic strains.

An example for sequential anaerobic/aerobic degradation of a recalcitrant compound is the mineralization of highly chlorinated PCBs or dioxins. Such compounds can be dechlorinated under very reducing conditions, resulting in predominantly mono- to tetrachlorinated congeners that could, in turn, be cometabolized by aerobic (oxygenase) enzymes. Aerobic ring cleavage produces monoaromatic intermediates that could be commensalistically mineralized by other aerobic microorganisms (Figure 3.28).

Another example of commensalism is when a specific degrader cannot synthesize a particular molecule required for the reaction but will degrade the target contaminant if a second organism synthesizes and excretes the needed cofactor, as illustrated in Figure 3.29. In this example, the "table scrap" is vitamin B_{12} (i.e., biotin), which is needed by strain 3-CL to degrade trichloroacetic acid. Vitamin B_{12} is provided by a *Streptomyces*, which is feeding on another substrate (Figure 3.29) (Slater and Lovatt 1984).

3.5.2. Syntrophism

Syntrophism is a nonobligatory association between two or more microbial populations that supply each other's nutritional requirements. Unlike commensalism,

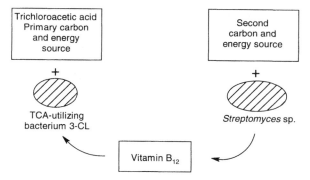

Figure 3.29. Commensalism based on vitamin production.

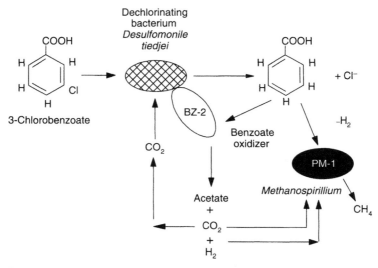

Figure 3.30. Consortium of microorganisms: a microbial web for syntrophic degradation of chlorinated aromatic compounds. (Tiedje and Stevens 1987.)

syntrophism is a mutually beneficial interaction. This is illustrated in Figure 3.30, where three different anaerobic microorganisms associate to mineralize a chlorinated aromatic compound (Tiedje and Stevens 1987). In this example, the first organism is *Desulfomonile tiedjei* that dechlorinates chlorobenzoate to benzoate. This organism is a dehalorespirer, which means that it can harvest energy from the dechlorination reaction. *Desulfomonile tiedjei* uses CO_2 as the carbon source, H_2 as the electron donor, and chlorobenzoate as the electron acceptor, which is reduced to benzoate. The second organism, BZ-2, is a fermenter that degrades benzoate, and it attaches to *D. tiedjei* to facilitate the transfer of benzoate. BZ-2 can ferment benzoate to CO_2 and H_2, which are used by *D. tiedjei* to grow. However, H_2 is produced faster than *D. tiedjei* can consume it, and the accumulation of H_2 inhibits the fermentation of benzoate due to thermodynamic constraints (see Section 3.5.3). The third microorganism, *Methanospirillium* PM-1 also feeds on H_2, preventing its accumulation and making it thermodynamically favorable for BZ-2 to mineralize benzoate.

The transfer of hydrogen that occurs between fermentative bacteria and organisms capable of chlorinated ethene respiration (i.e., dechlorination by halorespirers or chlorogenic bacteria) is very important for bioremediation. Fermentative bacteria are poorly suited to reductive metabolism of chlorinated ethenes but are capable of oxidizing a range of organic molecules under anaerobic conditions and releasing H_2 as a product. Halorespiring bacteria are well suited to the rapid reduction of chlorinated ethenes but require simple primary substrates for energy, including H_2. Thus, the ability of fermentative bacteria to produce H_2 for the halorespiring population allows many organic substrates to indirectly support bioremediation of chlorinated ethenes (Hughes et al. 2002).

3.5.3. Interspecies Hydrogen Transfer

The need for different microbial populations to collaborate is especially important under anaerobic conditions, where organic pollutants are degraded to nontoxic products such as acetate, CO_2, CH_4, and H_2 by the combined action of several different types of bacteria (White 1995). As illustrated in Figure 3.31, the anaerobic food chain consists of three stages. In the first stage, fermenters produce simple organic acids, alcohols, hydrogen gas, and carbon dioxide. Other members of the consortium oxidize these fermentation products in the second stage to CO_2, H_2, and acetate, such as sulfate reducers and organisms that use water-derived protons as the major or sole electron sink. The latter include the obligate proton-reducing acetogens, which oxidize butyrate, propionate, ethanol, and other compounds to acetate, H_2, and CO_2. Acetate can also be produced by homoacetogens, which are bacteria that utilize CO_2 and H_2 for this purpose (Madigan et al. 2003). Mineralization occurs in the third stage. This is accomplished by acetoclastic methanogens, which break down acetate into CO_2 and CH_4. Some sulfate reducers and other anaerobic microorganisms can also mineralize acetate and participate in the final stabilization stage (Atlas and Bartha 1998).

Interspecies hydrogen transfer is a critical link in the anaerobic food chain because it prevents the accumulation of fermentation products and enhances anaerobic mineralization. Specifically, hydrogen-producing fermentative and acetogenic bacteria are at a thermodynamic disadvantage if hydrogen accumulates (Conrad et al. 1985). For example, the fermentation of ethanol to acetate (i.e., acetogenesis,

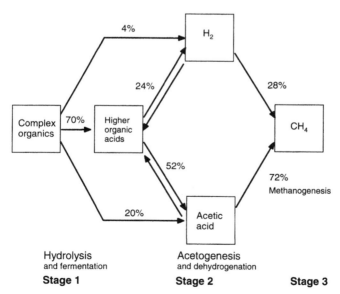

Figure 3.31. Electron flow through the methanogenic food web. (Adapted from McCarty and Smith 1986.)

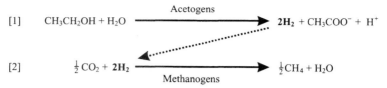

Figure 3.32. Interspecies hydrogen transfer.

Eq. [1] in Figure 3.32) is strongly inhibited by high hydrogen concentrations because this reaction is not thermodynamically feasible under standard conditions ($\Delta G^{\circ\prime} = +9.6$ kJ). Acetogenesis can proceed only if the H_2 produced by acetogens and other fermenters is removed (law of mass action) (Wu and Hickey 1996). Therefore, fermenters and acetogens associate mutualistically with hydrogen-consuming organisms (e.g., methanogens) to keep the H_2 levels low (Eq. [2] in Figure 3.32).

Methanogens and other hydrogen-consuming microorganisms are very important in anaerobic bioremediation because of their ecological relationship to other populations, making certain reactions possible by providing an improved thermodynamic environment. Figure 3.33 summarizes the effect of hydrogen partial pressure on the thermodynamic feasibility for the conversion of ethanol, propionate, acetate, and hydrogen during methane fermentation. The H_2 concentration must be low for acetogenesis and subsequent anaerobic mineralization to occur. On the other hand, the H_2 concentration must be sufficiently high to sustain the hydrogenotrophic species that remove it. For example, minimum H_2 requirements have been reported to be about 0.2 nM for iron reducers (10^{-6} atm), 1–1.5 nM for sulfate

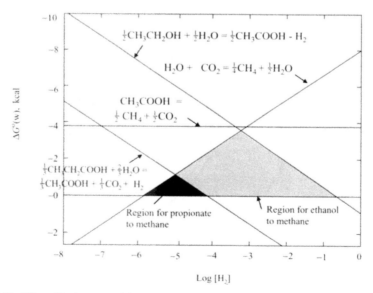

Figure 3.33. Effect of hydrogen partial pressure on the free energy of conversion of ethanol, propionate, acetate, and hydrogen during methane fermentation. (Adapted from McCarty and Smith 1986.)

reducers (10^{-5} atm), and 7–10 nM for methanogens ($10^{-4.5}$ atm) (Lovley and Goodwin 1988). Thus, there is a well-defined range of hydrogen partial pressures for which anaerobic mineralization of organic compounds is thermodynamically feasible. This is depicted as the shadowed regions in Figure 3.33.

Another example of a positive effect of interspecies hydrogen transfer is that which occurs between fermentative bacteria and organisms capable of chlorinated ethene respiration (i.e., dechlorination by halorespirers or chlorogenic bacteria). Fermentative bacteria are poorly suited to reductive metabolism of chlorinated ethenes but are capable of oxidizing a range of organic molecules under anaerobic conditions and releasing hydrogen (H_2) as a product. Halorespiring bacteria are well suited to the rapid reduction of chlorinated ethenes but require simple primary substrates for energy, including H_2. Thus, the ability of fermentative bacteria to produce H_2 for the halorespiring population allows many organic substrates to indirectly support bioremediation of chlorinated ethenes (Hughes et al. 2002).

3.6. BIOTRANSFORMATION OF METALS

Microorganisms need a wide variety of metals in trace quantities to grow. For example, metals such as zinc, cobalt, nickel, and molybdenum are used at the active sites of many enzymes to facilitate catalysis (e.g., by exerting electronic strain on the substrate) and/or electron transfer. Nevertheless, the quantity of metals assimilated for such metabolic purposes is relatively small and this is not an important attenuation mechanism compared to biotransformation reactions.

Whereas microorganisms cannot convert metals to different elements, they can catalyze oxidation or reduction reactions that affect the solubility and mobility of many metals. A wide variety of microorganisms can catalyze the reduction of redox-sensitive heavy metals, such as Cr(VI) and U(VI). This reaction may be directly or indirectly mediated by microbes. Microbes can be indirectly responsible for metal reduction by producing relatively strong reductants like H_2S from SO_4^{2-} and Fe(II) from Fe(III). In most cases, abiotic reduction using H_2S and Fe(II) is much slower than direct biotic reduction (Doong and Wu 1992; Kriegman-King and Reinhard 1994). A number of studies have shown that several microorganisms can directly (intracellularly) reduce Cr(VI) to Cr(III) using Cr(VI) as an electron acceptor during microbial respiration (Wang et al. 1989; Ishibashi et al. 1990; Yamamoto et al. 1993). Similarly, it has been shown that U(VI) can be reduced to U(IV) by microorganisms (Lovely and Phillips 1992a,b; Gorby and Lovely 1992; Thomas and Macaskie 1996). For the most part, metal reductases have been implicated in these studies. However, there is evidence that such metals can be used as electron acceptors for growth in energy-yielding reactions (Lovely et al. 1991). Many of these organisms (e.g., *Geobacter metallireducens* GS-15 and *Shewanella putrefaciens*) are coincidentally dissimilatory iron-reducing bacteria that can also use nitrate and Fe(III) as terminal electron acceptors.

Some microbial-mediated reductions and immobilization of heavy metals are not readily reversible and can be exploited for natural attenuation. For example, consider chromium(VI), which exists as an oxyanion, as bichromate ($HCrO_4^-$) below pH 6.5, chromate (CrO_4^{2-}) near pH 6.5, and dichromate ($Cr_2O_7^{2-}$) at concentrations greater than 10 mM. Chromium(VI) is mobile in groundwater and is a greater health hazard than chromium(III), which is a cation that tends to bind strongly to aquifer material. Many bacteria can use dissolved organic matter in the groundwater as an electron donor to reduce chromium(VI) to chromium(III), making it effectively immobile ($Cr(OH)_3$ precipitates). Some cells also obtain energy from Cr(VI) reduction, although most cells mediate this reaction cometabolically and do not harvest the energy. Note that Cr(VI) is toxic to most microorganisms at about 1 mg/L, but plasmids coding for resistance enable bacteria to tolerate up to 4000 mg/L. Once reduced, Cr(III) cannot be oxidized by microbes back to Cr(VI). When considering natural attenuation of Cr(VI), care should be taken that there are no oxidized forms of manganese in the aquifer matrix (e.g., MnO_2). Such minerals are known to oxidize and remobilize Cr(III) back to Cr(VI).

Another metal that can be immobilized by microbial reduction is uranium. Uranium(VI) (UO_2^{2+}, uranyl) is soluble when complexed by sulfate or phosphate and can be reduced by microorganisms to less mobile U(IV), which can precipitate as UO_2. Some bacteria also obtain energy from U(VI) reduction (e.g., *Desulfovibrio, Geobacter, Shewanella*).

Note that the reduction of a metal could be undesirable. For example, under anaerobic conditions, arsenic(V) (AsO_4^{3-}, arsenate) may also serve as an alternate electron acceptor and be reduced to *more toxic and more mobile* arsenic(III) (AsO_2^-, arsenite). This process is reversible because arsenite can be reoxidized to arsenate (and thus reimmobilized) under aerobic conditions. For example, the role of metal-reducing bacteria in arsenic release from Bengal delta sediments was investigated by Islam et al. (2004). These authors showed that anaerobic metal-reducing bacteria can play a key role in the mobilization of arsenic in sediments collected from a contaminated aquifer that is heavily used for drinking water. They also showed that, for the sediments in their study, arsenic release took place after Fe(III) reduction, rather than occurring simultaneously.

Manganese(IV) salts are reduced to *more toxic and more mobile* manganese(II) (Mn^{2+}) under anaerobic conditions. Similar to arsenic, this reaction is reversible.

3.7. BIODEGRADATION KINETICS

Monod kinetics (Monod 1949) are widely used to characterize biodegradation rates of organic contaminants. The popularity of Monod's equation is due to the fact that this empirical equation can describe biodegradation rates following zero- to first-order kinetics with respect to the target substrate concentration. Nevertheless, this is a hyperbolic equation that is difficult to incorporate into analytical fate-and-transport models because it precludes obtaining an explicit solution upon integration of the advection–dispersion reaction differential equation. Therefore, Monod's equation will be considered here mainly to provide insight into the physiologic and mass transport processes that affect biodegradation kinetics. Our attention will then shift to the first-order degradation model that is commonly used in analytical models.

Monod (1949) proposed a hyperbolic equation to describe the growth of bacterial cultures as a function of a limiting nutrient concentration:

$$\mu = -\frac{dX/dt}{X} = \frac{\mu_{max}C}{K_s + C}$$ [3.26]

where μ = specific growth rate; that is, growth rate normalized to the cell concentration (day^{-1})

X = microbial concentration (mg-cells/L)

μ_{max} = maximum specific growth rate (day^{-1})

C = limiting substrate concentration in the bulk liquid (mg-substrate/L)

K_s = half-saturation coefficient (mg-substrate/L), which is the substrate concentration corresponding to a specific growth rate equal to one-half of the maximum

$-\dfrac{dC/dt}{X}$ = specific substrate utilization rate, U (g-substrate/g-cells/day)

This equation was modified by Lawrence and McCarty (1970) to describe the rate at which microorganisms remove a target pollutant when this is the limiting substrate:

$$-\frac{dC}{dt} = \frac{kCX}{K_s + C}$$ [3.27]

where $-dC/dt$ = substrate utilization rate (mg-substrate/L/day)

k = maximum specific substrate utilization rate (g-substrate/g-cells/day)

All other terms are as previously defined. Note that K_s in this case corresponds to the substrate concentration that results in a specific substrate utilization rate equal to one-half of the maximum (i.e., $U = k/2$, Figure 3.34). Lower K_s values indicate a higher affinity for the substrate since this is conducive to higher growth and degradation rates at a given substrate concentration.

Note also that the growth rate is directly proportional to the degradation rate:

$$\frac{dX}{dt} = Y\left(-\frac{dC}{dt}\right)$$ [3.28]

where Y is the cell yield coefficient (g-cell produced/g-substrate degraded). Thus, the maximum specific growth and degradation rates are also proportional:

$$\mu_{max} = Yk$$ [3.29]

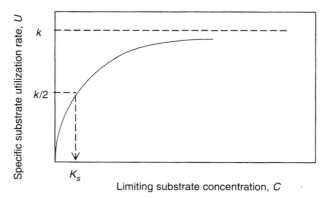

Figure 3.34. Graphic representation of Monod's equation.

Monod's equation is based on Michaelis–Menten enzyme kinetics, which applies to unireactant enzymes, rapid enzyme–substrate equilibrium, excess substrate, absence of mass transfer limitations, and constant temperature and pH. Such conditions are rarely found in bioremediation applications. Thus, the basic Monod equation has been modified to incorporate suboptimal conditions, such as substrate competition and inhibition by different mechanisms (Bellgardt 1991).

Merchuck and Ansejo (1995) interpreted Monod's equation by considering a dual-resistance model that provides a physical meaning to K_s. In this model, biodegradation involves two consecutive steps: (1) transport of the substrate from the bulk liquid to the cell surface, which occurs at a specific rate, μ_t; and (2) uptake and metabolism, occurring at a specific rate, μ_{max} (Figure 3.35).

Using an electrical circuit analogy, μ_t and μ_{max} are equivalent to *conductance*, and their inverses represent *resistance*. The overall resistance is the sum of each resistance in series:

$$\frac{1}{\mu} = \frac{1}{\mu_t} + \frac{1}{\mu_{max}}$$

[3.30]

This assumes that μ_{max} is constant (independent of substrate concentration) and is determined by either uptake or metabolism.

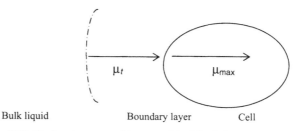

Figure 3.35. Dual-resistance model proposed by Merchuk and Ansejo (1995).

Assuming Fickian diffusion of the substrate from the bulk liquid to the cell surface (i.e., that this flux is driven by the corresponding concentration gradient), the decrease in dissolved substrate concentration is given by

$$-\frac{dC}{dt} = \frac{\text{cell surface area}}{\text{control volume}} h_s (C - C_c) \qquad [3.31]$$

where the first term represents the cell surface area concentration, h_s is an overall mass transfer coefficient [L/T], and C_c is the substrate concentration at the cell surface.

For spherical cells with diameter d_c, the cell volume is $\pi d_c^3 / 6$ and the cell surface area is πd_c^2. Thus, the cell surface area concentration is

$$\frac{\text{cell surface area}}{\text{control volume}} = \frac{6}{d_c} \frac{X}{\rho_c} \qquad [3.32]$$

where X is the cell concentration and ρ_c is the cell density.

Substituting into Eq. [3.31], we find

$$-\frac{dC}{dt} = \frac{6}{d_c} \frac{X}{\rho_c} h_s (C - C_c) \qquad [3.33]$$

For cylindrical cells (with spherical caps, and lengths of 3 d_c), the term $6/d_c$ should be replaced by $4.5/d_c$.

Assuming that μ_t is proportional to the substrate flux toward the cell, and recognizing that the proportionality constant between the substrate utilization and growth rates is Y (Eq. [3.28]), Eq. [3.33] implies that

$$\mu_t = \frac{6YX}{d_c \rho_c} h_s (C - C_c) \qquad [3.34]$$

Substituting into Eq. [3.30] and solving for μ, we find

$$\mu = \frac{\mu_{max} \left[\frac{6YX}{d_c \rho_c} h_s \right] (C - C_c)}{\mu_{max} + \left[\frac{6YX}{d_c \rho_c} h_s \right] (C - C_c)} \qquad [3.35]$$

If we assume that the substrate concentration at the cell surface is relatively small ($C_c \ll C$), Eq. [3.35] becomes

$$\mu = \frac{\mu_{max} \left[\frac{6YX}{d_c \rho_c} h_s \right] C}{\mu_{max} + \left[\frac{6YX}{d_c \rho_c} h_s \right] C} \qquad [3.36]$$

This is equivalent to Monod's equation [3.26] with

$$K_s = \frac{\mu_{max}}{\left[\frac{6YX}{d_c \rho_c} h_s\right]}$$

[3.37]

Equation [3.37] shows that K_s depends both on the metabolic capacity of the cell (Y and μ_{max}) as well as on mass transfer phenomena and substrate bioavailability (h_s). This equation also implies that smaller bacteria (e.g., oligotrophs) will tend to exhibit smaller K_s values and, thus, higher affinity for scarcely available substrates.

3.8. BIOAVAILABILITY

The rate at which microbes can degrade pollutants and the resulting treatment end points can be influenced by the availability of that pollutant to the microbe (i.e., bioavailability), which can be limited by mass transfer phenomena such as desorption from aquifer solids, dissolution from nonaqueous phase liquids (NAPLs), or diffusion from nanopores to name a few examples. Bioavailability is typically determined by physicochemical factors such as contaminant hydrophobicity, volatility, phase, and solubility, as well as reaction conditions, such as mixing intensity.

Two common observations suggest that mass transfer rates (and thus bioavailability) may control the overall rate of degradation: (1) laboratory assays often overestimate reaction rates at larger scales, and (2) mixing increases reaction rates (Figure 3.36).

There are three *Damkohler* (dimensionless) numbers that are often used to compare biodegradation rates with various mass transfer processes:

$$(1)\ D_1 = \frac{\text{biodegradation rate}}{\text{advection rate}} = \frac{k_{bio}L}{v}$$

[3.38]

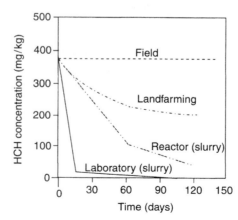

Figure 3.36. Biotransformation of hexachlorocyclohexane in systems ranging from laboratory slurries to field state. Note that mixing enhances the removal rate. (From Zhender 1991.)

where k_{bio} = pseudo-first-order biodegradation rate [T^{-1}]

v = pore-water velocity [L/T]

L = "characteristic length" in the flow direction, often taken as the plume length [L]

If $D_1 \gg 1$, slow advection limits overall degradation.

If $D_1 \ll 1$, slow biodegradation limits overall degradation.

(2) $D_2 = \dfrac{\text{biodegradation rate}}{\text{external mass transfer rate}} = \dfrac{k_{bio}}{k_1 a^{soil}}$ or $\dfrac{k_{bio}}{k_1 a^{NAPL}}$ [3.39]

where $k_1 a^{soil}$ = mass transfer rate coefficient from surface of aquifer solids (e.g., desorption) [T^{-1}]

$k_1 a^{NAPL}$ = mass transfer rate coefficient from NAPL surface (e.g., NAPL dissolution) [T^{-1}]

If $D_2 \gg 1$, slow external mass transfer limits overall degradation.

If $D_2 \ll 1$, slow biodegradation limits overall degradation.

(3) $D_3 = \dfrac{\text{biodegradation rate}}{\text{micropore diffusion rate}} = \dfrac{k_{bio} r^2}{D_{eff}}$ [3.40]

where r = grain size representing the maximum diffusion path length [L]

D_{eff} = effective intra-aggregate diffusion coefficient [L^2/T]

If $D_3 \gg 1$, slow micropore diffusion limits degradation.

If $D_3 \ll 1$, slow biodegradation limits overall degradation.

Successful interpretation of these dimensionless numbers depends entirely on the accuracy of the parameters used to calculate them. Because there is some uncertainty associated with these parameters, values for Damkohler numbers between 0.2 and 5 are not considered to be significantly different from 1 so that a firm conclusion can be reached as to which process governs overall degradation. Values for the various parameters can be obtained from the literature, from correlations developed by chemical engineers, from chemical properties, and from laboratory and field measurements (Bosma et al. 1997; Ramaswami and Luthy 1997).

The Best Equation

Best (1954) derived an equation that describes the kinetics of mass-transfer-limited biodegradation. First, assume Monod kinetics to describe the quantity of a pollutant that is transformed by a cell (q_c) after the contaminant gets to the cell surface, at a concentration C_c:

$$q_c = \frac{q_{max} C_c}{K_s + C_c}$$ [3.41]

Then assume that the resupply of the pollutant from a given source occurs by dissolution, desorption, or sorption-retarded diffusion at a rate that is proportional to the concentration gradient from a distant concentration (C_d) to the cell surface (C_c):

$$q_d = \kappa(C_d - C_c) \tag{3.42}$$

Here, the mass transfer rate to the cell surface (q_d) is described by a single (mass transfer) exchange coefficient, κ.

At steady state, $q_c = q_d$, and the two equations can be combined to predict the specific rate of pollutant removal (i.e., the Best equation):

$$q = q_{max} \left(\frac{C_d + K_s + q_{max}/\kappa}{2q_{max}/\kappa} \right) \left[1 - \sqrt{1 - \frac{4C_d q_{max}/\kappa}{(C_d + K_m + q_{max}/\kappa)^2}} \right] \tag{3.43}$$

where q = removal rate of pollutant per cell $[MT^{-1}]$

q_{max} = maximum conversion flux per cell $[MT^{-1}]$ (analogous to Monod's k)

C_d = distant (dissolved) pollutant concentration $[ML^{-3}]$

K_m = cell-surface pollutant concentration giving $q_{max}/2$ $[ML^{-3}]$

κ = exchange coefficient $[L^3T^{-1}]$

The permeability factor, κ $[L^3T^{-1}]$, is calculated as the product of a surface area $[L^2]$ and a mass transfer coefficient $[LT^{-1}]$, as illustrated in Table 3.5 (Bosma et al. 1997).

Bosma et al. (1997) modified the Best equation to incorporate macroscopic, gradient-driven, mass-transfer-dependent biodegradation in dimensionless form:

$$Q^* = \frac{1 + 1/B_n}{2(1 - C^*)} \left(1 - \sqrt{1 - 4C^* \frac{1 - C^*}{(1 + 1/B_n)}} \right) \tag{3.44}$$

Table 3.5. Expressions Used to Estimate the Exchange Coefficient κ

Mass Transfer Mechanism	Expression[a]
Linear diffusion	$D_{eff}A/\delta$
Radial diffusion, no advective flow	$D_{eff}4\pi R(R + \delta)/\delta$
Radial diffusion, $\delta \gg R$	$D_{eff}4\pi R$
Dissolution kinetics	$K_d S_{sw}$
Advective flow, monolayer of cells on particles with diffusion layer around them	$\eta A_p U/n$

[a]Symbols that are not explained in the text: R, cell radius (L); K_d, dissolution t = rate constant (LT^{-1}); A_{sw} solid–water contact surface area (L^2); η, collector efficiency; A_p, cross-sectional surface area of particles in porous media (L^2); U, flow velocity in porous media (LT^{-1}); n, number of bacterial cells per particle in porous media.

where

$$Q^* = \frac{q}{q_{max}} \tag{3.45}$$

$$C^* = \frac{C_d}{C_d + K_s + q_{max}/\kappa} \tag{3.46}$$

$$B_n = \frac{\kappa}{q_{max}/K_s} = \frac{\kappa K_m}{q_{max}} \tag{3.47}$$

In this model, Q^* is the normalized conversion (degradation) rate, which ranges from 0 to 1, and C^* represents the *bioavailable* contaminant concentration in dimensionless form. The term q_{max}/K_s is the microbial *specific affinity*, which is an asymptote of Monod's equation when $C_d \ll K_s$. Thus, high q_{max}/K_s values correspond to cells that are capable of fast degradation rates when the contaminant is present at low concentrations.

B_n is the *bioavailability number*, which can be considered to be the inverse of a Damkohler number to assess whether slow biodegradation ($B_n \gg 1$) or slow mass transfer to the cell surface ($B_n \ll 1$) controls the overall removal rate. Note that B_n indicates that bioavailability is a relative concept that relates the rate at which a pollutant can get to a cell to the rate at which the cell can degrade it. For example, B_n is lower for microbes with higher specific affinity (e.g., oligotrophs), which would tend to degrade a pollutant as fast as it becomes available. Nevertheless, under the same mass transfer limitations, a slower degrading strain could process the pollutant slower than it reaches the cell, in which case biodegradation would not be limited by bioavailability.

Equation [3.44] has two important limits. When mass transfer is fast compared to biodegradation (i.e., $B_n \gg 1$), the contaminant removal rate approaches hyperbolic (Monod) kinetics such as Eq. [3.41]. On the other hand, when biodegradation is fast compared to mass transfer (i.e., $B_n \ll 1$), there is a linear relationship between Q^* and C^*, and the degradation rate resembles the behavior of Eq. [3.42].

The concept of bioavailability can also help us understand how mass transfer limitations affect residual concentrations and treatment end points associated with biodegradation. Consider that substrate utilization per cell (q) can be separated into growth flux (q_g), which represents the metabolic flux of substrates that generates energy and building blocks for cell synthesis, and maintenance flux (q_m), which represents substrate consumption to prevent rapid decomposition of cell constituents and maintain cell viability.

$$q = q_g + q_m \qquad (\text{units} : [MT^{-1}]) \tag{3.48}$$

When the substrate supply rate (e.g., the diffusive flux out of soil pores, q_d) becomes equal to the maintenance substrate utilization rate (q_m), a threshold liquid concentration (C_t) will be reached below which a viable microbial community could not be sustained. Therefore, contaminant (i.e., substrate) degradation could not be sustained below this threshold concentration.

To illustrate how bioavailability affects C_t, assume again that mass transfer to the cell surface is directly proportional to the concentration gradient, and that the bulk liquid concentration has decreased already to the threshold level. Thus,

$$q_m = q_d = \kappa(C_t - C_c)$$ [3.49]

Where the concentration at the cell surface (C_c) is significantly smaller than C_t, then C_c can be ignored in Eq. [3.49] yielding

$$C_t = \frac{q_m}{\kappa}$$ [3.50]

Therefore, a slower diffusion rate out of nanopores or desorption from aquifer material or dissolution from a NAPL (i.e., decreased bioavailability) results in higher residual groundwater concentrations and less acceptable treatment end points.

3.9. APPLYING BIODEGRADATION KINETICS TO FATE AND TRANSPORT MODELING

Assessing the rate of degradation of organic pollutants is essential for predicting the extent to which contamination will spread and for estimating the time required to achieve acceptable levels. Analytical fate-and-transport models commonly describe biodegradation rates as a first-order decay regime with respect to the contaminant concentration (C):

$$\frac{dC}{dt} = -\lambda C$$ [3.51]

This first-order kinetic assumption is usually an appropriate simplification to describe biodegradation in aquifers, for two reasons. First, mass transfer limitations are often rate-limiting in porous media as the contaminants desorb, dissolve, and/or diffuse to the cell surface, and Fickian diffusion is a first-order process with respect to the contaminant concentration in the bulk liquid (e.g., Eq. [3.42]) (Simoni et al. 2001). Second, a decrease in substrate concentrations to levels that are below the corresponding Monod's half-saturation coefficient (K_s) favors first-order kinetics (Alvarez et al. 1991). In this case we can ignore C in the denominator from Eq. [3.29], and Monod's equation reduces to a linear (first-order) equation:

$$\frac{dC}{dt} = -\frac{kXC}{K_s + C} \approx -\left(\frac{kX}{K_s}\right)C \quad \text{(when } C \ll K_s\text{)}$$ [3.52]

Note that when mass transport is not rate-limiting, λ can be explained in terms of Monod parameters. A comparison of Eqs. [3.38] and [3.39] reveals that

$$\lambda = \frac{kX}{K_s}$$ [3.53]

This theoretical analysis indicates that the value of λ depends on (1) k (the maximum specific substrate utilization rate), which in turn depends primarily on the prevailing electron acceptor conditions, and on the type of microbe present; (2) K_s (the half-saturation coefficient), which was shown to be related to enzyme affinity, bioavailability, and mass transport limitations (Merchuk and Ansejo 1995); and (3) X (the active biomass concentration), which may not be constant and depends on environmental conditions and aquifer chemistry, including available substrates.

Therefore, λ is not necessarily a constant, but a coefficient that can vary in time and space due to microbial population shifts resulting from changes in aquifer chemistry. This parameter is often used as a volume-weighted average that includes an anaerobic biodegradation zone within the heart of the plume and an aerobic biodegradation zone at the edge of the plume. Despite this averaging approach, there is wide range of λ values that have been observed at different sites. For example, reported λ values for benzene range over several orders of magnitude from 0 to 0.087 day^{-1} (Howard 1990; Alvarez et al. 1991; Howard et al. 1991; Rifai et al. 1995; Aronson and Howard 1997). Therefore, λ should not be extrapolated from the literature. Rather, considerable care must be exercised in its determination to avoid overpredicting or underpredicting actual biodegradation rates and plume behavior. Methods to estimate λ are described in Chapter 7.

REFERENCES

Alexander, M. (1999). *Biodegradation and Bioremediation*, 2nd ed. Academic Press, San Diego, CA.

Alvarez, P.J.J., P.J. Anid, and T.M. Vogel (1991). Kinetics of aerobic biodegradation of benzene and toluene in sandy aquifer material. *Biodegradation* 2:43–51.

Aronson, D. and P.H. Howard (1997). Anaerobic Biodegradation of Organic Chemicals in Groundwater: A Summary of Field and Labloratory Studies. A report submitted to the American Petroleum Institute, Chemical Manufacturer's Association, National Council of the Paper Industry for Air and Stream Improvement, Edison Electric Institute, and American Forest and Paper Association.

Atlas, R.M., and R. Bartha (1998). *Microbial Ecology*, 4th ed. Benjamin/Cummings, Menlo Park, CA.

Beam H.W. and J.J. Perry (1974). Microbial degradation of cycloparaffinic hydrocarbons via cometabolism and commensalism. *J. Gen. Microbiol.* **82**:163–169.

Beller H.R. and A.M. Spormann (1998). Analysis of the novel benzylsuccinate synthase reaction for anaerobic toluene activation based on structural studies of the product. *J. Bacteriol.* **180**:5454–5457.

Bellgardt, H.-H. (1991). Cell models. In *Biotechnology*, 2nd ed. (completely revised), Vol. 4, *Measuring, Modeling and Control*, H.-J. Rehm and G. Reed (Eds.). VCH Publishers, New York.

Best, J.N. (1954). The inference of enzymatic properties from kinetic data obtained on living cells: I. Some kinetic considerations regarding an enzyme enclosed by a diffusion barrier. *J. Cell. Comp. Physiol.* 1–27.

Bosma, T.N.P., P.J.M. Middeldorp, G. Schraa, and A.J.B. Zehnder (1997). Mass transfer limitation of biotransformation: quantifying bioavailability. *Environ. Sci. Technol.*, **31**:248–252.

Chakrabarty A.M. (Ed.) (1982). *Biodegradation and Detoxification of Environmental Pollutants.* CRC Press, Boca Raton, FL.

Chapelle, F.H. (2001). *Ground-Water Microbiology and Geochemistry,* 2nd ed. John Wiley & Sons, Hoboken, NJ.

Conrad, R., T.J. Phelps, and J.G. Zeikus (1985). Gas metabolism evidence in support of the juxtaposition of hydrogen-producing and methanogenic bacteria in sewage sludge and lake sediments. *Appl. Environ. Microbiol.* **50**:595–601.

Corseuil, H. and W. Weber (1994). Potential biomass limitations on rates of degradation of monoaromatic hydrocarbons by indigenous microbes in subsurface soils. *Water Res.* **28**:1415–1423.

Cunningham, J.A., H. Rahme, G.D. Hopkins, C. Lebron, and M. Reinhard (2001). Enhanced in situ bioremediation of BTEX-contaminated groundwater by combined injection of nitrate and sulfate. *Environ. Sci. Technol.* **35**:1663–1670.

Dagley, S. (1984). Microbial degradation of aromatic compounds. *Dev. Indust. Microbiol.* **25**:53–65.

Da Silva, M.L. and P.J.J. Alvarez (2002). Effects of ethanol versus MTBE on BTEX migration and natural attenuation in aquifer columns. *ASCE J. Environ. Eng.* **128**(9):862–867.

Ellis, D.E., E.J. Lutz, J.M. Odom, R.J. Buchanan, C.L. Bartlett, M.D. Lee, M.R. Harkness, and K.A. Deweerd (2000). Bioaugmentation for accelerated in situ anaerobic bioremediation. *Environ. Sci. Technol.* **34**(11):2254–2260.

English, C.W. and R.C. Loehr (1991). Degradation of organic vapors in unsaturated soil. In *Bioremediation Fundamentals and Effective Application*, Proceedings of the 3rd Annual Symposium at the Gulf Coast Hazardous Substance Research Center, pp. 65–74.

Focht, D. (1988). Performance of biodegradative microorganisms in soil: xenobiotic chemicals and unexploited metabolic niches. In *Environmental Biotechnology*, G.S. Omenn (Ed.). Basic Life Sciences, Vol. 45. Plenum Press, New York.

Gale, E.F. (1952). *The Chemical Activities of Bacteria.* Academic Press, London.

Gerritse, J., V. Renard, T.M.P. Gomes, P.A. Lawson, M.D. Collins, and J.C. Gottschal (1996). *Desulfitobacterium* sp. strain PCE1, an anaerobic bacterium that can grow by reductive dechlorination of tetrachloroethene or ortho-chlorinated phenols. *Arch. Microbiol.* **165**(2):132–140.

Gibson, D.T. and V. Subramanian (1984). Microbial degradation of aromatic hydrocarbons. In *Microbial Degradation of Organic Compounds*, D.T. Gibson (Ed.). Marcel Dekker, New York, pp. 181–252.

Gibson, D.T., J.R. Koch, and R.E. Kallio (1968). Oxidative degradation of aromatic hydrocarbons by microorganisms. I. Enzymatic formation of catechol from benzene. *Biochemistry* **7**:2653–2662.

Gutfreund, H. (1976). Eilhelm Friedrich Kuhne: an appreciation. In *FEBS Letters: Enzymes: One Hundred Years.* H. Gutfreund (Ed.), Vol. 62 supplement, pp. 1–23.

Habe, H., J.S. Chung, J.H. Lee, K. Kasuga, T. Yoshida, H. Nojiri, and T. Omori (1991). Degradation of chlorinated dibenzofurans and dibenzo-*p*-dioxins by two types of bacteria having angular dioxygenases with different features. *Appl. Environ. Microbiol* **67**(8):3610–3617.

Haber, C.L., L.N. Allen, S. Zhao, and R.S. Hansen (1983). *Science* **221**:1147–1153.

Holliger, C.G. Schraag, A.J.M Stams, and A.J.B. Zehnder (1993). A highly purified enrichment culture couples the reductive dechlorination of tetrachloroethene to growth. *Appl. Environ. Microbiol.* **59**(9):2991–2997.

Howard, P.H. (Ed.) (1990). *Handbook of Environmental Fate and Exposure Data for Organic Chemicals—Volume II Solvents.* Lewis Publishers, New York.

Howard, P.H., R.S. Boethling, W.F. Jarvios, W.M. Meylan, and E.M. Michaelenko (1991). *Handbook of Environmental Degradation Rates.* Lewis Publishers, New York.

Hughes, J.B., K.L. Duston, and C.H. Ward (2002). *Engineered Bioremediation Technology.* Evaluation Report TE-02–03. Ground-Water Remediation Technologies Analysis Center, Pittsburgh, PA.

Hyman, M.R., C.R. Page, and D.J. Arp (1994). Oxidation of methyl fluoride and dimethyl ether by ammonia monooxygenase in *Nitrosomonas europea. Appl. Environ. Microbiol.* **60**:3033–3035.

Imai, T., H. Takigawa, S. Nakagawa, G.-J. Shen, T. Kodama, and Y. Minoda (1986). Microbial oxidation of hydrocarbons and related compounds by whole-cell suspensions of the methane-oxidizing bacterium H-2. *Appl. Environ. Microbiol.* **52**:1403–1406.

Islam, F.S., A.G. Gault, C. Boothman, D.A. Polya, J.M. Charnock, D. Chatterjee, and J.R. Lloyd (2004). Role of metal-reducing bacteria in arsenic release from Bengal delta sediments. *Nature* **430**:68–71.

Juliette, L.Y., M.R. Hyman, and D.J. Arp (1994). Inhibition of ammonia oxidation in *Nitrosomonas europea* by sulfur compounds: thioesters are oxidized to sulfoxides by ammonia monooxygenase. *Appl. Environ. Microbiol.* **59**:3718–3727.

Kim, M.H. and O.J. Hao (1999). Cometabolic degradation of chlorophenols by *Acinetobacter* species. *Water Res.* **33**(2):562–574.

King, R.B., G.M. Long, and J.K. Sheldon (1992). *Practical Environmental Bioremediation.* Lewis Publishers, Boca Raton, FL.

Kukor, J.J. and R.H. Olsen (1991). Genetic organization and regulation of a meta cleavage pathway for catechols produced from catabolism of toluene, benzene, phenol, and cresols by *Pseudomonas pickettii* PKO1. *J. Bacteriol.* **173**:4587–4594.

Lawrence, A. and P.L. McCarty (1970). Unified basis for biological treatment design and operation. *J. Sanitary Eng. Div. ASCE* **96**:757–778.

Lee, K. and D.T. Gibson (1996). Toluene and ethylbenzene oxidation by purified naphthalene dioxygenase from *Pseudomonas* sp. strain NCIB 9816-4. *Appl. Environ. Microbiol* **62**(9):3101–3106.

Lee, K., J.M. Brand, and D.T. Gibson (1995). Stereospecific sulfoxidation by toluene and naphthalene dioxygenases. *Biochem. Biophys. Res. Commun.* **212**(1):9–15.

Lee, M., J. Thomas, R. Borden, P. Bedient, C. Ward, and J. Wilson (1988). Biorestoration of aquifers contaminated with organic compounds. *CRC Crit. Rev. Environ. Control* **1**:29–89.

Li, S. and L.P. Wackett (1992). Trichloroethylene oxidation by toluene dioxygenase. *Biochem. Biophys. Res. Commun.* **185**(1):443–451.

Linkfield, T., J. Suflita, and J. Tiedje (1989). Characterization of the acclimation period before anaerobic dehalogenation of chlorobenzoates. *Appl. Environ. Microbiol.* **55**:2773–2778.

Little, C.D., A.V. Palumbo, S.E. Herbes, M.E. Lidstrom, T.L. Tyndall, and P.J. Gilmer (1988). Trichloroethylene biodegradation by a methane-oxidizing bacterium. *Appl. Environ. Microbiol.* **54**(4):951–956.

Lovanh, N. and P.J.J. Alvarez (2004). Effect of ethanol, acetate, and phenol on toluene degradation activity and *tod-lux* expression in *Pseudomonas putida* TOD102: an evaluation of the metabolic flux dilution model. *Biotechnol. Bioeng.* **86**(7):801–808.

Lovley, D.R. and S. Goodwin (1988). Hydrogen concentrations as an indicator of the predominant terminal electron accepting reactions in aquation sediments. *Geochim. Cosmochim. Acta* **52**:2993–3003.

Madigan, J.T., J.M. Martinko, and J. Parker (2003). *Brock Biology of Microorganisms*, 10th ed. Prentice Hall, Upper Saddle River, NJ.

Maymogatell, X.Y., T. Chien, J.M. Gossett, and S.H. Zinder (1997). Isolation of a bacterium that reductively dechlorinates tetrachloroethene to ethene. *Science* **276**(5318):1568–1571.

McCarty, P.L.M. and D.P. Smith (1986). Anaerobic wastewater treatment. *Environ. Sci. Technol.* **20**:1201–1206.

Merchuk, J.C. and J.A. Ansejo (1995). The Monod equation and mass transfer. *Biotechnol. Bioeng.* **45**:91–94.

Mohn, W.W. and J.M. Tiedje (1990). Strain dcb-1 conserves energy for growth from reductive dechlorination coupled to formate oxidation. *Arch. of Microbiol.* **153**(3):267–271.

Monod, J. (1949). The growth of bacterial cultures. *Annu. Rev. Microbiol.* **3**:371–394.

National Research Council (1993). *In situ Bioremediation: When Does It Work?* National Academy Press, Washington, DC.

Nelson, M.J.K., S.O. Montgomery, et al. (1988). Trichloroethylene metabolism by microorganisms that degrade aromatic compounds. *Appl. Environ. Microbiol.* **54**(2):604–606.

Oldenhuis, R., L. Kuijk, A. Lammers, D. Janssen, and B. Witholt (1989). Degradation of chlorinated and non-chlorinated aromatic solvents in soil suspensions by pure bacterial cultures. *Appl. Microbiol. Biotechnol.* **30**:211–217.

Paul, E. and F. Clark (1989). *Soil Microbiology and Biochemistry.* Academic Press, San Diego, CA.

Ramaswami, A. and R.G. Luthy, (1997). Measuring and modeling physicochemical limitations to bioavailability and biodegradation. In *Manual of Environmental Microbiology*, C.J. Hurst, G.R. Knudsen, M.J. McInerney, L.D. Stetzenbach, and M.V. Walter (Eds.). ASM Press, Washington, DC, Chap. 78.

Rasche, M.E., R.E. Hicks, M.R. Hyman, and D.J. Arp (1990). Oxidation of monohalogenated ethanes and *n*-chlorinated alkanes by whole cells of *Nitrosomonas europea*. *J. Bacteriol.* **172**:5368–5373.

Rasche, M.E., M.R. Hyman, and D.J. Arp (1991). Effects of soil and water content on methyl bromide oxidation by the ammonia-oxidizing bacteriums *Nitrosomonas europea*. *Appl. Environ. Microbiol.* **57**:2986–2994.

Richmond, S.A., J.E. Lindstrom, and J.F. Braddock (2001). Assessment of natural attenuation of chlorinated aliphatics and BTEX in subarctic groundwater. *Environ. Sci. Technol.* **35**(20):4038–4045.

Rieger, P.G. and H.J. Knackmuss (1995). Basic knowledge and perspectives on biodegradation of 2,4,6-trinitrotoluene and related nitroaromatic compounds in contaminated soil. In *Biodegradation of Nitroaromatic Compounds*, J.C. Spain (Ed.). Environmental Science Research, Vol. 49. Plenum Press, New York.

Rifai, H.S., R.C. Borden, J.T. Wilson, and C.H. Ward (1995). Intrinsic bioattenuation for subsurface restoration. In *Intrinsic Bioremediation*, R.E. Hinchee, et al. (Eds.). Battelle Press, Columbus, OH.

Robertson, B.R. and D.K. Button (1987). Toluene induction and uptake kinetics and their inclusion in the specific-affinity relationship for describing rates of hydrocarbon metabolism. *Appl. Environ. Microbiol.* **53**(9):2193–2205.

Salanitro, J.P., P. Johnson, G.E. Spinnler, P.M. Maner, H.L. Wisniewski, and C. Bruce (2000). Field scale demonstration of enhanced MTBE bioremediation through aquifer bioaugmentation and oxygenation. *Environ. Sci. Technol.* **34**(19):4152–4162.

Seeger, M., B. Camara, and B. Hofer (2001). Dehalogenation, denitration, dehydroxylation, and angular attack on substituted biphenyls and related compounds by a biphenyl dioxygenase. *J. Bacteriol.* **183**(12):3548–3555.

Sheremata, T. and J. Hawari (2000). Mineralization of RDX by the white rot fungus *Phanerochaete chrysosporium* to carbon dioxide and nitrous oxide. *Environ. Sci. Technol.* **34**:3384–3388.

Shields, M.S., S.O. Montgomery, P.J. Chapman, S.M. Cuskey, and P.H. Pritchard (1989). Novel pathway of toluene catabolism in the trichloroethylene-degrading bacterium G4. *Appl. Environ. Microbiol.* **55**:1624–1629.

Shields, M.S., S.O. Montgomery, S.M. Cuskey, P.J. Chapman, and P.H. Pritchard (1991). Mutants of *Pseudomonas cepacia* G4 defective in catabolism of aromatic compounds and trichloroethylene. *Appl. Environ. Microbiol.* **57**:1935–1941.

Simoni, S.F., A. Schafer, H. Harms, and A.J.B. Zehnder (2001). Factors affecting mass transfer limited biodegradation in saturated porous media. *J. Contam. Hydrol.* **50**(1-2):99–120.

Slater, J.H. and D. Lovatt (1984). Biodegradation and the significance of microbial communities. In *Microbial Degradation of Organic Pollutants*, D.T. Gibson (Ed.). Marcel Dekker, New York, pp. 439–485.

Tiedje, J.M. (1993). Bioremediation from an ecological perspective. In *In situ Bioremediation: When Does It Work?* National Research Council, National Academy Press, Washington, DC.

Tiedje, J.M. and T.O. Stevens (1987). The ecology of an anaerobic dechlorinating consortium. In *Environmental Biotechnology*. G.S. Ommen (Ed.). Plenum Press, New York.

Vogel, T.M., C.S. Criddle, et al. (1987). Transformations of halogenated aliphatic compounds. *Environ. Sci. Technol* **21**(8):722–736.

Wackett, L.P. and C.D. Hershberger (2001). *Biodegradation and Biocatalysis—Microbial Transformation of Organic Compounds*. ASM Press, Washington, DC.

Wackett, L.P., G.A. Brusseau, et al. (1989). Survey of microbial oxygenases: trichloroethylene degradation by propane-oxidizing bacteria. *Appl. Environ. Microbiol.* **55**(11):2960–2964.

Walter, M.V. (1999) Bioaugmentation. In *Manual of Environmental Microbiology.*, C.J. Hurst, G.R. Knudsen, M.J. McInerney, L.D. Stetzenbach, and M.V. Walter (Eds.). ASM Press, Washington, DC, Chap. 82.

Whited, G.M. and D.T. Gibson (1991). Toluene-4-monooxygenase, a three-component enzyme system that catalyzes the oxidation of toluene to *p*-cresol in *Pseudomonas mendocina* KR1. *J. Bacteriol.* **173**:3010–3016.

Williams, P.A. and K. Murray (1974). Metabolism of benzoate and the methylbenzoates by *Pseudomonas putida* (*arvilla*) mt-2: evidence for the existence of a TOL plasmid. *J. Bacteriol.* **120**:416–423.

Wolin, M.J. and T.L. Miller (1982). Interspecies hydrogen transfer: 15 years later. *ASM News* **48**:561–565.

Zhender, A.J.B. (1991). Mikrobiologische Reinigung eines kontaminierten Bodens, Beitrag des 9. Decheman-Fachgespräches, Umwetschultz Franfurt/Main, pp. 68–72.

Zehnder, A.J.B. and W. Stumm (1988). Geochemistry and biogeochemistry of anaerobic habitats. In *Biology of Anaerobic Organisms*, A.J.B. Zehnder (Ed.). Wiley-Liss, New York.

Zobell, C.E. (1946). Action of microorganisms on hydrocarbons. *Bacteriol. Rev.* **10**:149.

4

FUNDAMENTALS OF GROUNDWATER FLOW AND CONTAMINANT TRANSPORT PROCESSES

"Man masters nature not by force but by understanding."

—Jacob Bronowsky

"There is nothing softer and weaker than water, And yet there is nothing better for attacking hard and strong things. For this reason there is no substitute for it."

—Lao-Tzu (ca. 550 B.C.)

4.1. INTRODUCTION

Groundwater occurs in porous media such as soil, sediment, and rock beneath the ground surface. When all the pore space in a soil or rock is filled with water, the material is said to be *saturated*. When the pore space is less than completely filled with water, and partially filled with air, the geological medium is said to be *unsaturated*. As water from rainfall, snowmelt, rivers, and lakes percolates through the cracks and pores of soil and rock, it passes through a region called the *vadose zone* (Figure 4.1). *Groundwater* is a term used to describe water in the vadose and saturated zones. Water in the deeper portions of the unsaturated zone is unavailable for use by plants, which utilize mainly soil water in the shallow *root zone*.

Groundwater in the vadose zone is primarily held between grains under tension (negative pressure) by capillary forces. There are, within the vadose zone, regions where the pores are fully saturated. For example, pores can be fully saturated where stream water infiltrates and a perched aquifer may form above a stratum of low permeability. During heavy rainfall, the root zone may also reach full saturation when the infiltration capacity is reached. Pores are also fully saturated within the *capillary fringe*, but here, the water pressure is negative. The thickness of the capillary

Bioremediation and Natural Attenuation: Process Fundamentals and Mathematical Models
By Pedro J. J. Alvarez and Walter A. Illman   Copyright © 2006 John Wiley & Sons, Inc.

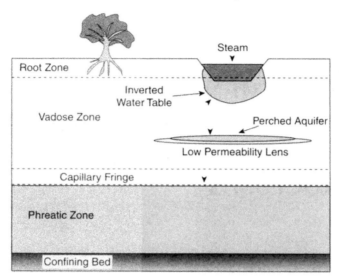

Figure 4.1. Conceptual model of the vadose and saturated zones.

fringe ranges from several centimeters for gravels up to 2 m for clayey materials. Water saturation, defined as the ratio between the volume of water and the volume of voids of the soil, decreases upward through the vadose zone from 100% at the water table to residual saturation above the capillary fringe (Figure 4.2). The *water table* or phreatic surface is, by definition, where the water pressure (p) is atmospheric ($p = p_{atm}$). Above the water table, water pressure becomes negative, while it increases with depth in the saturated zone due to hydrostatic forces.

The geological deposit, stratum, or succession of strata that contain and transmit groundwater are called *aquifers* (from Latin, *aqua* = water, *fer* = to bear). Water moving out of an aquifer is termed *discharge*, and water moving into an aquifer is termed *recharge* (Figure 4.3). Geologic strata interlayered with aquifers, but which tend to impede the movement of groundwater, are called *aquitards*. Aquifers

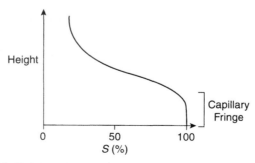

Figure 4.2. Variation of water saturation (S) with height in the vadose zone.

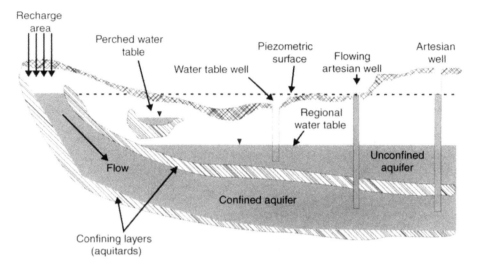

Figure 4.3. Confined and unconfined aquifers, aquitards, and various types of wells. (Adapted from Masters 1998.)

that are completely saturated and under pressure from overlying strata are termed *confined*. Unconfined aquifers, on the other hand, have the water table as an upper boundary. Unconfined aquifers are more vulnerable to contamination from activities on the ground because they are shallow, while confined aquifers are thought to be protected by the low permeability aquitards. However, contamination of a confined aquifer can occur through a constructed pathway such as a well or contamination resulting in the recharge area (Figure 4.3).

When a soluble contaminant dissolves into groundwater, it naturally spreads from the point of first encounter and begins to flow along with the groundwater (Figure 4.4a). The resulting shape of the dissolved contaminant mass in groundwater is called a *plume*. When the system is homogeneous and isotropic, the plume is usually ovoid, ellipsoidal, or tear-drop shaped in plan view. Physical heterogeneities such as clay lenses and multiple layers of different aquifer material often cause irregular plume shapes (Figure 4.4b). The presence of nonaqueous phase liquids (NAPLs) such as oils, gasoline, or industrial solvents also raises the level of complexity because the contaminants can migrate as a separate liquid phase, dissolved phase, and vapor phase. When the NAPL is heavier than water, it is called a dense nonaqueous phase liquid (DNAPL) and it tends to sink (e.g., chlorinated solvents; see Figure 4.4b). NAPLs that are lighter than water (LNAPL) tend to float on the water table (e.g., gasoline).

Contaminant plumes, whether composed of inorganic or organic chemicals, follow a natural life cycle beginning at the time of contaminant release, and ending, depending on the perspective of the evaluator, when the plume is diluted or

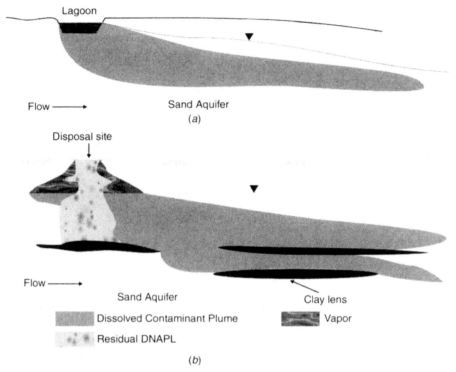

Flow ⟶

Sand Aquifer

(a)

Flow ⟶

Sand Aquifer

Clay lens

Dissolved Contaminant Plume   Vapor

Residual DNAPL

(b)

Figure 4.4. Groundwater contaminant plumes for (a) a simple system where percolation from a lagoon creates a dissolved plume, and (b) a complex system involving a DNAPL and physical heterogeneities. (Adapted from Domenico and Schwartz 1998.)

degraded to concentrations below concern, or attenuated, stabilized, and shrunk to insignificant dimensions. Active remediation can, of course, interrupt or accelerate the plume life cycle at any time or location.

Modeling of dissolved contaminant migration in groundwater involves two distinct processes: (1) groundwater flow and (2) contaminant transport. In the realm of groundwater flow, parameters characterizing water flow in the aquifer are of interest. In the realm of contaminant transport, parameters characterizing the aquifer pore network and interactions among various contaminants, the aquifer solid, and contaminant concentrations in water or air are of interest. In this chapter, the fundamental factors in both realms are discussed. This information forms the basis for the governing fate and transport equations described later in this chapter. Throughout the following discussion, the dimensions of introduced physical and chemical terms will be given as mass [M], length [L], and time [T]. Dimensions and common units for basic groundwater and contaminant transport parameters are provided in Table 4.1. Let us begin by considering the theoretical basis of groundwater flow.

Table 4.1. Symbols, Dimensions, and Units for Common Groundwater and Contaminant Transport Parameters

Parameter	Symbol	Dimensions	Units
Fluid pressure	p	$ML^{-1}T^{-2}$	N/m^2 or Pa
Fluid potential	Φ	L^2T^{-2}	m^2/s^2
Mass density	ρ	ML^{-3}	kg/m^3
Volumetric flow rate	Q	L^3T^{-1}	m^3/s
Specific discharge	q	LT^{-1}	m/s
Hydraulic head	h	L	m
Elevation head	z	L	m
Pressure head	ψ	L	m
Hydraulic conductivity	K	LT^{-1}	m/s
Permeability	k	L^2	m^2
Specific storage	S_s	L^{-1}	1/m
Total porosity	ϕ_T	L^0	—
Effective porosity	ϕ_e	L^0	—
Concentration	C	ML^{-3}	kg/m^3
Diffusion coefficient	D_d	L^2T^{-1}	m^2/s
Dispersivity	α	L	m
Hydrodynamic dispersion coefficient	D	L^2T^{-1}	m^2/s
Distribution coefficient	K_d	L^3M^{-1}	m^3/kg
Henry's Law constant	H	$ML^2T^{-2}mol^{-1}$	$kg^2/s^2 - mol$
Retardation factor	R	L^0	—

4.2. GROUNDWATER CONCEPTS

4.2.1. Hydraulic Head

Fluid flow through porous and fractured media is a mechanical process in which the driving forces causing fluid flow must overcome the resistance to it through viscous and frictional forces. Fluid flow is therefore accompanied by a transformation of mechanical energy to thermal energy through the mechanism of frictional resistance. The direction of flow in space must therefore be away from regions in which the mechanical energy per unit mass of fluid is higher toward regions in which it is lower. Therefore, the mechanical energy per unit mass at any point in a flow system can be defined as the work required to displace a unit mass of fluid from some datum to another point of interest. *The fluid potential for flow through porous media is therefore the mechanical energy per unit mass of fluid.* Figure 4.5 shows the components required to compute the mechanical energy of a unit mass of fluid from a reference state to the point of interest.

Consider a standard state of a unit mass of fluid at elevation $z = 0$; pressure is $p = p_0$ where p_0 is atmospheric pressure, and velocity is $v = 0$. A unit mass of fluid with density ρ_0 will occupy a volume V_0. We will now calculate the work required to raise the fluid from $z = 0$ to point X as shown in Figure 4.5. At this point the

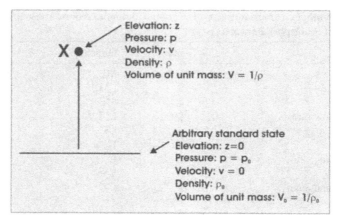

Figure 4.5. Calculation of mechanical energy of a unit mass of fluid. (Adapted from Freeze and Cherry 1979.)

fluid is at elevation z, with pressure p, velocity v, and a volume V of unit mass of fluid having density ρ. The work required to raise the mass of fluid from $z = 0$ to elevation z is

$$w_1 = mgz \qquad [4.1]$$

The work required to accelerate the fluid from velocity $v = 0$ to velocity v is

$$w_2 = \frac{mv^2}{2} \qquad [4.2]$$

And finally, the work required to change the fluid pressure from some initial pressure p_0 to p can be written as

$$w_3 = m \int_{p_0}^{p} \frac{V}{m} dp = m \int_{p_0}^{p} \frac{dp}{\rho} \qquad [4.3]$$

The mechanical energy per unit mass ($m = 1$) or the fluid potential Φ is the sum of the three terms (Eqs. [4.1]–[4.3]) of work done to the fluid:

$$\Phi = w_1 + w_2 + w_3 \qquad [4.4]$$

Substituting Eqs. [4.1] through [4.3] into Eq. [4.4] yields the Bernoulli equation, which is the classical equation in fluid mechanics describing the three components of energy loss (i.e., potential, kinetic, and elastic energy)

$$\Phi = gz + \frac{v^2}{2} + \int_{p_0}^{p} \frac{dp}{\rho} \qquad [4.5]$$

For groundwater flow through a porous medium, the second term is usually neglected because flow is considered to be slow. Only when the pore system consists of voids sufficient in size and connectedness to allow rapid flow as through a pipe is the velocity head of any consequence. In some fractured rocks and in karst terrain, the velocity head may be important. This equation can be additionally simplified as we are concerned primarily with flow of water which is treated to be incompressible. In this case, fluid density is not a function of pressure and therefore can be treated as a constant. Consequently, Eq. [4.5] can be simplified to the following form:

$$\Phi = gz + \frac{p - p_0}{\rho} \qquad [4.6]$$

Consider a manometer shown in Figure 4.6. At point X, the fluid pressure p is given by

$$p = \rho g \Psi + p_0 \qquad [4.7]$$

where Ψ is the height of the liquid above point X. Figure 4.6 shows that $\Psi = h - z$, where h is the hydraulic head and z is the elevation head. Hence,

$$p = \rho g(h - z) + p_0 \qquad [4.8]$$

Substituting Eq. [4.8] into [4.6] leads to

$$\Phi = gz + \frac{[\rho g(h - z) + p_0] - p_0}{\rho} = gh \qquad [4.9]$$

which simply states that the fluid potential at any point X in the porous medium is the product of the acceleration due to gravity and the hydraulic head. In groundwater hydrology, it is common to set $p_0 = 0$ and work in gauge pressure if p_0 is

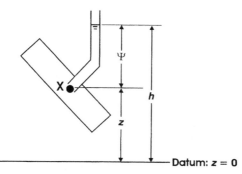

Figure 4.6. A laboratory manometer showing the relationships between hydraulic head, pressure head, and elevation head. (Adapted from Freeze and Cherry 1979.)

equal to the atmospheric pressure p_{atm}. Therefore, it is clear from Eqs. [4.6] and [4.9] that

$$\Phi = gz + \frac{p}{\rho} = g\left(z + \frac{p}{\rho g}\right) = gh \qquad [4.10]$$

and

$$h = z + \frac{p}{\rho g} = z + \Psi \qquad [4.11]$$

As shown in Eq. [4.11], hydraulic head (h) is the sum of the elevation (z) and pressure head (Ψ). The hydraulic head reflects the elevation of the top of a water column within the aquifer standing relative to some datum. To determine hydraulic head, the elevation of groundwater is measured in a monitoring well (Figure 4.7a). The screened interval or the intake is usually relatively long so the measured hydraulic head is considered to be a depth-integrated (or averaged) value. Alternatively, a piezometer (Figure 4.7b) can be used to measure the groundwater level. It consists of a small-diameter pipe extending from some point in an aquifer up to the ground surface and has an intake that is shorter in comparison to the monitoring well. Therefore, hydraulic head measurements obtained by means of a piezometer can be considered to be a point measurement. Measurement techniques for hydraulic head are described in Chapter 7.

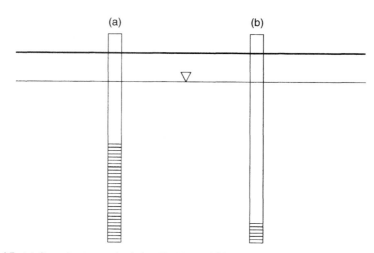

Figure 4.7. (a) Groundwater monitoring well with a long intake used to obtain depth-integrated water levels and water samples. (b) Piezometer with a short intake to obtain point-scale water levels and water samples.

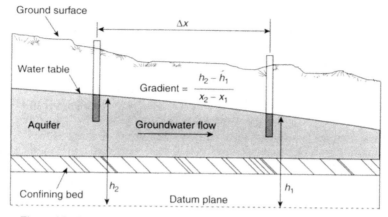

Figure 4.8. Hydraulic head (h) and hydraulic gradient in an unconfined aquifer.

The quantity of hydraulic head is expressed in units of length [L]. In unconfined aquifers, the head is the elevation of the water table (Figure 4.8). In confined aquifers, the head could be somewhat higher than the top of the aquifer (Figure 4.9).

Hydraulic gradient is a measure of the energy potential causing groundwater to flow between two points in an aquifer (Figure 4.8). The effective hydraulic gradient for an aquifer occurs in the direction of maximum head difference. The gradient is expressed as the difference in hydraulic head ($\Delta h = h_2 - h_1$) divided by the distance between measuring points ($\Delta x = x_2 - x_1$). Hydraulic gradients exist in horizontal or vertical directions. The total gradient is the sum of the x, y, and z gradient components, expressed as partial derivatives of hydraulic head with respect to the space coordinates; thus,

$$i_T = \frac{\partial h}{\partial x} + \frac{\partial h}{\partial y} + \frac{\partial h}{\partial z} \qquad [4.12]$$

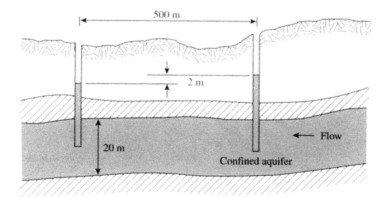

Figure 4.9. Hydraulic head (h) in artesian wells tapping a confined aquifer.

Groundwater flow is controlled by the three components of the hydraulic gradient, and contaminants in groundwater move mainly along the same downgradient direction. For sites that are far from points of recharge and discharge, the horizontal gradient will dominate groundwater movement. However, at sites near a pumping well, an infiltration field, or a river, vertical gradients can be important and can cause dissolved contaminant plumes to move downward. If no complicating factors (such as impermeable zones) occur in an unconfined aquifer, the horizontal hydraulic gradient as expressed by the water table tends to follow the landscape, flowing away from mounds beneath uplands, and flowing toward discharge zones in valleys. For groundwater under natural flow conditions, horizontal hydraulic gradients seldom exceed 0.05. Vertical gradients are normally less than the associated horizontal gradient except for locations very near high-capacity wells and when infiltration takes place in the vadose zone under unit gradient conditions.

4.2.2. Darcy's Law

Flow of groundwater through a saturated porous medium such as sand was described quantitatively first by a French hydraulic engineer named Henri Darcy in 1856. Using a laboratory experimental apparatus such as the one shown on Figure 4.10, he found that the discharged volume of water over time, $Q[L^3T^{-1}]$, that flowed through a porous medium was equal to the cross-sectional area of the flow system, $A[L^2]$, multiplied by the *hydraulic gradient*, $\Delta h/\Delta x$ $[LL^{-1}]$, where $\Delta h = h_2 - h_1$ and $\Delta x = x_2 - x_1$ is the distance between the intakes of the manometer, and multiplied by a proportionality constant, the *hydraulic conductivity*, $K[LT^{-1}]$, that is a characteristic constant of a porous medium under

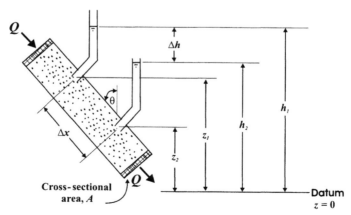

Figure 4.10. Darcy's experimental apparatus used to show the relationship among volumetric water discharge Q, hydraulic gradient $\Delta h/\Delta x$, and hydraulic conductivity K. (Adapted from Freeze and Cherry 1979.)

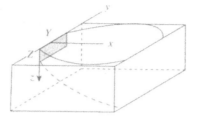

Figure 4.11. Cartesian system used to describe groundwater plumes in three dimensions.

consideration. This empirical relationship is known as Darcy's Law and is mathematically stated for the one-dimensional case as

$$Q = -KA \frac{\Delta h}{\Delta x} \qquad [4.13]$$

The physical parameters comprising the terms in Darcy's Law are elaborated below. In all subsequent discussion of groundwater or contaminant flow, the reader should keep in mind for reference a standard, three-dimensional Cartesian coordinate system (Figure 4.11). In this reference coordinate system, the direction of groundwater flow will always be along the positive x-axis. The downward, vertical direction will always be along the positive z-axis, and the horizontal transverse directions will always be along the positive and negative y-axes. With this in mind, we can speak of a one-dimensional flow or transport model with all processes occurring along the x-axis, denoted by a subscript x. Similarly, when we speak of a three-dimensional model, we refer to the three axial components of flow or transport and denote each with the appropriate subscript, x, y, or z.

4.2.3. Hydraulic Conductivity

Hydraulic conductivity (K) is an inherent characteristic of a porous medium that expresses the ease or difficulty of water flow through the medium. It has the dimension of length divided by time [LT^{-1}]. Other fluids such as air or petroleum have different liquid properties, and so they flow with different degrees of relative ease or difficulty through a given porous medium. Consequently, the hydraulic conductivity of a given porous medium will differ from the conductivity of air or some other fluid (liquid or gas). The parameter permeability (k) [L^2] is commonly used to describe the intrinsic capability of fluids to move through a porous or fractured medium, independent of fluid properties such as dynamic viscosity (μ) [$ML^{-1}T^{-1}$] and its density (ρ) [ML^{-3}].

The hydraulic conductivity is one of the parameters in natural sciences that exhibits tremendous variation. In fact, it varies over 13 orders of magnitude ranging from gravel deposits with $K = 1$ m/s (consideration for karst may yield higher K values!) down to unfractured metamorphic and igneous rocks with $K = 10^{-13}$ m/s. The hydraulic conductivity of geologic materials is also known to exhibit *spatial variability* and *directional dependence*. If hydraulic conductivity does not vary spatially from one point to the next (a rare condition indeed), the hydraulic conductivity of that particular formation is considered to be *homogeneous*. On the other hand, if the hydraulic conductivity varies from one point to the next, then the hydraulic conductivity of a formation is considered to be *heterogeneous*. If the hydraulic conductivity does not exhibit directional dependence, then it is considered to be *isotropic*, while it is *anisotropic* if it varies from one direction to the next. In many contaminant transport studies, hydraulic conductivity is treated as a constant because it can be costly to determine it at a large number of locations. This treatment of the hydraulic conductivity is justified in many studies where financial and time constraints preclude the determination of the spatial and directional dependence of hydraulic conductivity. However, one must recognize that it is the rule rather than the exception that, in many hydrogeologic conditions, the hydraulic conductivity exhibits directional and spatial dependence. For example, sedimentary aquifers typically have hydraulic conductivity variations between the horizontal (x) and vertical (z) directions that make the hydraulic conductivity an anisotropic quantity (Figure 4.12). In some aquifer materials, such as eolian sands, hydraulic conductivity differs in the x and y directions. For common considerations in contaminant transport, hydraulic conductivity is considered to be constant in the horizontal directions and is denoted K_h.

In the original, one-dimensional expression of Darcy, the hydraulic conductivity was recognized as a proportionality constant relating the hydraulic gradient ($dh/dx = \Delta h/\Delta x$) of a flow system to its *specific discharge* (q) [LT^{-1}] in Darcy's Law. Rearranging Eq. [4.13] gives

$$q = \frac{Q}{A} = -K\frac{dh}{dx} \qquad [4.14]$$

There is a negative sign in front of the right-hand side of the equation because flow takes place from higher to lower hydraulic gradient. Because hydraulic gradient is dimensionless, specific discharge, q, has the same units of velocity as K_h.

Figure 4.12. Anisotropy in K causing K_x to be larger than K_z.

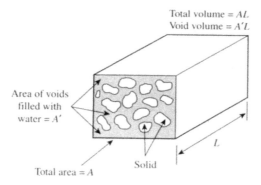

Figure 4.13. Cross-sectional area available for flow through porous medium (A).

Discharge, Q, of a porous medium is analogous to water flow in a pipe. The velocity, v_x, of water passing through a pipe with cross-sectional area A is

$$v_x = \frac{Q}{A} \qquad [4.15]$$

However, for a porous medium, the volume of water is flowing only through the pore system rather than through the entire cross-sectional area of the medium (Figure 4.13). Let us define the effective porosity (ϕ_e) as the interconnected pore fraction available for flow. Note that ϕ_e is also the fraction of the cross-sectional area of the porous medium through which water is actually flowing and excludes the portion of the cross-sectional area occupied by solid matrix. Thus, the product $\phi_e A$ represents the volume through which water flows ($A' = \phi_e A$). Adapting Eq. [4.15] to porous media by considering only the pore network shows that specific discharge, $Q/\phi_e A$, is equal to groundwater velocity, v_x. Substituting v_x for Q/A in Eq. [4.14] and considering that groundwater flows through interconnected pore space yields,

$$v_x = -\frac{K_h}{\phi_e}\frac{dh}{dx} \qquad [4.16]$$

Equation [4.16] is a simple, one-dimensional model commonly employed to provide rough estimates of groundwater velocity and travel time between points of interest. It permits the calculation of the groundwater velocity, v_x, through a geologic medium with effective porosity ϕ_e and hydraulic conductivity K_h, in response to hydraulic gradient dh/dx. Physically, it is the rate at which a particle of water moves along a flow path in an aquifer. Equation [4.16] characterizes the average groundwater velocity through the pore system of an aquifer. This average groundwater velocity is also known as the *seepage velocity*, the *average linear velocity*, or the *advective velocity*. Because it is an average, some water molecules will move faster than others in comparison to the calculated velocity. In a pipe or in a stream channel, the largest velocity of flowing water occurs near the center of the pipe or stream, away from the walls that cause friction. Likewise, the fastest groundwater flow occurs near the pore centers, and flow is slower adjacent to pore walls. To some extent, larger pore channels will transmit groundwater or any fluid faster than smaller pore channels will, but this depends on the connectivity of the pore channels. Even though there may be large pores, if they are not interconnected, they will not transmit any groundwater!

4.2.4. Heterogeneity in Hydraulic Conductivity and its Effect on Contaminant Transport

Groundwater contamination has become an ubiquitous environmental problem in many parts of the world and an active research topic since the 1970s. Significant advances have been achieved in our understanding of the complex processes of fluid flow and contaminant transport in natural geological media. It remains a formidable challenge, however, to accurately model and predict fluid flow and contaminant transport in subsurface environments. The main reasons for this are that geological media are inherently heterogeneous at different scales (from the pore, laboratory, field, and to the regional scales), and that the multiscale heterogeneity is difficult to fully characterize or describe. Figure 4.14 shows permeability and porosity data obtained from cores taken at 1 ft intervals in a sandstone aquifer in Illinois. It is clear from this figure that permeability varies erratically from one point to the next and that the variation spans over four orders of magnitude. Likewise, porosity varies erratically but the variation is considerably less, ranging from 3% to 20%.

In Figure 4.15, we see that different values of hydraulic conductivity can be defined for each of the investigated domains $[K_1 \neq K_2 \neq K_3]$ because of the variability in the geology, implying that hydraulic conductivity varies from one point to the next and that it is a measurement-scale-dependent quantity.

The effect of variation in hydraulic conductivity can be important in many contaminant transport studies. For example, a contaminant plume transported in a heterogeneous aquifer is strongly influenced by subsurface heterogeneity and thus exhibits an irregular shape (Figure 4.16) (see also Mackay et al. 1986; Killey and Moltyaner 1988; Garabedian et al. 1991; Boggs et al. 1992; Hubbard et al. 2001). This irregularity is caused mainly by the variation in hydraulic conductivity (K) and is difficult to predict with a model that treats the hydraulic conductivity to be deterministic or as a constant for certain units because hydraulic conductivity cannot be described spatially with certainty under many field conditions. Methods used to treat hydraulic conductivity and other hydraulic and transport parameters as spatially variable are beyond the scope of this textbook, but the reader is directed to Gelhar (1993), Dagan and Neuman (1997), and Zhang (2001) for further information on this exciting topic.

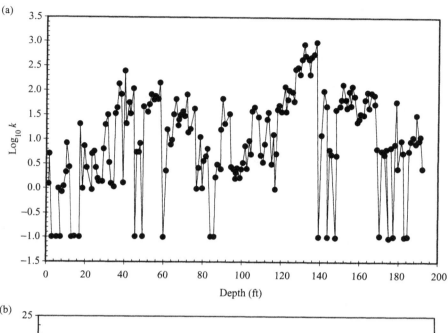

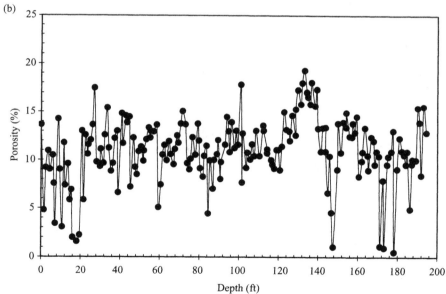

Figure 4.14. (a) Permeability and (b) porosity data obtained from cores taken at 1 ft intervals in a sandstone aquifer in Illinois. (Data from Bakr 1976.)

Figure 4.15. Highly heterogeneous fluvial deposits near an outcrop in Boise, Idaho with different values of hydraulic conductivity defined for each of the investigated domains $[K_1 \neq K_2 \neq K_3]$ implying that hydraulic conductivity varies from one point to the next and that it is a measurement-scale-dependent quantity.

4.2.5. Scale Effect in Hydraulic Conductivity

There is growing evidence that hydraulic conductivity (and permeability) determined for an aquifer tends to increase with the scale of the test in a variety of geological settings (Brace 1984; Bradbury and Muldoon 1990; Clauser 1992; Neuman 1994; Rovey and Cherkauer 1995; Illman et al. 1998; Schulze-Makuch and Cherkauer 1998; Illman and Neuman 2001, 2003; Vesselinov et al. 2001b; Hyun et al. 2002; Illman 2004). Such a trend is clearly visible when the hydraulic conductivity data from a large number of crystalline rock sites are plotted against its measurement scale ranging from the core to the regional scale (Figure 4.17). Observations of hydraulic conductivity scale variations at individual sites have been attributed to artifacts due to inadequate development of borehole intervals (Butler and Healey 1998a,b) and interpretive errors stemming from the use of disparate methods to gather and analyze field data (Zlotnik et al. 2000). Other possible causes of the hydraulic conductivity scale effect could be due to the conduct of well tests in composite structures such as the juxtaposition of various stratigraphic units and facies or the superposition of fractures and porous rock matrix (Rovey and Cherkauer 1995; Rovery 1998; Schulze-Makuch and Cherkauer 1998). Geostatistical (Vesselinov et al. 2001a,b) and theoretical (Hyun et al. 2002) analyses have shown that the primary cause of the scale effect in hydraulic conductivity arises from the use of a model that treats the hydraulic conductivity to be uniform when in fact the hydraulic conductivity is heterogeneous for a given aquifer. In fact, Vesselinov et al. (2001a,b) found that the hydraulic conductivity scale effect diminishes as one improves the degree to which heterogeneity is explicitly resolved by the model that one uses to estimate these parameters.

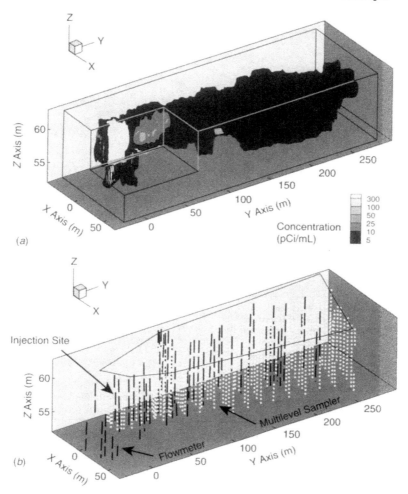

Figure 4.16. (a) Observed three-dimensional view of the tritium plume at 328 days after tracer injection showing its irregular shape. (b) Three-dimensional perspective view of the monitoring network used at the MADE site to capture the observed plume. (Adapted from Feehley et al. 2000.)

4.2.6. Derivation of Saturated Groundwater Flow Equation

Understanding advective transport of contaminants in the subsurface requires knowledge of fundamental flow processes. In order to facilitate this understanding, we will derive the saturated groundwater flow equation for both the steady and transient cases, beginning with the derivation of the continuity equation by writing a mass balance over a control volume with sides Δx, Δy, and Δz (Figure 4.18) through which the fluid is flowing. The derivations here are given in terms of rectangular coordinates. In some special cases, such as those involving fluid flow or solute transport to a well, cylindrical or spherical coordinates are more useful.

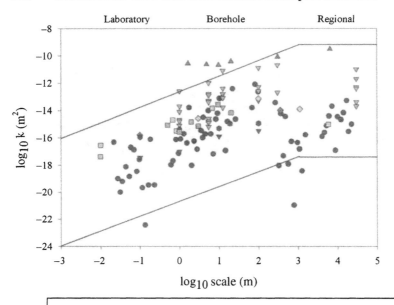

Figure 4.17. Permeability of crystalline rocks as a function of measurement scale.

The law of mass conservation states that there can be no net change in fluid mass in a control volume. In this control volume, fluid flow can take place through the entire volume, and in the x-direction the area of the face can be defined as $dy\ dz$. Similarly, fluid flux in the y- and z-directions takes place through an area defined by $dx\ dz$ and $dx\ dy$.

The mass inflow rate of fluid in the x-direction can be defined as

$$\rho_w q_x\ dy\ dz$$

where ρ_w =fluid density $[ML^{-3}]$, q_x = specific discharge in the x-direction $[LT^{-1}]$, and $dy\ dz$ = cross-sectional area $[L^2]$.

The mass outflow rate of fluid in the x-direction can be defined as

$$\left[\rho_w q_x + \frac{\partial}{\partial x}(\rho_w q_x)dx\right]dy\ dz$$

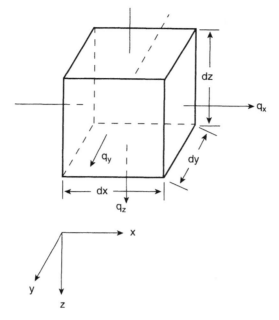

Figure 4.18. Control volume for the derivation of the fluid flow and advection–dispersion equations.

Similar expressions can be written for the y- and z-directions. The rate of mass accumulation due to the x-component of mass inflow minus mass outflow can be written as

$$-\frac{\partial(\rho_w q_x)}{\partial x} dx\ dy\ dz$$

Similar expressions can be written for the y- and z-directions. The sum of the three terms for the x-, y-, and z-directions yields

$$-\left[\frac{\partial}{\partial x}(\rho_w q_x) + \frac{\partial}{\partial y}(\rho_w q_y) + \frac{\partial}{\partial z}(\rho_w q_z)\right] dx\ dy\ dz$$

which is the total net mass accumulation within the control volume. The mass of water in the control volume (M) can be calculated by multiplying the density of water with porosity (n) by the control volume ($dx\ dy\ dz$):

$$\rho_w n\ dx\ dy\ dz$$

Therefore, the change in mass with respect to time can be obtained by taking the partial derivative of mass with respect to time as

$$\frac{\partial M}{\partial t} = \frac{\partial}{\partial t}(\rho_w n) dx\ dy\ dz$$

Conservation of mass requires that the difference between the rate of mass in and rate of mass out must be equal to the rate of mass accumulation; hence,

$$-\left[\frac{\partial}{\partial x}(\rho_w q_x) + \frac{\partial}{\partial y}(\rho_w q_y) + \frac{\partial}{\partial z}(\rho_w q_z)\right] dx\,dy\,dz = \frac{\partial}{\partial t}(\rho_w n) dx\,dy\,dz \qquad [4.17]$$

Dividing Eq. [4.17] by $dx\,dy\,dz$ and taking the limit as these dimensions approach zero, we obtain

$$-\left[\frac{\partial}{\partial x}(\rho_w q_x) + \frac{\partial}{\partial y}(\rho_w q_y) + \frac{\partial}{\partial z}(\rho_w q_z)\right] = \frac{\partial}{\partial t}(\rho_w n) \qquad [4.18]$$

which is the three-dimensional continuity equation. Physically, the equation describes the rate of change of fluid density at a fixed point resulting from the changes in mass flux. Under steady-state conditions, there is no mass storage, so Eq. [4.18] simplifies to

$$-\left[\frac{\partial}{\partial x}(\rho_w q_x) + \frac{\partial}{\partial y}(\rho_w q_y) + \frac{\partial}{\partial z}(\rho_w q_z)\right] = 0 \qquad [4.19]$$

Substitution of Darcy's Law for the x-, y-, and z-components yields

$$\frac{\partial}{\partial x}\left[K_x \frac{\partial h}{\partial x}\right] + \frac{\partial}{\partial y}\left[K_y \frac{\partial h}{\partial y}\right] + \frac{\partial}{\partial z}\left[K_z \frac{\partial h}{\partial z}\right] = 0 \qquad [4.20]$$

which is the partial differential equation that describes steady flow through a saturated anisotropic, heterogeneous medium. The above equation assumes that the xyz coordinate system is aligned with the principal axes of anisotropy in hydraulic conductivity. Otherwise, the left-hand side will have additional terms.

When the medium is homogeneous but hydraulic conductivity is anisotropic, we have

$$K_x \frac{\partial^2 h}{\partial x^2} + K_y \frac{\partial^2 h}{\partial y^2} + K_z \frac{\partial^2 h}{\partial z^2} = 0 \qquad [4.21]$$

For an isotropic, homogeneous medium $K_x = K_y = K_z = K$,

$$\frac{\partial^2 h}{\partial x^2} + \frac{\partial^2 h}{\partial y^2} + \frac{\partial^2 h}{\partial z^2} = 0 \qquad [4.22]$$

which is the Laplace equation. Equation [4.22] is one of the fundamental equations in mathematical physics and there are fundamental solutions for various boundary conditions (more on boundary conditions in Chapter 5). The solution to the equation is a function $h(x, y, z)$ that yields the hydraulic head distribution in three dimensions.

The saturated flow equation under transient conditions can be derived by expanding the right-hand side of the transient continuity equation [4.18] using the product rule of differentiation:

$$-\left[\frac{\partial}{\partial x}(\rho_w q_x) + \frac{\partial}{\partial y}(\rho_w q_y) + \frac{\partial}{\partial z}(\rho_w q_z)\right] = n\frac{\partial \rho_w}{\partial t} + \rho_w\frac{\partial n}{\partial t} \qquad [4.23]$$

Physically, the first term on the right-hand side describes the mass rate of water produced by an expansion of water due to the change in its density ρ. This first term is controlled by water compressibility $\beta[LT^2M^{-1}]$. The second term is the mass rate of water produced through the compaction of porous medium and it is evident through the partial derivative of porosity with respect to time. This second term is controlled by the compressibility of the porous medium, α.

The change in fluid mass within the control volume is therefore due to the change in porosity or change in fluid density with respect to time and they are both in *response* to *change* in *hydraulic head* with time. Therefore, one can write the following expression:

$$\frac{1}{\rho_w}\frac{\partial}{\partial t}(\rho_w n) = S_s\frac{\partial h}{\partial t} \qquad [4.24]$$

in which the unit volume of water produced by the two mechanisms for a unit decline in head is called the specific storage $(S_s)[L^{-1}]$ of a saturated aquifer and is given by

$$S_s = \rho_w(\alpha + n\beta) \qquad [4.25]$$

Inserting the x-, y-, and z-components of Darcy's Law and Eq. [4.24] into [4.23] yields

$$\frac{\partial}{\partial x}\left[K_x\frac{\partial h}{\partial x}\right] + \frac{\partial}{\partial y}\left[K_y\frac{\partial h}{\partial y}\right] + \frac{\partial}{\partial z}\left[K_z\frac{\partial h}{\partial z}\right] = S_s\frac{\partial h}{\partial t} \qquad [4.26]$$

which is the partial differential equation that describes *transient flow* in an *anisotropic, heterogeneous medium* when the *coordinate system is aligned* along the *principal coordinate system* (or axes of anisotropy). If the medium is isotropic and homogeneous, Eq. [4.26] takes on the simpler form

$$\frac{\partial^2 h}{\partial x^2} + \frac{\partial^2 h}{\partial y^2} + \frac{\partial^2 h}{\partial z^2} = \frac{S_s}{K}\frac{\partial h}{\partial t} \qquad [4.27]$$

where S_s/K is the diffusivity of the aquifer and Eq. [4.27] is known as the diffusion equation. The solution to the above equation is a function $h(x, y, z, t)$ that gives the hydraulic head distribution in three dimensions at any time.

4.3. CONTAMINANT TRANSPORT CONCEPTS

Contaminants in groundwater, whether miscible, soluble, or particulate, are transported by diffusion and advection. The processes involved in contaminant movement are called *transport* processes; the processes involved in reaction, decay, or volatilization of the contaminant are called *fate* processes. During advective movement, dispersion and diffusion act on the mass of contaminant causing it to spread and become dilute. Organic contaminants can be adsorbed onto matrix particles of carbon, clay, or iron–manganese oxides, thus reducing the flowing contaminant mass and in effect slowing down, or retarding, its advance relative to groundwater advection. Similarly, reaction of organic contaminants or destruction by biological processes will retard transport. Moreover, for a volatile contaminant being transported along the water table, vaporization into air-filled pores will reduce the concentration in groundwater.

Inorganic reactions that might take place in an aquifer include dissolution of salts, such as sodium chloride, and exchange of loosely bound ions between matrix particles and water. Sodium cations in clay, for example, may exchange with other cations dissolved in the groundwater. Radioactive solutes will decay according to their half-lives as they are transported. Other ionic inorganic substances may react chemically with groundwater or matrix minerals and precipitate new solid material while decreasing contaminant concentration in water. Each of these fate and transport factors is described in greater detail in the following sections.

4.3.1. Diffusion

Ogata (1970) defines molecular diffusion as "[t]he transport of mass in its ionic or molecular state due to differences in concentration of a given species in space...." Diffusion occurs in any system, gas, liquid, or solid, into which a substance is introduced in such a way that concentration differences are initially present. Figure 4.19

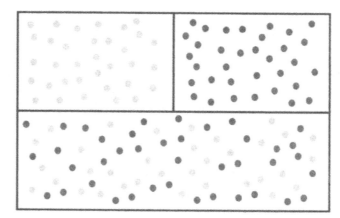

Figure 4.19. (a) Two substances separated by an impermeable barrier. (b) When the barrier between two substances is removed, the molecules will diffuse throughout the entire container.

shows two substances separated by an impermeable barrier. When the barrier between two substances is removed, the molecules will diffuse throughout the entire container. While the number of molecules in the container is the same as it was before the barrier was removed, the substances are now at lower concentrations. The rate of diffusion depends on the weight of the molecules—heavy molecules diffuse more slowly than light molecules.

The mchanisms causing diffusion are the random motions of dissolved molecules, ions, or suspended particles among the molecules of the containing phase and thermal kinetic energy of the solute. In an aquifer, diffusion occurs in the liquid phase, and also between the liquid and the solid phase of the matrix. In the present discussion, we are concerned only with diffusion in the liquid phase. A concentration gradient (dC/dx) detected in a fluid mass indicates the direction of contaminant movement as it diffuses from higher to lower concentration. Diffusion ceases when the concentration gradient becomes nonexistent or when the solutes encounter an impermeable boundary in which mixing is not allowed. It is important to recognize that mass transport can take place even without the fluid moving.

Diffusion into a homogeneous, still liquid is represented by the mathematical coefficient $D_d[L^2T^{-1}]$, which has a unique value for every ionic or molecular substance. Diffusion in a porous medium is not quite the same as diffusion in a homogeneous phase such as air or water. The grains or fracture walls of a porous medium form obstacles to diffusion along certain paths. To account for this effect, and the tortuosity of diffusion pathways around such obstacles, the diffusion coefficient for a substance in a homogeneous liquid phase is modified by a factor $D^*[L^0]$, which has a unique value for each porous medium. The bulk diffusion coefficient, $D_B[L^2T^{-1}]$, takes into account these nonideal effects and is expressed as

$$D_B = D^* D_d \qquad [4.28]$$

Table 4.2 lists values of D_d for selected anions and cations in water at 25°C and Table 4.3 lists diffusion coefficient values for organic compounds in air and water. Examination of Table 4.3 shows that the diffusion coefficient of various organic compounds is considerably larger in air than in water. Due to the tortuous nature of transport pathways in porous media, the effective diffusion coefficient can be a factor of 4–100 smaller, while in fractured rocks the effective diffusion coefficient can be several orders of magnitude lower in comparison to the diffusion coefficient in water (Schwartz and Zhang 2003).

Solute movement due to concentration gradient without fluid motion is due to molecular diffusion and can be described through Fick's First Law:

$$F = -D_B \frac{dC}{dx} \qquad [4.29]$$

where F is the mass flux of solute per unit area per unit time, C is solute concentration $[ML^{-3}]$, and dC/dx is concentration gradient $[ML^{-4}]$. Physically, Fick's First Law states that the mass of solute that diffuses through space is controlled

Table 4.2. Diffusion Coefficients in Water for Some Ions at 25°C

Cation	$D_d(10^{-6}cm^2/s)$	Anion	$D_d(10^{-6}cm^2/s)$
H^+	93.1	OH^-	52.7
Na^+	13.3	F^-	14.6
K^+	19.6	Cl^-	20.3
Rb^+	20.6	Br^-	20.1
Cs^+	20.7	HS^-	17.3
		HCO_3^-	11.8
Mg^{2+}	7.05		
Ca^{2+}	7.93	CO_3^{2-}	9.55
Sr^{2+}	7.94	SO_4^{2-}	10.7
Ba^{2+}	8.48	PO_4^{3-}	6.12
Ra^{2+}	8.89		
Mn^{2+}	6.88		
Fe^{2+}	7.19		
Cr^{3+}	5.94		
Fe^{3+}	6.07		

Source: Li and Gregory (1974).

Table 4.3. Diffusion Coefficients for Organic Compounds in Air (D_a) and Water (D_w)

Compound	$D_a(cm^2/s)$	$D_w(cm^2/s)$
Acetone	1.24×10^{-1}	1.14×10^{-5}
Aldrin	1.32×10^{-2}	4.86×10^{-6}
Anthracene	3.24×10^{-2}	7.74×10^{-6}
Benzene	8.80×10^{-2}	9.80×10^{-6}
Benzoic acid	5.36×10^{-2}	7.97×10^{-6}
Bromoform	1.49×10^{-2}	1.03×10^{-5}
Carbon disulfide	1.04×10^{-1}	1.00×10^{-5}
Carbon tetrachloride	7.80×10^{-2}	8.80×10^{-6}
Chlordane	1.18×10^{-2}	4.37×10^{-6}
Chlorobenzene	7.30×10^{-2}	8.70×10^{-6}
Chloroform	1.04×10^{-1}	1.00×10^{-5}
2-Chlorophenol	5.01×10^{-2}	9.46×10^{-6}
DDD	1.69×10^{-2}	4.76×10^{-6}
DDE	1.44×10^{-2}	5.87×10^{-6}
DDT	1.37×10^{-2}	4.95×10^{-6}
1,2-Dichlorobenzene	6.90×10^{-2}	7.90×10^{-6}
1,4-Dichlorobenzene	6.90×10^{-2}	7.90×10^{-6}
1,1-Dichloroethane	7.42×10^{-2}	1.05×10^{-5}
1,2-Dichloroethane	1.04×10^{-1}	9.90×10^{-6}
1,1-Dichloroethylene	9.00×10^{-2}	1.04×10^{-5}
cis-1,2-Dichloroethylene	7.36×10^{-2}	1.13×10^{-5}
trans-1,2-Dichloroethylene	7.07×10^{-2}	1.19×10^{-5}
Dieldrin	1.25×10^{-2}	4.74×10^{-6}
Endosulfan	1.15×10^{-2}	4.55×10^{-6}

Table 4.3. (*Continued*)

Compound	$D_a(\text{cm}^2/\text{s})$	$D_w(\text{cm}^2/\text{s})$
Endrin	1.25×10^{-2}	4.74×10^{-6}
Ethylbenzene	7.50×10^{-2}	7.80×10^{-6}
Heptachlor	1.12×10^{-2}	5.69×10^{-6}
γ-HCH (Lindane)	1.42×10^{-2}	7.34×10^{-6}
Hexachloroethane	2.50×10^{-3}	6.80×10^{-6}
Methyl bromide	7.28×10^{-2}	1.21×10^{-5}
Methylene chloride	1.01×10^{-1}	1.17×10^{-5}
Naphthalene	5.90×10^{-2}	7.50×10^{-6}
Nitrobenzene	7.60×10^{-2}	8.60×10^{-6}
Pentachlorophenol	5.60×10^{-2}	6.10×10^{-6}
Phenol	8.20×10^{-2}	9.10×10^{-6}
Pyrene	2.72×10^{-2}	7.24×10^{-6}
Styrene	7.10×10^{-2}	8.00×10^{-6}
1,1,2,2-Tetrachloroethane	7.10×10^{-2}	7.90×10^{-6}
Tetrachloroethylene	7.20×10^{-2}	8.20×10^{-6}
Toluene	8.70×10^{-2}	8.60×10^{-6}
1,2,4-Trichlorobenzene	3.00×10^{-2}	8.23×10^{-6}
1,1,1-Trichloroethane	7.80×10^{-2}	8.80×10^{-6}
1,1,2-Trichloroethane	7.80×10^{-2}	8.80×10^{-6}
Trichloroethylene	7.90×10^{-2}	9.10×10^{-6}
Vinyl chloride	1.06×10^{-1}	1.23×10^{-6}
m-Xylene	7.00×10^{-2}	7.80×10^{-6}
o-Xylene	8.70×10^{-2}	1.00×10^{-5}
p-Xylene	7.69×10^{-2}	8.44×10^{-6}

Source: Adapted from Schwartz and Zhang (2003).

by the concentration gradient and is proportional to the diffusion coefficient. The equation states that species C diffuses (moves relative to the mixture) in the direction of decreasing concentration, just as heat flows by conduction in the direction of decreasing temperature. The negative sign indicates that the transport is in the direction of decreasing concentration.

In systems where the concentration not only varies with space but also with time, Fick's Second Law, known also as the diffusion equation, is applicable:

$$\frac{\partial C}{\partial t} = D_B \frac{\partial^2 C}{\partial x^2} \qquad [4.30]$$

Here, Eq. [4.30] is written in one-dimensional form. Note the similarity between Eqs. [4.30] and [4.27] (the latter being written in three-dimensional form), which describes the diffusion of hydraulic head through porous materials. A general class of these diffusion problems has been solved by Crank (1956) and Carslaw and Jaeger (1959) in the heat conduction literature, many of which have been adapted for the analysis of groundwater flow and contaminant transport. Many solutions exist for various initial and boundary conditions (more on this in Chapter 5), but one of the most useful solutions is that for one-dimensional diffusion in an infinite medium (Crank 1956):

$$C(x,t) = C_0 \operatorname{erfc}\left(\frac{x}{2\sqrt{D_B t}}\right)$$ [4.31]

where C is the solute concentration at distance x from the source at time t since diffusion began, C_0 is the initial concentration, and $\operatorname{erfc}(y)$ is the complementary error function, where y is the argument of the function. This solution can be used to predict the concentration of solute at some distance and time from the source.

The complementary error function is a mathematical function that is related to the normal or Gaussian distribution. The relationship between the complementary error function and the error function is as follows:

$$\operatorname{erfc}(y) = 1 - \operatorname{erf}(y)$$ [4.32]

where

$$\operatorname{erf}(y) = \frac{2}{\sqrt{\pi}} \int_0^y e^{-t^2} dt$$ [4.33]

Because Eq. [4.33] cannot be solved analytically, one can use the following approximation:

$$\operatorname{erf}(y) = \sqrt{1 - \exp(-4y^2/\pi)}$$ [4.34]

The values of the complementary error function can also be obtained in Table 5.2 for a limited range of values. Alternatively, the value of the complementary error function can be obtained through a spreadsheet (e.g., MS EXCEL) and it can also be evaluated numerically using mathematical algorithms available from Press et al. (1992) and the International Mathematical and Statistical Library (IMSL).

Example 4.1. Prediction of Solute Concentration within a Clay Landfill Liner. As a consultant, you were asked to predict the concentration of chloride at some distance and time from the source in a clay liner such as that shown on Figure 4.20. Specifically, what will the concentration of chloride be within the

Figure 4.20. Schematic diagram of a landfill with a clay liner.

clay landfill liner 2 m from the landfill after 100 years of diffusion? The liner is 5 m thick and we would like to know what the ratio C/C_0 will be assuming a D_d of $1.0 \times 10^{-9}\, m^2/s$ and D^* of 0.5.

Step 1. Compute the effective diffusion coefficient from available data:

$$D_B = D^* D_d = (0.5)(1.0 \times 10^{-9}\, m^2/s) = 5 \times 10^{-10}\, m^2/s$$

Step 2. Convert 100 years to seconds:

$$100\, yr \times 365 \frac{days}{yr} \times 24 \frac{h}{day} \times 60 \frac{min}{hr} \times 60 \frac{s}{min} = 3.15 \times 10^9\, s$$

Step 3. Insert values into one-dimensional diffusion solution:

$$\frac{C}{C_0} = erfc\left(\frac{2\, m}{2\sqrt{(5 \times 10^{-10} m^2/s)(3.15 \times 10^9\, s)}} \right)$$
$$\frac{C}{C_0} = erfc(0.80) = 0.26$$

Therefore, one would expect to see a concentration of 26.0% of the original concentration in 100 yr at a distance of 2 m from the source. This example emphasizes the fact that diffusion is a very slow process and is predominantly a transport mechanism in low-permeability environments where groundwater flow is very slow or stagnant.

Transport by diffusion is neglected in most situations where groundwater flow and advective transport dominate, but there are certain situations in which diffusion cannot be neglected. Low groundwater velocities can be found in formations with low hydraulic conductivity or when the hydraulic gradient is exceedingly small. As we saw in Example 4.1, diffusion is an important transport mechanism for leachates in clay liners. Likewise, radionuclide transport in formations of very low permeability such as unfractured granite and shales also is another scenario that makes diffusion an important process. Diffusion can also be important in higher permeability materials in some cases. For example, one method to control the development of contaminant plumes from a known source zone (e.g., DNAPLs) is to place an artificial barrier around it. In this case, the hydraulic conductivity of the aquifer may be very high but because of the isolation, the hydraulic gradient within this isolated source zone may be exceedingly small. In this case, diffusion will be an important mechanism controlling transport of the organic contaminants within the source zone.

Example 4.2. Importance of Diffusion as a Transport Process. For which of the following scenarios 1 or 2 will transport due to diffusion be important? Discuss the solution in class.

> *Scenario 1*: Trichloroethene (TCE) and oil spill in an alluvial aquifer (see Figure 1.5).
> *Scenario 2*: TCE spill in fractured bedrock (see Figure 4.21).

4.3.2. Advection

Advection is mass transport due to flowing groundwater. For most contaminant transport problems, advection is considered to be the main process driving the movement of solutes from one location to another downgradient location. This implies that contaminants move at the same velocity as the flowing groundwater. Therefore, knowledge of groundwater flow patterns provides the necessary information for predicting solute transport under advection. Basic information required

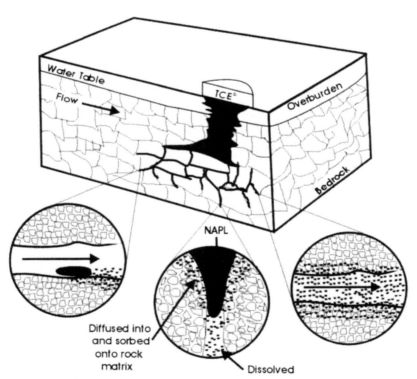

Figure 4.21. Schematic diagram showing a TCE spill in fractured bedrock. TCE is seen to penetrate deep into the formation due to its high density, low viscosity, and low interfacial tension. The dissolved phase travels with flowing groundwater. (Adapted from Pankow and Cherry 1996.)

includes the hydraulic gradient, hydraulic conductivity, and effective porosity. We saw earlier that groundwater flow can be described through Darcy's Law. From this, one can calculate the average linear velocity of groundwater through Eq. [4.16].

Example 4.3. Calculation of Advective Transport of Solute in an Unconfined Aquifer. A chemical spill occurs above a sloping, shallow unconfined aquifer consisting of medium sand with $K = 1$ m/day and a ϕ_e of 0.3. Several monitoring wells are drilled in order to determine the regional hydraulic gradient. The hydraulic head from a well drilled near the spill location yielded a value of 5 m. At a distance of 200 m down the slope another well yielded a hydraulic head of 1 m. Calculate how long it will take for the contaminants to travel 200 m.

Step 1. The regional hydraulic gradient can be estimated from two of the monitoring wells (it is advisable to use additional wells for the determination of hydraulic gradient, as we will see in Chapter 6).

$$\frac{dh}{dx} = \frac{5\,\text{m} - 1\,\text{m}}{200\,\text{m}} = 0.02$$

Step 2. Calculate the advective velocity using Eq. [4.16].

$$v_x = \frac{K}{\phi_e}\frac{dh}{dx} = \frac{1\,\text{m/day}}{0.3} \times 0.02 = 0.067\,\text{m/day}$$

Step 3. Calculate the time it will take for the contaminants to travel 200 m.

$$t = \frac{l}{v_x} = \frac{200\,\text{m}}{0.067\,\text{m/day}} = 3000\,\text{days}$$

Therefore, this calculation shows that the contaminants will travel over a distance of 200 m in 3000 days. It is important to remember that this calculation only gives the travel time of the contaminant between two points. When we need to predict the concentration of contaminants at some distance and time, we need to use analytical (Chapter 5) or numerical (Chapter 6) methods discussed later.

In order to compute and quantify the mass transported by flowing groundwater, one can define the one-dimensional mass flux per unit area due to advection $(F_x)[\text{ML}^{-2}\text{T}^{-1}]$ as

$$F_x = v_x \phi_e C \tag{4.35}$$

Physically, the mass flux due to advective transport represents the quantity of mass transported through a unit area per given time.

Example 4.4. Mass Flux Calculation of Agricultural Contaminants. Dissolved phosphate in a concentration of 5.0 mg/L is being advected with flowing

groundwater at a velocity of 0.1 m/day in a shallow, coarse sand and gravel unconfined aquifer with an effective porosity of 0.25. Groundwater from the aquifer discharges into a lake, where the nearby residents use it for drinking water and
recreational purposes. Calculate the total mass flux of phosphate into the lake if
the aquifer is 5.0 m thick and 500 m wide where it discharges.

Step 1. Convert concentration units given in mg/L to g/m^3.

$$C = 5.0 \frac{mg}{L} \times \frac{1000\,L}{1\,m^3} \times \frac{1\,g}{1000\,mg} = 5.0\,g/m^3$$

Step 2. One-dimensional mass flux (F_x) can be computed through substitution of
given data into Eq. [4.35]:

$$F_x = 0.1 \frac{m}{day} \times 0.25 \times 5.0 \frac{g}{m^3} = 0.125\,g/m^2\text{-day}$$

Step 3. Calculate the total mass flux of phosphate into the lake. In order to do
this, we must multiply the one-dimensional mass flux by the area with which
the lake comes into contact with the aquifer:

$$\text{Total mass flux} = F_x A = 0.125 \frac{g}{m^2\text{-day}} \times 5.0\,m \times 500\,m = 312.5\,g/day$$

For a lake with a volume of 500,000 m^3 (500 m by 200 m by 5 m), this flux is quite
significant. A simple calculation neglecting uptake and degradation processes of
phosphate shows that over 1 year, 114,062.5 g of phosphate will enter the lake.
The concentration of phosphate in this lake will be 0.23 mg/L (or ppm), which is well
above the 0.015 mg/L in which eutrophication of the lake will occur (Masters 1998).

4.3.3. Hydrodynamic Dispersion

The combined contaminant transport mechanisms of *mechanical dispersion* and
bulk diffusion comprise the process of *hydrodynamic dispersion*. This process
affects any substance introduced into solution or suspension in a fluid, whether
the fluid is air, water, or another liquid. As will be shown below for an advecting
contaminant mass, mechanical dispersion effects are orders of magnitude larger
than bulk diffusion effects. Only for a stationary or nearly stationary fluid is bulk
diffusion of concern.

The effects of hydrodynamic dispersion on a contaminant plume are dilution and
increased volume of contaminated water. Dilution occurs by spreading out the contamination in all three dimensions and, consequently, reducing contaminant concentration at points along the main flow path. The volume of contaminated water
increases as the contaminants spread to originally uncontaminated, peripheral areas.
Along the direction of flow (*longitudinally* with respect to plume shape), dispersion

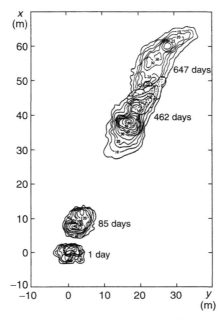

Figure 4.22. Vertically averaged concentration at 1 day, 85 days, 462 days, and 647 days after the slug injection of tracer into a shallow aquifer. (From Mackay et al. 1986.)

moves contamination ahead of the average groundwater flow velocity. Because of this, the velocity of contaminant movement will not necessarily be predicted accurately according to Darcy's Law alone. Perpendicular to the direction of flow (*transversely*) horizontally and vertically, dispersion moves contaminants out of the groundwater flow path. Figure 4.22 shows vertically averaged concentrations of a chloride plume during a natural gradient pulse injection tracer test conducted at the Borden site in Canada. Due to the process of hydrodynamic dispersion, we see that the plume spreads predominantly in the longitudinal direction as time passes.

4.3.4. Mechanical Dispersion

Ogata (1970) defines mechanical dispersion as "[t]he mixing mechanism that is present because of variations in the microscopic velocity within each flow channel and from one flow channel to another...." It is apparent from this definition that dispersion commences at the instant a contaminant mass begins moving in a porous medium. If unmodified by other effects, a contaminant can be expected eventually to disperse widely as a plume advances. Dispersion takes place across several scales of observation: from the intergranular scale at which a contaminant is subject to variations in flow velocity in the pore channels, through an intermediate scale subject to permeability heterogeneities of an aquifer, to a large scale crossing differing lithologies or structural features in a thick stratigraphic section.

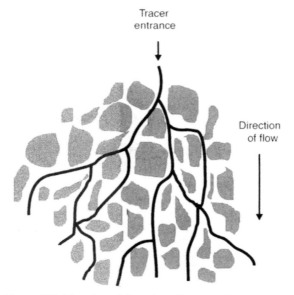

Tracer
entrance

Direction
of flow

Figure 4.23. Plan view of dispersion effects on a flowing volume.

Mechanical dispersion of a contaminant is caused by the variation of flow velocities in the pore network and random selection by contaminant-laden water of pore pathways around obstructions to flow (Figure 4.23). Obstructions range from individual sediment particles (i.e., silt or sand grains) through lenses of differing hydraulic conductivity (e.g., a clay lens within a sand layer), to large-scale stratigraphic changes (e.g., pinch-out of a sand between adjacent strata of differing lithology). Mechanical dispersion is represented as a mathematical coefficient, D_i', that is the product of groundwater advective velocity and an aquifer property known as *dispersivity*. The expression for longitudinal mechanical dispersion for unidirectional flows in the x-direction is

$$D_x' = \alpha_x v_x \qquad [4.36]$$

where D_x' is the coefficient of dispersion in the longitudinal (downgradient) sense $[L^2T^{-1}]$, α_x is the dispersivity in the x-direction $[L]$, and v_x is the groundwater velocity $[LT^{-1}]$ along the x-axis. Similarly, dispersion coefficients in the two directions transverse to the groundwater flow direction are given by

$$D_y' = \alpha_y v_x \qquad [4.37]$$

and

$$D_z' = \alpha_z v_x \qquad [4.38]$$

for horizontal transverse and vertical transverse mechanical dispersion, respectively. In general, longitudinal dispersivity (α_L) is much larger than transverse

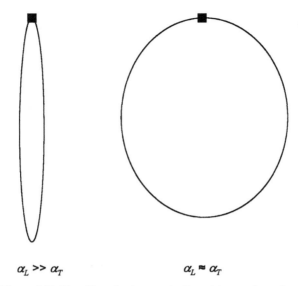

$$\alpha_L \gg \alpha_T \qquad\qquad\qquad \alpha_L \approx \alpha_T$$

Figure 4.24. The effect of anisotropy in dispersivity on plume shape.

dispersivity (α_T) by a factor of 10–20. Horizontal transverse dispersivity is greater than vertical transverse dispersivity by a factor of 2 or more. As α_L increases with respect to α_T, all other factors remaining constant, the shape of a plume becomes more elongate (Figure 4.24).

At the scale of fate-and-transport modeling, mechanical dispersion behaves in a similar way to bulk diffusion, so the two processes are combined mathematically. In one dimension, this is expressed as

$$D_x = D_x^{'} + D_B \qquad\qquad [4.39]$$

Determination of which process dominates in a flow system of interest is accomplished for granular aquifers by estimating the *Peclet number*, *P*, for the contaminant of interest and comparing this to the well-known relationship shown in Figure 4.25. The Peclet number is defined as follows

$$P = \frac{v_x d_0}{D_B} \qquad\qquad [4.40]$$

where v_x is average linear velocity, d_0 is the mean grain diameter, and D_B is the diffusion coefficient. Figure 4.25 shows that as the Peclet number increases, advective–dispersive processes become increasingly more important than molecular diffusion. We also see that the threshold for advective–dispersive transport to become more important is lower in the longitudinal direction than in the transverse direction of transport.

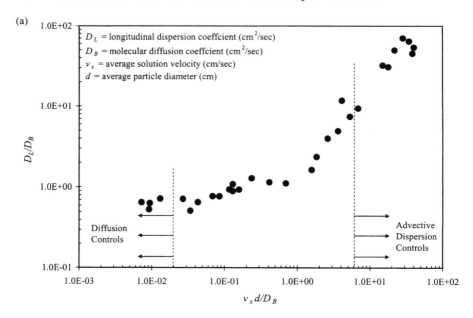

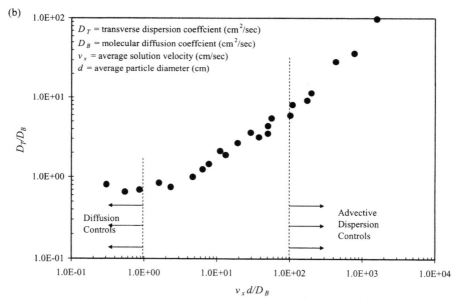

Figure 4.25. The ratio of (a) longitudinal dispersion coefficient and molecular diffusion coefficient and (b) transverse dispersion coefficient and molecular diffusion coefficient versus the Peclet number. Note the regions in which diffusion and advective dispersion processes are dominant. The zone betwen it is considered the transition zone. (Adapted from Perdins and Johnson 1963.)

4.3.5. Derivation of the Advection–Dispersion Equation for Solute Transport in Saturated Media

The derivation of the advection–dispersion equation (ADE) for solute transport in saturated porous media under steady-state flow conditions is based on the assumptions that the porous medium is (1) homogeneous, (2) isotropic, and (3) fully saturated with water and that, (4) groundwater flow is steady (i.e., no variation with time) and (5) Darcy's Law is valid. Analogous to the derivation of the saturated flow, consider a control volume (Figure 4.18) with sides dx, dy, and dz through which the solute flux is taking place. Two transport mechanisms control solute flux into and out of the control volume. The mass of solute per unit volume of the porous medium can be defined as $\phi_e C$; thus, advective transport in the x-direction can be defined as

$$v_x n_e C \, dy \, dz$$

Likewise, dispersive transport in the x-direction can be defined as

$$\phi_e D_x \frac{\partial C}{\partial x} dy \, dz$$

where v_x is the average linear velocity in the x-direction, C is the solute concentration (mass per unit volume of solution), $dy \, dz$ is the cross-sectional area of control volume or the representative elementary volume (REV) (Bear 1972) in the x-direction, and D_x is the hydrodynamic dispersion coefficient in the x-direction.

The total solute mass per unit cross-sectional area transported by means of advection and dispersion processes in the x-direction per unit time is F_x:

$$F_x = v_x \phi_e C - \phi_e D_x \frac{\partial C}{\partial x} \qquad [4.41]$$

in which the negative sign before the second term tells us that dispersive transport takes place from larger to smaller concentration. The dispersive transport usually is assumed to be Fickian. This means that the dispersive mass flux is directly proportional to the concentration gradient with the hydrodynamic dispersion coefficient being the constant of proportionality.

The total amount of solute entering the control volume is

$$F_x \, dz \, dy + F_y \, dz \, dx + F_z \, dx \, dy$$

while the total amount of solute leaving the control volume is

$$\left(F_x + \frac{\partial F_x}{\partial x} dx \right) dz \, dy + \left(F_y + \frac{\partial F_y}{\partial y} dy \right) dz \, dx + \left(F_z + \frac{\partial F_z}{\partial z} dz \right) dx \, dy$$

The partial derivatives of the mass flux indicate the change of solute mass in the indicated directions. The difference between the solute amount entering and leaving the control volume is

$$-\left(\frac{\partial F_x}{\partial x} + \frac{\partial F_y}{\partial y} + \frac{\partial F_z}{\partial z}\right) dx\ dy\ dz$$

The rate of mass change in the control volume is

$$\phi_e \frac{\partial C}{\partial t} dx\ dy\ dz$$

By the law of mass conservation, the rate of mass change in the control volume must be equal to the difference in the solute mass entering and leaving:

$$\frac{\partial F_x}{\partial x} + \frac{\partial F_y}{\partial y} + \frac{\partial F_z}{\partial z} = -\phi_e \frac{\partial C}{\partial t} \qquad [4.42]$$

Substituting Eq. [4.41] and corresponding expressions for the y- and z-directions, we obtain

$$\left[\frac{\partial}{\partial x}\left(D_x \frac{\partial C}{\partial x}\right) + \frac{\partial}{\partial y}\left(D_y \frac{\partial C}{\partial y}\right) + \frac{\partial}{\partial z}\left(D_z \frac{\partial C}{\partial z}\right)\right]$$
$$- \left[\frac{\partial}{\partial x}(v_x C) + \frac{\partial}{\partial y}(v_y C) + \frac{\partial}{\partial z}(v_z C)\right] = \frac{\partial C}{\partial t} \qquad [4.43]$$

which is the three-dimensional equation of mass transport for a conservative solute. It is important to note that the above equation does not factor in reactions of groundwater with the porous medium, chemical reactions among various constituents in the groundwater, and biological or radioactive decay.

If one assumes constant D_x, D_y, D_z in space, then Eq. [4.43] simplifies to

$$\left[D_x \frac{\partial^2 C}{\partial x^2} + D_y \frac{\partial^2 C}{\partial y^2} + D_z \frac{\partial^2 C}{\partial z^2}\right] - \left[\frac{\partial}{\partial x}(v_x C) + \frac{\partial}{\partial y}(v_y C) + \frac{\partial}{\partial z}(v_z C)\right] = \frac{\partial C}{\partial t} \quad [4.44]$$

It is important to recognize that the hydrodynamic dispersion coefficient (D) is a function of the flow direction such that even in an isotropic medium, $D_x \neq D_y \neq D_z$. The one-dimensional transport in homogeneous, isotropic porous media where v_x is uniform in space, can be described by

$$D_L \frac{\partial^2 C}{\partial x^2} - v_x \frac{\partial C}{\partial x} = \frac{\partial C}{\partial t} \qquad [4.45]$$

The two-dimensional transport in homogeneous, isotropic porous media with direction of groundwater flow that is parallel to the x-axis is

$$D_L \frac{\partial^2 C}{\partial x^2} + D_T \frac{\partial^2 C}{\partial y^2} - v_x \frac{\partial C}{\partial x} = \frac{\partial C}{\partial t} \qquad [4.46]$$

where D_L is the longitudinal hydrodynamic dispersion coefficient $[L^2T^{-1}]$ and D_T is the transverse hydrodynamic dispersion coefficient $[L^2T^{-1}]$. We will also later consider the ADE that contains chemical reactions such as sorption and degradation mechanisms.

4.3.6. Dilution

Dilution is another mechanism by which contaminant concentrations can be attenuated. For example, uncontaminated water naturally infiltrating from a surface water body or from precipitation displaces the plume downward and mixes with the contaminated groundwater. The resulting concentrations are therefore *diluted*. Dilution may also be induced as uncontaminated water flows into a plume from an injection well, or infiltrates from a surface stream or leaking conduit.

The effect of dilution on a contaminant plume may be estimated for a one-dimensional case if the recharge rate for the aquifer is known. Equation [4.47] is employed to estimate the concentration of a contaminant, C, at some distance, x, from the source, C_0, after recharge has affected the plume along the flow path (Wiedemeier et al. 1999, p. 152):

$$C = C_0 \exp\left(\frac{Wx}{(Ki)^2 z}\right) \qquad [4.47]$$

where W is the recharge rate $[LT^{-1}]$, K is the hydraulic conductivity, i is the hydraulic gradient, and z is the aquifer thickness. Dilution is not a factor explicitly evaluated in analytical models, and it is not a very significant factor in any but the most exceptional fate-and-transport situations, such as that with a plume migrating beneath a zone of high recharge rate. Equation [4.47] can be used to obtain a rough estimate in one dimension of concentration changes in a recharge zone.

4.3.7. Volatilization

At the water table and in the capillary zone, contaminated water comes in contact with air. Partitioning of the contaminant from water into air is termed *volatilization*. As a contaminant partitions from the aqueous phase in an attempt to find equilibrium with concentration in air, the aqueous concentration is reduced. Volatilization is thus another physical process that may diminish contaminant concentration as a plume flows. Volatilization is not a factor involving inorganic contaminants. Organic contaminants will volatilize to an extent dictated by their *Henry's Law constant*, H. Under equilibrium conditions, the concentrations of a contaminant of interest in water, C_{aq}, and air, C_{air}, can be calculated by the following formula:

$$C_{air} = HC_{aq} \qquad [4.48]$$

where C_{air} is the partial pressure of contaminant in the air $[ML^{-1}T^{-2}]$, and the dimension of C_{aq} is in mol-L^{-3}; thus, H has dimensions $[ML^2T^{-2}mol^{-1}]$. For practical purposes, only the portion of a contaminant plume that is adjacent to the water table is subject to volatilization. An insignificant amount of contaminant mass will volatilize from a plume unless the contaminant has extraordinarily high H value (e.g., acetone, vinyl chloride). As with dilution, the process of volatilization is typically ignored in fate-and-transport analysis but may appear in analytical models as a general attenuation factor. Henry's constant and other chemical parameters for commonly found compounds are listed in Appendix A.

4.3.8. Dissolution

Dissolution of nonaqueous phase liquids (NAPLs) from the source zone at a contaminated site is another important transport mechanism that will impact the concentration of the dissolved NAPL components in the groundwater as well as the duration of NAPL source zone persistence. Dissolution can be an important process also to biodegradation, which generally involves microbial uptake of contaminants from the water phase. Furthermore, biodegradation is most effective at low to moderate concentrations of contaminants. High concentrations found in NAPL source zones can be inhibitory and may not be effectively degraded by microorganisms.

The dissolution of a NAPL is primarily a function of the solubility of the compound, but groundwater velocity, the mass distribution of the NAPL in the subsurface, the interface (NAPL–water) contact area, the pore distribution of the medium, the aqueous phase diffusion coefficient(s) of the compound(s) in question, and the effects of other chemical constituents in the system are also important factors (Pankow and Cherry 1996). We note here that NAPLs can consist of a single compound or more commonly as a mixture of various compounds. The equilibrium aqueous phase concentration of each component in a NAPL mixture can be approximated using a solubility analog of Raoult's Law for vapor pressure (Pankow and Cherry 1996). It takes the following form:

$$C_m = X_m C_{\text{sat}} \qquad [4.49]$$

where C_m is the aqueous solubility of component m from the mixture, X_m is the mole fraction of component m in the NAPL mixture, and C_{sat} is the water solubility of the m component in its pure form. Therefore, one can readily calculate the aqueous solubility of component m from a mixture by looking up the water solubility of the compound in Appendix A and then calculating its mole fraction. Laboratory experiments by Banerjee (1984) and Broholm and Feenstra (1995) suggest that Eq. [4.49] is a reasonable approximation for mixtures of structurally similar hydrophobic organic liquids, such as mixtures of chlorinated solvents. However, for mixtures of dissimilar liquids and organic solids the relationship is more complex.

In the simplest form, dissolution or the rate of mass transfer $(N)[\text{MT}^{-1}]$ can be expressed mathematically as the product of the mass transfer coefficient $(K_c)[\text{LT}^{-1}]$, the concentration difference between the pure phase and the aqueous phase in the groundwater $(\Delta C)[\mathrm{ML^{-3}}]$, and the interfacial contact area $(A_s)[\mathrm{L^2}]$ between the two phases:

$$N = K_c \times \Delta C \times A_s \qquad [4.50]$$

The chief driving mechanism of mass transfer from the NAPL is the concentration gradient that develops across a boundary layer. This is usually defined as the difference between the effective solubilities of the component and the dissolved concentration in the bulk water flowing past the NAPL (Pankow and Cherry 1996). Several models have been developed to describe mass transfer coefficients for the dissolution of immiscible phase. The two most common models are the stagnant film model and the film penetration model (Bird et al. 1960). Below we briefly discuss the stagnant film model most commonly used in contaminant hydrogeology.

The three main assumptions that go into the stagnant film model are: (1) the entire resistance to mass transfer resides in a stagnant boundary layer of thickness δ adjacent to the interface, as shown in Figure 4.26; (2) the rate of equilibrium across the fluid–fluid interface is rapid; and (3) the concentration gradient within the stagnant film is linear. In this figure, C_{sat} is the concentration of the NAPL at the interface while C_w is the aqueous concentration of the NAPL in the bulk water. In this model, the overall rate of mass transfer is determined by molecular diffusion in the water phase coupled with advective flow (Levich 1962; Turitto 1975).

Recall that for one-dimensional diffusion, contaminant flux is described by Fick's First Law (Eq. [4.29]). The stagnant film model assumes that the concentration gradient is linear over the boundary layer thickness δ as shown in Figure 4.26 so that the flux (F) is given as

$$F = D\left(\frac{C_w - C_{\mathrm{sat}}}{\delta}\right) \qquad [4.51]$$

Therefore, Eq. [4.29] reduces to

$$F = K_c(C_w - C_{\mathrm{sat}}) \qquad [4.52]$$

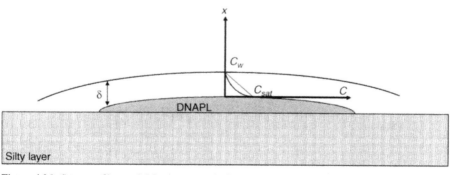

Figure 4.26. Stagnant film model for mass transfer by dissolution. (Adapted from Pankow and Cherry 1996.)

where $K_c = D/\delta$ is the mass transfer coefficient $[\text{LT}^{-1}]$. Equation [4.51] shows that the mass transfer coefficient is directly proportional to the diffusion coefficient. If the diffusive process is properly accounted for, the concentration gradient will look more like the curve shown in Figure 4.26.

In porous and fractured media, the mass transfer rate coefficient is commonly defined as a mass transfer rate per volume of porous medium, $F^*[\text{ML}^{-3}\text{T}^{-1}]$, by multiplying the mass transfer rate by the ratio of NAPL surface contact area A_s to the unit volume of the porous medium V. Hence,

$$F^* = F\frac{A_s}{V} = K_c(C_w - C_{\text{sat}})\frac{A_s}{V} = \kappa(C_w - C_{\text{sat}})$$ [4.53]

where κ is the "lumped" mass transfer coefficient $[\text{T}^{-1}]$ and is equal to $\kappa = K_c A_s/V$. Lumped mass transfer coefficients are commonly used in dissolution studies in porous media (e.g., Sleeps and Sykes 1989; Miller et al. 1990), but it is important to remember that such lumped parameters are considered to be fitting parameters and may not be able to describe more complex dissolution reactions.

Other models developed to describe dissolution include the film penetration model (Higbie 1935) and the random surface renewal model (Dankwerts 1951). The film penetration model attempts to consider the transient nature of the diffusive process between the interface and the bulk fluid phase with a given fluid element spending a characteristic penetration time that is constant. The random surface renewal model, on the other hand, considers a random distribution of time in which a fluid element contacts the interface participating in the mixing process. Details of these models are beyond the scope of this book but can be found in the respective references.

4.3.9. Adsorption and Retardation

Adsorption occurs as contaminant molecules come into contact with, and adhere to, certain types of particles in an aquifer. Adsorption is another process that removes the contaminant from the flowing mass in groundwater. Because of adsorption, concentrations of contaminant arriving at a point of interest will be lower than expected at a given time, and the contaminant is therefore said to be *retarded* to some degree in its transport. Retardation refers to a decrease in contaminant migration velocity (relative to the groundwater velocity) caused by sorption. This is illustrated using a trapping analogy in Figure 4.27.

Sorption-related retardation is quantified by the retardation factor (R_f), which is defined as the ratio of the groundwater velocity (v) to the contaminant migration velocity (v_c):

$$R_f = \frac{v}{v_c}$$ [4.54]

The tendency of a pollutant to be retarded by sorption depends on various factors and increases with the organic matter content of the soil and the hydrophobicity of

Migration With No Sorption

Migration With Sorption "Trap"

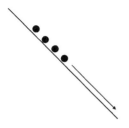

Assume that 100 balls reach the bottom during a given time, Δt

Balls do not reach the bottom until sorption trap is filled. If only 50 balls reach the bottom during the same time, Δt, then $R_f = 2$.

Figure 4.27. Schematic analogy of the effect of sorption on contaminant migration.

the substance. The retardation factor can be estimated a priori based on some properties of the contaminant and the aquifer material.

For inorganic, ionic molecules, adsorption occurs as cation exchange between dissolved ions in groundwater and loosely bound ions of certain silicate minerals such as clays and zeolites. Iron and manganese oxide particles are also capable of adsorbing dissolved or complexed metal ions. Organic molecules are chiefly adsorbed by organic carbon particles in an aquifer. Adsorption represents an equilibrium of contaminant concentration between the amount dissolved in groundwater and the amount adsorbed on aquifer particles. There is a maximum amount of contaminant that can be adsorbed on the available surfaces of particles. Moreover, the adsorption process is dependent on temperature and pH in the groundwater–aquifer system.

Adsorption parameters for contaminants of interest are determined by laboratory tests. For inorganic, ionic substances, the test determines the cation exchange capacity of a sample of aquifer material. For organic substances, the test determines the relationship between aqueous concentration and adsorbed concentration at a particular temperature. The concentration–adsorption relationship is known as the adsorption *isotherm* of the substance of interest.

There are several different types of adsorption isotherms (Figure 4.28). The isotherm most commonly employed in fate-and-transport modeling assumes the relationship between aqueous concentration and adsorbed concentration fits a linear mathematical model:

$$C_{\text{sorb}} = K_d C_{\text{aq}} \qquad [4.55]$$

where C_{sorb} is the concentration of contaminant sorbed on aquifer matrix $[\text{MM}^{-1}]$, C_{aq} is the contaminant concentration in water $[\text{ML}^{-3}]$. K_d is known as the *distribution coefficient* $[\text{L}^3\text{M}^{-1}]$, which is determined for the contaminant and aquifer

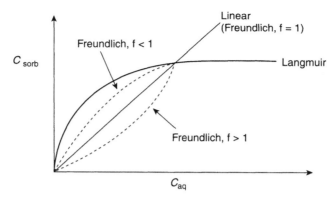

Figure 4.28. Diagram of three types of isotherms.

material of interest. K_d is a constant and is the ratio of contaminant mass sorbed on the aquifer matrix to the mass of contaminant dissolved in water. Discussion of K_d determination is found in Chapter 7.

The one-dimensional advection dispersion equation for a homogeneous, isotropic porous medium where v_x is uniform in space can be modified to include adsorption. The equation takes the following form:

$$D_x \frac{\partial^2 C}{\partial x^2} - v_x \frac{\partial C}{\partial x} - \frac{\rho_b}{\phi} \frac{\partial C_{\text{sorb}}}{\partial t} = \frac{\partial C}{\partial t} \qquad [4.56]$$

where ρ_b is the bulk density of the aquifer, and ϕ is the total porosity. Physically, the first term on the left-hand side represents solute dispersion, the second term is the advective term, and the third term describes the sorption of contaminants from the aqueous to the solid phase. The right-hand side describes the change in concentration with time. Substitution of Eq. [4.55] into [4.56] yields

$$D_x \frac{\partial^2 C}{\partial x^2} - v_x \frac{\partial C}{\partial x} - \frac{\rho_b}{\phi} \frac{\partial (K_d C)}{\partial t} = \frac{\partial C}{\partial t} \qquad [4.57]$$

Rearranging Eq. [4.57] yields

$$D_x \frac{\partial^2 C}{\partial x^2} - v_x \frac{\partial C}{\partial x} = \left(1 + \frac{\rho_b}{\phi} K_d\right) \frac{\partial C}{\partial t} \qquad [4.58]$$

where the term in parentheses on the right-hand side is the expression of the retardation factor for a linear isotherm:

$$R = 1 + \left(\frac{\rho_b}{\phi}\right) K_d \qquad [4.59]$$

There are two key limitations to the linear adsorption isotherm. One is that the linear adsorption model does not limit the amount of contaminant that it can adsorb onto the matrix despite the fact that there are a limited number of sorption sites. The other limitation is that when data are available for only a limited range of concentrations, a linear model may be fit to the sorption isotherm. A larger range of data may actually be fit to a model other than the linear sorption model, such as the Freundlich or Langmuir isotherms, which are nonlinear models. One key advantage is that the expression is simple mathematically, and for this reason it has been used in a large number of contaminant transport studies.

Two other isotherms used are the Freundlich,

$$C_{\text{sorb}} = K_d C_{\text{aq}}^f \qquad [4.60]$$

and the Langmuir,

$$C_{\text{sorb}} = \frac{M_s K_d C_{\text{aq}}}{1 + K_d C_{\text{aq}}} \qquad [4.61]$$

For the Freundlich isotherm, the exponent f has a positive value. For all values where $f \neq 1$, the relationship between sorbed and aqueous concentrations is nonlinear. When $f > 1$, the affinity of the contaminant for the adsorbing material is high, and when $0 < f < 1$, the affinity of the contaminant for the adsorbing material is low, relative to the affinity for the contaminant to remain in solution. When $f = 1$, the Freundlich isotherm is the same as the linear isotherm. For the Langmuir isotherm, the term M_s is the maximum mass of contaminant that the adsorptive material in an aquifer can hold $[\text{MM}^{-1}]$. Langmuir isotherms in general exhibit the effect of high contaminant affinity for the adsorbing material at low aqueous concentrations, and lower or invariant affinity at higher aqueous concentrations. The adsorption behavior of a chemical of interest in aquifer material is found by fitting C_{aq} versus C_{sorb} data determined in the laboratory with various isotherm types and determining which mathematical expression best fits the data.

Many different approaches have been published for estimation of K_d. Several of these are discussed in Chapter 7. The distribution coefficient is of interest in fate-and-transport modeling because it controls the *retardation* of a dissolved contaminant flowing in groundwater. Equation [4.62] shows the effect of retardation, R, on contaminant velocity.

$$v_{x,\text{benzene}} = \frac{K}{(R_{\text{benzene}} \phi_e)} \frac{dh}{dx} \qquad [4.62]$$

Equation [4.62] assumes constant conditions along the flow path and over any time interval. No processes other than retardation (i.e., dispersion, reaction) are considered in this simple, one-dimensional model. Transport rates calculated with this equation must, therefore, be used with caution.

Because the density and porosity terms in Eq. [4.59] are always positive, the equation shows that the retardation factor for a chemical of interest will be greater than or equal to 1, and that for a nonretarded chemical $R = 1$, implying $K_d = 0$. In terms of the physical process, this implies the adsorbed mass of contaminant is zero. Thus, when considering adsorption alone, the retarded contaminant velocity in groundwater will be slower than or, at most, equal to the advective velocity of groundwater.

In some instances, the retardation factor also can be less than 1, when the contaminants move faster relative to groundwater due to colloid transport or due to ion exclusion effects.

The contaminant migration velocity (v_c) can be estimated by incorporating the concept of retardation into Darcy's Law (Eq. [4.16]):

$$v_c = \frac{K}{R_f \phi_e} \frac{dh}{dx} = \frac{K}{[1 + \frac{\rho_b}{\phi} Kd] \phi_e} \frac{dh}{dx} \qquad [4.63]$$

In most analytical models of organic chemical fate and transport, a linear isotherm is assumed, and the retardation factor is estimated using Eq. [4.59]. Equation [4.63] can then be employed to estimate contaminant travel time as shown in the following example.

Example 4.5. Estimation of Contaminant Travel Time. A site has groundwater contaminated by a chemical with retardation factor of 3. This means that the mean rate of contaminant migration in the groundwater is 1/3 the rate of the groundwater itself. Assume further that K_h is 2 m/day, ϕ_e is 0.15, and dh/dx is 0.004. How long will it take for this contaminant to travel 100 m?

The groundwater velocity according to Eq. [4.16] is thus $v = 2(0.004)/0.15 = 0.053$ m/day, and the mean contaminant advection velocity, v_c, according to Eq. [4.63] is $0.053/3 = 0.018$ m/day. This velocity can be used to estimate travel times between two points, or distance traveled after a certain time.

$$\text{Travel time} = \text{Distance}/v_c \qquad [4.64]$$

In this example, time $= (100\,\text{m})/(0.018\,\text{m/day}) = 5555.6$ days.

Equation [4.64] assumes constant conditions along the flow path and over any time interval. No processes other than advection and retardation (e.g., dispersion, reaction) are considered in this simple, one-dimensional model. Therefore, caution must be taken when using this equation to estimate the time required for a contaminant to migrate from one point to another. For example, if hydrodynamic dispersion is considered, the travel time would correspond to the center of mass of the plume but, due to longitudinal dispersion, the first detection of the contaminant at the point of interest would likely occur earlier than predicted by Eq. [4.64].

The retardation factor for a Freundlich isotherm is as follows:

$$R = 1 + fC^{f-1}K_d\left(\frac{\rho_b}{\phi}\right) \qquad [4.65]$$

where f is the exponent from Eq. [4.60], and C is the aqueous concentration of contaminant. For the Langmuir isotherm, the retardation factor has the form

$$R = 1 + K_d\left(\frac{\rho_b}{\phi}\right)\left[\frac{M_sK_d}{(1 + K_dC)^2}\right] \qquad [4.66]$$

where the term M_s is as given in Eq. [4.61].

Retardation factors are not necessarily constant and can increase with time until some steady state is reached. This trend has been observed in field studies in a sandy aquifer, where R_f increased over 600 days from about 2 to 6 for perchloroethylene (PCE) and from about 4 to 9 for dichlorobenzene (Roberts et al. 1986). Presumably, this is due to two concentration-dependent phenomena. As contaminant concentrations decrease over time due to dilution and degradation, (1) there is less competition among contaminants and other potential sorbates for sorption sites on the aquifer material; and (2) lower contaminant concentrations favor nonlinear (stronger) adsorption mechanisms onto glassy sites (e.g., charcoal fragments) compared to linear absorption mechanisms into spongelike organic matter (e.g., humic substances).

4.3.10. Degradation Processes

The advection–dispersion equation can contain a reaction (sink) term (r_c) that represents different potential sources or sinks. The one-dimensional advection–dispersion equation for a homogeneous, isotropic porous medium, where v_x is uniform in space, can be modified to include the reaction term as

$$D_x\frac{\partial^2 C}{\partial x^2} - v_x\frac{\partial C}{\partial x} - \frac{\rho_b}{\phi}\frac{\partial C_{\text{sorb}}}{\partial t} \pm r_c = \frac{\partial C}{\partial t} \qquad [4.67]$$

The most common degradation processes are discussed in greater detail in Chapters 2 and 3. These are:

Microbial Degradation. Biodegradation is one of the most important mechanisms to remove organic pollutants from contaminated aquifers. Examples include oxidative transformations, reductive transformations, and hydrolytic and other biotransformations not involving redox processes.

Natural Exponential Decay. Several contaminants experience endogenous decay, such as radioactive elements and bacteria. Such decay often follows first-order kinetics (i.e., exponential decay), where the rate of decay (r_c) of a pollutant with concentration C is given by

$$r_c = -\kappa C \qquad [4.68]$$

Thus,

$$C(t) = C_0 e^{-\kappa t} \qquad [4.69]$$

where $C(t)$ is concentration at time t, C_0 is concentration at time $t = 0$, and κ is the first-order decay coefficient [T^{-1}].

The use of a first-order decay term facilitates the incorporation of degradation processes in mass balances, which are often expressed in differential equation form. First-order sink terms are additive and facilitate the integration of the differential equation to obtain analytical solutions (Chapter 5). However, it should be kept in mind that (with the exception of radioactive decay, where κ is an intrinsic property of the decaying element) exponential decay is often an empirical approximation to complex phenomena that are often affected by multiple site-specific variables (e.g., mass transfer limitations, temperature, microbial community structure). Therefore, κ values should not be extrapolated from the lab to the field or from one site to another.

Hydrolytic Reactions. Hydrolysis was defined in Chapter 2 as the degradation of a molecule that reacts with water. This process can also be described by first-order kinetics, and it can be important for the degradation of some natural polymers (e.g., proteins, polysaccharides, and glycerides) and priority organic pollutants.

Photochemical Reactions. Solar radiation in the ultraviolet range can cause the breakdown of many organic compounds, including plastics. However, due to the absence of light in the subsurface, photolysis is not an important process in aquifers. Photolysis can be an important degradation mechanism of some synthetic organic chemicals in surface waters and in the atmosphere.

Oxidation–Reduction Reactions. As discussed in Chapter 2, these reactions involve the transfer of electrons from an electron donor (i.e., the reductant) to an electron acceptor (i.e., the oxidant). Whereas oxidation–reduction reactions of organic pollutants are often mediated by microorganisms, some abiotic reactions can be important in environmental systems. One example is the oxidation of ferrous iron (Fe^{2+}) by molecular oxygen:

$$4Fe^{2+} + O_2 + 4H \rightleftharpoons 4Fe^{3+} + 2H_2O$$

The rate of this reaction depends on the pH, the oxidation–reduction potential of the reaction medium, and the ferrous iron concentration. Thus, if reaction conditions are relatively constant, the rate can also be approximated by a first-order expression. Since the oxidized iron form, ferric iron (Fe^{3+}), is prone to precipitate, this reaction is often exploited for removing dissolved iron from groundwater prior to domestic use.

4.3.11. Effect of Different Processes on the Attenuation of Contaminant Migration

The effect of different processes on the migration of a groundwater contaminant plume will be illustrated with the following example. A contaminant is migrating toward a drinking water well (Figure 4.29a) and the concentration at the well is plotted as a function of time under different scenarios (Figure 4.29b). The y-axis is the concentration observed at the receptor's location relative to a constant, continuous-source concentration (C_0). The x-axis is the elapsed time relative to the mean travel time for groundwater (t/τ). If migration occurs due to advection

(a) Hypothetical Realease Scenario

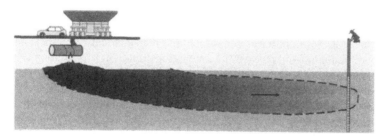

(b) Observed Concentration at the Receptor Location

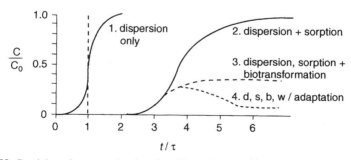

Figure 4.29. Breakthrough curves showing the effects of dispersion (d), sorption (s), constant-rate biodegradation (b), and increasing-rate biodegradation due to adaption. C/C_0 is the concentration observed at the receptor's location relative to a constant source concentration, and t/τ is time relative to the mean travel time for groundwater. The retardation factor for the sorbing substance in this example is approximately 3.7. (Adapted from Schnoor 1996.)

alone, an ideal plug-flow behavior will be observed, and the concentration at the well will increase suddenly up to C_0 after a time corresponding to the groundwater travel time. If hydrodynamic dispersion occurs, some molecules will travel faster, resulting in earlier detection, and it will take longer for the concentration to reach C_0. However, it would take the same time as for plug flow $(t/\tau = 1)$ for the average concentration of the contaminant "wave" $(C/C_0 = 0.5)$ to reach the well. Sorption will retard the migration, resulting in a longer time for detection and increased spreading. However, if dilution is ignored (as is the case for one-dimensional models), C_0 will eventually be reached. Only biodegradation (or another degradation reaction) results in a decrease in the asymptotic concentration observed at the well. Note that bacterial growth and acclimation increase the rate of degradation, which results in a further decrease in the concentration at the well. If biodegradation is sufficiently fast, the pollutant may never reach the well.

In Chapter 5, the mathematical description of the problem is completed by discussing initial and boundary conditions relevant to typical transport scenarios. The mathematical problem consisting of the governing equation and the initial and boundary conditions forms a boundary value problem. We then discuss the advantages and disadvantages of various solution techniques used to solve this boundary value problem. After that, analytical models of solute transport and reaction processes are applied to several examples to show how they can predict contaminant concentrations in both space and time. Such predictions are needed for risk assessment of groundwater contamination, management of plumes, and design of active remediation and passive (i.e., natural attenuation) remediation schemes. A sensitivity analysis using an analytical model is performed to study the effect of varying key parameters on contaminant transport.

Analytical models are useful in producing simple and quick answers to contaminant transport problems but are limited to a few idealized cases. No book on quantitative treatment of subsurface contaminant transport can be complete without a discussion of numerical techniques. Numerical techniques are more flexible, allowing for solutions to more complex situations such as irregular boundaries, spatial variability in physical and chemical properties, and time-varying boundary conditions. Some background on commonly used numerical techniques are provided and an example problem on the transport and biodegradation of a BTEX plume is studied using RT3D (reactive transport in 3 dimensions), a numerical model for simulating three-dimensional multispecies, reactive transport in groundwater.

REFERENCES

Alvarez, P.J.J., P. J. Anid, and T. M. Vogel (1991). Kinetics of aerobic biodegradation of benzene and toluene in sandy aquifer material. *Biodegradation* **2**:43–51.

Alvarez, P.J.J., Y.-K. Zhang, and N. Lovanh (1998). Evaluation of the Iowa Tier-2 Model Based on the Benzene Plume Dimensions. Iowa Comprehensive Petroleum Underground Storage Tank Fund Board, Des Moines, IA, December.

ASTM (1998). *1998 Annual Book of ASTM Standards: Standard Guide for Remediation of Ground Water by Natural Attenuation at Petroleum Release Sites* (Designation: E 1943-98). American Society for Testing and Materials, West Conshohocken, PA, pp. 875–917.

Bakr, A.A. (1976). Stochastic analysis of the effects of spatial variations of hydraulic conductivity on groundwater flow. Ph.D. dissertation, New Mexico Institute of Mining and Technology, Socorro, NM.

Banerjee, S. (1984). Solubility of organic mixtures in water. *Environ. Sci. Technol.* **18**:587–591.

Barker, J.F., G.C. Patrick, and D. Major (1987). Natural attenuation of aromatic hydrocarbons in a shallow sand aquifer. *Groundwater Monitoring Rev.* **7**:64–71.

Bear, J. (1972). *Dynamics of Fluids in Porous Media.* Dover Publications, New York.

Bedient, P.B., H.S. Rifai, and C.J.Newell (1994). *Ground Water Contamination. Transport and Remediation.* PTR Prentice Hall, Upper Saddle River, NJ.

Bird, R.B., W.E. Stewart, and E.N. Lightfoot (1960). *Transport Phenomena.* John Wiley & Sons, Hoboken, NJ.

Boggs, J.M., S.C. Young, L.M. Beard, L.W. Gelhar, K.R. Rehfeldt, and E.E. Adams (1992). Field study of dispersion in a heterogeneous aquifer, 1. Overview and site description. *Water Resour. Res.* **28**(12):3281–3291.

Borden, R.C. and P.B. Bedient (1986). Transport of dissolved hydrocarbons influenced by oxygen-limited biodegradation: 1. Theoretical development. *Water Resour. Res.* **22**:1973–1982.

Brace, W.F. (1984). Permeability of crystalline rocks: new in situ measurements. *J. Geophys. Res.* **89**(B6):4327–4330.

Bradbury, K.R. and M.A. Muldoon (1990). Hydraulic conductivity determinations in unlithified glacial and fluvial materials. In *Ground Water and Vadose Zone Monitoring*, ASTM STP 1053, D.M. Nielson and A.I. Johnson (Eds.), pp. 138–151.

Broholm, K. and S. Feenstra (1995). Laboratory measurements of the aqueous solubility of mixtures of chlorinated solvents. *Environ. Toxicol. Chem.* **14**:9–15.

Butler, J.J. Jr. and J.M. Healey (1998a). Relationship between pumping-test and slug-test parameters: scale effect or artifact? *Ground Water* **36**(2):305–313.

Butler, J.J. Jr. and J.M. Healey (1998b). Discussion of papers: authors' reply. *Ground Water* **36**(6):867–868.

Carslaw, H.S. and J.C. Jaeger (1959). *Conduction of Heat in Solids*, 2nd ed. Oxford Science Publications, New York.

Chen, Y.M., L.M. Abriola, P.J.J. Alvarez, P.J. Anid, and T.M. Vogel (1992). Biodegradation and transport of benzene and toluene in sandy aquifer material:model–experiment comparisons. *Water Resour. Res.* **28**:1833–1847.

Chiang, C.Y., J.P. Salanitro, E.Y. Chai, J.D. Colthart, and C.L. Klein (1989). Aerobic biodegradation of benzene, toluene, and xylene in a sandy aquifer—data analysis and computer modeling. *Ground Water* **27**:823–834.

Clauser, C. (1992). Permeability of crystalline rocks. *Eos Trans. Am. Geophys. Union* **73**(21): 233.

Cline, O.V., J.J. Delfino, and P.S. Rao (1991). Partitioning of aromatic constituents into water from gasoline and other complex mixtures. *Environ. Sci. Technol.* **25**:914–920.

Crank, J. (1956). *The Mathematics of Diffusion*, 2nd ed. Oxford Science Publications, New York.

Dagan, G. and S.P. Neuman (Eds.) (1997). *Subsurface Flow and Transport: The Stochastic Approach*, Proceedings of the Second Kovacs Colloquium, International Hydrological Programme of UNESCO, Paris, France, 26–28 January.

Dankwerts, P.W. (1951). Significance of liquid-film coefficients in gas absorption. *Ind. Eng. Chem. Process Design Dev.* **43**:1460–1467.

De Joselin and G. de Jong (1958). Longitudinal and transverse diffusion in granular deposits. *Trans. Am. Geophys. Union* **39**(1):67.

Domenico, P.A. (1987). An analytical model for multidimensional transport of a decaying contaminant species. *J. Hydrol.* **91**:49–58.

Domenico, P.A. and F.W. Schwartz (1998). *Physical and Chemical Hydrogeology*, 2nd ed. John Wiley & Sons, Hoboken, NJ.

EPA (1980). Acid Rain, EPA-600/9-79-036. NO_x, SO_x and VOC:Methodology and Results. Argonne National Lab, Argonne, IL.

Feehley, C.E., C. Zheng, and F.J. Molz (2000). A dual-domain mass transfer approach for modeling solute transport in heterogeneous aquifers: application to the macrodispersion experiment (MADE) site. *Water Resour. Res.* **36**(9):2501–2515.

Freeze, R.A. and J.A. Cherry (1979). *Groundwater.* Prentice Hall, Upper Saddle River, NJ.

Garabedian, S.P., D.R. LeBlanc, L.W. Gelhar, and M.A. Celia (1991). Large-scale natural gradient test in sand and gravel, Cape Cod, Massachusetts. 2. Analysis of spatial moments for a nonreactive tracer. *Water Resour. Res.* **27**(5):911–924.

Gelhar, L.W. (1993). *Stochastic Subsurface Hydrology.* Prentice-Hall, Englewood Cliffs, NJ.

Gelhar, L.W., C. Welty, and K.R. Rehfeldt (1992). A critical review of data on field-scale dispersion in aquifers. *Water Resour. Res.* **28**(7):1955–1974.

Guimerà, J., L. Vives, and J. Carrera (1995). A discussion of scale effects on hydraulic conductivity at a granitic site (El Berrocal, Spain). *Geophys. Res. Lett.* **22**:1449–1452.

Guzman, A.G., A.M. Geddis, M.J. Henrich, C.F. Lohrstorfer, and S.P. Neuman (1996). Summary of Air Permeability Data from Single-Hole Injection Tests in Unsaturated Fractured Tuffs at the Apache Leap Research Site:Results of Steady-State Test Interpretation, NUREG/CR-6360, U.S. Nuclear Regulatory Commission, Washington, DC.

Hanor, J.S. (1993). Effective hydraulic conductivity of fractured clay beds at a hazardous waste landfill, Louisiana Gulf Coast. *Water Resour. Res.* **29**:3691–3698.

Higbie, R. (1935). The rate of absorption of a pure gas into a still liquid during short periods of exposure. *Trans. Am. Inst. Chem. Eng.* **31**:365–385.

Howard, P.H. (Ed.) (1989). *Handbook of Environmental Fate and Exposure Data for Organic Chemicals.* Lewis Publishers, Chelsea, MI.

Hsieh, P.A. (1998). Scale effects in fluid flow through fractured geologic media. In *Scale Dependence and Scale Invariance in Hydrology*, G. Sposito (Ed.). Cambridge University Press, Cambridge, UK, pp. 335–353.

Hsieh, P.A., S.P. Neuman, G.K. Stiles, and E.S. Simpson (1985). Field determination of the three-dimensional hydraulic conductivity tensor of anisotropic media. 2. Methodology and application to fractured rocks. *Water Resour. Res.* **21**:1667–1676.

Hubbard, S.S., J. Chen, J. Peterson, E.L. Majer, K.H. Williams, D.J. Swift, B. Mailloux, and Y. Rubin (2001). Hydrogeological characterization of the South Oyster bacterial transport site using geophysical data.

Hyun, Y., S.P. Neuman, V.V. Vesselinov, W.A. Illman, D.M. Tartakovsky, and V. Di Federico (2002). Theoretical interpretation of a pronounced permeability scale-effect in unsaturated fractured tuff. *Water Resour. Res.* **38**(6):1092.

Illman, W.A. (2004). Analysis of permeability scaling within single boreholes. *Geophys. Res. Lett.* **31**(5):1029.

Illman, W.A. and S.P. Neuman (2001). Type-curve interpretation of a cross-hole pneumatic test in unsaturated fractured tuff. *Water Resour. Res.*, **37**(3):583–604.

Illman, W.A. and S.P. Neuman (2003). Steady-state analyses of cross-hole pneumatic injection tests in unsaturated fractured tuff. *J. Hydrol.* **281**:36–54.

Illman, W.A., D.L. Thompson, V.V. Vesselinov, G. Chen, and S.P. Neuman (1998). Single- and Cross-Hole Pneumatic Tests in Unsaturated Fractured Tuffs at the Apache Leap Research Site:Phenomenology, Spatial Variability, Connectivity and Scale, NUREG/CR-5559, U.S. Nuclear Regulatory Commission.

Killey, R.W.D. and G.L. Moltyaner (1988). Twin lake tracer tests:setting methodology and hydraulic conductivity distribution. *Water Resour. Res.* **24**(10):1585–1612.

Klecka, G.M., J.W. Davis, D.R. Gray, and S.S. Madsen (1990). Natural bioremediation of organic contaminants in groundwater: Cliffs-Dow Superfund site. *Ground Water* **28**:534–543.

Levich, V.G. (1962). *Physicochemical Hydrodynamics*. Prentice Hall, Englewood Cliffs, NJ.

Li, Y.-H. and S. Gregory (1974). Diffusion of ions in sea water and in deep-sea sediments. *Geochem. Cosmochim. Acta* **38**:703–714.

Longhurst J.W. (Ed.) (1989). *Acid Deposition Sources. Effects and Controls*. British Library Technical Communication, London.

Mackay, D.M., D.L. Freyberg, P.V. Roberts, and J.A. Cherry (1986). A natural gradient experiment in a sand aquifer. 1. Approach and overview of plume movement. *Water Resour. Res.* **22**(13):2017–2030.

Martínez-Landa, L., J. Carrera, J. Guimerà, E. Vazquez-Suñé, L. Vives, and P. Meier (2000). Methodology for the hydraulic characterization of a granitic block. In *Calibration and Reliability in Groundwater Modeling:Coping with Uncertainty*, ModelCARE 99, F. Stauffer, W. Kinzelbach, K. Kovar, and E. Hoehn (Eds.), IAHS Publication 265. IAHS Press, Wallingford, Oxfordshire, UK, pp. 341–345.

Masters, G.M. (1998). *Introduction to Environmental Engineering and Science*, 2nd ed. Prentice Hall, Englewood Cliffs, NJ.

Miller, C.T., M.M. Poirier-McNeill, and A.S. Mayer (1990). Dissolution of trapped nonaqueous phase liquids:mass transfer characteristics. *Water Resour. Res.* **26**:2783–2796.

Neuman, S.P. (1994). Generalized scaling of permeabilities: validation and effect of support scale. *Geophys. Res. Lett.* **21**(5):349–352.

Newell, C.J. and J.A. Connor (1998). Characteristics of dissolved hydrocarbon plumes: results of four studies. In *Proceedings of the 1998 Petroleum Hydrocarbons and Organic Chemicals in Ground Water: Prevention, Detection, and Remediation*, Conference and Exposition, Houston, Texas, pp. 51–59.

NRC (1986). *Acid Deposition. Long Term Trends*. National Research Council, National Academy Press, Washington, DC.

Ogata, A. (1970). *Theory of Dispersion in Granular Medium*. U.S. Geological Survey Professional Paper 411-I.

Pankow, J.F. and J.A. Cherry (1996). *Dense Chlorinated Solvents and Other DNAPLs in Groundwater: History, Behavior, and Remediation*. Waterloo Press, Guelph, Canada.

Perkins, T.K. and O.C. Johnson (1963). A review of diffusion and dispersion in porous media. *Soc. Petroleum Eng. J.* **3**:70–84.

Postel, S. (1984). *Air Pollution, Acid Rain and the Future of Forest*. World Watch Paper 58, World Watch Institute, Washington, DC.

Press, W.H., S.A. Teukolsky, W.T. Vetterling, and B.P. Flannery (1992). *Numerical Recipes in FORTRAN—The Art of Scientific Computing*, 2nd ed. Cambridge University Press, Cambridge, UK.

Rasmussen, T.C., D.D. Evans, P.J. Sheets, and J.H. Blanford (1993). Permeability of Apache Leap tuff: borehole and core measurements using water and air. *Water Resour. Res.* **29**:1997–2006.

Rifai, H.S., R.C. Borden, J.T. Wilson, and C.H. Ward (1995). Intrinsic bioattenuation for subsurface restoration. In *Intrinsic Bioremediation*, R.E. Hinchee, F.J. Brockman, and C.M. Vogel (Eds.). Battelle Press, Columbus, OH, Vol. 3, No. 1, pp. 1–29.

Roberts, P.V., M.N. Goltz, and M.M. Douglas (1986). A natural gradient experiment on solute transport in a sand aquifer. 3. Retardation estimates and mass balances for organic solutes. *Water Resour. Res.* **22**(13):2047–2058.

Rovey, C.W. II and D.S. Cherkauer (1995). Scale dependency of hydraulic conductivity measurements. *Ground Water* **33**:769–780.

Rovey, C.W. II (1998). Digital simulation of the scale effect in hydraulic conductivity. *J. Hydrol.* **6**:216–225.

Schnoor, J.L. (1996). *Environmental Modeling: Fate and Transport of Pollutants in Water, Air, and Soil*. Wiley InterScience, Hoboken, NJ.

Schulze-Makuch, D. and D.S. Cherkauer (1998). Variations in hydraulic conductivity with scale of measurements during aquifer tests in heterogenous, porous carbonate rock. *Hydrogeol. J.* **6**:204–215.

Sleeps, B.E. and J.F. Sykes (1989). Modeling the transport of volatile organics in variably saturated media. *Water Resour. Res.* **25**:81–92.

Tchobanoglous, G. and E.D. Schroeder (1987). *Water Quality*. Addison–Wesley Publishing, Reading, MA.

Turitto, V.T. (1975). Mass transfer in annuli under conditions of laminar flow. *Chem. Eng. Sci.* **30**:503–509.

Vesselinov, V.V., S.P. Neuman, and W.A. Illman (2001a). Three-dimensional numerical inversion of pneumatic cross-hole tests in unsaturated fractured tuff: 1. Methodology and borehole effects. *Water Resour. Res.* **37**(12):3001–3018.

Vesselinov, V.V., S.P. Neuman, and W.A. Illman (2001b). Three-dimensional numerical inversion of pneumatic cross-hole tests in unsaturated fractured tuff: 2. Equivalent parameters, high-resolution stochastic imaging and scale effects. *Water Resour. Res.* **37**(12): 3019–3042.

Wild, A. (1993). *Soil and the Environment: An Introduction*. Cambridge University Press, Cambridge, UK.

Zhang, D. (2001). *Stochastic Methods for Flow in Porous Media: Coping with Uncertainties*. Academic Press, San Diego, CA.

Zheng, C. and S.M. Gorelick (2003). Analysis of solute transport in flow fields influenced by preferential flowpaths at the decimeter scale. *Ground Water* **41**(2):142–155.

Zlotnik, V.A., B.R. Zurbuchen, T. Ptak, and G. Teutsch (2000). Support volume and scale effect in hydraulic conductivity:experimental aspect. In *Theory, Modeling, and Field Investigation in Hydrogeology: A Special Volume in Honor of Shlomo P. Neuman's 60th Birthday*, D. Zhand and C. L. Winter (Eds.) Geological Society of America. Boulder, CO, Special Paper 348, pp. 191–213.

Zoetman, B.C.J., E. De Greef, and F.J.J. Brikmann (1981). Persistence of organic chemicals in groundwater, lessons from soil pollution incidents in The Netherlands. *Sci. Tot. Environ.* **21**:187–202.

5

FATE AND TRANSPORT EQUATIONS AND ANALYTICAL MODELS FOR NATURAL ATTENUATION

"Models are to be used, not believed."

—*Principles of Econometrics*, Henri Theil

5.1. THEORETICAL BACKGROUND

Most fate-and-transport models are based on mass balances that incorporate processes such as advection, dispersion, chemical reactions, and biodegradation of the target contaminants as a function of time. Such an expression was derived in Chapter 4 from first principles and is given here in compact form using del notation as:

$$\frac{\partial C}{\partial t} = \frac{1}{R_f}[D\,\nabla^2 C - v\,\nabla C - \lambda C] \qquad [5.1]$$

where C is the concentration of a dissolved contaminant at a given point in the aquifer at time t, R_f is the retardation factor for instantaneous, linear sorption, D is the hydrodynamic dispersion coefficient tensor, v is the groundwater velocity, λ (lambda) is the first-order decay coefficient, and ∇ is the derivative operator, where $\nabla = \partial/\partial x + \partial/\partial y + \partial/\partial z$ in a three-dimensional Cartesian system. Physically, this expression means that the change in contaminant concentration with time is a function of the amount of plume spreading in all directions (first term on the right-hand side of Eq. [5.1], dispersive transport term), the amount of contaminant that migrates with groundwater in bulk flow (second term, advective

Bioremediation and Natural Attenuation: Process Fundamentals and Mathematical Models
By Pedro J. J. Alvarez and Walter A. Illman   Copyright © 2006 John Wiley & Sons, Inc.

transport term), and the amount of contaminant that is degraded (last term on the right, sink term).

The hydrodynamic dispersion coefficient given in Eq. [5.1] assumes that D is constant and independent of distance. However, the value of the dispersion coefficient is scale-dependent (e.g., Lallemande-Barres and Peaudecerf 1978; Pickens and Grisak 1981; Neuman 1990; Zhang and Neuman 1990; Gelhar 1993), and changes in D that are associated with the scale of experiments or the travel distance of a contaminant are difficult to determine. Therefore, D is usually assumed to be constant in many models, which may introduce an error in model simulations.

For the sink term, the decay coefficient (λ) consists of processes such as hydrolysis, photolysis (unlikely if the pollutant is underground), chemical redox reactions, adsorption, volatilization, and biodegradation (Domenico and Schwartz 1998). However, most studies have shown that the decay coefficient for benzene (and other BTEX compounds) is primarily due to biodegradation (Chiang et al. 1989; Rifai et al. 1995). For example, Chiang et al. (1989), through data analysis and computer modeling, demonstrated that the contribution of volatilization to the dissolved contamination attenuation was only 5% at one site. It can be seen that, except in the case of very shallow groundwater, volatilization is not expected to contribute significantly to the overall attenuation of the contaminant plume. Therefore, we neglect volatilization and the decay rate is assumed to be a measure of biodegradation. We stress that in the form given in Eq. [5.1], degradation takes place in the aqueous phase only.

5.1.1. Initial and Boundary Conditions

The solution to Eq. [5.1] can be obtained through a variety of analytical and numerical techniques. Regardless of the solution technique used, in order to obtain a unique solution to a flow and transport problem such as that for a domain Ω with boundary Γ shown in Figure 5.1, one must specify the material properties of the domain and the forcing functions consisting of initial and boundary conditions and source/sink terms. In a steady-state problem, an initial condition is not required. However, for a transient problem one must specify an initial condition in addition to boundary conditions.

The initial condition is used to specify the initial distribution of a contaminant within the model domain. For example, the initial condition for a one-dimensional problem takes the following form:

$$C(x, 0) = C_0(x) \tag{5.2}$$

which states that at time zero throughout the modeled region, the concentration of the contaminant has an initial concentration distribution of $C_0(x)$. A special case of Eq. [5.2] is when $C_0(x)$ is set to zero. This means that the solute concentration is zero everywhere throughout the model domain. The latter initial condition is commonly used to evaluate scenarios involving the contamination of an aquifer due to a placement of some contaminant source.

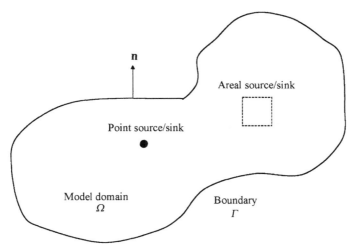

Figure 5.1. Hypothetical aquifer with domain Ω, boundary Γ, and regions that act as point and areal source/sinks.

Boundary conditions are used to specify fixed values of concentrations, the gradient (or a mix of the two) at the physical boundary Γ of the domain Ω that is to be modeled. Boundary conditions control the way the domain of interest communicates with areas outside the domain.

There are three principal types of boundary conditions. The first boundary condition, also known as the Dirichlet condition, specifies the concentration along the boundary over some time:

$$C(x, t) = c(x) \qquad [5.3]$$

over some boundary Γ for $t > 0$. Here, $c(x)$ is the concentration that is specified along the boundary Γ. The latter also can be varied over time if the boundary condition is a time-varying boundary condition. Depending on the specified concentration, the Dirichlet boundary condition can act to add or remove solute mass from the modeled domain.

A second type of boundary condition, also known as the Neumann condition, involves the specification of the concentration gradient normal to the boundary, denoted with a unit normal vector in Figure 5.1. Here, as in the Dirichlet boundary condition, the concentration gradient can vary with space and time. It takes the form

$$-D_x \frac{\partial C}{\partial x} = g(x) \qquad [5.4]$$

where $g(x)$ represents the dispersive flux that is normal to the boundary Γ. When the dipersive flux on the right-hand side of Eq. [5.4] is set to zero, the boundary condition is called the no-flux boundary condition. In this case, there is no concentration gradient normal to the boundary; thus, the mass cannot leave or enter the region Ω.

The third type of boundary condition is called the Cauchy or mixed boundary condition. This boundary condition is used to specify both the concentration and its gradient along the boundary. It takes the following form:

$$-D_x \frac{\partial C}{\partial x} + v_x C = k(x) \qquad [5.5]$$

where $k(x)$ is a known function that represents the dispersive and advective flux normal to the boundary Γ. When the boundary is impermeable, then $k(x) = 0$, and there is no advective–dispersive flux of solutes through the boundary.

If the concentration flux across the boundary is proportional to the concentration difference between the boundary and the medium surrounding Ω, so that it is given by

$$H(C - C_0)$$

where C_0 is the concentration of the medium and H is a constant, then the boundary condition reads

$$D \frac{\partial C}{\partial x} + H(C - C_0) = 0 \qquad [5.6a]$$

or

$$\frac{\partial C}{\partial x} + h(C - C_0) = 0 \qquad [5.6b]$$

where $h = H/D$. As $h \to 0$, this tends to the no-flux condition, and as $h \to \infty$ it approaches the prescribed boundary condition given by Eq. [5.2] (Carslaw and Jaeger 1959). Similar boundary conditions can be written for two- and three-dimensional problems.

We have seen that there are three main types of boundary conditions. Examples of the three boundary conditions with different loading scenarios are summarized in Table 5.1. We emphasize that the choice of specifying a boundary condition for a particular problem is one of the most important components of modeling groundwater flow and solute transport. The misspecification of a boundary condition can have a large effect on concentration profiles within the modeled domain and this could lead to large errors in predictions of contaminant concentrations. We next illustrate how the three boundary conditions that we discussed can be specified.

Consider a situation in which there is a river that releases nitrate at a constant concentration over space and time into an aquifer confined within a valley that is considered to be impermeable (Figure 5.2a). The aquifer discharges into a pristine lake with endangered fish and reptiles that need to be protected. We would like to predict how fast the contaminants will migrate through the aquifer and reach the lake. To simplify the problem, we treat this problem as a two-dimensional aquifer in plan view and only consider flow within the aquifer. Figure 5.2b shows the

Table 5.1. Examples of Boundary Conditions with Different Loading Scenarios

Name	Type	Form[a]
Constant concentration	Fixed concentration	$C(0,t) = C_0$
Pulse-type loading with constant concentration	Fixed concentration	$C(0,t) = \begin{cases} C_0, 0 < t \leq t_0 \\ 0, t > t_0 \end{cases}$
Exponential decay with source concentration $\to 0$	Fixed concentration	$C(0,t) = C_0 e^{-\alpha t}$
Exponential decay with source concentration $\to C_a$	Fixed concentration	$C(0,t) = C_a + C_b e^{-\alpha t}$
Constant flux with constant input concentration	Variable flux	$\left(-D\frac{\partial C}{\partial x} + vC\right)_{x=0} = vC_0$
Pulse-type loading with constant input fluxes	Variable flux	$\left(-D\frac{\partial C}{\partial x} + vC\right)_{x=0} = \begin{cases} vC_0, 0 < t \leq t_0 \\ 0, t > t_0 \end{cases}$

[a]$C_0, C_a, C_b,$ = various constant concentrations; α = decay constant; t_0 = time at which concentration changes due to pulse loading.

Source: Adapted from Domenico and Schwartz (1998).

corresponding boundary conditions assigned on each of the four boundaries for the transport problem. On the left border of the flow domain, the river fully penetrates the aquifer and releases nitrate at a concentration C_0. Therefore, Γ_1 is treated as a Dirichlet boundary. The aquifer is bounded on Γ_2 and Γ_3 by impermeable bedrock.

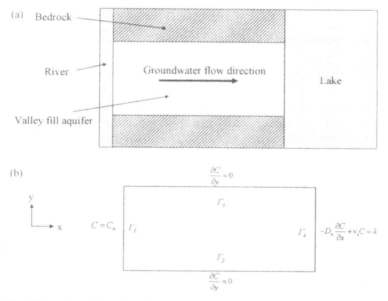

Figure 5.2. (a) Plan view of the schematic diagram showing a river recharging contaminant-laden water into a narrow valley aquifer confined by impermeable bedrock on the left and right sides. This aquifer then subsequently discharges into a pristine lake. (b) Boundary conditions for the transport problem formulated for the aquifer.

Therefore, boundaries Γ_2 and Γ_3 are assigned Neumann conditions in which the dispersive flux is specified as zero [i.e., $g = 0$]. Boundary Γ_4 separates the aquifer and the lake. In this case, we treat this as a Cauchy boundary as the advective and dispersive fluxes must be specified.

5.1.2. Sources and Sinks

Source and sink terms appear in both the governing flow and transport equations. For the flow problem, the source/sink term represents a mechanism in which water is added or removed from the system. The source and sink terms are broadly classified as point or areal (Figure 5.1). These terms in the governing transport equation represent solute mass that is dissolved in groundwater and is added to or removed from the system through source and sink terms appearing in the governing flow equation. Examples of point sources or sinks include wells that are either recharging or discharging, buried drains, and localized recharge resulting from contaminant spill events. Examples of physical features that are modeled as areal sources or sinks include recharge that takes place over a large area due to precipitation or irrigation, evapotranspiration, impoundments of contaminants such as sewage lagoons and mill tailings piles, and surface water features including wetlands, lakes, and oceans. A third class of source and sink terms could be added to represent curvilinear features such as rivers and canals that are present within the model domain. In many cases, however, these features are used as natural boundaries and treated as boundary conditions instead of internal source and sink terms.

5.1.3. Analytical Versus Numerical Solutions

The governing partial differential equation, the initial conditions and the boundary conditions form a boundary value problem. As with many other boundary problems that appear in applied physics, it can be solved analytically or numerically. Analytical solutions are exact and are generally simple to implement. Closed form solutions are usually available, which can be used readily to make preliminary predictions of fate and transport. However, it is generally limited to simple geometries and requires that aquifer and transport parameters be uniform throughout the modeled region. There are stochastic analytical solutions that treat variabilities in hydraulic and transport parameters, but we do not deal with these solutions as they are beyond the scope of this book.

We present in this chapter analytical solutions to various cases as they are often used in screening studies (Tiers 1 and 2). These analytical solutions are generally limited to steady, uniform flow and should not be used for groundwater flow or solute transport problems in strongly anisotropic or heterogeneous aquifers. These models additionally should not be applied under nonuniform flow conditions. One example of this is when there are strong seasonal effects in the direction of hydraulic gradients. Another example is when there are pumping or injection systems that can create a complicated nonuniform flow field. Furthermore, these models should not be applied where there are large dominant high-permeability features (e.g., fractures) in fractured or karst aquifers, or where vertical flow gradients affect contaminant transport. It should be kept in mind that analytical models are best utilized for order-of-magnitude estimations, since a number of potentially important processes are treated in the models in an approximate manner or sometimes ignored totally.

Numerical solutions, on the other hand, treat the boundary value problem as a system of algebraic equations or alternatively simulate transport by tracking a larger number of particles in a known velocity field. The main advantage over analytical solutions is that it is flexible and can handle complex geometries. This flexibility allows one to incorporate complex geological features and spatial variability in flow and transport parameters. It can also incorporate spatial variation in the initial condition and both space and time variations in boundary conditions. However, numerical solutions can be prone to discretization and roundoff errors as well as numerical dispersion.

5.2. ANALYTICAL MODELS FOR CONTAMINANT TRANSPORT AND REACTION PROCESSES

Consider a situation in which there is a leaking underground storage tank (LUST) or a dense nonaqueous phase liquid (DNAPL) spill that forms a source zone. In most groundwater contamination problems, we are interested in predicting the concentration of the dissolved phase contaminants that may appear at a receptor or point of compliance (POC) some distance downgradient from the source zone. In order to make predictions of contaminant concentrations with space and time, we will examine several analytical models that are widely used.

The situation generating the contaminant plume from the source zone is shown diagrammatically in Figure 5.3. The dashed lines emanating from the vertical source area represent an ideal contaminant plume undergoing advection and dispersion at three successive times. This is a classical case in which a hydrogeologist or a civil engineer is asked to predict the contaminant concentration at some location away from the source zone at some time. In order to accomplish this, an appropriate analytical model described below or from the literature can be employed to estimate site-specific risk due to migration of chemicals of concern. These equations can be used to calculate contaminant concentrations at points of interest in a groundwater plume. Use of a spreadsheet program such as Excel makes testing of many scenarios for a site possible in a short amount of time. The trick is to select a model that best represents the conditions in the field. This is important as incorrect selection of the model can have a devastating effect on the outcome of the site investigation (see the case study in Section 5.5.1).

As in any mathematical representation of natural processes, it is important that initial and boundary conditions and simplifying assumptions be clearly understood. Many models are based on solutions to Eq. [5.1] that incorporate different initial and boundary conditions. These solutions commonly assume the following:

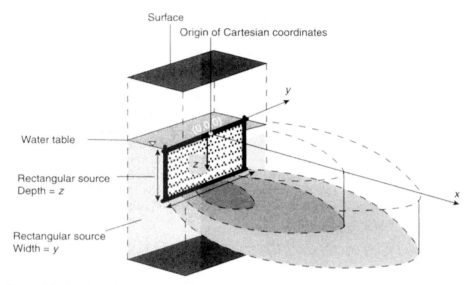

Figure 5.3. Geometry of the model given in Eq. [5.7]. The center of the rectangular source corresponds to $x = y = z = 0$. (Adapted from Newell et al. 1996.)

1. Contaminant concentration at time zero is zero (i.e., $C(x, y, z, 0) = 0$).
2. The aquifer is homogeneous and isotropic; that is, K is constant in all directions.
3. The groundwater flow field is uniform; that is v_x is constant and $v_y = v_z = 0$.
4. Groundwater flow is fast enough so that molecular diffusion in the hydrodynamic dispersion term can be neglected.
5. Dispersion is Fickian, and longitudinal dispersivity can be approximated with a constant apparent dispersivity (α_x) and the dispersion coefficient is proportional to the groundwater velocity; that is $D_x = \alpha_x v_x$.
6. Adsorption is a reversible process at equilibrium, represented by a linear isotherm.
7. Biodegradation kinetics is first-order with respect to the contaminant concentration.
8. The biodegradation rate coefficient, λ, is constant, which is a very influential assumption.

5.2.1. Multidimensional Transport from a Finite, Planar, Continuous Source of Contamination Under Transient Conditions

From the above assumptions, the analytical solution for multidimensional transport can be obtained for Eq. [5.1]. Because the solution is multidimensional, it incorporates both longitudinal and transverse dispersion in the horizontal and vertical directions. This solution assumes that the domain is infinite. That is, there are no physical boundaries that can affect the transport process. It is also assumed that there is a rectangular planar source with width Y and height Z at $x = 0$. The planar source is perpendicular to groundwater flow, releasing dissolved constituents. In this case, the source zone concentration is set to be constant with time, although there are models that allow for time variation in source zone concentration. Conceptually, the source zone is situated a short distance behind the vertical source plane defined by YZ (see Figure 5.3). The solution for this situation reads (Domenico and Schwartz 1998):

$$C(x, y, z, t) = \left(\frac{C_0}{8}\right) \exp\left[\frac{x}{2\alpha_x}\left(1 - \sqrt{\frac{1 + 4\lambda\alpha_x}{v_c}}\right)\right] \cdot \text{erfc}\left[\frac{x - v_c t\sqrt{\frac{1+4\lambda\alpha_x}{v_c}}}{2\sqrt{\alpha_x v_c t}}\right]$$

$$\cdot \left[\text{erf}\left(\frac{y + Y/2}{2\sqrt{\alpha_y x}}\right) - \text{erf}\left(\frac{y - Y/2}{2\sqrt{\alpha_y x}}\right)\right] \qquad [5.7]$$

$$\cdot \left[\text{erf}\left(\frac{z + Z}{2\sqrt{\alpha_z x}}\right) - \text{erf}\left(\frac{z - Z}{2\sqrt{\alpha_z x}}\right)\right]$$

where C = contaminant concentration, C_0 = initial contaminant concentration at the source, x = distance downgradient of source, y = distance from the centerline of the source, z = vertical distance from the groundwater surface to the measurement point, Y = source width, Z = source depth, α_x = longitudinal dispersivity, α_y = horizontal transverse dispersivity, α_z = vertical transverse dispersivity, λ = site-specific first-order decay coefficient, t = time, v_c = contaminant velocity in groundwater, $\text{erf}(x)$ = error function, and $\text{erfc}(x)$ = complementary error function = $1 - \text{erf}(x)$.

The analytical solution includes one or more of the following functions:

$\text{erf}(\beta)$	Error function
$\text{erfc}(\beta)$	Complementary error function
$\exp(\beta)$	Exponential function

The variable β represents any mathematical argument inside the parentheses. Table 5.2 gives $\text{erf}(\beta)$ values for $0 < \beta < 3$. Figure 5.4 shows a graphical representation of the error and complementary error functions. Alternatively, it is convenient to use computer spreadsheets (e.g., Excel, Quattro-Pro) that automatically calculate the error function for positive values of β. Error function values for negative β or complementary error functions values can be estimated based on the following relationships:

$$\text{erf}(-\beta) = -\text{erf}(\beta)$$
$$\text{erfc}(\beta) = 1 - \text{erf}(\beta)$$
$$\text{erfc}(-\beta) = 1 + \text{erf}(\beta)$$

Equation [5.7] is most appropriate when the source (spill or leakage from an underground storage tank) is known to be continuous and the contaminant

Table 5.2. Values of $erf(\beta)$ and $erfc(\beta)$ for Positive Values of β

β	$erf(\beta)$	$erfc(\beta)$	β	$erf(\beta)$	$erfc(\beta)$
0	0	1			
0.05	0.056372	0.943628	1.10	0.880205	0.119795
0.10	0.112463	0.887537	1.20	0.910314	0.089686
0.15	0.167996	0.832004	1.30	0.934008	0.065992
0.20	0.222703	0.777297	1.40	0.952285	0.047715
0.25	0.276326	0.723674	1.50	0.966105	0.033895
0.30	0.328627	0.671337	1.60	0.976348	0.023652
0.35	0.379382	0.620618	1.70	0.983790	0.016210
0.40	0.428392	0.571608	1.80	0.989091	0.010909
0.45	0.475482	0.524518	1.90	0.992790	0.007210
0.50	0.520500	0.479500	2.00	0.995322	0.004678
0.55	0.563323	0.436677	2.10	0.997021	0.002979
0.60	0.603856	0.396144	2.20	0.998137	0.001863
0.65	0.642029	0.357971	2.30	0.998857	0.001143
0.70	0.677801	0.322199	2.40	0.999311	0.000689
0.75	0.711155	0.288845	2.50	0.999593	0.000407
0.80	0.742101	0.257899	2.60	0.999764	0.000236
0.85	0.770668	0.229332	2.70	0.999866	0.000134
0.90	0.796908	0.203092	2.80	0.999925	0.000075
0.95	0.820891	0.179109	2.90	0.999959	0.000041
1.00	0.842701	0.157299	3.00	0.999978	0.000022

concentration varies as a function of time at a given sampling point. It applies for contaminant migration in the saturated zone with horizontal dispersion along the y-axis, vertical dispersion along the z-axis, and longitudinal dispersion along the flow direction, the x-axis. This is a transient solution that allows one to calculate the concentration of contaminants at any time within the modeled region in three dimensions. This means that the user can assess whether a prespecified concentration limit

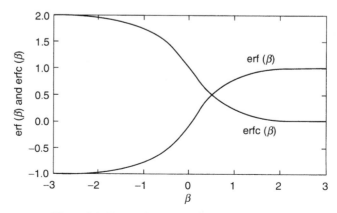

Figure 5.4. Error and complementary error functions.

might be exceeded *at some* point of compliance (POC) and also *when* this limit may be exceeded. For example, if concentrations are predicted to be higher than prede-termined levels at a receptor in the near future, remedial actions may be needed to be taken immediately. However, if the concentration is predicted to stay below the regulatory limit for many years, then time may be available to evaluate an alterna-tive remedial approach. Consideration of transient plume behavior can be important when establishing remedial goals, setting monitoring frequencies, evaluating the plume status, and determining when steady-state models may be appropriately applied (ASTM 1998).

Sometimes it is desirable to estimate contaminant concentrations along the plume centerline at the water table $C(x, 0, 0, t)$, subject to vertical and horizontal dispersion. The solution for this situation, assuming a constant rectangular source, is given by (ASTM 1998)

$$C(x,0,0,t) = \left(\frac{C_0}{2}\right) \exp\left[\frac{x}{2\alpha_x}\left(1 - \sqrt{1 + \frac{4\lambda\alpha_x}{v_c}}\right)\right] \cdot \text{erfc}\left(\frac{x - v_c t \sqrt{\frac{1 + 4\lambda\alpha_x}{v_c}}}{2\sqrt{\alpha_x v_c t}}\right)$$
$$\cdot \left[\text{erf}\left(\frac{Y}{4\sqrt{\alpha_y x}}\right) \cdot \text{erf}\left(\frac{Z}{2\sqrt{\alpha_z x}}\right)\right] \tag{5.8}$$

5.2.2. Multidimensional Transport from a Finite, Planar, Continuous Source of Contamination Under Steady-State Conditions

As part of its life cycle, a plume expands and reaches a maximum size upon achiev-ing a steady state (i.e., when contaminant concentrations at any point within the plume do not change with time). Under these conditions, the contaminant concen-tration does not change with time and thus the left-hand side of Eq. [5.1] is equal to zero. The steady-state conditions can be determined by analyzing historical data to make sure the plume is stable through time.

A steady-state form of Eq. [5.7] can readily be obtained when the argument of the complementary error function approaches -2 [i.e., $\text{erfc}(-2) = 2$ from Figure 5.4]. The solution reads

$$C(x, y, z) = \left(\frac{C_0}{4}\right) \exp\left[\frac{x}{2\alpha_x}\left(1 - \sqrt{1 + \frac{4\lambda\alpha_x}{v_c}}\right)\right]$$
$$\cdot \left[\text{erf}\left(\frac{y + Y/2}{2\sqrt{\alpha_y x}}\right) - \text{erf}\left(\frac{y - Y/2}{2\sqrt{\alpha_y x}}\right)\right] \cdot \left[\text{erf}\left(\frac{z + Z}{2\sqrt{\alpha_z x}}\right) - \text{erf}\left(\frac{z - Z}{2\sqrt{\alpha_z x}}\right)\right]$$
$$\tag{5.9}$$

Steady state is reached when the contaminant is released from the source at the same rate that it is attenuated. This is an important situation to model because it represents the largest distance to which a contaminant will migrate and, thus, the worst case scenario.

Domenico (1987) also proposed an equation to estimate the steady-state concentration along the plume centerline at the water table:

$$C(x,0,0) = C_0 \exp\left[\frac{x}{2\alpha_x}\left(1 - \sqrt{1 + \frac{4\lambda\alpha_x}{v_c}}\right)\right] \cdot \left[\text{erf}\left(\frac{Y}{4\sqrt{\alpha_y x}}\right) \cdot \text{erf}\left(\frac{Z}{2\sqrt{\alpha_z x}}\right)\right] \quad [5.10]$$

Comparing Eqs. [5.8] and [5.10], we observe that one-half of the steady-state concentration is reached when the complementary error function (erfc) in Eq. [5.8] equals 1 (i.e., when erfc equals 1, Eq. [5.8] is one-half of Eq. [5.10]). This occurs when

$$\text{erfc}\left(\frac{x - v_c t \sqrt{1 + \frac{4\lambda\alpha_x}{v_c}}}{2\sqrt{\alpha_x v_c t}}\right) = 1 \quad [5.11]$$

This relationship can be used to obtain a rough estimate of the time required to achieve the steady state. First, we need to assume that the time required to reach one-half of the steady-state concentration is roughly the same time required to reach the steady-state concentration (ASTM 1998). This counterintuitive assumption is based on the relatively high variability that is commonly observed in contaminant concentrations at any given point in an aquifer, which makes a 50% difference in concentrations measured at a monitoring well often statistically insignificant. Equation [5.11] can be rearranged to explicitly calculate the time required to reach steady state (t_{ss}):

$$t_{ss} = \frac{x}{v_c\sqrt{1 + \frac{4\lambda\alpha_x}{v_c}}} \quad [5.12]$$

This formula is similar to Eq. [4.64], except for the square root term. This term is always greater than or equal to 1 and yields t_{ss} values that are shorter than the travel time estimated by considering advection and retardation alone (Eq. [4.59]). Other processes such as biodegradation and dispersion (considered in Eq. [5.12]) attenuate the plume migration and contribute to reaching the steady state faster.

5.2.3. Multidimensional Transport from a Finite, Planar, Decaying Source of Contamination Under Transient Conditions

For a decaying contaminant source (which is more representative of a contaminant source undergoing remediation by either engineered solutions or natural weathering), with the same assumptions and source geometry, the analytical solution to Eq. [5.1] is (Newell et al. 1996)

$$
C(x,y,z,t) = \left(\frac{C_0 e^{-kt}}{8}\right) \exp\left[\frac{x}{2\alpha_x}\left(1 - \sqrt{1 + \frac{4\lambda\alpha_x}{v_c}}\right)\right] \cdot \mathrm{erfc}\left(\frac{x - v_c t\sqrt{\frac{1+4\lambda\alpha_x}{v_c}}}{2\sqrt{\alpha_x v_c t}}\right)
$$

$$
\cdot \left[\mathrm{erf}\left(\frac{y + Y/2}{2\sqrt{\alpha_y x}}\right) - \mathrm{erf}\left(\frac{y - Y/2}{2\sqrt{\alpha_y x}}\right)\right] \cdot \left[\mathrm{erf}\left(\frac{z + Z}{2\sqrt{\alpha_z x}}\right) - \mathrm{erf}\left(\frac{z - Z}{2\sqrt{\alpha_z x}}\right)\right]
$$

$$[5.13]$$

Here, the transient condition is still applied; however, the source is either being removed or undergoing remediation.

Often, the decrease in contaminant concentration at the source is due to NAPL dissolution and washout as uncontaminated groundwater flows through the source. In this case, the source decay coefficient (k_s) is given by

$$
k_s = \frac{QC_0}{M_0}
$$

$$[5.14]$$

where Q is groundwater flow rate $(L^3 T^{-1})$, C_0 is the contaminant concentration in groundwater leaving the source (ML^{-3}), and M_0 is the initial mass of the contaminant in the source zone (M). Another form of the source decay coefficient takes the form

$$
k_s = \frac{\ln 2}{t_{1/2}}
$$

$$[5.15]$$

where $t_{1/2}$ is the half-life of the contaminant.

5.2.4. Multidimensional Transport from an Instantaneous Point Source of Contamination Under Transient Conditions

Some contamination events are best modeled conceptually as an instantaneous point source, such as chemical spills following a delivery truck or train accident, or other spills of large magnitude where the free products and source zone are removed by emergency response teams. Several analytical models exist for this case. One is the parallelepiped finite source model of Hunt (1978) and the point source model of Baetslé (1969). Since the source resulting from an instantaneously spill event can be viewed as a point in many contamination problems, we provide its solution here. Such events result in elliptical plumes that migrate and disperse along the groundwater flow direction (Figure 5.5).

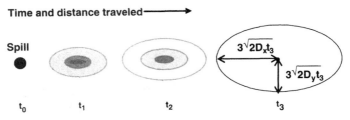

Figure 5.5. Plan view of the migration of a plume that originated from an instantaneous point source.

Baetslé (1969) provided the following equation for instantaneous point source plumes:

$$C(x,y,z,t) = \left(\frac{C_0 V_0}{8(\pi t)^{3/2}(D_x D_y D_z)} \right) \exp\left(-\frac{(x-v_c t)^2}{4D_x t} - \frac{y^2}{4D_y t} - \frac{z^2}{4D_z t} - \lambda t \right)$$

[5.16]

where D_x, D_y, and D_z are the dispersion coefficients along the x-, y-, and z-directions; C_0 is the initial point source concentration; and V_0 is the spilled volume. Thus, the product $C_0 V_0$ represents the contaminant mass that is released. The solution to the nondecaying case can be obtained simply by setting $\lambda = 0$. Recall that the hydrodynamic dispersion coefficient along the i direction is related to the dispersivity (α_i) and the advective velocity (v_x) as $D_i = \alpha_i v_x$.

Under ideal conditions (e.g., homogeneous isotropic aquifers with constant and uniform groundwater flow), the contaminant concentration follows a Gaussian (normal) distribution, and the maximum concentration occurs at the center of the plume, when $x = vt$ and $y = z = 0$ (Figure 5.5):

$$C_{\max} = \frac{C_0 V_0 e^{-\lambda t}}{8(\pi t)^{1.5}\sqrt{D_x D_y D_z}}$$

[5.17]

To estimate the dimensions of an ellipsoid plume, it is useful to recognize that, by definition (De Josselin and de Jong, 1958), the standard deviations for a spreading solute are

$$\sigma_x = \sqrt{2D_x t}$$

[5.18a]

$$\sigma_y = \sqrt{2D_y t}$$

[5.18b]

$$\sigma_z = \sqrt{2D_z t}$$

[5.18c]

Because 99.7% of the area of a normal (Gaussian) distribution is contained within three standard deviations, it follows that the ellipsoid plume can arbitrarily be defined from its center of mass as having dimensions $3\sigma_x$, $3\sigma_y$, and $3\sigma_z$.

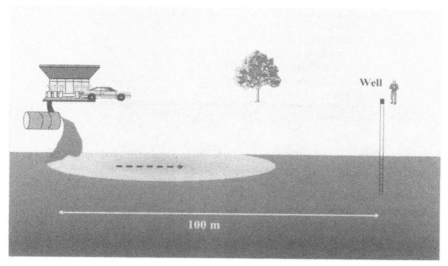

Figure 5.6. Diagram of hypothetical site used in contamination Scenarios 1, 2, 3, and 4.

5.3. APPLICATION OF ANALYTICAL MODELS IN CONTAMINANT HYDROGEOLOGY

To illustrate use of the analytical models described in the previous section, four different contamination scenarios will be considered using a hypothetical release site. The hydrogeological parameters and receptors at risk are the same for each scenario (Figure 5.6). Table 5.3 summarizes the site-specific parameters.

Table 5.3. Site-Specific Parameters for Contamination Scenarios

Depth to Water Table	2 m
Hydraulic gradient (dh/dx)	0.005
Hydraulic conductivity (K)	0.5 m/day
Total porosity (ϕ)	30%
Effective porosity (ϕ_e)	15%
Bulk density (ρ_b)	1.86 kg/L
Longitudinal dispersivity (α_x)	10 m
Horizontal dispersivity (α_y)	2 m
Vertical dispersivity (α_z)	0.5 m
Biodegradation rate coefficient (λ)	0.0007 day^{-1}
Source width (Y)	5 m
Source depth (Z)	3 m
Volume released	1000 L
Maximum initial benzene concentration	20,000 μg/L
Partitioning coefficient for benzene (K_d)	0.038 L/kg

Table 5.4. Parameters Calculated from Table 5.3 Values

Retardation factor for benzene (R_f)	1.23
Advective velocity for benzene (v)	0.013 m/day
Longitudinal dispersion coefficient (D_x)	0.13 m²/day
Horizontal dispersion coefficient (D_y)	0.026 m²/day
Vertical dispersion coefficient (D_z)	0.00625 m²/day

Table 5.3 values can be used to calculate other parameters that are needed to use the analytical models in these examples (Table 5.4):

Retardation factor, $$R_f = 1 + \frac{K_d \rho_b}{\phi} = 1 + \frac{0.038(1.86)}{0.30} = 1.24$$

Advective velocity for benzene, $$v_c = \frac{K}{R_f \phi_e} \frac{dh}{dx} = \frac{0.5(0.005)}{(1.24)(0.15)}$$
$$= 0.013 \, \text{m/day} = 4.9 \, \text{m/year}$$

Neglecting diffusion, the longitudinal, horizontal, and vertical hydrodynamic dispersion coefficients can be calculated as

$$D_x = \alpha_x v_c = 10(0.013) = 0.13 \, \text{m}^2/\text{day}$$
$$D_y = \alpha_y v_c = 2(0.013) = 0.026 \, \text{m}^2/\text{day}$$
$$D_z = \alpha_z v_c = 0.5(0.013) = 0.00625 \, \text{m}^2/\text{day}$$

Many frequently asked questions about contaminant releases cannot be solved for explicitly with analytical models, and answering them is accomplished through implicit solutions that require iterative approaches. Due to the repetitive and cumbersome nature of iterating relatively complex equations, it is recommended to use iterative tools included in electronic spreadsheets, such as SOLVER and GOAL SEEK for Microsoft Excel, or their equivalent OPTIMIZER and SOLVE FOR, respectively, for Corel Quattro Pro.

Scenario 1: Continuous Planar Source with Constant Source Concentration

A release that could be considered continuous and causing a constant source concentration would be such as one occurring at an undetected pipe leak in the subsurface. After an initial buildup of contaminant to the point that liquid gasoline occurs in soil near the source, migration will bring the liquid in contact with groundwater. The continual release of gasoline, slow though it may be, maintains a liquid phase at the water table as benzene dissolves in the groundwater and flows away. The released mass in this case is not known, but the benzene concentration in groundwater at the source can be measured.

Risk assessment must address human ingestion of benzene through drinking water pumped from contaminated wells. For this hypothetical site, the well in danger of being contaminated is located 100 m directly downgradient from the source on the x-axis (Figure 5.6). Therefore, we can use Eqs. [5.8] and [5.10] with the hydrogeologic parameters in Tables 5.3 and 5.4. The standard for allowable benzene in drinking water is $5\,\mu g/L$. Questions to be answered with the aid of the analytical model include:

1. How far will the benzene plume (defined by the $5\,\mu g/L$ contour) migrate?
2. To what level should the source concentration be removed in order to prevent exposing the receptor to benzene concentrations greater than $5\,\mu g/L$?
3. How long will it take for the plume to reach the steady state if site conditions remain constant?

To answer the first question, we should keep in mind that the maximum extension of a plume (for a constant source) occurs when the steady state is achieved. Thus, we can use Eq. [5.8] to determine the distance (x) from the source that corresponds to a benzene concentration of $5\,\mu g/L$ (i.e., determine the x value for $C/C_0 = 5/20,000 = 0.00025$). Since the x variable is included in the exponential and error function terms, we cannot solve for an explicit algebraic expression for x. To answer question 1 it is convenient to enter the steady-state solution (Eq. [5.10]) with the values in Tables 5.3 and 5.4 in a spreadsheet. Care should be taken that the variables have compatible units. To determine the x value corresponding to $C/C_0 = 0.00025$, we can iterate with the GOAL SEEK tool, varying the value of x until the desired C/C_0 value is reached. This procedure gives $x = 115\,m$. This implies that the drinking water well (which is located 100 m downgradient from the source) will be impacted. Alternatively, we could program a spreadsheet to calculate the benzene concentration (using Eq. [5.10]) for various x values (in 1 m increments) and read directly the x value that corresponds to $C = 5\,\mu g/L$. The same answer would be obtained $(x = 115\,m)$.

The second question can be answered explicitly without the need to iterate. We need to calculate the value of C_0 that results in a benzene concentration equal to or smaller than $5\,\mu g/L$ at the well, for the worst case scenario (i.e., when the plume reaches its maximum extension, during steady state). For safety, we should allow for a buffer zone upgradient of the well. Using a margin of safety of 20 m, the x distance of interest for the $5\,\mu g/L$ contour will be 80 m. Using these values in Eq. [5.10] gives

$$C(80,0,0,\infty) = C_0\,\exp\left[\frac{80}{2(10)}\left(1 - \sqrt{1 + \frac{4(0.0007)10}{0.013}}\right)\right]$$
$$\cdot\,\text{erf}\left(\frac{5}{4\sqrt{(2)80}}\right)\text{erf}\left(\frac{3}{2\sqrt{(0.5)80}}\right)$$

Solving, $C_0 = 3492\,\mu g/L$. Thus, the well would be safe from contamination if the source is cleaned to a level that results in C_0 lower than this value.

The third question can be answered using Eq. [5.12], keeping in mind that the maximum extension of the $5\,\mu g/L$ contour would be 115 m:

$$t_{ss} = \frac{x}{v_c\sqrt{1 + \frac{4\lambda\alpha_x}{v_c}}} = \frac{115}{0.013\sqrt{1 + \frac{4(0.0007)10}{0.013}}} = 4863 \text{ days} = 13.3 \text{ years}$$

Scenario 2: Continuous Source with Decaying Source Concentration

This scenario is typical when contamination is discovered shortly after the release occurred, and source removal and remediation activities have begun. This would result in a decaying source concentration as indicated by Eq. [5.13]. Monthly monitoring of the source concentration could yield a good estimate for the source decay coefficient (k_s given as the slope of $\ln(C_0)$ versus time; see Chapter 7 for further details). Other examples of sources with decaying concentration include radioactive wastes and dilution and dissolution of the source by the surrounding, uncontaminated groundwater. Similar to the first scenario, we will consider the risk associated with human ingestion of benzene through drinking water pumped from a nearby well. The hydrogeologic parameters given in Tables 5.3 and 5.4 are applicable to this example. Furthermore, we will consider that biostimulation of the source zone is working, and benzene concentrations are decreasing in the groundwater that comes out of the source with a half-life of 180 days. The questions we seek to answer with the help of analytical models include:

1. What is the highest benzene concentration that will occur in groundwater in contact with a potable water main that is located 8 m downgradient of the source? (The water main runs perpendicular to the plume flow direction, and the critical point is at $x = 8$, $y = z = 0$.)
2. When will this concentration be reached?

First, we need to calculate the source decay coefficient, $k_s = \ln(2)/180 = 0.038\,\text{day}^{-1}$. To answer both questions we will use a model that determines the contaminant concentration along the centerline (Eq. [5.8]), with a slight modification—we will add a decay term for the source, similar to that in Eq. [5.13]:

$$C(x,0,0,t) = \left(\frac{C_0 e^{-k_s t}}{2}\right) \exp\left[\frac{x}{2\alpha_x}\left(1 - \sqrt{1 + \frac{4\lambda\alpha_x}{v_c}}\right)\right]$$
$$\cdot \text{erfc}\left(\frac{x - v_c t\sqrt{1 + \frac{4\lambda\alpha_x}{v_c}}}{2\sqrt{\alpha_x v_c t}}\right)\left[\text{erf}\left(\frac{Y}{4\sqrt{\alpha_y x}}\right)\right]\left[\text{erf}\left(\frac{Z}{2\sqrt{\alpha_z x}}\right)\right]$$

Using the above value for k_s and the parameters in Tables 5.3 and 5.4, for $x = 8\,\text{m}$

$$C(8,0,0,t) = \left(\frac{20,000e^{-0.038t}}{2}\right) \exp\left[\frac{8}{2(10)}\left(1 - \sqrt{1 + \frac{4(0.0007)10}{0.013}}\right)\right]$$

$$\cdot \text{erfc}\left(\frac{8 - 0.013t\sqrt{1 + \frac{4(0.0007)10}{0.013}}}{2\sqrt{10(0.013)t}}\right)\left[\text{erf}\left(\frac{5}{4\sqrt{2(8)}}\right)\right]\left[\text{erf}\left(\frac{3}{2\sqrt{0.5(8)}}\right)\right]$$

Simplifying, we get

$$C(8,0,0,t) = 1796.3\ e^{-0.038t}\text{erfc}\left(\frac{8 - 0.0237t}{0.735\sqrt{t}}\right)$$

Iterating with a spreadsheet, we can determine that the maximum value for $C(8,0,0,t)$ is $18.8\,\mu\text{g/L}$ corresponding to $t = 62$ days. This concentration exceeds the drinking water standard ($5\,\mu\text{g/L}$) and it could represent a threat to public health if contaminated groundwater infiltrated into the water main, even though it runs $8\,\text{m}$ away from the source.

Scenario 3: Continuous Radioactive Source

Concentrations of dissolved radioactive pollutants are typically expressed as bequerels (Bq), curies (Ci), or picocuries (pCi) per liter. These are all measures of nuclear disintegration per time. One curie is 3.7×10^{10} disintegrations per second, 1 pCi is equal to 10^{-12}Ci, and 1 Bq (SI units) is equal to 2.7×10^{-11}Ci. Analytical models describing radioactive plumes in groundwater can be used with these concentration units instead of the more traditional mass/volume units used for dissolved organic pollutants.

In this example, we will consider groundwater contamination by a landfill leachate that contains strontium (^{90}Sr), initially at $500\,\mu\text{Ci/L}$. The half-life for ^{90}Sr is 28 years, which corresponds to a source decay coefficient (k_s) of $0.000068\,\text{day}^{-1}$. ^{90}Sr is not subject to biodegradation or sorption, and it is expected to mix readily with the groundwater. The question we seek to answer is:

How long will it take for ^{90}Sr to contaminate the well, located $100\,\text{m}$ downgradient? (The allowable limit for radionuclides in drinking water is 1 pCi/L).

Since we are dealing with a decaying source, we can use Eq. [5.13] with two modifications. First, ^{90}Sr is removed by radioactive decay rather than by biodegradation, so we need to substitute k_s for λ. Second, we will ignore retardation because

[90]Sr does not sorb significantly, and we will replace the contaminant velocity (v_c) with the groundwater velocity (v).

$$C(x, y, z, t) = \left(\frac{C_0 e^{-k_s t}}{8}\right) \exp\left[\frac{x}{2\alpha_x}\left(1 - \sqrt{1 + \frac{4k_s \alpha_x}{v}}\right)\right] \text{erfc}\left(\frac{x - vt\sqrt{1 + \frac{4k_s \alpha_x}{v}}}{2\sqrt{\alpha_x vt}}\right)$$

$$\cdot \left[\text{erf}\left(\frac{y + Y/2}{4\sqrt{\alpha_y x}}\right) - \text{erf}\left(\frac{y - Y/2}{4\sqrt{\alpha_y x}}\right)\right]\left[\text{erf}\left(\frac{z + Z}{4\sqrt{\alpha_z x}}\right) - \text{erf}\left(\frac{z - Z}{4\sqrt{\alpha_z x}}\right)\right]$$

For the central (x) axis, this equation can be simplified to a form similar to that used in Scenario 2.

$$C(x, 0, 0, t) = \left(\frac{C_0 e^{-k_s t}}{2}\right) \exp\left[\frac{x}{2\alpha_x}\left(1 - \sqrt{1 + \frac{4k_s \alpha_x}{v}}\right)\right]$$

$$\cdot \text{erfc}\left(\frac{x - vt\sqrt{1 + \frac{4k_s \alpha_x}{v}}}{2\sqrt{\alpha_x vt}}\right)\left[\text{erf}\left(\frac{Y}{4\sqrt{\alpha_y x}}\right)\right]\left[\text{erf}\left(\frac{Z}{2\sqrt{\alpha_z x}}\right)\right]$$

For this scenario, expressing concentrations in pCi/L, we obtain

$$C(100, 0, 0, t) = \left(\frac{5 \times 10^8 e^{-0.000068 t}}{2}\right) \exp\left[\frac{100}{2(10)}\left(1 - \sqrt{1 + \frac{4(0.000068)10}{0.017}}\right)\right]$$

$$\cdot \text{erfc}\left(\frac{100 - 0.017t\sqrt{1 + \frac{4(0.000068)10}{0.017}}}{2\sqrt{10(0.017)t}}\right)\left[\text{erf}\left(\frac{5}{4\sqrt{3(100)}}\right)\right]$$

$$\cdot \left[\text{erf}\left(\frac{3}{2\sqrt{0.5(100)}}\right)\right]$$

This can be simplified as follows:

$$C(100, 0, 0, t) = 3.96 \times 10^6 e^{-0.000068 t} \, \text{erfc}\left(\frac{100 - 0.018t}{0.816\sqrt{t}}\right)$$

This equation can be solved by iteration using a spreadsheet, varying t until the desired value for $C = 1$ pCi/L is obtained. This occurs when $t = 822$ days. Interestingly, the travel time for 100 m, subject only to advection and adsorption (as predicted by Eq. [4.64]) would be $100/0.017 = 6000$ days, which is more than seven times longer than calculated above. This illustrates the importance to consider longitudinal dispersion, which causes low-concentration contours to migrate faster than the center of mass, resulting in earlier arrival times at the receptor's location than predicted by Eq. [4.64].

Scenario 4: Gasoline Spill as an Instantaneous Point Source

Large, sudden spills that could be considered as instantaneous point sources include delivery truck accidents and overfilling underground storage tanks when gas stations are serviced. In these cases, the amount of contamination released is approximately known. Often, emergency response teams rapidly remove much of the contamination, excavating the contaminated soil or extracting the free phase that floats on the water table. However, significant residual contamination could remain in the aqueous phase, which is prone to migration as illustrated by Figure 5.6.

For this example, we will consider a gasoline spill equivalent to 1000 L of benzene (based on the corresponding content of benzene in gasoline) with an initial concentration of $20,000\,\mu g/L$ and the parameters described in Tables 5.3 and 5.4. Equations [5.16] and [5.17] can be used to answer the following question:

How long will it take for benzene to disappear from this site (i.e., to decrease below $5\mu g/L$ as a result of natural attenuation)?

In this case, it is convenient to use Eq. [5.17] with $C_{max} = 5\,\mu g/L$. It should be kept in mind that this equation is based on simplifying assumptions (e.g., homogeneous aquifers with constant, uniform velocity), and that the maximum concentration occurs in the center of the plume when $y = z = 0$, and $x = vt$.

$$C_{max} = \frac{C_0 V_0 e^{-\lambda t}}{8(\pi t)^{1.5}\sqrt{D_x D_y D_z}}$$

$$5 = \frac{20,000(1000)e^{-0.0007t}}{8(\pi t)^{1.5}\sqrt{(0.13)(0.026)(0.00625)}}$$

Iterating, we obtain $t = 5520$ days, which is more than 15 years.

5.4. SENSITIVITY ANALYSIS

Analytical models are increasingly being used to evaluate the fate and transport of groundwater pollutants, to assess the degree of natural attenuation of contaminant plumes, and to predict the potential region of influence of a chemical spill. However, simulation results from different modeling efforts may differ considerably from each other and from the actual plume data, depending on the value of the model parameters used. To ensure simulation accuracy, it is important to ascertain that site-specific conditions are consistent with the model's simplifying assumptions, and to use representative parameter values. The determination of site-specific model parameters is discussed in Chapter 7. This section will focus on determining which fate and transport parameters are most influential regarding model predictions. The identification of the most "important" parameters is essential to prioritize data collection and estimation efforts. Influential parameters will be identified by conducting a sensitivity analysis for a simple model: the steady-state, centerline, advection, dispersion, and biodegradation equation with constant rectangular source (Eq. [5.10]).

$$C(x,0,0) = C_0 \exp\left[\frac{x}{2\alpha_x}\left(1 - \sqrt{1 + \frac{4\lambda\alpha_x}{v_c}}\right)\right]\mathrm{erf}\left(\frac{Y}{4\sqrt{\alpha_y x}}\right)\mathrm{erf}\left(\frac{Z}{2\sqrt{\alpha_z x}}\right)$$

This model will be used to predict the maximum benzene plume length, defined as the steady-state distance from the source to the 5 µg/L contour (i.e., the drinking water standard). Simulations will be run using typical ranges of model parameters from the literature. The sensitivity analysis will consist of increasing each parameter value by 50% (one at a time) and comparing the resulting changes in plume length. This exercise will also be performed by varying the model parameters within typical (wider) ranges reported in the literature (Alvarez et al. 1998).

5.4.1. Effect of 50% Increase in Model Parameter Values on Plume Length

Using the baseline values depicted in Table 5.5, the simulated baseline plume length was 195 m (640 ft). This is longer than the reported average plume length of 31 m (101 ft) for a study of 271 plumes in California (Rice et al. 1995), although this study used the 10 µg/L contour line to define plume length. Our baseline plume length is within the range of 4.4–920 m (8–3020 ft) reported for a compilation of 604 studies by Newell and Connor (1998).

Table 5.5 shows that the biodegradation rate coefficient (λ) and the groundwater velocity (v) are the most influential parameters in determining the simulated plume length. The plume length decreases by about 24% (from the baseline length) as the value of the biodegradation coefficient increases by 50%. This illustrates that faster biodegradation results in shorter plumes. The simulated plume length increases as the groundwater velocity increases, which is also logical since a faster groundwater velocity will cause the plume to move farther from the source. The hydrogeologic and source characteristics also affect the simulated plume dimensions. For example, the organic fraction (f_{oc}), the porosity (ϕ), and the bulk density (ρ_b) affect the

Table 5.5. Sensitivity Analysis for Eq. [4.31]—Change in Plume Length After Increasing Variable by 50%

Variable	Baseline Value	Change in Plume Length (%)
λ (per day)	0.0005	−24
C_o (ppb)	25000	+7
Z (m)	6	+7
Y (m)	10	+7
α_x (m)	10	−1
f_{oc}	0.01	−17
ϕ	0.3	+17
ρ_b(g/cm^3)	1.86	−17
v (m/day)	0.044	+33

retardation factor and, thus, contaminant migration velocity and plume dimensions. Table 5.5 also shows that increasing the source initial concentration, its width, or its depth by 50% increases the simulated plume length, but only by about 7%. Thus, whereas larger sources (in terms of concentration or source area) are conducive to longer plumes, these effects are less influential on plume length than variations in biodegradation rate coefficient and groundwater velocity. Source dimensions and source mass are probably more influential determinants of plume duration before natural attenuation decreases pollutant concentrations to acceptable levels.

5.4.2. Change in Plume Length Based on Varying Parameters Within Wider (Typical) Ranges

When model parameters are varied over a wider (but realistic) range of values from the literature (Table 5.6), groundwater velocity and the biodegradation rate coefficient continue to exert the greatest influence on the simulated plume length. The simulated plume length increased by more than 1500% upon increasing the groundwater velocity within the range under consideration $(0.0022-1.12\,\text{m/day})$ with all other parameters held constant at the baseline values in Table 5.5. The biodegradation rate coefficient (λ), on the other hand, was increased from 0.00025 to $0.1\,\text{day}^{-1}$, which resulted in a decrease in plume length of 97%. Even though percent change in plume length due to groundwater velocity changes was much higher than that from λ variation, the ratio of the maximum to minimum plume length due to the variation in λ (29) was higher than that from the variation in groundwater

Table 5.6. Simulation Results from Varying Parameters Within Typical Ranges

Variables	Range of Value	Range of Plume Length[a] (m)	Percent Change in Plume Length[b]	Max/Min[c]
λ (per day)[d]	0.00025–0.1	319 to 11	−97	29.0
C_0 (ppb)[e]	12500–100000	172 to 244	+42	1.4
Z (m)	1.5–4.5	172 to 209	+22	1.2
Y (m)	5–15	172 to 209	+22	1.2
α_x (m)[f]	1–1000	232 to 120	−48	1.9
α_y (m)[f]	1–100	195 to 120	−38	1.6
α_z (m)[f]	1–25	100 to 57	−43	1.8
f_{oc}	0.005–0.015	253 to 162	−36	1.6
ϕ	0.15–0.45	141 to 229	+62	1.6
$\rho_b(\text{g/cm}^3)$	0.93–2.79	253 to 162	−36	1.6
v (m/day)[g]	0.0022–1.12	123 to 2040	+1580	16.6

[a]The plume length is determined at the contaminant concentration of 5 ppb (MCL for benzene).
[b]percent change in plume length is calculated from the lower to higher parameter value.
[c]The ratio of the maximum over the minimum plume length.
[d]From Rifai et al. (1995).
[e]From Cline et al. (1991).
[f]From Gelhar et al. (1992).
[g]From Rifai (personal communication, 1998).

velocity (16.6) (Table 5.6). This suggests that λ is likely to be the most influential parameter that determines the plume length. Therefore, special care should be taken in determining this parameter accurately, as discussed in Chapter 7.

The results summarized in Table 5.6 also show that other parameters can exert a significant influence on the outcome of the model simulation. For example, the longitudinal dispersivity and the porosity of the subsurface soil can cause the overall plume length to vary by more than 50%. This illustrates that failure to use representative, site-specific hydrogeologic parameters could significantly hinder the accuracy of the simulation.

5.5. RELIABILITY AND LIMITATIONS OF ANALYTICAL MODELS

We have shown that analytical models can be very useful tools to evaluate contaminant behavior in groundwater and to characterize potential exposure pathways in risk assessment efforts. Analytical models can also be used to assess the age and stability of contaminant releases. One common approach is to use the available data on the spatial distribution of the target contaminant(s) with appropriate hydrogeologic parameters to calibrate the model. Using standard data fitting techniques, such as nonlinear regression, groundwater professionals can estimate site-specific migration and decay rates and assess the age of a release. However, analytical models generally have limited predictive ability, usually with an order-of-magnitude accuracy at best for complex sites. This is so because groundwater flow and microbial behavior do not always follow the model's simplifying assumptions discussed in Section 5.2. Therefore, hydrogeologists and environmental engineers should ascertain that the groundwater flow and contaminant behavior are consistent with the principles utilized in their models. Unfortunately, this cardinal rule is easy to forget. Since model simulations are only as good as their input and assumptions, there is a potential for inadvertent misuse of models in liability or risk assessment.

Deviations from model assumptions are commonly due to heterogeneity of the aquifer material, which causes spatial variability in model parameters such as hydraulic conductivity, porosity, sorption capacity, and biodegradation kinetics (due to changes in availability of nutrients or electron acceptors needed for microbial activity). Other deviations from ideal behavior include seasonal changes in hydraulic gradient and associated variability in groundwater flow velocity and direction, artificial gradients caused by pumping wells, and variable source flux with undefined shape. Such complexities cannot be incorporated in analytical models, which feature their relative simplicity as both their main advantage and disadvantage. Therefore, analytical model results should be interpreted with caution. If site heterogeneities are large, but well known, it will probably be more appropriate to use a numerical model that discretizes the site into small elements or blocks and assigns appropriate parameters to different domains. When dealing with uncertainty and variability of model parameters, it is good practice to generate simulations using a reasonable range of all parameters that are not well defined for the site of interest. This type of analysis allows the modeler to demonstrate worst case and best case scenarios and to provide valuable input for risk assessment and control decisions.

The use of analytical fate-and-transport models is likely to increase with the adoption of risk-based corrective action (RBCA) in the environmental protection rules of numerous countries. Given the anticipated increase in modeling activity to evaluate the potential of natural attenuation and determine the need for remedial action, there is a need to ensure that current modeling practices are adequate for protecting public welfare. The purpose of this section is to caution the reader against the inappropriate use of analytical fate-and-transport models. A case study will be used to illustrate common faults and misjudgments associated with modeling petroleum product releases to groundwater. This case study was abstracted from settled litigation, and it emphasizes the fate of benzene, which is often the gasoline constituent of greatest concern because of its potential to cause leukemia.

5.5.1. Case Study—General Background

This case involves a groundwater contamination by a gasoline release from a leaking underground storage tank (LUST). The facility under consideration had two different operators. Company X operated this facility until 1997. The current operator, Company Y, replaced X's underground storage tank system in 1997. The new system passed all annual tightness tests, and there are no reports of releases or overfills during Company Y's operation. In 2000, gasoline contamination was found by Company Y in an underlying silty bed; a site assessment ensued. Benzene was detected above its maximum contaminant level (MCL) of $5\,\mu g/L$ in nearby monitoring wells. Nevertheless, no hydrocarbons were detected in the fill sand that surrounds the new UST system. This suggested that contamination occurred prior to the installation of the new UST system in 1997, while Company X operated the facility. Therefore, a cost recovery claim was filed against Company X.

In litigation, Company X argued that it was not liable for this contamination and hired a consultant to support its defense against the cost recovery claim. The consultant used modeling analyses to determine the likelihood that contamination occurred while Company X operated the facility, prior to 1997. Based on a fate-and-transport model, they concluded that "... *assuming a pulse source, a single release of benzene dissipates to below the MCL in less than a year. Therefore, any contamination from before 1997 would either have migrated off-site or degraded by the time samples were collected in the year 2000.*" As discussed below, this conclusion is questionable because of intrinsic limitations on the applicability of their analytical model and their choice of benzene biodegradation rate coefficient in their fate-and-transport simulations.

5.5.2. Intrinsic Limitations of the Analytical Model

To simulate the fate of benzene, the consultant used an analytical solution to the three-dimensional advection–dispersion equation that considers an instantaneous pulse source, local equilibrium with linear partitioning by sorption, and first-order biodegradation kinetics (Eq. [5.16]). While this model is relatively easy to use and requires minimum input of site-specific parameters, there are several intrinsic assumptions that should be fulfilled to ascertain its appropriateness and ensure the validity of its output. Nevertheless, data limitations (and perhaps also budgetary constraints) precluded the consultant from determining whether the contamination scenario and the groundwater flow acted in a manner consistent with the principles utilized in the model. Specific limitations are discussed below:

1. *Contamination source.* The consultant concluded that the observed benzene contamination must have occurred after 1997 because any prior contamination should have migrated off-site by 2000. The simulated contamination scenario assumed a single instantaneous pulse source. LUST contamination, however, often resembles a constant source (e.g., Eq. [5.7]) because tanks may leak for an extended period of time, and desorption of hydrocarbons from contaminated soil constitutes another source of sustained groundwater contamination. Therefore, the pulse source assumption underestimates the time required for contamination to dissipate by physical processes. A constant source over a stipulated period may be a more representative way of modeling LUST contamination. It should also be pointed out that the pulse source analytical solution requires knowledge of the volume of contaminant released. This is a common uncertainty associated with LUST contamination which introduces error in the time required for a plume to dissipate.

2. *Steady flow.* The model used by the consultant is applicable only to steady flow fields (i.e., at any point in space, the flow does not vary in direction or velocity with respect to time). Often, however, the direction and velocity of groundwater flow changes (at least seasonally). In the case under consideration, there was not enough data over time and space to evaluate the validity of the steady flow assumption. Unaccounted fluctuations in groundwater flow direction and velocity could result in significant error.

3. *Uniform flow.* The simulations generated for this case assumed a uniform flow field (i.e., straight and parallel velocity vectors), which intrinsically assumes homogeneity of the porous medium. Nevertheless, the site under consideration is heterogeneous. Stratigraphic heterogeneities can result in unaccounted for preferential flow pathways that transport contaminants faster than predicted, and "dead spots" that hinder contaminant advection and dissipation.

4. *Steady-state plume.* Steady state was not an assumption made in this case. Nevertheless, this assumption deserves attention because it is commonly inherent to some commonly used fate-and-transport analytical models. By definition, a steady-state plume is one where, at any point, contaminant concentrations do not change with time. This occurs only under two ideal conditions: (a) the plume does not migrate and does not degrade, at least within the time frame of the investigation (i.e., the "trivial" solution), or (b) there is a constant source, and the migration rate equals the decay rate so that there is no net expansion or recession. Often, steady state is assumed for simplicity without establishing the validity of this assumption.

Validation of the steady-state assumption can be costly because of extensive data requirements over sufficient time and space. Nevertheless, disregarding the validity of this assumption constitutes inappropriate modeling practice that could lead to significant error.

5.5.3. Modeling of Benzene Biodegradation

Biodegradation is widely recognized as an important mechanism by which benzene is eliminated from LUST plumes. Whereas very small gasoline releases could dissipate solely by physical processes such as dilution, one must consider biodegradation to contemplate the possibility that benzene should completely disappear from typical LUST plumes. Mass balance studies have demonstrated that "passive" biodegradation (i.e., no biostimulation by oxygen or nutrients addition) is a significant attenuation mechanism in benzene transport (e.g., Zoetman et al. 1981; Chiang et al. 1989; Klecka et al. 1990; Chen et al. 1992). Although anaerobic bioremediation is possible, in such cases, biodegradation rates are typically controlled by oxygen diffusion from the atmosphere (Borden and Bedient 1986).

Biodegradation is the only "true" sink considered by most analytical models. Adsorption, advection, and dispersion do not remove benzene from the aquifer, and volatilization is often (conservatively) ignored in analytical models because it is not a major benzene removal mechanism from dissolved LUST plumes (Chen et al. 1992). Therefore, choosing an appropriate biodegradation model and a reasonable decay rate coefficient is critical to the defensibility of the modeling results.

As we have discussed in Chapter 3, biodegradation rates are often modeled using Monod kinetics (Eq. [3.26]) because Monod's equation has a mechanistic (enzymological) basis and considers the active microbial concentration. Therefore, most sophisticated (numerical) fate-and-transport models use Monod's equation. This equation, however, is hyperbolic and does not yield an explicit analytical solution for the contaminant concentration as a function of time and space. Therefore, simpler (analytical) fate-and-transport models use empirical kinetic expressions, such as first-order kinetics (i.e., the rate is proportional to the contaminant concentration) and zero-order kinetics (i.e., the rate is constant and thus independent of the contaminant concentration). Although benzene biodegradation rates in aquifers have been reported to follow zero-order kinetics (Barker et al. 1987), first-order rates are more common. This is probably due to the fact that benzene is often present at trace concentrations, and Monod's equation reduces to a first-order expression whenever the target contaminant is present at levels much lower than the half-saturation Monod coefficient, K_s (Alvarez et al. 1991). This condition was met in the case under consideration, and the choice of first-order biodegradation kinetics was appropriate. The selected value of the first-order biodegradation coefficient, however, was inappropriate.

Following common practice, the consultant estimated a first-order biodegradation rate coefficient (λ) of 0.0462 day^{-1} (i.e., a half-life of 15 days) using site-specific data and "validated" this coefficient by comparison to values reported in the literature. Nevertheless, the literature values that they considered (Howard et al. 1991) reflect laboratory measurements. Benzene degradation in the field is usually slower because it is often limited by lower temperatures and mass transfer limitations for nutrients and electron acceptors (e.g., the rate at which molecular oxygen diffuses into the plume). In situ rate coefficients are typically one order of magnitude lower than their selected value. A paper by Rifai et al. (1995) presents a summary of first-order decay coefficients for benzene that have been measured at twelve sites under "passive" conditions. These literature values have an average of 0.0046 day^{-1} (i.e., a half-life of 149 days) with a common range of 0–0.0085 day^{-1} and a geometric (log) mean of only 0.0018 day^{-1} (i.e., a half-life of 375 days). To put the magnitude of these coefficients in perspective, let us consider the time required for the benzene concentration to drop from 10 mg/L to the detection limit (1 µg/L) in a batch system. In this example, the time required would be 0.55 year using the consultant's estimated coefficient. This is consistent with the consultant's claim that contamination prior to 1997 would have been degraded by 2000. Nevertheless, a time of 14 years would be required if one uses the geometric mean of the literature values reported by Rifai et al. (1995). Therefore, common literature values did not support the consultant's claim of fast (intrinsic) biodegradation. The discrepancy between the estimated and the commonly reported first-order decay coefficients cast doubt on the accuracy of the consultant's fate-and-transport simulations.

The dimensions of the contamination source were not relevant in this case because the simulated scenario assumed a pulse source. When a constant source is assumed, however, the area of the source can be an important factor in estimating the biodegradation coefficient. Care must be taken not to exaggerate the area of the source; this would overestimate the contaminant release rate (i.e., flux times area), and thus, it would overestimate the decay coefficient necessary to simulate the observed contaminant distribution. Similarly, if steady state is assumed, an overestimation of the source area (and thus the release rate) results in an overestimation of the decay coefficient necessary to equalize the migration and degradation rates. An overestimation of the decay coefficient is conducive to underestimating the potential health risk associated with the release because model simulations would underestimate the contaminant concentration at a distant receptor.

The consultant did not evaluate whether their model was adequately calibrated with the selected decay coefficient (i.e., goodness-of-fit). Furthermore, a sensitivity analysis was not conducted to assess the effect of varying the degradation coefficient within its statistical error. Consequently, it could not be ruled out that the selected biodegradation coefficient overestimated the biodegradation rate and underestimated the age of the release. Indeed, when a model relies on multiple parameters, the best-fitting parameter combination is not necessarily precise or reasonable. To prevent this problem, groundwater professionals should consider adopting compatibility constraints that limit the input of decay coefficients. These limits of acceptability could be based on comparison with values that have been measured at similar sites, and the relative standard error of the estimated decay coefficient that is necessary to meet case-specific precision requirements. Sensitivity analyses encompassing a reasonable range of reported values should also be conducted as routine practice.

5.5.4. Conclusion

The modeling work presented by the consultant could not rule out the possibility that the observed contamination occurred prior to 1997, while Company X operated the facility. This conclusion is based on two facts: (1) there was not sufficient data to ensure that groundwater contamination and flow acted in a manner consistent with the principles utilized in their mathematical model; and (2) the instantaneous pulse source assumption and a relatively high first-order decay coefficient are conducive to underestimating the age of the release. Therefore, the modeling defense by Company X was ineffectual.

In general, groundwater professionals should fully justify that modeling assumptions accurately reflect site conditions, that site-specific data used in modeling are valid, and that literature values abstracted for input are realistic and applicable. Without such detailed justification, risk from exposure to petroleum contamination will likely be erroneously assessed and responsible parties, operators, and consultants might find themselves in expensive litigation.

REFERENCES

Alvarez, P.J.J., P.J. Anid, and T.M. Vogel (1991). Kinetics of aerobic biodegradation of benzene and toluene in sandy aquifer material. *Biodegradation* **2**:43–51.

Alvarez, P.J.J., Y.-K. Zhang, and N. Lovanh (1998). Evaluation of the Iowa Tier-2 Model Based on the Benzene Plume Dimensions. Iowa Comprehensive Petroleum Underground Storage Tank Fund Board, Des Moines, IA, December.

ASTM (1998). *1998 Annual Book of ASTM Standards: Standard Guide for Remediation of Ground Water by Natural Attenuation at Petroleum Release Sites* (Designation: E 1943-98). American Society for Testing and Materials, West Conshohocken, PA, pp. 875–917.

Baetslé, L.H. (1969). Migration of radionuclides in porous media. In *Progress in Nuclear Energy Series XII, Health Physics*, A.M.F. Duhamel (Ed.). Pergamon Press, Elmsford, NY, pp. 707–730.

Barker, J.F., G.C. Patrick, and D. Major (1987). Natural attenuation of aromatic hydrocarbons in a shallow sand aquifer. *Groundwater Monitor. Rev.* **7**:64–71.

Bedient, P.B., H.S. Rifai, and C.J.Newell (1994). *Ground Water Contamination. Transport and Remediation.* PTR Prentice Hall, Upper Saddle River, NJ.

Borden, R.C. and P.B. Bedient (1986). Transport of dissolved hydrocarbons influenced by oxygen-limited biodegradation: 1. Theoretical development. *Water Resour. Res.* **22**:1973–1982.

Carslaw, H. S. and J. C. Jaeger (1959). *Conduction of Heat in Solids*, 2nd ed. Oxford Science Publications, New York.

Chen, Y.M., L.M. Abriola, P.J.J. Alvarez, P.J. Anid, and T.M. Vogel (1992). Biodegradation and transport of benzene and toluene in sandy aquifer material: model–experiment comparisons. *Water Resour. Res.* **28**:1833–1847.

Chiang, C.Y., J.P. Salanitro, E.Y. Chai, J.D. Colthart, and C.L. Klein (1989). Aerobic biodegradation of benzene, toluene, and xylene in a sandy aquifer—data analysis and computer modeling. *Ground Water* **27**:823–834.

Cline, O.V., J.J. Delfino, and P.S. Rao (1991). Partitioning of aromatic constituents into water from gasoline and other complex mixtures. *Environ. Sci. Technol.* **25**:914–920.

De Josselin and G. de Jong (1958). Longitudinal and transverse diffusion in granular deposits. *Trans. Am. Geophys. Union* **39**(1):67–74.

Domenico, P.A. (1987). An analytical model for multidimensional transport of a decaying contaminant species. *J. Hydrol.* **91**:49–58.

Domenico P.A. and F.W. Schwartz (1998). *Physical and Chemical Hydrogeology*, 2nd ed. John Wiley & Sons, Hoboken, NJ.

Gelhar, L.W. (1993). *Stochastic Subsurface Hydrology*. Prentice-Hall. Englewood Cliffs, N.J.

Gelhar, L.W., C. Welty, and K.R. Rehfeldt (1992). A critical review of data on field-scale dispersion in aquifers. *Water Resour. Res.* **28**(7):1955–1974.

Howard, P.H. (Ed.) (1989). *Handbook of Environmental Fate and Exposure Data for Organic Chemicals*. Lewis Publishers, Chelsea, MI.

Howard, P.H., R.S. Boethling, W.F. Jarvios, W.M. Meylan, and E.M. Michaelenko (1991). *Handbook of Environmental Degradation Rates*. Lewis Publishers, New York.

Hunt, B.W. (1978). Dispersive sources in uniform groundwater flow. *J. Hydraul. Div. ASCE* **99**:13–21.

Klecka, G.M., J.W. Davis, D.R. Gray, and S.S. Madsen (1990). Natural bioremediation of organic contaminants in groundwater: Cliffs-Dow Superfund site. *Ground Water* **28**:534–543.

Lallemande-Barres, A. and P. Peaudecerf (1978). Recherche des relations entre la valeur de la dispersivité macroscopique d'un milieu aquifère, ses autres caractéristiques et les conditions de mesure. *Bull. Bur. Rech. Geol. Min.*, Sect. 3:Ser. 2, p. 4.

Neuman, S.P. (1990). Universal scaling of hydraulic conductivities and dispersivities in geologic media. *Water Resour. Res.* **26**(8):1749–1758.

Newell, C.J. and J.A. Connor (1998). Characteristics of dissolved hydrocarbon plumes: results of four studies. In *Proceedings of the 1998 Petroleum Hydrocarbons and Organic Chemicals in Ground Water: Prevention, Detection, and Remediation*. Conference and Exposition, Houston, Texas, pp. 51–59.

Newell, C.J., R.K. McLeod, and J.R. Gonzales (1996). BIOSCREEN *Natural Attenuation Decision Support System, User's Manual Version 1.3*, EPA/600/R-96/087, United States Environmental Protection Agency, Office of Research and Development, Washington, DC.

Pickens, J.F. and G.E. Grisak (1981). Scale dependent dispersion in a stratified granular aquifer. *Water Resour. Res.* **17**(4):1191–1211.

Rice, D.W., R.D. Grose, J.C. Michaelsen, B.P. Dooher, D.H. MacQueen, S.J. Cullen, W.E. Kastenberg, L.G. Everett, and M.A Marion (1995). California leaking underground fuel tank (LUFT) historical case analysis. Environmental Protection Department, Nov. 16.

Rifai, H.S., R.C. Borden, J.T. Wilson, and C.H. Ward (1995). Intrinsic bioattenuation for subsurface restoration. In *Intrinsic Bioremediation*, R.E. Hinchee, F.J. Brockman, and C.M. Vogel (Eds.). Battelle Press, Columbus, OH, Vol. 3, No. 1, pp. 1–29.

Zhang, Y.-K. and S.P. Neuman (1990). A quasi-linear theory of non-Fickian and Fickian subsurface dispersion, 2. Application to anisotropic media and the Borden site. *Water Resour. Res.* **26**(5):903–913.

Zoetman, B.C.J., E. De Greef, and F.J.J. Brikmann (1981). Persistence of organic chemicals in groundwater, lessons from soil pollution incidents in The Netherlands. *Sci. Tot. Environ.* **21**:187–202.

6

NUMERICAL MODELING OF CONTAMINANT TRANSPORT, TRANSFORMATION, AND DEGRADATION PROCESSES

"Computer models do not make the job of decision making any easier. They make it harder. They enforce rigorous thinking and expose fallacies in mental models we have always been proud of. We think it is worth it. We think it pushes our mental models to be a bit closer to reflecting the world as it is."
—*Groping in the Dark: The First Decade of Global Modelling*, Donella Meadows

6.1. WHY USE NUMERICAL MODELS TO PREDICT THE FATE AND TRANSPORT OF CONTAMINANTS?

This chapter provides an introduction to the basic theory and application of contaminant transport modeling by numerical methods. After reading this chapter, the reader should be able to select an appropriate numerical model for the situation under consideration, run the code, and calibrate it. The use of numerical methods to solve groundwater contaminant transport problems has become a widely used technique because of the intensified interest in groundwater quality and the rapid development of computing technology, which has made numerical simulations available to a large number of hydrogeologists and civil engineers. It has become indispensable in answering questions regularly posed by stakeholders and regulators regarding groundwater quality at a certain point of compliance (POC) or how a contaminant plume may evolve over time. If properly used, numerical models can provide answers to the following questions (Zheng and Bennett 2002):

- What solute concentration will eventually be attained at a given location?
- When will the solute concentration at a certain point decrease to a given level due to the placement of an active or passive remediation scheme?

Bioremediation and Natural Attenuation: Process Fundamentals and Mathematical Models
By Pedro J. J. Alvarez and Walter A. Illman   Copyright © 2006 John Wiley & Sons, Inc.

- Will a remedial design achieve a targeted reduction in concentration by a certain time (cleanup time)?

- How can one reconstruct the history of a contamination event (backward in time simulations) to estimate the concentration to which populations have been exposed, and the time periods of that exposure?

In order to answer these types of questions, qualitative answers will not be scientifically and legally defensible. A detailed site investigation combined with the development of analytical and/or numerical models will be required. As discussed previously, analytical models are useful in producing simple and quick answers to contaminant transport problems when an order-of-magnitude level of accuracy is required, but numerical techniques are more flexible, allowing for solutions to more complex situations such as irregular boundaries, spatial variability in physical and chemical properties, space–time variation in forcing functions (initial and boundary conditions and source/sink terms), and complex biogeochemical processes. Groundwater flow and contaminant transport modeling can help the decision maker by providing insight into the relative impacts of various alternatives on groundwater and solute transport, in particular, when intrinsic bioremediation is chosen as a remedial alternative where rigorous monitoring of the site is required.

Whether an analytical or numerical model is used as a tool to investigate the problem also depends on the level of risk-based corrective action (RBCA) implemented at a particular site. RBCA is a decision making process for the assessment of and response to subsurface contamination, based on the protection of human health and environmental resources (see Chapter 1 for additional details). The RBCA program was devised for state regulatory programs for petroleum release sites (ASTM 1995) and was extended for nonpetroleum release sites in 1998 (Wiedemeier et al. 1999). In a Tier 2 RBCA investigation, the use of analytical models is considered to be adequate, while for a Tier 3 RBCA investigation, appropriate numerical models are used to evaluate the risks from a number of pathways to receptors. Tier 3 assessments can be implemented when modeling of a site with Tier 2 techniques, with a simple analytical solution, shows the existence of contaminant pathways to at-risk receptors. In some cases, a Tier 3 effort might involve merely determining a few additional facts, such as construction detail of an apparently threatened drinking water well to show the well and aquifer are not actually threatened. But a numerical model is often chosen when environmental risk at a contaminant site is characterized as high after a Tier 2 assessment. One of the options available for proceeding with obligatory corrective action is to do more realistic modeling of the site at the Tier 3 level to determine whether the oversimplifications inherent in the Tier 2 model are causing a false high-risk condition. In many cases, Tier 3 analysis might need a greater effort, such as gathering additional site-specific data that are required to run an analytical or a numerical computer code of a fate-and-transport model of chemicals in soils and aquifers. The hydrogeological conditions in those cases may be complex, for example, involving a pumping well or a multilayered aquifer. A computer software package suited for the situation to be modeled has to be chosen in order to simulate the fate and transport of the contaminants from contaminant source zones at those sites. Modeling assessments at the Tier 3 level are advantageous for site owners and operators in situations where expensive corrective action would be required based on Tier 2 results, and where the Tier 2 model with default values for fate and transport parameters does not reasonably reflect the hydrogeological conditions.

We have seen in previous chapters that the subsurface migration of contaminants is influenced by a variety of transport, transformation, and attenuation processes, with many of them that are coupled with one another making the solution to the boundary value problem more complex and often unachievable by analytical methods. Therefore, the need for a numerical model also arises from its ability to capture many of the processes that we have seen in Chapters 2, 3, and 4. There are many numerical models available with widely varying capabilities (see Appendix C). Many of them are available in the public domain, as they were developed with federal research grants, but others are proprietary. Some of these proprietary codes were developed by making the free codes easier to use. That is, extensive pre- and postprocessing capabilities have been added so that construction of the model has become considerably easier and the postprocessing of results from a numerical simulation can be done within the modeling package, which allows one to quickly visualize results. In other words, without an extensive knowledge of how the code was written or how the input and output files were prepared, one can readily set up a numerical simulation, run the code, and visualize the results.

Making the codes easy to use is a good thing for the profession, as more users can readily simulate groundwater flow and transport; but it also shows that many of the codes can be operated without the detailed understanding of theory behind and applied to field problems, which gives rise to potential misuse. Codes that are misapplied can give garbage-in garbage-out (GIGO) answers. Therefore, it is imperative that a modeler understands the theory behind the code, its capabilities (i.e., what the code is solving), and the limits to the algorithms used to solve the problem.

It is additionally important for the modeler to select the numerical model that best fits the situation that s/he is interested in modeling. It is important to keep in mind the limitations of the code and recognize that no code is suitable for all situations. Therefore, the modeler should know some basics about the techniques on which the solutions are computed (i.e., finite difference, finite element, and method of characteristics). Some solution techniques are suitable for certain classes of problems, while others do better in other situations. Every situation encountered is different, necessitating a very good working knowledge of computer models. It is better to understand how the code works before relying on it to solve problems that can affect the outcome of hydrogeologic investigations. One analogy to this is that one can easily learn to ride a car without understanding how the car works. However, what if you were asked to cross the Sahara Desert in Africa by yourself without a mechanic? Would you be inclined to go across the desert without an intimate knowledge of how the car runs and methods of troubleshooting if it breaks down? The same applies to running numerical models to solve contaminant transport problems.

In many cases we will not write our own numerical model to solve the problems at hand. Instead, we will use an off-the-shelf code. However, there are situations in which the modeler may have to modify an existing code or write a new code to solve the problem at hand.

The numerical approach that we will discuss here refers to the use of a computer to obtain an approximate solution to the boundary value problem that we are interested in solving. Sometimes, this takes on the form of solving the advection–dispersion equation (ADE) by assuming a constant groundwater velocity or a coupled groundwater flow-and-transport model. When the solute concentration yielded by solution of the transport equation causes negligible variation in water density or parameters such as hydraulic conductivity and porosity, the flow equation and solute transport equation can be solved independently. In this case, the flow equation (Eq. [4.26]) is solved prior to the transport simulation first yielding a value of hydraulic head at every point in the simulation domain, and this can be used to calculate the velocity distribution of the domain for all time periods and used as input to the transport code. Such a decoupled approach is implemented in common numerical models such as MOC (Konikow and Bredehoeft 1978; Konikow et al. 1996), MT3D (Zheng 1990; Zheng and Wang 1999), and RT3D (Clement 1997). If, on the other hand, solute transport affects the water density and/or parameters of the model significantly, the flow and transport equations must be solved simultaneously for each time step. Descriptions of these types of models can be found in Zheng and Bennett (2002).

Solute concentrations in many field problems are low enough so that fluid density is assumed to be constant. There are situations in which density dependence may play a major role in solute transport. For example, the injection of brine into an aquifer may cause the salt plume to sink, so that a density-dependent flow-and-transport model should be used in this particular case. If one were to model the infiltration of dense nonaqueous phase liquids (DNAPLs) such as trichloroethylene (TCE) or perchloroethylene (PCE) in the subsurface, one would have to consider the density dependence of the DNAPL in relation to water and model it as a separate phase, making it a two-phase flow problem. If one were to consider the infiltration of a DNAPL in the unsaturated zone, it should be treated as a three-phase problem because of the presence of air in addition to the water and the DNAPL. In some situations, chemical reactions may cause the pore structure to enlarge through dissolution or become smaller through precipitation of minerals, causing the permeability and porosity to change with time. In such situations (such as the modeling of karst terrain or biological growth in porous media), a flow and transport code that updates the parameter values as the simulation marches forward in time may have to be used to account for the changes in the physical and chemical properties of the aquifer itself. Yet in another situation, the population dynamics of microbes in a particular aquifer may vary due to the consumption of contaminants. In such a situation, the population dynamics of the microbial community may have to be built into the code to correctly represent the system under study.

We begin by describing the common numerical approaches implemented in many of the numerical models. The discussion presented here is considered to be elementary but should provide sufficient background for the reader interested in the inner workings of a numerical model. More detailed descriptions of the approaches presented here can be found in Smith (1985), Huyakorn and Pinder (1983), and Zheng and Bennett (2002). It is strongly recommended for the reader who will be modeling to become intimately familiar with further details of the algorithms described in the references.

After a survey on numerical methods, we will discuss the process of setting up and running a numerical model in a step by step fashion. The section begins with a discussion of collection and assessment of site data, in particular, what data are needed to build a numerical model to assess monitored natural attenuation. We then discuss the importance of formulating a conceptual model for the features, processes, and events for the particular field situation under study. We also discuss the procedures involved in calibrating and validating a model before it can be used for predictive purposes. We then present several examples of nonreactive and reactive transport simulations to describe how numerical flow-and-transport models are built and used. For this purpose, we use the codes MT3D and RT3D to illustrate the solute transport and several degradation processes in a confined aquifer. We conclude the chapter by discussing three case studies that describe the numerical modeling process for the transport and attenuation of BTEX compounds and chlorinated solvents.

6.2. THE NUMERICAL MODELING APPROACH: FUNDAMENTALS

Recall from Chapter 5 that a boundary value problem for a region of interest consists of a governing partial differential equation and boundary conditions. For a transient problem, an initial condition is needed additionally. We have seen that an analytical solution can be obtained for certain simple cases, but in many situations, numerical approaches are necessary to accommodate the complexities of flow and transport situations encountered in the field. Numerical methods are powerful techniques that allow us to obtain solutions to such boundary value problems for both fluid flow and solute transport. Nevertheless, a numerical model is just another model that attempts to approximate the real situation. Whether a model represents the real situation adequately is a topic that we will cover in the section on model validation (see Section 6.3).

Broadly speaking, the two main numerical approaches used for solving groundwater flow and transport equations are the *finite difference* and *finite element methods*. Both approaches replace the governing partial differential equation for groundwater flow and transport by a system of algebraic equations. We will later see how this is done with a simple, steady-state flow problem in two dimensions. But before doing so, we will discuss the basic mechanics of constructing a numerical model.

The approaches discussed here require that the modeled region be subdivided or discretized into finite-sized blocks in a mesh or grid network. These blocks can

(a)

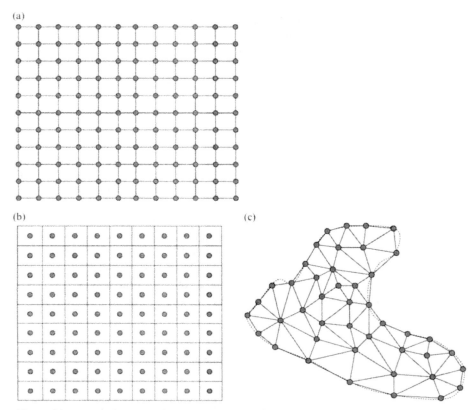

(b)   (c)

Figure 6.1. (a) A mesh centered finite difference grid, (b) a block centered finite difference grid, and (c) a finite element grid consisting of triangular finite elements. Solid gray circles denote nodes and the dashed curve delineates the irregular boundary of the modeled region.

come in a variety of shapes but a typical finite difference approach utilizes a regular discretization, in which an aquifer is subdivided into a series of rectangular blocks (Figure 6.1a,b). In the mesh centered approach, the nodes coincide with mesh. On the other hand, in the block centered approach, the nodes are placed in the center of each grid block. The popular flow code MODFLOW (Harbaugh and McDonald 1996) is based on this latter approach. In a two-dimensional model, each model cell is assumed to have a thickness b. Thus, each of the grid blocks represents a volume of the aquifer, $b\Delta x\Delta y$. In a three-dimensional model consisting of aquifers and confining beds, individual units are subdivided vertically into cells of a specified thickness. The size of the grid block is kept small relative to the overall extent of the aquifer. The spacing between rows and columns varies, but for simplicity the example grids (Figure 6.1a,b) assume a constant spacing (i.e., $\Delta x = \Delta y$). Associated with the grid blocks are nodes that represent the points where the unknown hydraulic heads are calculated.

The finite difference grid consists of grid blocks that are square or rectangular; thus, they may not conform to aquifer boundaries. In contrast, the discretization of the region is considerably more flexible with the finite element approach (Figure 6.1c). In Figure 6.1c, we see that the original aquifer boundary is delineated with a dashed curve. The discretization of the aquifer with triangular finite elements appears to capture the boundary quite accurately.

Later, we will discuss the advantages and disadvantages of finite difference and finite element approaches. This is important because, in some situations, one form of discretization and numerical approach is preferred over the other. In either case, the hydrogeological and chemical properties of the blocks/nodes are constant but in some situations those can be made to vary internally. An algebraic equation is written for each node and a global solution to the boundary value problem is obtained by assembling the system of equations and obtaining a solution to it through the use of techniques in linear algebra.

Before a solution to a contaminant transport model can be obtained, the solution to the flow model must be computed to obtain the values of hydraulic head for the entire solution domain. One can then use Darcy's Law to calculate the velocity vectors that determine magnitude and direction of groundwater flow, which is the primary driving mechanism for contaminant transport. Other natural attenuation mechanisms such as dispersion and biogeochemical transformation mechanisms can then be added on later. However, in order to understand the basics of numerical modeling, we will start by obtaining the numerical approximation of a two-dimensional steady flow problem in a homogeneous isotropic medium.

6.2.1. Finite Difference Approach

We first illustrate the finite difference approach using a steady-state, two-dimensional groundwater flow equation. The partial differential equation that describes steady flow in a two-dimensional, homogeneous aquifer is the Laplace equation:

$$\frac{\partial^2 h}{\partial x^2} + \frac{\partial^2 h}{\partial y^2} = 0 \qquad [6.1]$$

The solution to the above equation by finite differences involves the replacement of the second-order partial derivative with an approximation. This allows us to obtain the finite difference equation for an interior node in the nodal grid used to discretize the flow region. Recall from calculus that the definition of the partial derivative of hydraulic head which is a function of two variables (x, y) with respect to the space coordinate x takes the form

$$\frac{\partial h}{\partial x} = \lim_{\Delta x \to 0} \frac{h(x + \Delta x, y) - h(x, y)}{\Delta x} \qquad [6.2]$$

Analytically, one can take the limit of the above expression, but it is impossible to do this on a computer. Instead, we approximate the limit by assigning Δx to be the distance between the nodes in the finite difference grid.

For any value of y, for example, y_0, one can expand $h(x, y_0)$ using a Taylor's expansion about the point (x_0, y_0) in the following manner:

$$h(x, y_0) = h(x_0, y_0) + (x - x_0)\left(\frac{\partial h}{\partial x}\right)(x_0, y_0) + \frac{(x - x_0)^2}{2!}\left(\frac{\partial^2 h}{\partial x^2}\right)(x_0, y_0) + \cdots$$

$$[6.3]$$

To obtain the *forward difference approximation*, let $x = x_0 + \Delta x$ and truncate the higher order terms:

$$\left(\frac{\partial h}{\partial x}\right)_{(x_0, y_0)} = \frac{h(x_0 + \Delta x, y_0) - h(x_0, y_0)}{\Delta x}$$

$$[6.4]$$

where the abandoned higher order terms represent the *truncation error* of the finite difference approximation.

A *backward difference approximation* may be obtained by substituting $x = x_0 - \Delta x$:

$$\left(\frac{\partial h}{\partial x}\right)_{(x_0, y_0)} = \frac{h(x_0, y_0) - h(x_0 - \Delta x, y_0)}{\Delta x}$$

$$[6.5]$$

The approximation for the second degree partial derivative of hydraulic head with respect to $x\,(\partial^2 h/\partial x^2)$ can readily be obtained by writing the finite difference equation for $\partial^2 h/\partial x^2$ in terms of $\partial h/\partial x$ as

$$\left(\frac{\partial^2 h}{\partial x^2}\right)_{(x_0, y_0)} = \frac{\frac{\partial h}{\partial x}(x_0 + \Delta x, y_0) - \frac{\partial h}{\partial x}(x_0, y_0)}{\Delta x}$$

$$[6.6]$$

Substitution of Eq. [6.5] into [6.6] yields

$$\left(\frac{\partial^2 h}{\partial x^2}\right)_{(x_0, y_0)} = \frac{h(x_0 + \Delta x, y_0) - 2h(x_0, y_0) + h(x_0 - \Delta x, y_0)}{(\Delta x)^2}$$

$$[6.7]$$

A similar expression can be obtained for $\partial^2 h/\partial y^2$ as

$$\left(\frac{\partial^2 h}{\partial y^2}\right)_{(x_0, y_0)} = \frac{h(x_0, y_0 + \Delta y) - 2h(x_0, y_0) + h(x_0, y_0 - \Delta y)}{(\Delta y)^2}$$

$$[6.8]$$

Substitution of Eqs. [6.7] and [6.8] into Eq. [6.1] yields the finite difference approximation of Laplace's equation:

$$\frac{h(x_0 + \Delta x, y_0) - 2h(x_0, y_0) + h(x_0 - \Delta x, y_0)}{(\Delta x)^2}$$
$$+ \frac{h(x_0, y_0 + \Delta y) - 2h(x_0, y_0) + h(x_0, y_0 - \Delta y)}{(\Delta y)^2} = 0$$

$$[6.9]$$

For a square grid, we have $\Delta x = \Delta y$ and so Eq. [6.9] can be simplified to

$$\frac{1}{(\Delta x)^2}[h(x_0 + \Delta x, y_0) + h(x_0 - \Delta x, y_0)$$

$$+ h(x_0, y_0 + \Delta y) + h(x_0, y_0 - \Delta y) - 4h(x_0, y_0)] = 0 \qquad [6.10]$$

Letting (x_0, y_0) be the nodal point (i, j), Eq. [6.10] can be rewritten to read

$$h_{ij} = \tfrac{1}{4}\big(h_{i+1,j} + h_{i-1,j} + h_{i,j+1} + h_{i,j-1}\big) \qquad [6.11]$$

which simply states that in a homogeneous, isotropic medium under steady flow conditions, the hydraulic head at any node is the average of the four surrounding values. Similar equations can be developed for the boundary and corner nodes. One can develop these equations for all the nodes in the grid. In fact, if a numerical model contains N nodes, there are N finite difference equations that need to be solved. Within the N finite difference equations, there are N unknown values of hydraulic head. Therefore, we have a set of N linear, algebraic equations for N unknowns, which could be solved by techniques described in any linear algebra textbook. The objective here is to obtain a solution to the linear system $\mathbf{Ax} = \mathbf{b}$ in the form $\mathbf{x} = \mathbf{A}^{-1}\mathbf{b}$, where $\mathbf{A}$ is a matrix and $\mathbf{x}$ and $\mathbf{b}$ are vectors. If there are only a few nodes, one would be able to obtain the solution to the problem via Cramer's method or Gaussian elimination. For a large system of equations, this has to be done with a computer and there are several direct and iterative methods of obtaining solutions to the system of linear equations. The direct method such as Gaussian elimination yields the exact solution to the linear system in a fixed finite number of operations provided that $\mathbf{A}$ is nonsingular and there is no large accumulation of round-off error. Iterative methods, on the other hand, start from a first approximation of the unknowns that is improved successively until a sufficiently accurate solution is obtained. Some examples of these iterative techniques are the successive overrelaxation and conjugate gradient methods. Details on these approaches can be found in various linear algebra textbooks.

A similar procedure is used to solve the advection–dispersion equation with finite differences.

6.2.2. Method of Characteristics

The method of characteristics (MOC) is somewhat different from the finite difference technique discussed in the previous section. One can think of this technique by visualizing a number of solute particles being transported through a given flow field and observing changes in chemical concentration as the plume moves downgradient. Consider the two-dimensional, transient advection–dispersion equation

$$D_x \frac{\partial^2 C}{\partial x^2} + D_y \frac{\partial^2 C}{\partial y^2} - v_x \frac{\partial C}{\partial x} - v_y \frac{\partial C}{\partial y} + F = \frac{\partial C}{\partial t} \qquad [6.12]$$

The above equation describes the transient transport of a solute in a two-dimensional steady velocity field with F being the source and sink term. The MOC relies on expressing the partial derivatives in Eq. [6.12] as total or material time derivatives. We briefly discuss the differences between those two derivatives as they are important when studying transport phenomena.

Suppose we stand on a bridge and observe the number of cars that are passing beneath it with time. This amounts to observing how the concentration of cars changes with time at a fixed position in space. Therefore, in words, $\partial C/\partial t$ is the partial derivative of concentration (of cars) with respect to time at a given position (i.e., fixed x, y, and z). Solute transport simulations done by solving Eq. [6.12] are said to be Eulerian, which has a fixed coordinate system. Unfortunately, direct solution of the Eulerian form of the advective transport equation is marred by a serious problem, called the *numerical dispersion* of concentrations (Zheng and Bennett 2002). Spreading of particles that arises from numerical dispersion is a computational artifact and is completely separate from physical dispersive processes.

On the other hand, instead of standing on the bridge and observing a stream of cars passing by, you decide to drive in a car and count the number or the concentration of cars (around your car) changing with time. This involves driving around in different directions and so you must consider the velocity of your car, which is being used to conduct the survey, when determining the number of cars observed. This amount to observing how the concentration of cars changes with time in a moving framework, reference to as Lagrangian coordinates. In this case, the change in concentration with time is a total or a material derivative of $C(x, y, t)$, which is written

$$\frac{dC}{dt} = \frac{\partial C}{\partial t} + \frac{\partial C}{\partial x}\frac{dx}{dt} + \frac{\partial C}{\partial y}\frac{dy}{dt} \qquad [6.13]$$

where

$$v_x = \frac{dx}{dt} \qquad [6.14a]$$

$$v_y = \frac{dy}{dt} \qquad [6.14b]$$

Note that the second and third terms on the right-hand side of Eq. [6.13] correspond to the advection term in Eq. [6.12]. Substituting Eq. [6.13] into [6.12], we get

$$D_x \frac{\partial^2 C}{\partial x^2} + D_y \frac{\partial^2 C}{\partial y^2} + F = \frac{dC}{dt} \qquad [6.15]$$

where Eqs. [6.14a,b] and [6.15] are considered to be the characteristics of Eq. [6.12]. Solution by the method of characteristics consists of two main steps:

1. Solving for the convection terms (Eqs. [6.14a,b]).
2. Solving for the diffusive-dispersion terms (Eq. [6.15]).

In other words we are not solving for the advection–dispersion equation directly but instead we solve the equivalent equations of the advection–dispersion equation. The equivalent equations are the characteristics of the advection–dispersion equation. The first step of the MOC begins with the discretization of the solution domain (block centered grid system). Then, a set of moving particles distributed uniformly in the solution domain with initial coordinates $(x_{p,0}, y_{p,0})$ and initial concentration $C_{p,0}$, where p is an index identifying the moving points and 0 defines the initial concentration, is introduced into the simulation domain. These particles are traced within the stationary coordinates of the grid used to conduct the simulation. Particles are traced through a velocity field with v_x and v_y as we know the hydraulic heads at all stationary points within the grid system from the solution to the flow problem. Using the velocities that apply to the center of blocks, v_x and v_y can be calculated at any point (x, y) within the block by linear interpolation. The next step is to determine the new positions of the moving particles according to

$$\left(\frac{dx}{dt}\right)^{n+1}_p = v_x\left(x_p^n, y_p^n\right) \qquad [6.16a]$$

$$\left(\frac{dy}{dt}\right)^{n+1}_p = v_y\left(x_p^n, y_p^n\right) \qquad [6.16b]$$

where n and $n + 1$ are successive time steps and

$$v_x = \frac{x_p^{n+1} - x_p^n}{\Delta t} \qquad [6.17]$$

$$\Delta t = t^{n+1} - t^n \qquad [6.18]$$

The new coordinates of the moving particles can readily be computed by

$$x_p^{n+1} = x_p^n + \Delta t \cdot v_x\left(x_p^n, y_p^n\right) \qquad [6.19a]$$

$$y_p^{n+1} = y_p^n + \Delta t \cdot v_y\left(x_p^n, y_p^n\right) \qquad [6.19b]$$

After moving the particles, the coordinates of the new positions of the moving particles $\left(x_p^{n+1}, y_p^{n+1}\right)$ are tested to see in which block the moving particles lie. Now a temporary concentration $\left(C_{ij}^*\right)$ in each block is determined by taking the average of the concentration of the moving point concentration $\left(C_p^n\right)$ within each block. After doing this, the change in concentration due to dispersion is calculated by solving the finite difference form of Eq. [6.12]. In this case assume that $F = 0$.

$$\frac{dC}{dt} = \frac{\Delta C_{ij}^n}{\Delta t} = D_x \frac{\partial^2 C^*}{\partial x^2} + D_y \frac{\partial^2 C^*}{\partial y^2} \qquad [6.20]$$

where

$$\frac{\partial^2 C^*}{\partial x^2} \cong \frac{C_{i-1,j}^{*n} - 2C_{ij}^{*n} + C_{i+1,j}^{*n}}{(\Delta x)^2} \qquad [6.21a]$$

$$\frac{\partial^2 C^*}{\partial y^2} \cong \frac{C_{i,j-1}^{*n} - 2C_{ij}^{*n} + C_{i,j+1}^{*n}}{(\Delta y)^2} \qquad [6.21b]$$

We then update the concentration of the moving particles using the following expression:

$$C_p^{n+1} = C_p^n + \Delta C_{ij}^n \qquad [6.22]$$

The points lying within a block undergo the same change in concentration due to dispersion. Finally, the concentrations at the stationary grid points are corrected for the new time step by

$$C_{ij}^{n+1} = C_{ij}^{*n} + \Delta C_{ij}^n \qquad [6.23]$$

which completes the step from t^n to t^{n+1}. The procedure is repeated for each subsequent time step. Additional details on the methodology can be found in Konikow and Bredehoeft (1978), Goode (1990), and Zheng and Bennett (2002).

6.2.3. Finite Element Approach

The finite element approach is another technique that is used by groundwater hydrologists to numerically simulate fluid flow and solute transport. It is less frequently used than the finite difference approach because of its numerical complexity. However, the finite element approach permits a more flexible arrangement of nodes that can achieve higher accuracies in solutions. This higher accuracy comes at the expense of higher numerical complexity and more computational effort (Gray 1984). The discretization of the solution domain with triangular or quadrilateral elements illustrates how one can define the boundaries of irregularly shaped aquifers (Figure 6.1c), unlike the finite difference approach. Another key advantage of the finite element approach is that nodal points can be placed to coincide with wells and hydrological features such as rivers or other surface water bodies. As the mathematics are beyond the scope of this book, we suggest references such as Zienkiewicz (1977), Huyakorn and Pinder (1983), and Zheng and Bennett (2002) for further information on this fascinating topic.

6.2.4. Forward Versus Inverse Problems

In conducting modeling studies, one may encounter different classes of problems including forward and inverse problems. Here, we briefly describe the difference between the two. Consider the partial differential equation that describes transient saturated flow in a geologic medium. If the medium is isotropic but heterogeneous, Eq. [4.26] takes on the simpler form

$$\frac{\partial}{\partial x}\left[K\frac{\partial h}{\partial x}\right] + \frac{\partial}{\partial y}\left[K\frac{\partial h}{\partial y}\right] + \frac{\partial}{\partial z}\left[K\frac{\partial h}{\partial z}\right] = S_s\frac{\partial h}{\partial t} \qquad [6.24]$$

If K and S_s are defined as *parameters* or *primary variables*, then h is a *state variable, secondary variable*, or *system response*. A *forward problem* refers to solving the flow equation for the hydraulic head in space and time with known primary variables (primary information), and for given initial and boundary conditions. On the other hand, an *inverse problem* refers to estimating values of the primary variables from information about excitations to the system and its response (secondary information) to those excitations. We will only deal with the solution to forward numerical models. Readers interested in solving inverse problems should consult the text by Sun (1994) and review articles by Yeh (1986), McLaughlin and Townley (1996), and Kitanidis (1997).

6.3. DESCRIPTION OF THE MODELING PROCESS

We have thus far discussed the theoretical basis of numerical methods used to solve fluid flow and solute transport equations. We now discuss the steps involved in building a numerical model for an actual site to evaluate the natural attenuation of contaminants. Many tasks are involved in the construction of a numerical model. In general, one can follow the steps outlined in the flowchart in Figure 6.2. Details of the procedure are described in the following sections. Additional discussion of the modeling steps and a broader perspective on the philosophy of numerical modeling can be found in Anderson and Woessner (1992).

Step 1. Define the Purpose of Modeling

The first step in a modeling project, whether it be a Tier 3 assessment involving numerical modeling or a monitored natural attenuation design for a petroleum

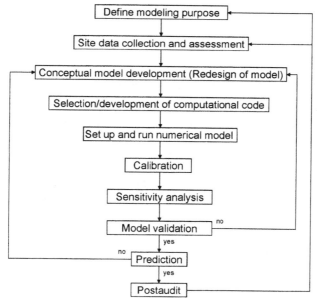

Figure 6.2. Steps in modeling groundwater flow and transport.

release, is to establish the purpose of using a groundwater model. The main purpose of constructing a site numerical model for natural attenuation is to help one quantitatively analyze the natural attenuation process at a given site and to predict contaminant concentrations at some distance and time. At this stage, it is of paramount importance to quantify the cost–benefits of building a model before actually investing a considerable amount of financial resources and time. Prior to proposing a modeling effort, a groundwater professional must have a clear understanding, in consultation with his or her client, of what is needed at a site to obtain, for example, a "No Action Required" (NAR) designation, or to demonstrate that no high-risk condition exists, or to demonstrate that natural attenuation is a viable corrective action alternative. A numerical model may be used, for example, to show what effect a new pumping well might have on a nearby contaminant plume, or what effect shutting down a water well might have on plume migration, or how source removal might enhance plume degradation. Depending on the purpose of the model, the modeler may simply use available site data to build a scoping model or build a comprehensive model that can involve the collection of an extensive set of site data that will be used to design and calibrate the numerical model.

Before building a numerical model, one may ask some questions to clarify its purpose(s): What is the purpose of modeling the natural attenuation process? How important is the problem? Would an analytical model suffice? If there are no important receptors nearby the plume, then a detailed numerical model may not be necessary. However, if there is a receptor such as a municipal well that is close to the contaminant source zone then a numerical model may become necessary. If so, how important is the model in quantifying the transport of contaminants and their natural attenuation? What level of accuracy is needed?

Sometimes, during and after constructing a model, the purpose of the modeling can change. This can be due to unexpected movement of contaminants that can threaten municipal wells and changing of regulations.

Step 2. Collection and Assessment of Site Data

In building a defensible model one must collect a broad range of site and regional data that are relevant to the context of the problem for which a solution is sought. Data is needed to construct an initial site model and for purposes of model calibration. Additional site data may be needed for model confirmation or validation. Collected data must be evaluated for its sufficiency to support models considered in the site-scale numerical model for natural attenuation. It is of utmost importance to evaluate the technical basis for data on design features, physical phenomena, geology, hydrology, microbiological processes, and geochemistry used in the site-scale model. This basis may include the collection of data that is available in the literature, a combination of techniques including laboratory column or microcosm experiments and site-specific field measurements. Sufficient data should be available to adequately define relevant parameters and to support models, assumptions, and forcing functions (initial conditions, boundary conditions, and source/sink terms) necessary for developing detailed and site-scale models. The data should also be sufficient to assess the degree to which natural attenuation affects compliance with state and federal standards and/or regulations.

The construction of a detailed numerical model is usually undertaken after a comprehensive set of site data is available. Data needed to construct a model are summarized in Table 6.1 and include hydrogeological, geochemical, and biological information. In addition, site and regional data are needed to correctly specify the initial and boundary conditions.

Table 6.1. Site Data Needed to Build a Numerical Model for Flow and Transport

General Site Data

- Chronology/history of operations/activities leading to the contamination event
- List of known contaminants and degradation products
- Quantity (mass and/or volume) of the contaminants disposed of at the site
- Summary of known impacts of the site activities on the hydrologic system and background water quality
- Surrounding land and water uses
- Meteorological data

Description of Hydrogeologic Units

- Hydrostratigraphic cross sections/maps
- Hydrogeologic units that constitute the aquifer(s)
- Description of perched aquifers (areal/volumetric extent)
- Description of the vadose zone (thickness, extent)
- Other geologic characteristics (presence of layers, continuity, faults)

Data on the Hydraulic and Transport Properties of Hydrogeologic Units

- Hydraulic conductivity
- Unit thickness
- Hydraulic head contour maps (of each layer)
- Background horizontal and vertical hydraulic gradients and temporal variations to determine flow directions
- Vertical hydraulic gradients and inter-aquifer flow within and between multiple aquifer systems
- Effective porosity
- Storativity
- Longitudinal, vertical, and horizontal transverse dispersivity
- Retardation factors
- Degradation rates
- Unsaturated flow and transport parameters of the vadose zone

Data on Regional Recharge Rates and Groundwater/Surface Water Interactions with Nearby Streams, Rivers, or Lakes

- Areal recharge rates
- Water fluxes to and from rivers, aquifers, and surface water bodies
- Surface water bodies (e.g., stream flow rates, dimensions of nearby surface water bodies)
- Concentration of hazardous constituents in surface water bodies

Table 6.1. (*Continued*)

Contaminant Source Zone Characteristics

- Identification of contaminant source locations
- Design and materials for engineered barriers
- Spatial and temporal distribution of seepage fluxes within the source zone
- Mass, concentration and physical/chemical characteristics of contaminants
- Information on whether the contaminants are in the nonaqueous phase, sorbed or are dissolved in groundwater

Data on Geochemical Conditions and Water Quality

- Background (baseline) groundwater quality
- Delineation of the contaminant plume extent
- Characterization of subsurface geochemical properties (pH, E_h, major/trace cations and anions, temperature)
- Identification of attenuation mechanisms and estimation of attenuation rates

Existing Site Cleanup Data

- Information on remediation technique deployed at the site (active and/or passive technique)
- Information on pumping, injection, and sampling wells (coordinates, depths, completion diagrams, flow rates)
- Pumping/injection rates and rate history for each well
- Mass of hazardous constituents recovered to date

Parameter values, assumed ranges, probability distributions, and/or bounding assumptions used in the modeling of natural attenuation should be technically defensible and should reasonably account for uncertainties and variabilities. The modeler should be able to provide the technical bases for each parameter value, ranges of values, or probability distributions used in the modeling of the groundwater cleanup. If site parameter values are not available (due to cost and time constraints) the modeler should provide sufficient bases for parameter values and representative parameter values taken from the literature (and their references) and the bounds and statistical distributions for hydrologic and transport parameters that are important to the modeling of the natural attenuation and groundwater cleanup.

Step 3. Development of a Site Conceptual Model

After sufficient data has been collected and interpreted, the next step is to develop a conceptual model of the site that is to be studied. The National Research Council (2001) defines a conceptual model as an evolving hypothesis identifying the important features, processes, and events controlling fluid flow and contaminant transport of consequence at a specific field site in the context of a recognized problem. According to Neuman and Wierenga (2003), it is a mental construct or hypothesis accompanied by verbal, pictorial, diagrammatic, and/or tabular interpretations and representations of site conditions as well as corresponding flow and transport dynamics. In constructing the site conceptual model, the modeler should examine the descriptions of features and physical phenomena, and the descriptions of the geological, hydrological, geotechnical, microbiological, and geochemical aspects of the contaminants, the site, and the underlying vadose zone and aquifer. The modeler should verify that the descriptions are adequate and that the conditions and assumptions used in building the conceptual model are realistic or reasonably conservative and supported by the body of data presented in the descriptions. The modeler should assess the technical bases for these descriptions and for incorporating them in the numerical model of the site (ASTM guidelines should be consulted for this purpose). Questions that may be asked in developing a conceptual model include, but are not limited to:

- Can the groundwater flow or contaminant transport be characterized as a one-, two-, or three-dimensional system?
- In what direction is groundwater moving?
- Is the hydrogeologic system composed of more than one aquifer, and is vertical flow between aquifers important?
- How does recharge to the aquifer occur? Does it occur through direct recharge of precipitation through the vadose zone into the aquifer, or through leakage from a river, drain, lake, or infiltration pond, or a combination of these?
- Is groundwater exiting the aquifer by seepage to a river or lake, flow to a drain, or extraction through a well?
- Is evapotranspiration important?
- Does it appear that the aquifer hydrogeological characteristics remain relatively uniform, or do geologic data show significant variation over the site?
- Have the boundary conditions been defined around the perimeter of the model domain, and do they have a hydrogeological or geochemical basis?
- Do groundwater flow or contaminant source conditions remain constant, or do they change with time?
- Is the contaminant plume growing, stable, or shrinking?
- Are the biogeochemical processes taking place fully known?

There may be other questions related to site-specific conditions that one may need to ask in order to construct a defensible conceptual model of the site. It should be stressed that this conceptualization step must be completed and described in the model documentation report.

It is of paramount importance to recognize that we will, of course, not be able to capture all the transport and transformation mechanisms at the site. The important aspect about correctly conceptualizing site processes is to make sure that one captures the important mechanisms that affect contaminant transport in a significant way and that one implements them correctly. Deciding which processes are most important relies heavily on the judgment of the hydrogeologist/engineer. The best conceptual models (and corresponding site numerical models) are produced when all the information available to the modeler is assimilated, capturing the most important and relevant processes.

It is also important to remember that conceptual modeling includes proposing several alternate models, ranking them, and discarding those that are in conflict with available data (Neuman and Wierenga 2003). This means that the mathematical models (analytical or numerical) produced from the conceptual model should be evaluated and one should strive to document the degree of conservatism in modeling the natural attenuation process, and the level of conservatism presumed by the modeler must be commensurate with the data and conceptual model uncertainty. To automate this approach, Neuman (2003) proposed the maximum likelihood Bayesian model averaging (MLBMA) technique, which is a comprehensive strategy for constructing alternative conceptual–mathematical models of flow-and-transport models, selecting the best among the competing models and assessing their joint predictive uncertainty. The approach relies on Monte Carlo simulations done with various models to compute statistics to assess predictive uncertainty and model performance.

Step 4. Selection of a Suitable Computational Code

The next step is to select a computational code that best represents the proposed conceptual model. In a way, one can think of this process as translating the conceptual model into a numerical mathematical model. It implies that one identifies certain processes that may govern flow and transport at a site, their symbolic mathematical representation, and their numerical approximation. Appendix C lists a number of codes that may be used to evaluate natural attenuation processes at a given site.

It is important to have a good grasp of the mathematical approach implemented in a given numerical model. Without this understanding there is potential danger in applying the code as a "black box," leading to a potential misuse of the code. Every effort should be made to understand the physical, chemical, and biological processes considered in a given model and its numerical implementation.

An important component of the code selection process is whether the selected code is right for the situation considered. The code should be consistent with the study goals and should also consider the availability of site data. In going through the selection process, the modeler should follow the "principle of parsimony," which states that a simpler model is preferred over a more complex model when the simple model is able to model the system as well as the complex model given a set of data. There is the temptation to construct a more complex model for a given situation but a more complex model usually requires a larger number of parameters, which are often not available.

Finally, in selecting a computational code, one should be concerned with the accuracy of the code through comparison against published analytical solutions (verification), should question whether the code includes a water or chemical mass balance computation, and should see if the code has been used in other studies. That is, one should research whether the code has a proven track record (Anderson and Woessner 1992).

Step 5. Setting Up and Running the Numerical Model

After selecting a suitable computational code for the problem of interest, a site numerical model is built by selecting the dimension of the problem (whether it is a one-, two-, two-, or three-dimensional grid), its grid size, and number of nodes to represent the problem domain. At this stage the simulation domain is designed and the layers are incorporated into the numerical model. The size of the simulation domain is often larger than the region of interest so that boundary effects do not affect the solutions of the interior.

The next task is to define zones that will be modeled. Each zone will have its own hydrogeological, geochemical, and microbiological properties and assigned values. For a stochastic model, the parameter values at each node may vary from one node to the next. As the site hydrogeology is inherently heterogeneous and we do not know the parameters at each of these nodes with certainty, we prefer to build a stochastic model that incorporates aquifer complexities and spatial variability in various properties. A stochastic model allows one to incorporate parameter uncertainties with the uncertainty represented by probability or related quantities such as statistical moments. However, due to the lack of data, in a majority of situations, constant parameter values are assigned for each layer or a zone defined within a layer, but there may be numerous layers and zones that may be incorporated into the numerical model.

The locations of source/sink terms (e.g., rivers, drains, wells, recharge, spills, and evapotranspiration and its corresponding strengths) and boundary conditions must also be specified. If a transient model is solved, the initial condition of the model must be additionally specified. In many cases, one could obtain the initial condition of the transient model through simulation with a steady-state model. The output of the steady-state model is then used as an input to the transient model. Boundary conditions are in many cases difficult to define so that the model is constructed and positioned such that natural boundaries are utilized. One example of this is a watershed boundary that acts as a no-flow boundary. Another example is a surface water body, such as a lake, pond, or river, which can act as a constant hydraulic head and concentration boundary.

After the features of the models are defined, they are translated to an input file. Running of codes requires an input file that must be prepared according to the specification of the computational code. Commercially available codes such as Groundwater Modeling Systems (GMS) (Environmental Modeling Systems, Inc.) and Visual MODFLOW (Waterloo Hydrogeologic, Inc.) make this task much easier by providing a graphical user interface (GUI) that speeds up the setting up of the model. If such a GUI is not available, the modeler should seek example input and output files prepared by the author of the model. The input file can be prepared by a text editor and the compiled executable will read the input file and perform the calculations according to the set of instructions and parameters provided in the input file. Examples are usually provided, so the modeler can take this and learn how to use the code. The output file can be plotted with a utility provided with the software or can be plotted using third party codes.

After a code is found to work properly, one can apply it to situations of interest. At this stage it is important to check the water and chemical mass balance. This is a mass balance involving the amount of water and/or contaminant entering and leaving a modeled system.

Step 6. Calibration of Numerical Models

After the model has been "correctly" set up and run, it should be calibrated against available site data. Calibration is a process of selecting model parameters to achieve a good match between the simulated and measured hydraulic heads, or other relevant data such as concentration data within some acceptable criteria. The main parameters for flow model calibration include hydraulic conductivity (or transmissivity), layer thickness, recharge, and boundary conditions. For the calibration of the transport model, one should consider adjusting the source zone characteristics (size, flux, etc.), dispersivity, sorption, and biodegradation parameters.

Calibration of the model applies to both steady-state and transient simulations. Model calibration for steady-state simulations involves matching hydraulic head or contaminant concentrations obtained at the steady state for the field conditions being modeled. Transient model calibrations involve matching the change in hydraulic head with time with field test results. Transient data may be obtained from a pumping test, or if the hydraulic head values vary considerably with time at a given site, such seasonal variations in hydraulic head may also be used. Calibration of transient models may be accomplished without simulating steady-state flow conditions, but this involves additional complexity. The whole process of model calibration requires that field conditions at a site be properly characterized and represented in the calibrated model. The lack of proper site characterization data may result in a calibrated model that is not representative of actual field conditions.

Commonly, one must first obtain a calibrated flow model before a transport simulation can be run and calibrated. During this phase, the modeled hydraulic head values are first used to calibrate the flow model. In many cases, the calibration of the model is accomplished with a trial and error procedure in which model parameters are systematically varied until a good fit between the simulated and observed values can be obtained. The first case study described later in this chapter provides an example of how one calibrates a model and how the results are presented. Automated approaches based on the maximum likelihood approach such as through the use of PEST (Doherty et al. 1994) or UCODE (Poeter and Hill 1999) can potentially speed up calibrations, but the details to these approaches are beyond the scope of this book.

After the flow model is calibrated, a transport model is run with the velocity fields obtained from the calibrated head fields. The modeled concentration values are then used as the basis for model calibration. It is of utmost importance to set criteria in advance of the calibration exercise. There are no "rules of thumb" for what constitutes a good calibration. However, the main goal of model calibration is to minimize the difference between simulated and measured hydraulic head or concentration values. At a minimum, for a contaminated site such as for a leaking underground storage tank (LUST) site, the calibration procedure should involve the following data:

- Hydraulic head and its gradient
- Groundwater flow direction
- Contaminant concentrations
- Contaminant migration rates
- Contaminant migration directions

The comparisons between simulated and actual field data should be presented in maps, tables, or graphs. Each modeler and model reviewer will need to use their professional judgment in evaluating the calibration results. For initial assessments, it is possible to obtain useful results from models that are not calibrated. The application of uncalibrated models can be very useful as a screening tool or in guiding data collection activities.

Evaluation of the calibrated model can be assessed through the criteria based on (1) mean error, (2) mean absolute error, or (3) root mean square error (Anderson and Woessner 1992):

$$\frac{1}{n}\sum_{i=1}^{n}(C_m - C_s)_i \qquad [6.25]$$

$$\frac{1}{n}\sum_{i=1}^{n}|(C_m - C_s)_i| \qquad [6.26]$$

$$\left[\frac{1}{n}\sum_{i=1}^{n}(C_m - C_s)_i^2\right]^{1/2} \qquad [6.27]$$

where n is the number of points were comparisons are made, C_m is the measured concentration at some point i, and C_s is the simulated concentration at the same point (Anderson and Woessner 1992). Of the three error estimates, Anderson and Woessner (1992) point to the root mean square error as the best quantitative measure if the errors are normally distributed. The mean error (Eq. [6.25]) is not preferred because large positive and negative errors can cancel each other out. Thus, a small error estimate may hide a poor model calibration. Other requirements are sometimes applied in addition to error estimates such as quantitatively correct flow directions and hydraulic head gradients.

A reasonably good calibration for the head at a particular monitor well might be a RMS (root mean square) error of 5%. For a target contaminant such as benzene at a LUST site, a calibrated model should be able to reproduce the concentration at the well within 500 μg/L of latest observed sample results (Zhang and Heathcote 2001).

Step 7. Sensitivity Analysis

The modeler is also expected to conduct sensitivity analyses to:

- Identify aquifer flow and transport parameters that are expected to significantly affect the site model outcome.
- Test the degree to which the performance of natural attenuation and groundwater cleanup may be affected if a range of parameter values must be used as input to the model due to sparsity of, or uncertainty in, available data.
- Test for the need for additional data.

Determination of model sensitivity to the input parameters is a key step that should be performed for every model application. This is because the model that has been calibrated may be influenced by uncertainty owing to the inability to define the exact spatial (and temporal) distribution of parameter values (and processes) in the solution domain. There are uncertainties of the magnitude and form of the initial/boundary conditions and stresses imposed on the system.

A sensitivity analysis involves the systematic adjustment of (1) parameters such as hydraulic conductivity, storativity, dispersivity, degradation rate constants, and mass transfer rates and (2) forcing functions such as the recharge rate, evapotranspiration, and fluxes from surface water bodies over the range that is physically and chemically reasonable. The magnitude of changes in the hydraulic head and concentration values from the calibrated solution is the measure of the sensitivity of the solution to that particular parameter (Anderson and Woessner 1992). The sensitivity of one model parameter versus other parameters is also frequently demonstrated.

The standard procedure for conducting a sensitivity analysis of a groundwater flow-and-transport model is to first obtain a calibrated model for a given site. This calibrated model and its corresponding parameters form the "base" case. The sensitivity analysis can then be conducted by adjusting certain hydrogeological, chemical, and biodegradation parameters systematically over a reasonable parameter range, and the resulting variations in hydraulic head distributions or contaminant concentrations are reported in a tabular or graphical manner. Parameters are usually adjusted by 50% larger or smaller than the base case parameters, but these changes are often made up to an order of magnitude and even larger depending on the situation and the parameter that is being tested. Examples of sensitivity analysis are presented in the three case studies described later.

Sensitivity analysis is also important in determining the amount and frequency of future data collection activities. Model parameters that have been found to be sensitive at a particular site would perhaps require additional characterization, as opposed to data for which the model is relatively insensitive.

Step 8. Model Validation

It is relatively easy to construct models and to run them. With automated approaches, calibration also becomes a relatively easy exercise. However, a calibrated model does not mean that the system has been adequately represented. Model validation involves the establishment of greater confidence in the model by conducting simulations of the system under conditions in which the data has not been used for calibration purposes. One should note though that multiple data sets on pumping and tracer tests are usually not available.

For example, one can calibrate a model using one set of pumping test data. If the calibrated model with this first pumping test data can do a good job of predicting system response with a second set of pumping test data (e.g., conducted using another well), one can have greater confidence in the calibrated model. On the other hand, if the parameters need to be adjusted to match the response of the second pumping test, the process becomes a second calibration and additional data sets become necessary to continue with the validation exercise. Model validation is complete when the validation data are matched against simulated values resulting from the calibrated parameter values that have not been significantly varied.

Once the model validation process is completed, it can be used in a predictive mode. It is important to remember that even if the model has been validated, this does not mean that it will yield correct predictions since the model has been calibrated for certain forcing functions (initial and boundary conditions and source/sink terms). In many cases, data needs to be collected over a long period of time and perhaps the model may need to be updated periodically. For example, depletion of a sorption site can cause retardation to decrease over time, resulting in the plume moving faster. In another example, the geochemical and microbiological mechanisms may change over time at a site; hence, the reaction kinetics also become time dependent. Relying on a numerical model with sensitive parameters obtained through calibration using old data can seriously underestimate the travel time of the plume if site conditions have changed significantly, and so the modeler should revisit the problem periodically and update the model as more/new data becomes available.

Step 9. Prediction

Predictive simulations are conducted using a validated model and its parameter sets. The simulations are run into the future to quantify the system response to future events and stresses imposed on the system. For example, a validated model may be used to evaluate different remediation alternatives, such as pump-and-treat, hydraulic containment, or intrinsic remediation to assist with risk assessment. Predictions can range from several years for a small spill from an underground storage tank to several decades to centuries involving a DNAPL spill and its corresponding natural attenuation. Some predictive simulations for fluid flow and solute transport are run much longer: in the case of the evaluation of high-level nuclear waste storage at Yucca Mountain, predictive simulations are run for over 10,000 years as the radioactive contaminants remain hazardous for long periods due to small decay constants.

In conducting predictive simulations, quantitative estimates of future stresses need to be made. For example, the recharge rate measured for present conditions is x mm/yr. But with global warming, the recharge rate may significantly vary over the next few decades to y mm/yr. This could significantly affect the outcome of the measured hydraulic head values and the calibrated model using recharge rate x mm/yr may not be able to predict the hydraulic head levels at another recharge rate (y mm/yr).

In another example, the current decay constant for natural attenuation of BTEX may be x day^{-1}. As the plume migrates from one region to the next, the geochemical characteristics of the aquifer may be very different, causing a large difference in the decay constant to y day^{-1} because of the manner in which microbes could behave in a slightly different geochemical environment. This could ultimately affect the concentration levels of contaminants and could significantly vary from simulated values.

Uncertainty in a predictive simulation arises from uncertainty in the calibrated model and our inability to make accurate estimates of future stresses. One way to assess predictive uncertainty for deterministic and/or stochastic models is to conduct a Monte Carlo simulation. The method is straightforward and applies to a very broad range of fluid flow-and-transport models. However, it can be time consuming as a large number of simulations need to be run to obtain meaningful uncertainty estimates. Further discussion on this important and fascinating topic can be found in Zhang (2002) and Zheng and Bennett (2002).

Step 10. Postaudit

A postaudit of the calibrated model should be conducted several years after the modeling study is completed (Anderson and Woessner 1992). Sufficient time should pass before a postaudit is conducted so that the model response can vary significantly from the calibrated model. A postaudit performed too soon after the initial calibration may lead to the conclusion that the prediction came close to estimating the observed value when in fact not enough time elapsed to allow the system to move sufficiently far from the calibrated values (Anderson and Woessner 1992).

When a postaudit is conducted, new field data are collected to assess whether the prediction is correct. In the case of simulating natural attenuation, concentration data along the plume may be collected to see if indeed the plume has reduced in size after several years. The amount of data collected and its collection frequency depend on the level of confidence in the model results and the associated level of risk to the downgradient receptors.

Step 11. Redesign of Model

Typically, the postaudit will lead to new insights into system behavior, which may lead to changes in the conceptual model or model parameters. A change in conceptual model can lead to the use of an alternative computational code to solve the same problem. Where appropriate and when natural attenuation mechanisms are highly uncertain, the modeler should use alternative site models, which involves the reformulation of the conceptual model and corresponding adjustment to the numerical model used. Alternative modeling approaches consistent with available data and current scientific understanding are investigated where necessary, and results and limitations are appropriately factored into the evaluation of the natural attenuation process. The modeler must provide sufficient evidence that relevant site features have been considered, that the models are consistent with available data and current scientific understanding, and that the effects of natural attenuation on contaminant degradation have been evaluated. Specifically, one must adequately consider alternative modeling approaches where necessary to incorporate uncertainties in site parameters and ensure they are propagated through the modeling (this is why a stochastic or probabilistic approach in some cases is necessary). Uncertainty in data interpretation is considered by analyzing reasonable conceptual models that are supported by site data, or by demonstrating through sensitivity studies that the uncertainties have little impact on the natural attenuation process.

6.4. NUMERICAL MODELS USED TO EVALUATE NATURAL ATTENUATION

We have seen how a numerical model is constructed for a given site. In this section, we review the most common numerical models used in the industry and the research community to model groundwater flow and contaminant transport, transformation, and attenuation mechanisms. These descriptions are intended to provide a general overview of each of these models. One should consult the original publications for further information on each of these models, particularly when it is considered for application to a particular site.

6.4.1. MODFLOW

MODFLOW is one of the most commonly used computer codes for groundwater modeling. It is a modular, three-dimensional, block centered, finite difference flow code developed by the U.S. Geological Survey (McDonald and Harbaugh 1988) to simulate fluid flow in saturated geological media. MODLFOW was designed such that the user can select a set of modules to be used during a given simulation. Each module deals with a specific feature of the hydrologic system (e.g., well, recharge, and rivers). MODFLOW is used to calculate the site hydraulic head and groundwater velocity distributions subsequently used in the contaminant transport model.

MODFLOW can be used to simulate fully three-dimensional systems and quasi-three-dimensional systems, in which the flow in aquifers is horizontal and flow through confining beds is vertical. The code can also be used for a two-dimensional mode for simulating flow in a single layer, or two-dimensional flow in a vertical cross section. An aquifer can be confined, unconfined, or a mixed confined/unconfined. Flow from external stresses, such as flow to wells, areal recharge, evapotranspiration, flow to drains, and flow through riverbeds, can also be simulated. This code has been validated with many real case applications.

6.4.2. MT3D

The MT3D code (Zheng 1990) is widely used for simulating transport of one contaminant species at a time (e.g., benzene) in groundwater. It is a modular, three-dimensional transport code for simulating advection, dispersion, and chemical reactions of contaminants in groundwater systems. Its solute transport simulation is based on the flow field generated with MODFLOW. MT3D can simulate radioactive decay or biodegradation, and linear and nonlinear sorption processes of Freundlich or Langmuir isotherms.

MT3D includes five files that control various aspects of groundwater contaminant transport (basic, advection, dispersion, source/sink, and reaction). The package contains basic information for the model, including grid size and timing information, which are required for all model runs. The advection file controls the mathematical solution scheme to be used to represent contaminant movement in groundwater. MT3D contains four different solution methods: (1) method of characteristics (MOC); (2) modified method of characteristics (MMOC), a particle tracking method that combines "backward" particle tracking with an interpolation scheme to reduce the computational burden of a simulation, but does not handle sharp contamination fronts very well; (3) hybrid method of characteristics (HMOC), which combines the standard MOC and the MMOC methods, alternating back and forth depending on the presence of sharp concentration fronts; and (4) upstream finite difference, a method not based on particle tracking, but similar to the forward difference scheme. The dispersion file controls the amount of dispersion introduced by the model at each step in a simulation. The source/sink mixing file controls sources and sinks of concentration due to wells, drains, recharge, rivers/streams, general head boundaries, and evapotranspiration. MT3D requires only the concentration information for these processes, as the flow characteristics of the sources/sink are contained in the "head and flow file" previously created by MODFLOW. The sources/sinks, however, must correspond to sources and sinks entered into the MODFLOW model. Finally, the reaction file controls sorption of contaminants by selected isotherms, and first-order radioactive decay or biodegradation.

6.4.3. MT3DMS

The MT3DMS code (Zheng and Wang 1999) is an expanded version of MT3D that can simulate transport of multiple contaminant species at one time. Additional enhancements over MT3D include a new advection solver method called the total variation diminishing (TVD) scheme. This solver allows the user to select from three different mathematical methods, depending on the requirements of the system being modeled. Other enhancements are an implicit generalized conjugate gradient method, a nonequilibrium sorption option that allows more accurate retardation modeling if certain parameters are known, and the multiple-species structure that will accommodate add-on reaction packages, such as those in RT3D.

6.4.4. RT3D

The RT3D code solves the coupled partial differential equations that describe reactive transport of multiple mobile and/or immobile solutes in a three-dimensional aquifer. The code is a generalized multispecies version of the single-species transport code MT3D (Zheng 1990). The code was originally developed to support contaminant transport modeling efforts at natural attenuation demonstration sites (Clement and Johnson 1998; Lu et al. 1999; Clement et al. 2000) but has also been used to model several laboratory and pilot-scale active bioremediation experiments (Johnson et al. 1998).

RT3D involves a much more detailed modeling of reactive contaminant transport than does MT3DMS. This code was based on the 1997 version of MT3D (DOD Version 1.5) but has extended capabilities of MT3DMS with the addition of several reaction packages. The code requires the MODFLOW groundwater flow code for computing groundwater head distributions. RT3D can accommodate multiple sorbed and aqueous phase species with any one of seven predefined reaction frameworks, or any other novel framework that the user may define. This allows, for example, natural attenuation processes or an active remediation system to be evaluated and simulation to be made to consider contaminants such as heavy metals, explosives, petroleum hydrocarbons, and/or chlorinated solvents. The preprogrammed reaction packages include (Clement 1997):

1. Two-species instantaneous reaction (hydrocarbon and oxygen).
2. Instantaneous hydrocarbon biodegradation using multiple electron acceptors (O_2, NO_3^-, Fe^{2+}, SO_4^-, CH_4).
3. Kinetically limited hydrocarbon biodegradation using multiple electron acceptors (O_2, NO_3^-, Fe^{2+}, SO_4^-, CH_4).
4. Kinetically limited reaction with bacterial transport (hydrocarbon, oxygen, and bacteria).
5. Nonequilibrium sorption/desorption (can also be used for nonaqueous phase liquid dissolution).
6. Reductive, anaerobic biodegradation of PCE, TCE, DCE, and VC.
7. Reductive, anaerobic biodegradation of PCE, TCE, DCE, and VC combined with aerobic biodegradation of DCE and VC.

6.4.5. BIOPLUME III

The BIOPLUME III software package was written by Rifai et al. (1997). It is based on the U.S. Geological Survey Method of Characteristics (MOC) Model (Konikow and Bredehoeft 1978). It simulates the biodegradation of organic contaminants using a number of aerobic and anaerobic electron acceptors: oxygen, nitrate, iron(III), sulfate, and carbon dioxide. The model also considers the natural processes of advection, dispersion, sorption, and ion exchange. It solves the transport equation six times per computer run for the fate and transport of the hydrocarbons and the electron acceptors/reaction by-products. Three different kinetic expressions can be used to simulate the aerobic and anaerobic biodegradation reactions. These include first-order decay, instantaneous reaction, and Monod kinetics. The principle of superposition is used to combine the hydrocarbon plume with the electron acceptor plume.

BIOPLUME III assumes that biodegradation occurs sequentially in the following order:

$$\text{Oxygen} \rightarrow \text{Nitrate} \rightarrow \text{Iron(III)} \rightarrow \text{Sulfate} \rightarrow \text{Carbon dioxide}$$

The model can also be used to simulate bioremediation of hydrocarbons in groundwater by injecting electron acceptors (except for iron(III)) and can also be used to simulate air sparging for low injection flow rates. Finally, the model can be used to simulate advection, dispersion, and sorption without including biodegradation.

There are two limitations imposed by the biodegradation expressions incorporated in BIOPLUME III. First, the model does not account for selective or competitive biodegradation of individual hydrocarbon species. This means that hydrocarbons are generally simulated as a lumped organic, which represents the sum of benzene, toluene, ethyl benzene, or xylene. If a single component is to be simulated, the user would have to determine how much electron acceptor would be available for the component in question. Second, the conceptual model for biodegradation used in BIOPLUME III is a simplification of the complex biologically mediated redox reactions that occur in the subsurface. The main differences between BIOPLUME III and the RT3D code are:

- BIOPLUME III is a two-dimensional code while RT3D can handle three dimensions.
- BIOPLUME III is based on the USGS MOC transport code, which has only one transport solver—the method of characteristics; whereas RT3D is based on MT3DMS code, which has the finite difference and TVD solvers in addition to the MOC.
- BIOPLUME III can simulate two types of multispecies reactions, instantaneous reactions, and sequential electron donor–acceptor reactions, while in addition to those two, RT3D can simulate other types of reactions, including a user-defined reaction.

6.5. MODELING THE NATURAL ATTENUATION OF HYDROCARBONS AND DNAPLs

Numerical models of flow and transport can be used to study the effectiveness of natural attenuation at a particular site by incorporating the combined effects of physical, geochemical, and microbiological attenuation mechanisms. Numerous situations can be studied in a relatively rapid fashion, allowing the user to conduct a sensitivity analysis of parameters for a given case. Once the sensitive parameters are found for a given situation, characterization efforts can focus around collecting the data that can affect the results of a site investigation. Numerical models can also be used to design measures for active remediation if natural attenuation is deemed at the site to be insufficient. For example, it can be used to design the cleanup operation by pump and treat, primarily by identifying optimal well configurations for constructing efficient capture zones. Optimization of well configuration may result in the expedient reduction of hazardous constituent concentrations and could result in significant cost reduction for the cleanup operation.

In this section, we illustrate the use of a numerical model by working on several simple examples using MT3D and RT3D. We encourage the reader to try out these simple situations before using the model for realistic situations. The purpose of this exercise is to show how one can build and run a model using a widely used commercial software to simulate an idealized spill of various contaminants that are non-reactive or reactive at a hypothetical site to study its evolution with space and time. If the contaminant is reactive, then we will observe the changes in the geochemistry observed in the aquifer. If the contaminant additionally degrades over some distance away from the spill location, then the transport of the degradation products will also be studied in relation to the original contaminant.

Once we understand how the code works for these simplified cases, we will be ready to apply it to a more complex, real field situation. The application of the code to real field situations is beyond the scope of this text (although the techniques are directly applicable to them); instead, we will examine three representative case studies that are based on the codes described here.

6.5.1. Mathematical Background

Both MT3D and RT3D use flow fields computed in MODFLOW and the corresponding flow velocities. A three-dimensional groundwater flow equation with appropriate initial and boundary conditions gives values of hydraulic head at all points in the solution domain:

$$S_s \frac{\partial h}{\partial t} = \frac{\partial}{\partial x_i}\left(K_i \frac{\partial h}{\partial x_i}\right) + q_s \qquad [6.28]$$

The corresponding flow velocities are computed through Darcy's Law:

$$v_i = -\frac{K_i}{n}\frac{\partial h}{\partial x_i} \qquad [6.29]$$

where $h =$ the hydraulic head [L], $S_s =$ the specific storage coefficient [L^{-1}], $K_i =$ principal components of the hydraulic conductivity tensor [LT^{-1}], $v_i =$ pore water velocity [LT^{-1}], and $n =$ soil porosity. Equations [6.28] and [6.29] imply that the principal components of the hydraulic conductivity tensor is aligned with the principal coordinate system. Once the flow field is computed via MODFLOW, one can compute the transport of aqueous and solid phases of contaminants. In order to accomplish this, we write the macroscopic equations for a multidimensional system that describe the transport of aqueous and solid phase species, respectively.

The transport equation for the aqueous phase species is (Clement 1997)

$$\frac{\partial C_k}{\partial t} = \frac{\partial}{\partial x_i}\left(D_{ij}\frac{\partial C_k}{\partial x_j}\right) - \frac{\partial}{\partial x_i}(v_i C_k) + \frac{q_s}{n}C_{S_k} + r_c - r_a + r_d \qquad [6.30]$$

where $n = 1, 2, \ldots, m$ is the total number of species, m = total number of aqueous phase or mobile species, C_k = the aqueous phase concentration of the kth species $[ML^{-3}]$, D_{ij} = the hydrodynamic dispersion tensor, q_s = the volumetric flux of water per unit volume of aquifer representing sources and sinks $[T^{-1}]$, C_s = the concentration of source/sink $[ML^{-3}]$, r_c = the reaction rate that describes the mass of the species removed or produced per unit volume per unit time $[ML^{-3}T^{-1}]$, and r_a = attachment (or adsorption) rate $[ML^{-3}T^{-1}]$, and r_d = detachment (or desorption) rate that describe the kinetic exchange of the transported species between aqueous and solid phases $[ML^{-3}T^{-1}]$.

The transport equation for the solid phase species is (Clement 1997)

$$\frac{d\tilde{C}_l}{dt} = \tilde{r}_c \qquad [6.31]$$

where $l = 1, 2, \ldots, (n - m)$, and $(n - m)$ is the total number of solid phase or immobile species, $\tilde{C}_l$ = the solid phase concentration of the lth species $[MM^{-1}]$, and $\tilde{r}_c$ = the reaction rate at the solid phase $[MM^{-1}T^{-1}]$.

In order to solve the reactive transport equations [6.30] and [6.31], RT3D uses a reaction operator-split (OS) numerical scheme to solve any number of transport equations that may be coupled via nonlinear reaction expressions. Using this approach, Eq. [6.30], which describes the transport of the mobile species, can be divided into four distinct equations of the following form:

$$\frac{\partial C_k}{\partial t} = \frac{\partial}{\partial x_i}(v_i C_k) \qquad [6.32]$$

$$\frac{\partial C_k}{\partial t} = \frac{\partial}{\partial x_i}\left(D_{ij}\frac{\partial C_k}{\partial x_j}\right) \qquad [6.33]$$

$$\frac{\partial C_k}{\partial t} = \frac{q_s}{n}C_{S_k} \qquad [6.34]$$

$$\frac{\partial C_k}{\partial t} = r_c - r_a + r_d \qquad [6.35]$$

where Eq. [6.32] describes advection, Eq. [6.33] describes dispersion, Eq. [6.34] describes source/sink mixing, and Eq. [6.35] is the reaction equation. The transport of immobile species is solved with only Eq. [6.35]. Figure 6.3 illustrates the logical steps of numerical procedures implemented in RT3D.

6.5.2. Background on Simple Examples Solved

We now illustrate the capabilities of RT3D by running the code for a number of simple example problems modified after the tutorials in *Groundwater Modeling Systems–Tutorials*, Vol. 16 (Environmental Modeling Research Laboratory 2003). In case 1, we illustrate the transport of a nonreactive species using MT3D to study the effect of dispersion. In cases 2 and 3 we compare the transport of BTEX subjected to instantaneous aerobic degradation and degradation with multiple electron

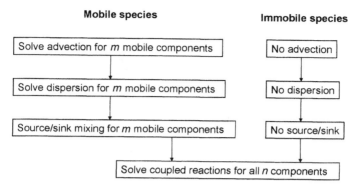

Figure 6.3. Flowchart describing the numerical solution scheme implemented in RT3D. (Adapted from Clement et al. 1998.)

acceptors with RT3D. Comparison of transport distances can be made with the first three cases. In case 4, we study the natural attenuation and transport of PCE and its degradation products using RT3D. Finally, in case 5, we study the effects of rate-limited sorption on a pump-and-treat cleanup operation. A brief description of the reactions and its mathematical implementation are discussed for each of the cases. Further details to the reaction processes can be found in Chapters 2 and 3.

A sample problem has been devised using parameters and conditions that are representative of a typical leaking underground storage tank. The purpose of these simulations is to illustrate the importance of certain variables and processes that are critical in natural attenuation processes discussed in the previous chapters and to illustrate the importance of certain parameters on the outcome of the simulation results. Results from the simulation are discussed to illustrate which parameters can have a large impact on natural attenuation. Here, we use the codes MT3D and RT3D implemented in the computer software Groundwater Modeling System (GMS) (Brigham Young University 1997) to simulate groundwater contaminant transport, transformation, and attenuation processes. The GMS software provides a graphical user interface for several commonly used groundwater flow and transport codes.

Consider an underground storage tank leaking hydrocarbons into a sandy confined aquifer. The simulation domain for cases 1 through 3 consists of a single layer of a confined aquifer with dimensions of 210 m by 310 m with 50 m grid blocks shown on Figure 6.4. Hydraulic tests conducted within the aquifer yielded a transmissivity value of $300 \, \text{m}^2/\text{day}$. Analysis of cores taken from the site revealed a porosity of 0.3. The hydraulic head on the left boundary is 100 m while on the left boundary it is 98 m. The upper and lower boundaries are treated as no-flow boundaries. The underground storage tank is simulated with a well situated at the location $i = 85 \, \text{m}$ and $j = 105 \, \text{m}$ with the leakage simulated by injection of contaminated water at a volumetric flow rate of $4 \, \text{m}^3/\text{day}$. The contaminants enter the aquifer as a continuous source through the well. The longitudinal dispersivity (α_L) was estimated to be 5 m while the transverse dispersivity (α_T) was estimated to be

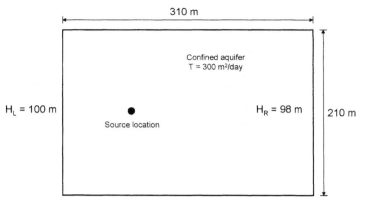

Figure 6.4. Plan view of the model domain dimensions, boundaries, and source location.

one-third of α_L. Table 6.2 summarizes the flow and transport parameters for cases 1 through 3.

Case 1: Transport of a Nonreactive Contaminant

We first simulate the transport of a nonreactive contaminant. The results from this simulation are used as a benchmark that will be compared against cases involving the natural attenuation of BTEX in the same aquifer. Here, we use the code MT3D for the simulation of the transport of a nonreactive contaminant. But before we do this, we must first compute the flow field and corresponding seepage velocities. We use the flow code, MODFLOW, to compute the flow field for all cases presented here. Running a steady-state simulation for a confined aquifer yields the hydraulic head field shown in Figure 6.5. As this is a steady-state simulation in the confined aquifer, the hydraulic head field is uniform and flow takes place uniformly from the left to the right.

We next simulate the transport of a nonreactive contaminant in a confined aquifer through continuous injection from a well at a concentration of 1000 mg/L. The simulation takes place for 1000 days over 10 time steps. For the advection package,

Table 6.2. Flow and Transport Parameters for Cases 1 through 4

Parameter	Value
H_R	98 m
H_L	100 m
T	300 m^2/day
ϕ	0.3
a_L	5 m
a_T	1.7 m

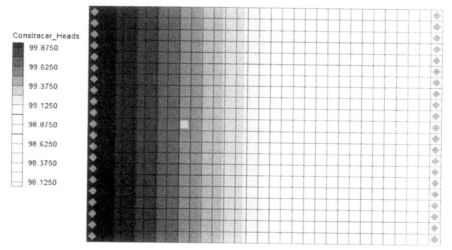

Figure 6.5. Hydraulic head contour generated using MODFLOW.

the modified method of characteristics (MMOC) is used. Figure 6.6a shows a contour plot of the nonreactive contaminant in the aquifer after 500 days of injection into the aquifer while Figure 6.6b shows the distribution of the contaminant after 1000 days. The figures show that the contaminants move from the left to right with longitudinal and transverse dispersion spreading the contaminants away from the nodes along the well injecting the contaminants. In the absence of dispersion, the contaminants will be transported only by advection and should be confined to the nodes along the flow path where the well is placed. It is evident from Figure 6.6b that, after 1000 days, the contaminants have reached the right boundary.

Case 2: Instantaneous Aerobic Degradation

The instantaneous reaction model was proposed by Borden and Bedient (1986) as a simple model to approximate the biodegradation of organic chemicals in groundwater. The reaction simulated in this package is similar to those incorporated into BIOPLUME II (Rifai et al. 1997) and the results obtained from the two codes are compared in Clement et al. (1998).

The instantaneous reaction model assumes that microbial biodegradation kinetics is fast in comparison to oxygen transport. It additionally assumes that microbial growth and oxygen utilization can be simulated as an instantaneous reaction between the organic contaminant and oxygen. The reaction model treated below simulates the instantaneous degradation of fuel hydrocarbons under aerobic conditions based on (e.g., benzene)

$$C_6H_6 + 7.5O_2 \rightarrow 6CO_2 + 3H_2O \qquad [6.36]$$

Equation [6.36] implies that the mass ratio of benzene to oxygen (F) is 3.08.

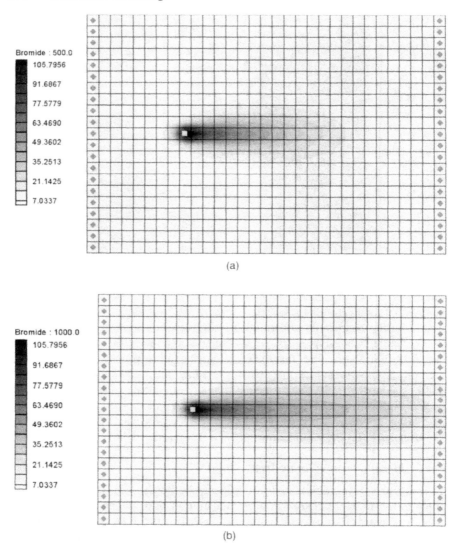

Figure 6.6. Distribution of a nonreactive contaminant (bromide) (a) after 500 days of injection and (b) after 1000 days of injection.

The transport equations solved are (Clement 1997)

$$\frac{\partial}{\partial x_i}\left(D_{ij}\frac{\partial C_{HC}}{\partial x_j}\right) - \frac{\partial}{\partial x_i}(v_i C_{HC}) + \frac{q_s}{\phi}C_{HC-s} + r_{HC} = R_{HC}\frac{\partial C_{HC}}{\partial t} \qquad [6.37]$$

$$\frac{\partial}{\partial x_i}\left(D_{ij}\frac{\partial C_{O_2}}{\partial x_j}\right) - \frac{\partial}{\partial x_i}(v_i C_{O_2}) + \frac{q_s}{\phi}C_{O_2-s} + r_{O_2} = R_{O_2}\frac{\partial C_{O_2}}{\partial t} \qquad [6.38]$$

where the subscripts HC and O_2 indicate hydrocarbon and oxygen, respectively, while r_{HC} and r_{O_2} are the hydrocarbon and oxygen removal rates. Instantaneous removal of hydrocarbon (H) and oxygen (O) is controlled by the following algorithm, where t refers to the particular time step. The reaction algorithm takes the form (Rifai et al. 1988)

$$H(t+1) = H(t) - \frac{O(t)}{F} \qquad [6.39]$$

and

$$O(t+1) = 0 \qquad [6.40]$$

when $H(t) > O(t)/F$.
    For the oxygen

$$O(t+1) = O(t) - H(t) \cdot F \qquad [6.41]$$

and

$$H(t+1) = 0 \qquad [6.42]$$

when $O(t) > H(t) \cdot F$. With this algorithm, either the hydrocarbon or oxygen concentration in a given grid cell will be reduced to zero at each time step, depending on which component is stoichiometrically limiting in the prior time step. The main advantages of this approach over other approaches are that the governing equations are simplified and that the data requirements are small. In particular, this model does not require kinetic data but it is limited to situations where the biodegradation rate is fast in comparison to the groundwater flow rate.
    We now illustrate the degradation of BTEX compounds with the instantaneous reaction model using the grid and parameters utilized in case 1. Consider a continuous spill of BTEX with a uniform concentration of 2100 mg/L simulated through a well injecting at a rate of 4 m^3/day. The fate and transport of BTEX (electron donor) and oxygen (electron acceptor) plumes in a two-dimensional aerobic aquifer are modeled using an explicit reaction algorithm described above. The initial concentration of the oxygen throughout the model domain is set to 10 mg/L and BTEX concentration is set to 0 mg/L. These same concentration values are set at the boundary and throughout the model domain except at the injection well.
    Results from the simulation after 500 and 1000 days for the BTEX and oxygen are shown in Figures 6.7 and 6.8, respectively. We note that the longitudinal and transverse extent of the BTEX plume is considerably smaller than that of the oxygen plume, as the reaction seems to deplete oxygen from a larger portion of the aquifer.

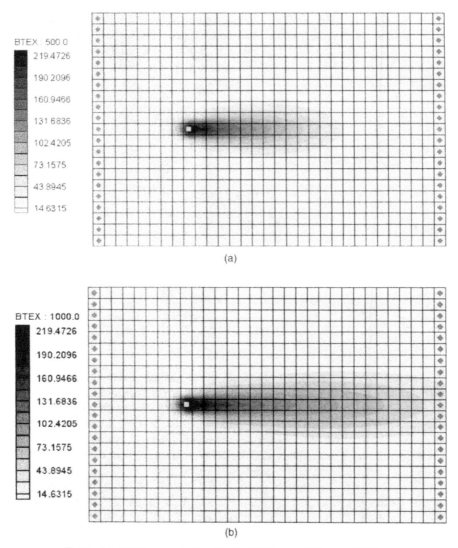

BTEX : 500.0
219.4726
190.2096
160.9466
131.6836
102.4205
73.1575
43.8945
14.6315

(a)

BTEX : 1000.0
219.4726
190.2096
160.9466
131.6836
102.4205
73.1575
43.8945
14.6315

(b)

Figure 6.7. BTEX concentration (a) after 500 days and (b) after 1000 days.

This approach has several important implications. Results of the simulations show that oxygen transport is the limiting process for biodegradation and that transverse mixing is the major source of oxygen to the plume. The reaction mechanism is considered to be valid when the BTEX is rapidly degrading in relation to the transport of oxygen. The modeling approach may not be appropriate at sites where the degradation takes place at a very slow rate. In such a case, the kinetics of the reaction process will take a more significant role in limiting the degradation of BTEX.

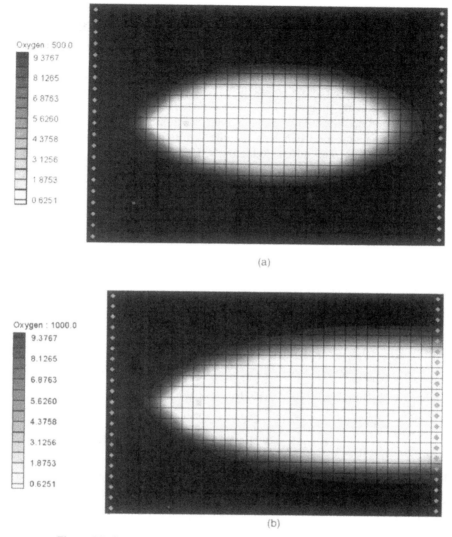

Figure 6.8. Oxygen concentration (a) after 500 days and (b) after 1000 days.

Case 3: BTEX Degradation with Multiple Electron Acceptors Under Rate-Limited Conditions

In case 2, we dealt with a case in which BTEX degrades under aerobic conditions. Indeed, in shallow environments in which oxygen is abundant, aerobic respiration is the first reaction that will cause biodegradation. However, once the oxygen in the aquifer becomes depleted due to the consumption by the bacteria feeding on the BTEX, anaerobic conditions will become more important in the interior of the BTEX plume. Therefore, we consider here the case in which BTEX undergoes degradation with multiple electron acceptors under rate-limited conditions. Details of the biogeochemical processes can be found in Chapter 3.

In this case, we will again model the continuous leak of BTEX from an underground storage tank but instead of the instantaneous aerobic degradation model, we assume the biodegradation of BTEX compounds via five different aerobic/anaerobic degradation pathways under rate-limited conditions. The five processes considered in this reaction model include aerobic respiration (AR), denitrification (D), iron(III) reduction (IR), sulfate reduction (SR), and methanogenesis (M). It is assumed that all biogeochemical reactions occur within the aqueous phase mediated by the existing subsurface microbes (see Chapter 3 for details). The assumption that reactions take place only in the aqueous phase is a simplification because the reactions do also take place on the solid phase but we take this to be conservative. The reaction is assumed to proceed in the following sequence:

$$AR \rightarrow D \rightarrow IR \rightarrow SR \rightarrow M \qquad [6.43]$$

Using benzene as an example, the chemical reactions take the following form (Clement 1997):

$$C_6H_6 + 7.5O_2 \rightarrow 6CO_2 + 3H_2O \qquad [6.44]$$

$$6NO_3^- + 6H^+ + C_6H_6 \rightarrow 6CO_2 + 6H_2O + 3N_2 \qquad [6.45]$$

$$30Fe(OH)_3 + 60H^+ + C_6H_6 \rightarrow 6CO_2 + 78H_2O + 30Fe^{2+} \qquad [6.46]$$

$$3.75SO_4^{2-} + 7.5H^+ + C_6H_6 \rightarrow 6CO_2 + 3H_2O + 3.75H_2S \qquad [6.47]$$

$$C_6H_6 + 4.5H_2O \rightarrow 2.25CO_2 + 3.75CH_4 \qquad [6.48]$$

Since the concentrations of Fe^{3+} and CO_2 are not readily measurable under normal field conditions, these terms are replaced with the assimilative capacity for iron reduction and methanogenesis, defined as

$$[Fe^{3+}] = [Fe^{2+}_{max}] - [Fe^{2+}] \qquad [6.49]$$

$$[MC] = [CO_2] = [CH_{4,max}] - [CH_4] \qquad [6.50]$$

where $[Fe^{2+}_{max}]$ and $[CH_{4,max}]$ are the maximum possible aquifer levels of these species that represent the aquifer's maximum capacity for iron reduction and methanogenesis. Note that the concentration of CO_2 used here is the CO_2 evolved while the hydrocarbon is destroyed via methanogenesis, which may be thought of as the "methanogenic capacity" (MC) of the aquifer. Using these relations, Fe(III) reduction and the methanogenesis process may be related back to measurable Fe^{2+} and CH_4 concentration levels.

The coupled transport equations solved including equation [6.37] take the following form:

$$\frac{\partial}{\partial x_i}\left(D_{ij}\frac{\partial C_{O_2}}{\partial x_j}\right) - \frac{\partial}{\partial x_i}(v_i C_{O_2}) + \frac{q_s}{\phi}C_{O_2-s} + r_{O_2} = R_{O_2}\frac{\partial C_{O_2}}{\partial t} \qquad [6.51]$$

$$\frac{\partial}{\partial x_i}\left(D_{ij}\frac{\partial C_{NO_3}}{\partial x_j}\right) - \frac{\partial}{\partial x_i}(v_i C_{NO_3}) + \frac{q_s}{\phi}C_{NO_3-s} + r_{NO_3} = R_{NO_3}\frac{\partial C_{NO_3}}{\partial t} \qquad [6.52]$$

$$\frac{\partial}{\partial x_i}\left(D_{ij}\frac{\partial C_{Fe^{2+}}}{\partial x_j}\right) - \frac{\partial}{\partial x_i}(v_i C_{Fe^{2+}}) + \frac{q_s}{\phi}C_{Fe^{2+}-s} + r_{Fe^{2+}} = R_{Fe^{2+}}\frac{\partial C_{Fe^{2+}}}{\partial t} \qquad [6.53]$$

$$\frac{\partial}{\partial x_i}\left(D_{ij}\frac{\partial C_{SO_4}}{\partial x_j}\right) - \frac{\partial}{\partial x_i}(v_i C_{SO_4}) + \frac{q_s}{\phi}C_{SO_4-s} + r_{SO_4} = R_{SO_4}\frac{\partial C_{SO_4}}{\partial t} \qquad [6.54]$$

$$\frac{\partial}{\partial x_i}\left(D_{ij}\frac{\partial C_{CH_4}}{\partial x_j}\right) - \frac{\partial}{\partial x_i}(v_i C_{CH_4}) + \frac{q_s}{\phi}C_{CH_4-s} + r_{CH_4} = R_{CH_4}\frac{\partial C_{CH_4}}{\partial t} \qquad [6.55]$$

The rate-limited reaction is assumed to be first-order, which means that the rate of hydrocarbon decay is directly proportional to the hydrocarbon concentration that is available at a given node. The following reaction kinetic framework is used to model the degradation of hydrocarbon via different electron acceptor pathways (Clement 1997):

$$r_{HC,O_2} = -k_{HC,O_2}[HC]\frac{[O_2]}{K_{O_2}+[O_2]} \qquad [6.56]$$

$$r_{HC,NO_3} = -k_{HC,NO_3}[HC]\frac{[NO_3]}{K_{NO_3}+[NO_3]}\cdot\frac{K_{HC,NO_3}}{K_{HC,NO_3}+[O_2]} \qquad [6.57]$$

$$r_{HC,Fe^{2+}} = -k_{HC,Fe^{3+}}[HC]\frac{[Fe^{3+}]}{K_{Fe^{3+}}+[Fe^{3+}]}\cdot\frac{K_{i,O_2}}{K_{i,O_2}+[O_2]}\cdot\frac{K_{i,NO_3}}{K_{i,NO_3}+[NO_3]} \qquad [6.58]$$

$$r_{HC,SO_4} = -k_{HC,SO_4}[HC]\frac{[SO_4]}{K_{SO_4}+[SO_4]}\cdot\frac{K_{i,O_2}}{K_{i,O_2}+[O_2]}\cdot\frac{K_{i,NO_3}}{K_{i,NO_3}+[NO_3]}$$
$$\cdot\frac{K_{i,Fe^{3+}}}{K_{i,Fe^{3+}}+[Fe^{3+}]} \qquad [6.59]$$

$$r_{HC,CH_4} = -k_{HC,CH_4}[HC]\frac{[CO_2]}{K_{CH_4}+[CO_2]}\cdot\frac{K_{i,O_2}}{K_{i,O_2}+[O_2]}\cdot\frac{K_{i,NO_3}}{K_{i,NO_3}+[NO_3]}$$
$$\cdot\frac{K_{i,Fe^{3+}}}{K_{i,Fe^{3+}}+[Fe^{3+}]}\cdot\frac{K_{i,SO_4}}{K_{i,SO_4}+[SO_4]} \qquad [6.60]$$

where r_{HC,O_2} is the rate at which hydrocarbon is destroyed by utilizing oxygen, r_{HC,NO_3} is the rate at which hydrocarbon is destroyed by utilizing nitrate, $r_{HC,Fe^{3+}}$ is the rate at which hydrocarbon is destroyed by utilizing Fe^{3+}, r_{HC,SO_4} is the rate at which hydrocarbon is destroyed by utilizing sulfate, r_{HC,CH_4} is the rate at which hydrocarbon is destroyed by producing methane, $[O_2]$ is the oxygen concentration $[ML^{-3}]$, k_{HC,O_2} is the first-order rate constant $[T^{-1}]$, K_{O_2} is the Monod half-saturation constant $[ML^{-3}]$ and K_{i,O_2} is the oxygen inhibition constant $[ML^{-3}]$ while a similar nomenclature is utilized for all other reactions. We see that, by setting all the half-saturation constants to a small value, it is possible to simulate zero-order dependency with respect to electron donor and hence a first-order degradation model with respect to hydrocarbon. One can also set all inhibition constants to a small value to force the reactions to occur in a sequential fashion (clement 1997).

The total rate of hydrocarbon destruction, via all the above described processes, is the sum of each individual rate and is given as

$$r_{HC} = r_{HC,O_2} + r_{HC,NO_3} + r_{HC,Fe^{2+}} + r_{HC,SO_4} + r_{HC,CH_4} \qquad [6.61]$$

The corresponding rates of electron acceptor utilization can be obtained through the product of the rates of hydrocarbon destruction and the yield coefficient (Y) for each reaction:

$$r_{O_2} = Y_{O_2/HC} r_{HC,O_2} \qquad [6.62]$$

$$r_{NO_3} = Y_{NO_3/HC} r_{HC,NO_3} \qquad [6.63]$$

$$r_{Fe^{2+}} = -Y_{Fe^{2+}/HC} r_{HC,Fe^{2+}} \qquad [6.64]$$

$$r_{SO_4} = Y_{SO_4/HC} r_{HC,SO_4} \qquad [6.65]$$

$$r_{CH_4} = -Y_{CH_4/HC} r_{HC,CH_4} \qquad [6.66]$$

The yield values for the case when BTEX represents the contaminants present in the groundwater are $Y_{O_2/HC} = 3.14$, $Y_{NO_3/HC} = 4.9$, $Y_{Fe^{2+}/HC} = 21.8$, $Y_{SO_4/HC} = 4.7$, and $Y_{CH_4/HC} = 0.78$ (Wiedemeier et al. 1995). The yield value can be thought of the mass ratio of electron acceptors removed or metabolic by-products produced to total BTEX degraded.

The problem that we solve follows the previous two examples. As in case 2, we simulate the leakage of BTEX from an underground storage tank and transport at uniform groundwater velocity. As it moves downgradient it undergoes degradation with multiple electron acceptors. We use the default values for BTEX, Fe^{2+}, and methane and input nonzero values for oxygen (10 mg/L), nitrate (5.0 mg/L), and sulfate (10.0 mg/L). We assume the same porosity and dispersivities for the aquifer as in the previous examples. Here again, we assume that BTEX leaks from the tank at the same volumetric flow rate with a concentration of 2100 mg/L.

The inhibition coefficients are all assumed to be 0.01 mg/L and all Monod constants are treated to be 0.5 mg/L. These are the default values in RT3D. The yield values are the same as those described in the previous section (default values in the model). Reaction constants for this model are noted in Table 6.3.

Results after running the simulation for 1000 days are shown for BTEX (Figure 6.9), oxygen (Figure 6.10), nitrate (Figure 6.11), iron (Figure 6.12), sulfate (Figure 6.13), and methane (Figure 6.14). As in the case of instantaneous aerobic

**Table 6.3. Reaction Constants for BTEX
Degradation with Multiple Electron Acceptors**

Constant	Value
$[Fe^{2+}_{max}]$	25.0 mg/L
$[CH_{4,max}]$	30.0 mg/L
k_{HC,O_2}	0.1 day^{-1}
k_{HC,NO_3}	0.008 day^{-1}
$k_{HC,Fe^{3+}}$	0.0005 day^{-1}
k_{HC,SO_4}	0.00025 day^{-1}
k_{HC,CH_4}	0.0001 day^{-1}

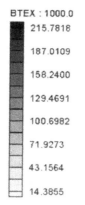
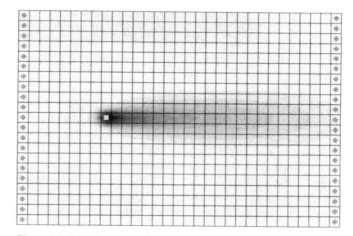

Figure 6.9. BTEX concentration after 1000 days.

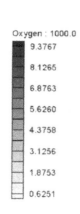
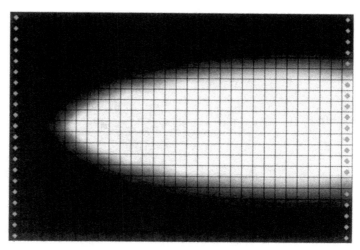

Figure 6.10. Oxygen concentration after 1000 days.

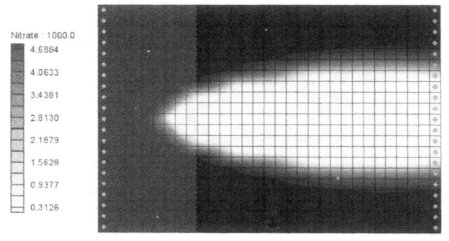

Nitrate : 1000.0

4.6884

4.0633

3.4381

2.8130

2.1879

1.5628

0.9377

0.3126

Figure 6.11. Nitrate concentration after 1000 days.

degradation, we note that the longitudinal and transverse extent of the BTEX plume is considerably smaller than that of the zone of oxygen depletion since the reaction seems to deplete oxygen from a larger portion of the aquifer. The zone of nitrate depletion is also larger than the BTEX plume but it is smaller than the zone of oxygen depletion. The size of the iron plume is on the same order as the nitrate plume while the sulfate and methane plumes are considerably smaller. These comparisons may not be strictly valid as the contours vary from one constituent to the next; however, the varying extent of the plume for each species can be

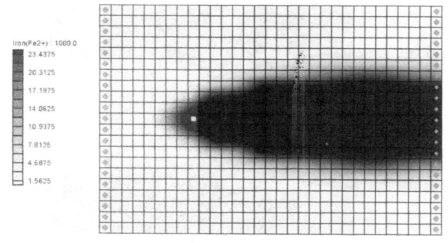

Iron(Fe2+) : 1000.0

23.4375

20.3125

17.1875

14.0625

10.9375

7.8125

4.6875

1.5625

Figure 6.12. Fe^{2+} concentration after 1000 days.

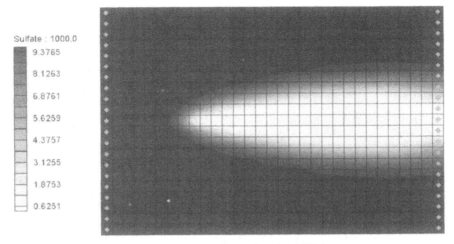

Figure 6.13. Sulfate concentration after 1000 days.

described qualitatively. It would be beneficial to conduct a sensitivity analysis on selected transport parameters to determine whether such a trend is visible under widely varying conditions.

Comparison of the simulation results with electron acceptor zones in the subsurface shows that the spatial distribution of terminal electron-accepting processes match that observed in the field (Wiedemeier et al. 1999) and conceptualized (Figure 3.13).

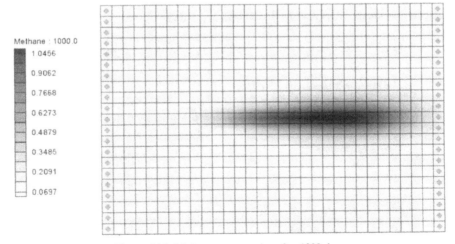

Figure 6.14. Methane concentration after 1000 days.

Case 4: Sequential Anaerobic Degradation of PCE

Dense nonaqueous phase liquids (DNAPLs) are prevalent at a large number of sites throughout the world. The high densities, low interfacial tensions, and low viscosities of halogenated organic solvents can lead to deep DNAPL penetration (Pankow and Cherry 1996). In porous media, much of the DNAPL mass remains in the groundwater as persistent source zones. The variable release history and geologic heterogeneity make the distribution of DNAPLs in the source zone complex, where DNAPLs exist as residuals or as pools of pure phase. Without remediation, these source zones can contribute to long-term groundwater contamination for decades to centuries. Therefore, the spatial distribution, mass, and composition of DNAPLs present in both the source zone and as dissolved phase plumes need to be characterized so that efficient remediation schemes can be designed. The more common DNAPLs are PCE (tetrachloroethylene) and TCE (trichloroethene). These compounds are considered to be very difficult to remediate by active remediation technologies, but recent research has shown the potential for natural attenuation.

In case 4, we simulate the sequential degradation of PCE in an anaerobic subsurface environment. It involves the dechlorination of PCE by first-order degradation kinetics and we assume further that degradation all the way to vinyl chloride (VC) is anaerobically favorable. PCE degradation takes the following pathway:

$$\text{PCE} \rightarrow \text{TCE} \rightarrow \text{DCE} \rightarrow \text{VC} \qquad [6.67]$$

Laboratory and field estimates of first-order rate coefficients for selected compounds are reported by Wiedemeier et al. (1999) and reproduced in Table 6.4.

The large spread in the magnitude of these rate constants indicates that any remediation design or management decision based on natural attenuation needs to be analyzed thoroughly using a contaminant transport model that considers such a reaction process (Clement et al. 1998).

The following set of equations describes the reactive transport of PCE and its daughter products (Clement et al. 1998):

$$\frac{\partial}{\partial x_i}\left(D_{ij}\frac{\partial C_{\text{PCE}}}{\partial x_j}\right) - \frac{\partial}{\partial x_i}(v_i C_{\text{PCE}}) + \frac{q_s}{\phi}C_{\text{PCE}-s} - K_{\text{PCE}}C_{\text{PCE}} = \frac{\partial C_{\text{PCE}}}{\partial t} \qquad [6.68]$$

$$\frac{\partial}{\partial x_i}\left(D_{ij}\frac{\partial C_{\text{TCE}}}{\partial x_j}\right) - \frac{\partial}{\partial x_i}(v_i C_{\text{TCE}}) + \frac{q_s}{\phi}C_{\text{TCE}-s} + Y_{\text{TCE/PCE}}K_{\text{PCE}}C_{\text{PCE}}$$

$$-K_{\text{TCE}}C_{\text{TCE}} = \frac{\partial C_{\text{TCE}}}{\partial t} \qquad [6.69]$$

$$\frac{\partial}{\partial x_i}\left(D_{ij}\frac{\partial C_{\text{DCE}}}{\partial x_j}\right) - \frac{\partial}{\partial x_i}(v_i C_{\text{DCE}}) + \frac{q_s}{\phi}C_{\text{DCE}-s} + Y_{\text{DCE/TCE}}K_{\text{TCE}}C_{\text{TCE}}$$

$$-K_{\text{DCE}}C_{\text{DCE}} = \frac{\partial C_{\text{DCE}}}{\partial t} \qquad [6.70]$$

$$\frac{\partial}{\partial x_i}\left(D_{ij}\frac{\partial C_{\text{VC}}}{\partial x_j}\right) - \frac{\partial}{\partial x_i}(v_i C_{\text{VC}}) + \frac{q_s}{\phi}C_{\text{VC}-s} + Y_{\text{VC/DCE}}K_{\text{DCE}}C_{\text{DCE}}$$

$$-K_{\text{VC}}C_{\text{VC}} = \frac{\partial C_{\text{VC}}}{\partial t} \qquad [6.71]$$

Table 6.4. Mean and Recommended First-Order Rate Coefficients for Selected Chlorinated Solvents

Compound	Mean of Field/In Situ Studies			Recommended First-Order Rate Coefficients				Comments
	First-Order Rate Coefficient (day^{-1})	Half-life (day^{-1})	Number of Studies Used for Mean	Low End		High End		
				Coefficient (day^{-1})	Half-life (day^{-1})	Coefficient (day^{-1})	Half-life (day^{-1})	
Tetrachloroethylene (PCE)	0.0029	239	16	0.00019	3647	0.0033	210	Lower limit was reported for a field study under nitrate reducing conditions.
Trichloroethylene (TCE)	0.0025	277	47	0.00014	4950	0.0025	277	Lower limit was reported for a field study under unknown redox conditions.
Vinyl chloride (VC)	0.0079	88	19	0.00033	2100	0.0072	96	Lower limit was reported for a field study under methanogenic/sulfate-reducing conditions.
1,1,1-Trichloroethane (TCA)	0.016	43	15	0.0013	533	0.01	69	Range not appropriate for nitrate-reducing conditions. Expect lower limit to be much less.
1,2-Dichloroethane	0.0076	91	2	0.0042	165	0.011	63	Range reported from a single field study under methanogenic conditions.
Carbon tetrachloride (CT)	0.37	1.9	9	0.0037	187	0.13	5	Range not appropriate for nitrate-reducing conditions. Expect lower limit to be much less.
Chloroform	0.030	23	1	0.0004	1733	0.03	23	Only one study available. Biodegradation under nitrate-reducing conditions expected.
Dichloromethane	0.0064	108	1	0.0064	108	—	—	Rate constant reported from a single field study under methanogenic conditions.
Trichlorofluoromethane	—	—	—	0.00016	4331	0.0016	433	All studies with very low concentration of this compound.
2,4-Dichlorophenol	0.014	50	2	0.00055	1260	0.027	26	Range may not be appropriate for nitrate-reducing conditions.

Source: Data originally from Aronson and Howard (1997); adapted from Wiedemeier et al. (1999).

where C_{PCE}, C_{TCE}, C_{DCE}, and C_{VC} are the concentration of the respective contaminants in mg/L; K_{PCE}, K_{TCE}, K_{DCE}, and K_{VC} are first-order degradation rates; and $Y_{TCE/PCE}$, $Y_{DCE/TCE}$, and $Y_{VC/DCE}$ are yield coefficients whose values can be computed from stoichiometric relations to be 0.79, 0.74, and 0.64, respectively (Clement et al. 1998). This kinetic model ignores sorption, but it may be incorporated if desired. Utilizing the operator splitting procedure, the kinetics of the degradation mechanism can be represented through the following set of differential equations:

$$\frac{dC_{PCE}}{dt} = -K_{PCE}C_{PCE} \qquad [6.72]$$

$$\frac{dC_{TCE}}{dt} = Y_{TCE/PCE}K_{PCE}C_{PCE} - K_{TCE}C_{TCE} \qquad [6.73]$$

$$\frac{dC_{DCE}}{dt} = Y_{DCE/TCE}K_{TCE}C_{TCE} - K_{DCE}C_{DCE} \qquad [6.74]$$

$$\frac{dC_{VC}}{dt} = Y_{VC/DCE}K_{DCE}C_{DCE} - K_{VC}C_{VC} \qquad [6.75]$$

The kinetic constants input by the user are listed in Table 6.5, while the yield constants are fixed internally in the reaction package and are tabulated in Table 6.6.

The problem geometry, boundary conditions, and flow and transport parameters are similar to the previous three cases. Simulations are run to 1000 days over 10 time steps to examine the fate and transport of PCE and its degradation products. The spatial distribution of PCE, TCE, DCE, and VC plumes at 1000 days are shown on Figures 6.15–6.18. One observation to note is that the location of maximum concentration for different degradation products moves in the direction of flow. The plume also appears to be increasingly spread out as PCE becomes increasingly degraded. These observations are due to the nonsorptive, sequential nature of the

Table 6.5. Kinetic Constants for the Sequential Anaerobic Degradation of PCE

Constant	Value (day^{-1})	Description
K_{PCE}	0.005	PCE anaerobic constant
K_{TCE}	0.003	TCE anaerobic constant
K_{DCE}	0.002	DCE anaerobic constant
K_{VC}	0.001	VC anaerobic constant

Table 6.6. Yield Constants for the Sequential Anaerobic Degradation of PCE

Constant	Value (mg/mg)	Description
$Y_{TCE/PCE}$	0.7920	TCE/PCE stoichiometric yield
$Y_{DCE/TCE}$	0.7377	DCE/TCE stoichiometric yield
$Y_{VC/DCE}$	0.6445	VC/DCE stoichiometric yield

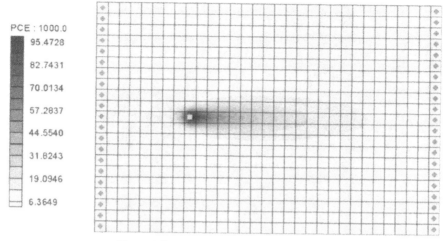

Figure 6.15. PCE concentration after 1000 days.

reaction kinetics assumed for this case. In this simulation, the PCE plume is contained within the model domain when $t = 1000$ days (Figure 6.15) and it does not reach the right boundary, which is treated as the receptor location. At longer times, or with different parameters, such conditions may be violated. In fact, we see that the degradation products (TCE, DCE, and VC) all reach the right boundary at $t = 1000$ days (Figures 6.16–6.18). Therefore, sensitivity analyses should be performed with expected minimum and maximum values of parameters to consider the worst and best case scenarios. Such sensitivity analyses can readily be conducted with a model such as RT3D.

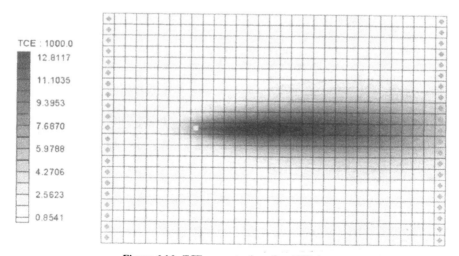

Figure 6.16. TCE concentration after 1000 days.

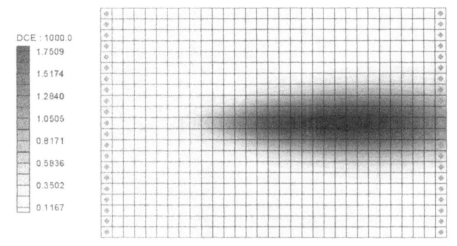

Figure 6.17. DCE concentration after 1000 days.

Case 5: Rate-Limited Sorption Reaction

The transport of various contaminants can be dependent on their sorption characteristics. Under most groundwater flow conditions, the partitioning of contaminants between the aqueous and solid phases can be assumed to be in local equilibrium. This implies that the mass transfer of contaminants between the solid and aqueous phases is fast in comparison to groundwater flow. In most field situations, this idealization seems to hold. In these situations the widely used retardation approach for modeling sorption provides an adequate description of the transport process. However, when the reaction takes place much slower relative to the groundwater

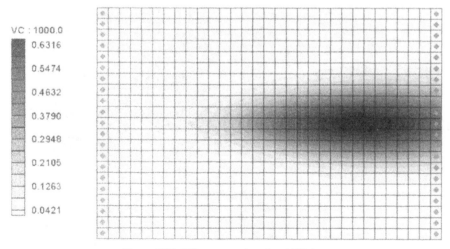

Figure 6.18. VC concentration after 1000 days.

velocity, this assumption may not be valid (Haggerty and Gorelick 1994). One condition in which the local equilibrium assumption can fail is when pumping is taking place as part of an active remediation process. In such a situation, contaminant concentration monitored in the extraction well may exhibit a "plume tailing effect," in which the concentration declines at a very slow rate which causes aquifer remediation to take an excessively long period. This may be one cause of why pump-and-treat operations can take decades to centuries to remediate some sites. Another effect seen is the "rebounding effect" in which the aqueous concentrations of the contaminant decrease with time but once pumping is stopped the concentration rebounds. These effects cannot be readily simulated with the standard retardation approach because the mass transfer reactions are oversimplified.

To better simulate the sorption reaction under these conditions, a mass transfer model that considers the transfer of contaminants between the aqueous and solid phases is needed. Such a model is referred to as a rate-limited mass transfer model in which the rate of exchange is controlled by the mass transfer coefficient. Here we follow the approach outlined in Haggerty and Gorelick (1994) and write the transport equations for the aqueous and solid phases, respectively:

$$\frac{\partial}{\partial x_i}\left(D_{ij}\frac{\partial C}{\partial x_j}\right) - \frac{\partial}{\partial x_i}(v_i C) + \frac{q_s}{\phi}C_s = \frac{\partial C}{\partial t} + \frac{\rho_b}{\phi}\frac{\partial \tilde{C}}{\partial t} \qquad [6.76]$$

$$\xi\left(C - \frac{\tilde{C}}{\lambda}\right) = \frac{\rho_b}{\phi}\frac{\partial \tilde{C}}{\partial t} \qquad [6.77]$$

where C is the contaminant concentration in the mobile phase $[ML^{-3}]$, $\tilde{C}$ is the contaminant concentration in the immobile phase (contaminant mass per unit mass of porous media $[MM^{-1}]$, ρ_b is the bulk density of the soil, ϕ is porosity, ξ is the first-order mass transfer rate coefficient $[T^{-1}]$, and λ is the linear partitioning coefficient $[L^3M^{-1}]$. When mass transfer between the aqueous and solid phases is high, the left-hand side of Eq. [6.77] becomes dominant in comparison to the storage term on the right-hand side. In this case, Eq. [6.77] can be rewritten as

$$\tilde{C} = \lambda C \qquad [6.78]$$

which is the linear sorption model. Substitution of this relationship into Eqs. [6.76] and [6.77] yields the retardation form of the transport equation [4.58]. In contrast, when the mass transfer coefficient is set very low, the contaminants are sorbed onto the solid phase and are considered to be fully sequestered.

The rate-limited sorption reaction discussed above was implemented in RT3D to track a single mobile and immobile species. Utilizing the reaction operator splitting, the model reduces to

$$-\xi\left(C - \frac{\tilde{C}}{\lambda}\right) = \frac{dC}{dt} \qquad [6.79]$$

$$\frac{\phi\xi}{\rho_b}\left(C - \frac{\tilde{C}}{\lambda}\right) = \frac{d\tilde{C}}{dt} \qquad [6.80]$$

which are incorporated into the RT3D model.

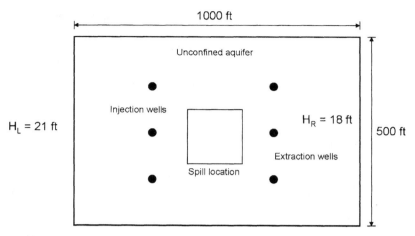

Figure 6.19. Model domain showing the location of spill, injection wells, and extraction wells. (Adapted from Clement and Jones 1998.)

The synthetic example that is shown in Figure 6.19 is based on an example in Clement and Jones (1998). The site is a 1000 ft × 500 ft section of an unconfined aquifer with background flow gradient from left to right. A spill at the center of the site has created a contaminant plume as shown on Figure 6.19. A pump-and-treat system consisting of three injection (each at $Q = 900$ ft^3/day) and three extraction wells (each at $Q = -900$ ft^3/day) is used to clean the contaminant spill. The aqueous concentration of contaminant level throughout the plume is assumed to be at 300 mg/L. Table 6.7 summarizes the flow and transport parameters used in this example.

Note that one can readily calculate the retardation coefficient value of 1.5 using $(R = 1 + \rho_b \lambda / \phi)$. Assuming that equilibrium conditions exist before starting the pump-and-treat system, the initial soil-phase contaminant concentration levels, $\tilde{C} = \lambda C$, can be estimated to be at 3.0×10^{-5} (mg of contaminant/mg of soil).

Table 6.7. Flow and Transport Parameters for Cases 1 through 4

Parameter	Value
h_R	18 ft
h_L	21 ft
K	10 ft/day
ϕ	0.3
α_x	5 m
α_y	1.7 m
λ_d	1.0×10^{-7} L/mg
ρ_b	1.6×10^6 mg/L

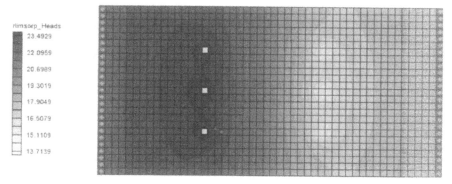

Figure 6.20. Hydraulic head contours due to a pump-and-treat operation under steady-state flow conditions.

The objective of this treatment system is to clean both dissolved and solid-phase contamination. The model will simulate the effectiveness of the system under different mass transfer conditions. A 300 day simulation will be performed. One can readily vary the mass transfer coefficient values to simulate retardation conditions (using $\xi = 0.1$ day^{-1}), intermediate conditions (using $\xi = 0.002$ day^{-1}), and sequestered conditions (using $\xi = 0.0001$ day^{-1}). The student working on this example should conduct a sensitivity analysis of the mass transfer coefficient and compare the results to see the variation in the number of days it will take to clean up the spill.

Figure 6.20 shows the hydraulic head distribution computed by MODFLOW. The elevated zones of hydraulic head (injection cones) are visible on left side of the domain, while the drawdown cones are visible on the right side of the domain. The aqueous phase concentrations of the contaminants are plotted after 50 days (Figures 6.21) and 400 days (Figure 6.22), showing that the plume is being displaced from the left to right with this active remediation scheme.

The modeling cases that we have seen should give some ideas on how to construct and run a numerical model. It is also of paramount importance to recognize

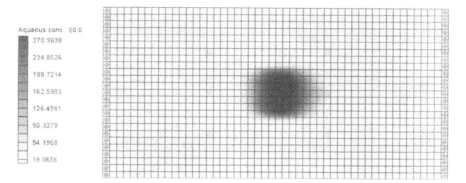

Figure 6.21. Aqueous phase concentration of contaminant after 50 days.

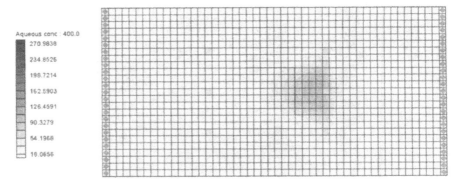

Figure 6.22. Aqueous phase concentration of contaminant after 400 days.

that certain parameters can have a larger effect on the transport behavior of contaminants. We now examine three case studies in which the codes MT3D and RT3D have been used to study the transport of BTEX compounds and chlorinated solvents in real field cases.

6.6. NUMERICAL MODELING CASE STUDIES

Case Study 1: RBCA Tier 3 Modeling of a Leaking Underground Storage Site (LUST) Site in Western Iowa

The first case study concerns the modeling of BTEX transport at a LUST site in western Iowa. The numerical model was built for this site by Zhang and Heathcote (2001) and their results are summarized here. The main objectives of modeling groundwater flow and contaminant transport at this site were to determine (1) what the regional flow in the area is, (2) what effect the residential wells have on the flow regime, and (3) to determine the distribution of the benzene and TEX plume, in which the latter is simulated as xylene in the two-layer aquifer.

The town of Climbing Hill is located in western Iowa with the surficial geology consisting of glacial deposits. The town has several small, private water supply wells that produce from thin, sandy layers at depths of about 75 ft. The producing zone is stratigraphically at the contact between Pre-Illinoian till and overlying loess. Other wells are drilled into a sandstone layer beneath a shale layer in the Cretaceous Age Dakota Formation that directly underlies the Pre-Illinoian till. Petroleum contamination has shown up in several of the water wells screened in the shallower sandy interval. However, contamination has not been found in water wells screened in the Dakota Formation. A known source zone is located near the east end of town but the times and amounts of petroleum releases are uncertain.

The water table is between 20 and 30 ft deep in the loess and slopes westward. Annual recharge to the water table is not precisely known but is in the range of

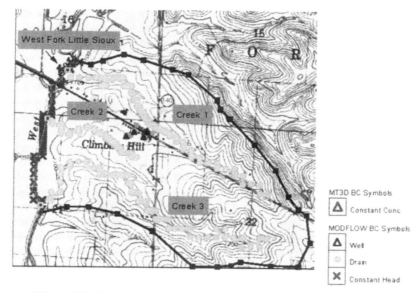

MT3D BC Symbols

△ Constant Conc

MODFLOW BC Symbols

△ Well

⚛ Drain

✖ Constant Head

Figure 6.23. The simulation domain with boundary conditions at Climbing Hill.

2–6 inches (0.051–0.152 m/y) or 7–25% of annual precipitation. Existing monitoring wells are completed in the loess, while residential wells are completed in the sand layer underlying the loess.

The simulation domain is a small groundwater basin (Figure 6.23). The groundwater basin is bounded by a topographic high in all directions except to the west. The west boundary is formed by the West Fork Little Sioux River that flows from north to south. The river elevation varies from 329.8 to 328.4 m along the domain boundary. The river is assumed to be well connected to the aquifers and thus is modeled as constant-head boundary (the crosses in Figure 6.23). There are three unnamed creeks (Creeks 1, 2, and 3) inside the domain, which are modeled with the drain package in MODFLOW.

A three-dimensional, two-layer model was constructed to examine hydraulic behavior and contaminant transport at Climbing Hill. The first layer is for the loess and the second layer is for the thin, sandy layer used as a water source. Figure 6.24 is a cross section that depicts the site hydrogeology. Three wells screened in the thin sand layer are included in Figure 6.24. These wells are intended to simulate residential water demand and so are assigned small pumping rates of 18 m³/day (20 gpm for about 4 h/day). The West Fork Little Sioux River is simulated as a constant-head boundary and the three creeks are simulated with the drain package in MODFLOW. The river stage and the bottom elevation and conductance of the three creeks are listed in Table 6.8.

The loess is simulated with coverage Layer 1. This layer is treated as an unconfined, horizontal aquifer with bottom elevation at 326.1 m. The thin sandy layer is modeled by coverage Layer 2 with a uniform thickness of 2.1 m and is also

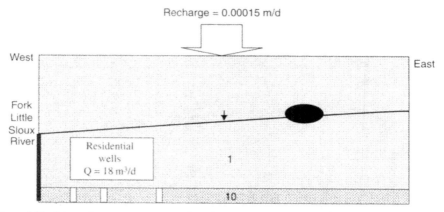

Figure 6.24. Schematic east–west cross section of the Climbing Hill conceptual model (not to scale). Black triangle denotes water table; black oval denotes petroleum source. Numbers within various fields are hydraulic conductivity (m/day) for the unit indicated.

horizontal. Both layers are treated as homogeneous. The hydraulic conductivity for Layer 1 is one order of magnitude smaller than that of Layer 2 (Table 6.9). Hydraulic conductivity is assumed to be isotropic; that is, $K_h = K_v$.

The flow model is calibrated against the long-term average of the observed hydraulic head values at eleven monitoring wells by changing the hydraulic conductivity, the net recharge rate, and the conductance of the creeks. The calibration

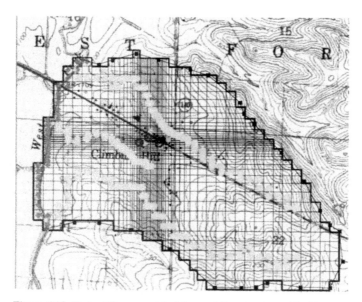

Figure 6.25. Finite difference grid of the modeling domain at Climbing Hill.

Table 6.8. Input Parameters for the Rivers and Creeks

Parameter	West Fork Little Sioux River	Creek 1	Creek 2	Creek 3
Simulated in MODFLOW as:	Constant head	Drain	Drain	Drain
River stage (m)	328.4–329.8	NA	NA	NA
River or drain bottom elevation (m)	NA	329.7–371.8	329.2–347.5	328.4–374.9
River or drain conductance (m/day)	NA	5.0	5.0	5.0

target is set to be within 0.5 m of the observed water levels at eleven monitor wells, and the results are listed in Table 6.10. All errors are much smaller than 0.5 m. The root mean square error is 0.19 m. The calibrated conductance and hydraulic conductivity values are listed in Tables 6.8 and 6.9, respectively, and the calibrated net recharge rate is 0.00015 m/day (2.1 in./yr). The calibrated steady-state head

Table 6.9 Input Parameters of the Layers for Groundwater Flow Modeling

Parameter	Layer 1	Layer 2
Aquifer type	Unconfined	Confined
Top elevation (m)	400	326.1
Bottom elevation (m)	326.1	324.0
Horizontal conductivity, K_h (m/day)	1.0	10
Vertical conductivity, K_v (m/day)	1.0	10
Net recharge rate (m/day)	0.00015	NA

Table 6.10. Calibration Results for Hydraulic Head Values at the Observation Wells

Well Number	X (m)	Y (m)	I	J	K	Observed Head (m)	Simulated Head (m)	Error (m)
MW 7	246265	4692177	17	25	1	334.40	334.37	− 0.03
MW 12	246302	4692217	14	29	1	334.40	334.66	0.26
MW 18	246396	4692151	20	39	1	335.50	335.28	− 0.22
MW 101	246433	4692151	20	41	1	335.40	335.55	0.15
MW 103	246373	4692101	26	37	1	335.00	335.15	0.15
MW 105	246265	4692128	23	25	1	334.60	334.42	− 0.18
MW 106	246348	4692179	17	31	1	335.00	334.94	− 0.06
MW 112	246299	4692178	17	29	1	334.40	334.58	0.18
MW 113	246255	4692128	23	24	1	334.50	334.36	− 0.14
MW 115	246396	4692136	22	39	1	335.60	335.29	− 0.31
MW 118	246337	4692178	17	33	1	335.10	334.86	− 0.24

Root mean square error = 0.19

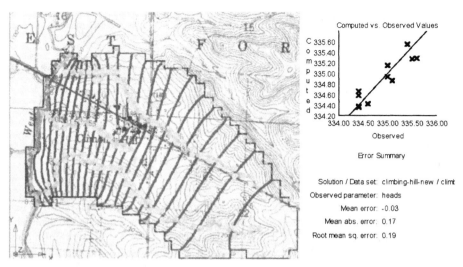

Figure 6.26. Steady-state hydraulic head contours with calibration results at Climbing Hill.

contours are illustrated in Figure 6.26 along with a scatter plot and error summary of the observed versus modeled head at the four monitoring wells.

Calibrated results of the flow model indicate hydraulic conductivities chosen for the two layers are reasonable (Table 6.10). The recharge value is within the reasonable range for western Iowa. Groundwater flow direction is westward through Climbing Hill in the shallower aquifer.

A sensitivity analysis for the Climbing Hill model was conducted by doubling the values of hydraulic conductivities and recharge rate. The simulation results are given in Table 6.11, where it is seen that all modeled head values are higher than

Table 6.11. Comparison of Observed Versus Modeled Head Values with Both Hydraulic Conductivity and Recharge Rate Doubled

Well Number	X (m)	Y (m)	I	J	K	Observed Head (m)	Simulated Head (m)	Error (m)
MW 7	246265	4692177	17	25	1	334.40	335.5	0.65
MW 12	246302	4692217	14	29	1	334.40	335.28	0.88
MW 18	246396	4692151	20	39	1	335.50	335.92	0.42
MW 101	246433	4692151	20	41	1	335.40	336.15	0.75
MW 103	246373	4692101	26	37	1	335.00	335.83	0.83
MW 105	246265	4692128	23	25	1	334.60	335.11	0.51
MW 106	246348	4692179	17	31	1	335.00	335.59	0.59
MW 112	246299	4692178	17	29	1	334.40	335.26	0.86
MW 113	246255	4692128	23	24	1	334.50	335.05	0.55
MW 115	246396	4692136	22	39	1	335.60	335.93	0.33
MW 118	246337	4692178	17	33	1	335.10	335.52	0.42

Root mean square error = 0.64

Table 6.12. Source Locations and Concentrations at the LUST Site 8LTU14

LUST Site	X (m)	Y (m)	I	J	K	Benzene Concentration (ppb)	Xylene Concentration (ppb)
8LTU14	246395	4692151	20	39	1	14,000	6,200

observed ones. This result is due to the dominance of the low-permeability Layer 1 on effective hydraulic conductivity of the hydrostratigraphic sequence. The RMSE increased from 0.19 to 0.64 with the largest error of 0.88 m at MW 12.

The contaminants of concern are benzene and xylene released from LUST site 8LTU14 (the triangle in Figure 6.23). The benzene plume and xylene plume are simulated with MT3DMS in GMS v. 3.1 based on the steady-state groundwater flow condition. Four packages—Basic, Advection, Dispersion, and Chemical Reactions—are used in MT3DMS. The method of characteristics (MOC) is selected in the Advection package.

The simulation domain for contaminant transport is the same as that for groundwater flow shown in Figure 6.23, with no solute flux across any boundary. The LUST site is treated as an internal constant concentration source with benzene at 14,000 ppb and xylene at 6200 ppb (Table 6.12). These concentrations are the highest reported for the two hydrocarbons from monitoring well samples at the source LUST site (8LTU14). The constant source condition represents a conservative scenario even though the times and amounts of petroleum releases from the sources are uncertain.

The other parameters needed in this simulation are effective porosity (n_e), dispersivity (α), distribution coefficient (K_d), and biodegradation rate (λ). These parameters have not been determined from aquifer samples, so assumptions were made based on available data and the borehole log descriptions. The value for effective porosity is estimated to be 0.2. The value for longitudinal dispersivity (α_L) is estimated based on the minimum plume length (between 8LTR14 and the farthest contaminated monitor well) (EnecoTech 2000) of about 700 ft. The estimation formula of Neuman (1990) yields a value of 73 ft (22 m). However, a value of 15 m is used for α_L because part of large-scale heterogeneity (i.e., layering) that contributes to dispersion has been considered explicitly in this numerical model. Transverse dispersivity (α_T) was taken as 0.75 m and molecular diffusion was neglected since it is much smaller than pore-scale mechanical and macro dispersion. Uncertainty in dispersivity is addressed in the sensitivity analysis. The distribution coefficient for benzene is 0.081 cm^3/g and that for xylene is 0.177 cm^3/g. These values for K_d were selected to give retardation factors of 1.5 for benzene and 4.5 for xylene, consistent with behaviors of these compounds in field studies (Wiedemeier et al. 1995). The biodegradation rate was set as 0.0001 day^{-1} for both layers and both contaminants. The values of these parameters are listed in Table 6.13.

A calibration effort in this case would involve systematically adjusting the values of effective porosity (n_e), dispersivities (α_L, α_T), biodegradation rate (λ), and

Table 6.13. Input Parameters for Contaminant Transport Modeling

Parameter	Layer 1	Layer 2
Effective porosity, n_e	0.2	0.2
Longitudinal dispersivity, α_L(m)	15	15
Transverse dispersivity, α_T (m)	0.75	0.75
Biodegradation rate, λ(day^{-1})	0.0001	0.0001
Bulk density (g/cm^3)	1.86	1.86
Distribution coefficient, K_d(cm^3/g) for benzene	0.081	0.081
Distribution coefficient, K_d(cm^3/g) for xylene	0.177	0.177

distribution coefficients (K_d) in successive simulations, and comparing the results against the observed concentrations at the monitoring wells. The transport model has not been fully calibrated. As the model now stands, the gross plume shapes from the simulations can be compared with mapped contamination from the field data.

Figure 6.27a illustrates the benzene plume in both layers after 10 years using the parameters listed in Table 6.13. Figure 6.27b is a close view of the benzene plume shown in Figure 6.27a. Figure 6.28a is the comparable xylene plume and Figure 6.28b is a close view of Figure 6.28a. The maximum length and width of the modeled benzene plume in the two layers are listed in Table 6.14. The plume boundary is set at the concentration of 100 ppb due to the accuracy of the numerical scheme used in MT3DMS. It is seen that both the length and width of the plume increase much faster at earlier years than later years and the plume, especially its width, becomes almost stable after seven or eight years. Actual site monitoring data show that neither benzene nor xylene have ever been detected in the loess layer as far downgradient as MW 7. Nor have they been detected in MW 114 (not listed, located south of MW 113), which provides a constraint on plume spreading. Benzene but not xylene has been detected in residential wells #4 and #11, indicating the contaminant plumes have entered the sand layer downgradient from the source. The lack of contaminant detection in well #6 is a constraint on plume spreading in the sand layer. With a calibrated model, evolution of the benzene and xylene plumes could be simulated with greater confidence, and predictions about plume behavior could be made.

In summary, the model simulations showed that groundwater flow direction is westward through Climbing Hill in the shallower aquifer. Low-capacity residential water wells have little effect on the local groundwater gradient. The location of many of these wells is in the direct, down gradient position from the contamination source. This fact was the main condition responsible for contamination reaching the water wells. The shale layer separating the shallow, contaminated water source from an underlying sand in the upper portion of the Dakota Formation appears to be an effective protection of the Dakota Formation.

The stratigraphic configuration of a higher permeability layer beneath a lower permeability layer, together with water wells pumping from the higher permeability

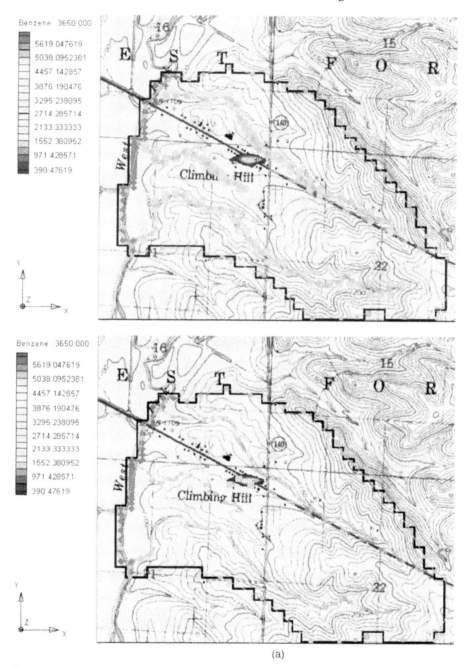

(a)

Figure 6.27. (a) Benzene concentration at 3650 days in layer 1 (*top*) and in layer 2 (*bottom*). (b) Close view of benzene concentrations at 3650 days in layer 1 (*top*) and in layer 2 (*bottom*).

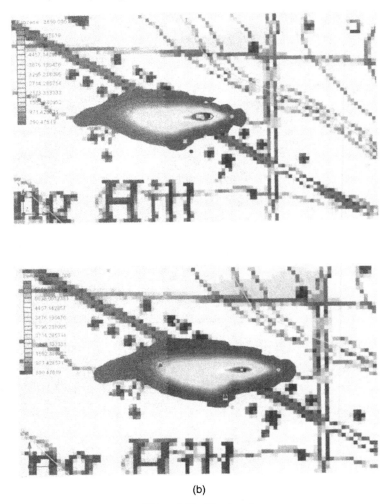

(b)

Figure 6.27. (*Continued*)

layer, creates a condition for vertical flow from the water table toward the under-lying layer. Petroleum constituents dissolved in groundwater will be transported downward from the water table into the higher permeability layer. The transport model needs to be further calibrated and its sensitivity to the parameter changes needs to be explored, although initial modeling results of the benzene and xylene plumes seem to agree reasonably well with observed concentration. If biodegrada-tion is not a discriminating or particularly influential parameter, benzene can rea-sonably be expected to migrate at three times the rate of xylene, suggesting that releases of heavier products than gasoline will be strongly retarded.

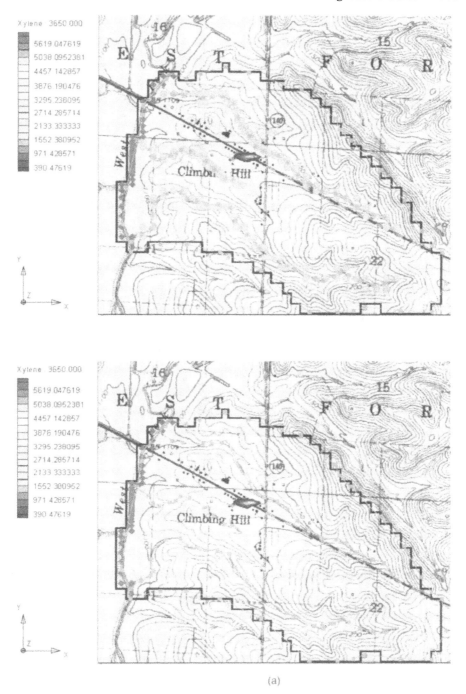

(a)

Figure 6.28. (a) Xylene concentrations at 3650 days in layer 1 (*top*) and in layer 2 (*bottom*). (b) Close view of xylene concentrations at 3650 days in layer 1 (*top*) and in layer 2 (*bottom*).

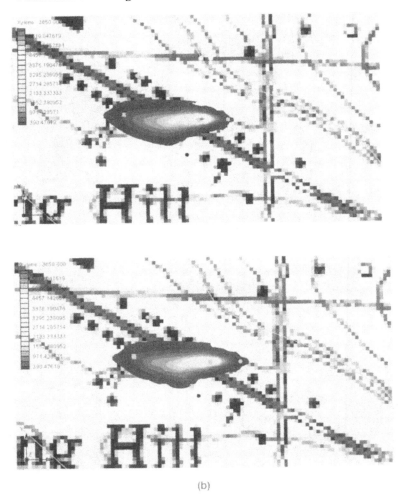

(b)

Figure 6.28. (*Continued*)

Case Study 2: Modeling of Natural Attenuation of BTEX Compounds at Hill Air Force Base, Utah

The second case study involves the work of Lu et al. (1999), who modeled the degradation of BTEX compounds at the Hill Air Force Base in Utah. The objectives of the study are (1) to simulate the fate and transport of BTEX and their associated electron acceptors, (2) to present a modeling framework to simulate the natural attenuation of BTEX compounds using RT3D, and (3) to assess the relative sensitivity of model parameters with respect to BTEX removal via natural attenuation (Lu et al. 1999).

Table 6.14. Length and Width of the Benzene Plume at Different Times

Time (year)	Layer 1		Layer 2	
	Length	Width	Length	Width
1	125	45	144	55
2	168	55	172	66
3	185	71	188	71
4	186	74	201	73
5	205	78	207	78
6	216	85	226	84
7	228	88	232	90
8	236	92	244	91
9	244	92	252	92
10	252	92	258	92

Hill AFB is located at the base of the Wasatch Mountains and the geology consists of fluvial-deltaic deposits. The site hydrogeology consists of an unconfined aquifer with poorly to moderately sorted silty fine-grained sands. Overlying the sands are clayey silt to silty clay intervals, and underlying the sand is a sequence of thinly interbedded clay to silty clay that acts to prevent migration of contaminants from overlying strata. Figure 6.29 shows the location of the facility, which is where the leaking underground storage tanks are located, and Figure 6.30 is the cross section showing the hydrogeology of the site. At the site, a 1000 gallon

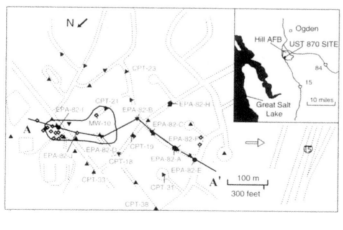

- ● slug test well location
- ▶ flow calibratrion well location
- ◇ monitoring well location
- ▼ monitoring flow calibration well location
- ▲ piezometer   ⌒ LNAPL plume (July/August 1993)   ➝ regional flow direction

Figure 6.29. Plan view of the POL site at Hill AFB with the locations of observation wells. (Adapted from Wiedemeier et al. 1995.)

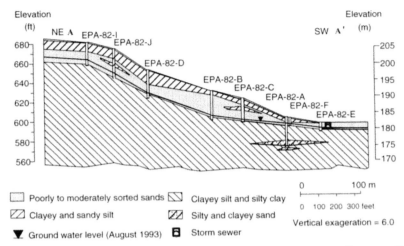

Figure 6.30. Hydrogeologic section A–A for the POL site at Hill AFB. *Note*: Datum at 4000 ft (1219.2 m) MSL. (Adapted from Wiedemeier et al. 1995.)

underground storage tank was used to store a variety of jet fuel. Excavation of the tank revealed contamination of soil and groundwater at the site. The actual source of contamination was not known with certainty nor is the total volume of fuel that leaked into the subsurface. Sampling of groundwater throughout the site found mobile LNAPLs in several monitoring wells and piezometers. Figure 6.31a is an isopach map showing the distribution and measured thickness of mobile LNAPLs at the site on December 28, 1993. Twenty monitoring wells were installed to monitor the extent of dissolved groundwater contamination in the area. The presence of a certain geochemical species, its relative concentration, and its spatial distribution (Figure 6.32a–d) support the hypothesis that the BTEX is attenuating naturally by means of multiple terminal electron-accepting processes (Wiedemeier et al. 1995). We see from this figure that, in the upgradient portion of the aquifer, BTEX concentration is high while DO, nitrate, and sulfate have low concentrations. Concentrations of Fe(II) and methane are relatively high in this area. In contrast, outside the BTEX plume, concentrations of DO, nitrate, and sulfate are higher while Fe(II) and methane concentrations are low. Figure 6.32 shows that the BTEX plume is undergoing natural attenuation through microbial activity in which oxygen, nitrate, Fe(III), and sulfate are being consumed as electron acceptors, while methane is being produced through methanogenesis.

BTEX transport at the site was simulated using the RT3D code from August 1993 through July 1994. Flow was assumed to be at steady state and was modeled using MODLFOW (with a grid of 20 rows and 30 columns). Values of selected model parameters are provided in Table 6.15. Constant head boundaries were placed on the left and right boundaries, which coincided with the direction of groundwater flow. The upper and lower boundaries were treated to be no-flow boundaries. As for the boundary conditions of the transport simulations, no mass flux was assigned at the upper and lower boundaries while a specified concentration

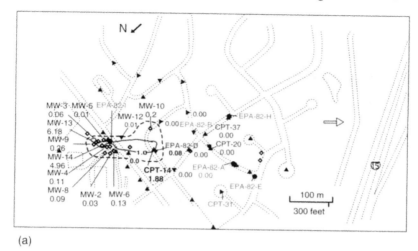

(a)

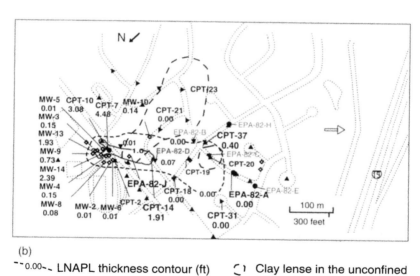

(b)

`--0.00--` LNAPL thickness contour (ft)   ⊂⊃ Clay lense in the unconfined
                                                  aquifer identified by drilling

Figure 6.31. Approximate LNAPL plume map (feet). (a) Approximate LNAPL plume from August 1, 1993 to January 9, 1994 (data shown were measured on December 28, 1993). (b) Approximate LNAPL plume from January 10 to April 25, 1994 (data shown measured on January 20, 1994). (Data from Wiedemeier et al. 1995.)

was assigned at the left boundary and a specified mass flux boundary was assigned at the right boundary.

As stated earlier, there were no data available concerning the amount of hydrocarbons spilled at the site. In order to define the initial condition of the model runs, the distribution of the plume from the July–August 1993 data set was used. The model was run and initially the flow system was calibrated against available

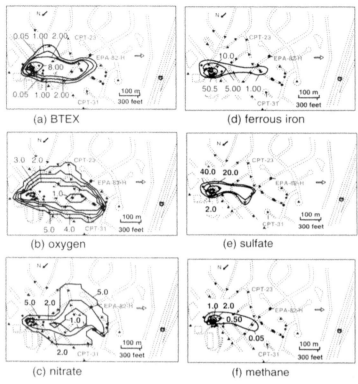

Figure 6.32. Distributions of interpolated concentrations (in mg/L): (a) BTEX, (b) dissolved oxygen (DO), (c) nitrate, (d) ferrous iron (Fe^{2+}), (e) sulfate, and (f) methane. The measurements were made in July–August 1993. (Adapted from Wiedemeier et al. 1995.)

hydraulic head data (Figure 6.33). Calibration was conducted by varying the hydraulic conductivity values to match the observed hydraulic head distribution by hand. These results are summarized in Table 6.16.

Simulation results show that the observed and modeled results are in good agreement (Figures 6.34 and 6.35). In particular, the model was able to predict the overall pattern of the plumes and the position of the plume fronts quite well. Mass balance results show that there was an 8.5% discrepancy, but considering the lack of information on the source volumes, the agreement between the observed and simulated plume movement can be considered to be quite good.

The initial values of all the first-order biodegradation constants were obtained from the literature (Rifai et al. 1995). After the flow field has been calibrated, those values were also calibrated by manual trial and error with the values noted in Table 6.17. At this site Lu et al. (1999) found that anaerobic degradation is the dominant mechanism with respect to the degradation of BTEX. Anaerobic processes accounted for 70% of BTEX degradation: 23% by denitrification, 18% by Fe(III) reduction, 25% by sulfate reduction, and 4% by methanogenesis. Aerobic degradation is attributed to the remaining 30% of the BTEX mass.

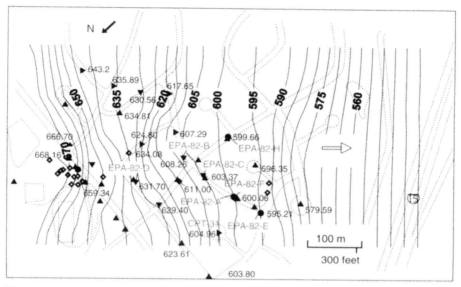

Figure 6.33. Calibrated flow field with observed head (in feet) as measured in August 1993 for POL facility site at Hill AFB.

Table 6.15. Model Parameters for the POL Facility Site at Hill AFB

Parameter	Value
K	19 ft/day (5.79 m/day)
ϕ	0.25
α_L	27 ft (8.23 m)
α_T / α_L	0.01
Discretization in:	
x-direction	110 ft (33.53 m)
y-direction	85 ft (25.91 m)
Extent of model in:	
x-direction	3300 ft (1005.84 m)
y-direction	1700 ft (518.16 m)
Background concentration (mg/L)	
BTEX	0
Oxygen	6.0
Nitrate	17.0
Ferrous iron	0.001
Sulfate	100
Methane	0.0001
Maximum Fe(II) concentration	50.5
Maximum methane concentration	2.05

Source: Adapted from Lu et al. (1999).

Table 6.16. Comparison of Measured and Simulated Piezometric Head[a] Values in the Observation Wells for July/August 1993 (units in feet)

Well Number	Measured Head (h_m)	Simulated Head (h_s)	$h_m - h_s$	$(h_m - h_s)^2$
EPA-82-B	608.28	608.35	-0.07	0.0049
EPA-82-C	603.37	604.10	-0.73	0.5329
EPA-82-D	631.70	632.69	-0.89	0.7921
EPA-82-E	595.21	594.92	0.29	0.0841
EPA-82-F	600.06	599.70	0.36	0.1296
EPA-82-H	599.66	599.27	0.14	0.0196
CPT-2	659.34	659.20	-0.32	0.1023
CPT-11	666.70	666.79	-0.09	0.0081
CPT-12	643.21	643.51	-0.30	0.0900
CPT-13	630.56	630.76	-0.20	0.0400
CPT-15	607.29	607.62	-0.33	0.1089
CPT-17	623.61	623.72	-0.11	0.0121
CPT-18	629.10	628.96	0.14	0.0196
CPT-21	624.80	624.32	0.48	0.2304
CPT-23	617.65	617.82	-0.17	0.0289
CPT-26	579.59	579.65	-0.06	0.0036
CPT-27	596.35	596.20	0.15	0.0225
CPT-31	604.83	604.75	0.08	0.0064
CPT-41	635.89	635.91	-0.02	0.0004
MW-10	634.15	634.14	0.01	0.0001
MW-12	653.76	653.65	0.11	0.0121
MW-14	668.16	668.30	-0.14	0.0196
			Mean $= -0.049$	Mean $= 0.102$

[a]Head is relative head, datum set at 40,000 ft (1219.2 m) MSL. $N = 22$, RMS $= 0.319$.
Source: Adapted from Lu et al. (1999).

A sensitivity analysis for total BTEX mass and BTEX plume front shows that the total mass is least sensitive to changes in values of dispersivity. The results of the simulations were more sensitive to changes in hydraulic conductivity, the overall rate reaction constant, and recharge (listed in decreasing sensitivity) (Table 6.18). Anaerobic processes were found to be most sensitive to changes in reaction rate constants of sulfate and Fe(III) reductions, methanogenesis, and denitrification in decreasing order. The BTEX front location was found to be strongly influenced by the saturated thickness of the aquifer, hydraulic conductivity, and longitudinal dispersivity. However, the front location was most sensitive to the reaction rate constant of oxygen, nitrate, sulfate, and Fe(II) and least sensitive to methane.

Case Study 3: Modeling of Natural Attenuation of Chlorinated Ethane Compounds at the Dover Site

Clement et al. (2000) modeled the fate and transport of PCE and TCE and their degradation products (dichloroethene, vinyl chloride, and ethane) due to a series of chemical releases at the Area-6 site at the Dover Air Force Base in Delaware.

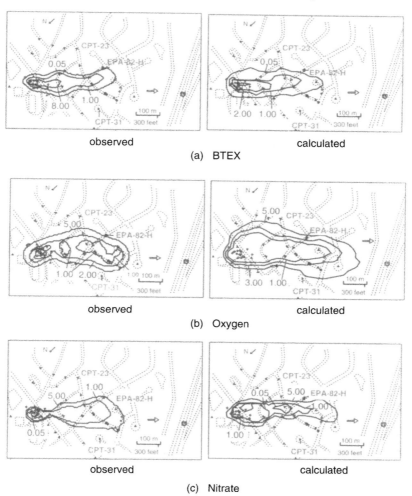

observed   calculated

(a) BTEX

observed   calculated

(b) Oxygen

observed   calculated

(c) Nitrate

Figure 6.34. Observed (July 1994) and calculated plumes (concentration in mg/L): (a) BTEX, (b) dissolved oxygen (DO), and (c) nitrate. (Observed plumes adapted from Wiedemeier et al. 1995.)

This case study shows how the sequential anaerobic degradation model implemented in RT3D can effectively model the combined transport of these contaminants and their degradation products. Since contamination problems such as the one presented here typically involve the transport and interaction of multiple contaminants, models that can simulate reactive transport of multiple species are very useful. The main objective of this study was to develop a flow-and-transport model that predicts the transport of chlorinated solvents and their degradation products at the Dover site with hopes of generalizing the modeling approach to other sites.

The site consists of a mixture of silty, sandy, and gravelly material that can be separated lithologically into two distinct zones. A silty sand that is low in permeability with a thickness of 5–10 ft composes the shallow layer. Underlying this layer is a

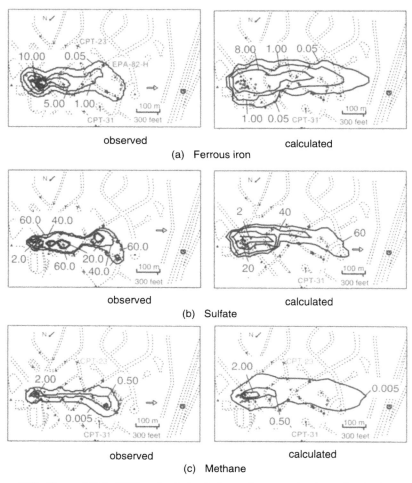

observed   calculated

(a) Ferrous iron

observed   calculated

(b) Sulfate

observed   calculated

(c) Methane

Figure 6.35. Observed (July 1994) and calculated plumes (concentration in mg/L): (a) ferrous iron (Fe^{2+}), (b) sulfate, and (c) methane. (Observed plumes adapted from Wiedemeier et al. 1995.)

Table 6.17. Estimated First-Order Biodegradation Rate Constants for the POL Facility Site at Hill AFB

Parameter	Value (day^{-1})
Aerobic degradation constant	0.051
Denitrification rate constant	0.031
Iron reduction rate constant	0.005
Sulfate reduction rate constant	0.004
Methanogenic degradation rate constant	0.002

Source: Adapted from Lu et al. (1999).

Table 6.18. Results of Parameter Sensitivity Analysis for Total BTEX Mass and BTEX Plume Front

Sensitivity Coefficient	Hydraulic Conductivity	Recharge	Longitudinal Dispersivity	Retardation Factor	DO	Nitrate	Ferrous Iron	Sulfate	Methane
						Degradation Rate Constants			
Total BTEX mass	−0.202	−0.047	−0.036	−0.058	−0.031	−0.047	−0.084	−0.111	−0.054
Front of BTEX plume	+0.563	−0.088	−0.419	−0.077	−0.331	−0.287	−0.188	−0.265	−0.003

Source: Adapted from Lu et al. (1999).

271

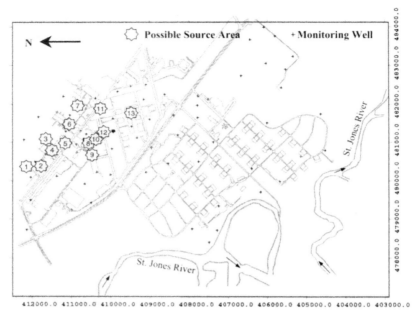

Figure 6.36. Site map of Dover Air Force Base, Delaware.

sandy gravelly deposit approximately 10–15 ft thick. Beneath the high-permeability unit is a clayey layer. The site numerical model reflects the hydrogeological conditions and the high-permeability unit is treated as a separate layer, which is bounded by a low-permeability silty sand layer at the top and a clayey layer at the bottom.

Site data indicate that TCE is the most abundant contaminant with maximum soil-phase concentration that is 40 times greater than PCE (Ball et al. 1997). However, there is no evidence of a continuous free-phase of TCE at the site. Figure 6.36 is a site map showing the possible source areas and locations of monitoring wells. Samples taken from groundwater monitoring wells downgradient of the suspected source zone (Klier et al. 1998) revealed the presence of high concentrations of degradation products of the disposed chlorinated solvents such as *cis*-DCE and VC, traces of ethane, and elevated levels of chloride ions (Figure 6.37). The presence of these degradation products suggests that natural attenuation of chlorinated solvents is taking place by biotic means. Previous studies at other sites have shown that anaerobic reductive dechlorination can be a dominant attenuation mechanism for the biotransformation of chlorinated solvents (Wiedemeier et al. 1996; Wu et al. 1998).

The model assumes that the kinetics of the degradation reactions is first-order with respect to the contaminant concentration. Processes such as the growth and decay of bacteria and the limiting availability of carbon, oxygen, and other biochemical intermediates were not simulated with the current model. This approach is considered to be valid under most natural field conditions where microbial populations are not increasing or decreasing with time (Wiedemeier et al. 1996). The

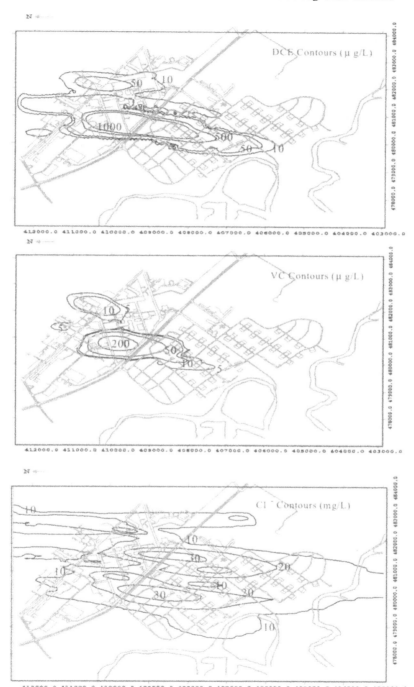

Figure 6.37. Field measured DCE, VC, and chloride plume contours.

approach may not be strictly valid when active bioremediation techniques are implemented to increase the bacteria population by means of injecting bacteria or nutrients, which can lead to increased populations. The site was conceptualized to have four distinct reaction zones. The reaction rates were varied from one zone to the next while the rates are constant spatially and temporally within a given zone over the entire simulation period.

Groundwater simulations were first run under steady-state conditions using MODFLOW to compute the hydraulic head distribution. A finite difference grid with a uniform cell size with an anisotropy factor of 0.75 was used in all simulations. A trial and error procedure was used to adjust the transmissivity values to fit the observed hydraulic head distribution. Reactive transport was simulated over

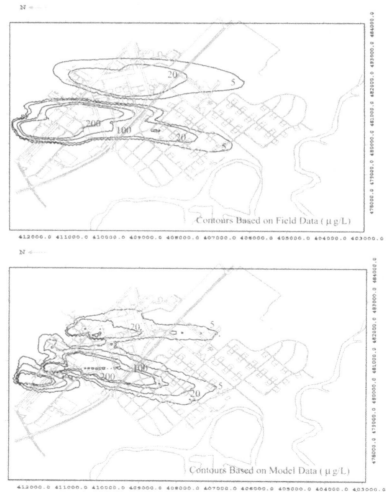

Figure 6.38. Comparison of field-measured and model-predicted PCE plume contours.

a period of 40 years by assuming retardation factors for PCE, TCE, and DCE (Ei et al. 1999; West et al. 1999) based on laboratory and field data at the site. Dispersivity values were based on literature data (Gelhar et al. 1992). The 40 year period was divided into four equal stress periods to reflect the temporal changes in contaminant release from the source zone. The mass released was determined through the model calibration process. Reaction rate constants used in the simulations were bounded by those reported in the literature (Wiedemeier et al. 1996). Model simulations were run by varying the reaction parameters and source strengths until simulated concentration values fit the measured profiles.

The concentration contours for the chlorinated solvents (PCE and TCE) computed by the reactive transport model are shown in Figures 6.38 and 6.39,

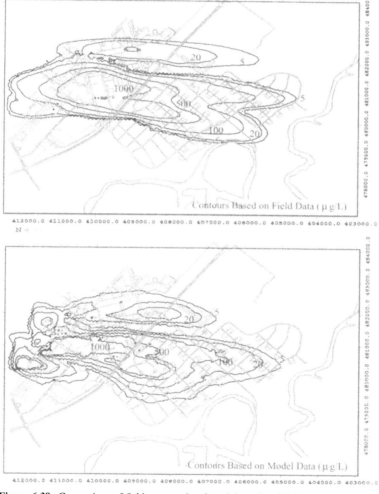

Figure 6.39. Comparison of field-measured and model-predicted TCE plume contours.

respectively, and are compared against the observed data. The concentration contours for the degradation products (DCE, VC, ETH, and Cl⁻) are also plotted in Figures 6.40 and 6.41 (compare them against measured data in Figure 6.37). Results show that the comparisons between the simulated and the observed values are quite good for the chlorinated solvent plumes and their degradation products.

The calibrated degradation rates and contaminant source mass release rates are tabulated in Tables 6.19 and 6.20, respectively. It was found that most model-estimated rate constants were within the range of values reported in the literature. The dissolved phase masses of the chlorinated solvents and their degradation products were also calculated and compared against the estimated masses based on available field data. Comparison of the two (in Table 6.21) shows very good

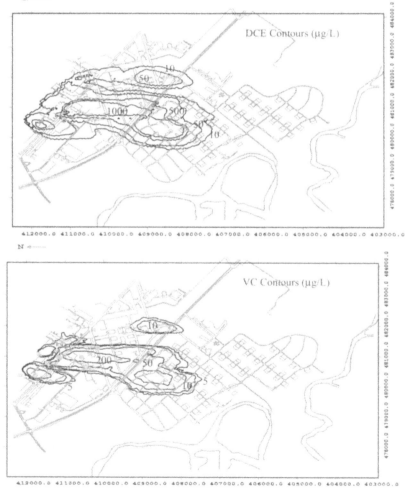

Figure 6.40. Model-predicted DCE and VC plume contours.

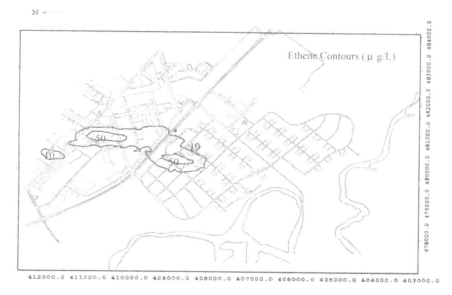

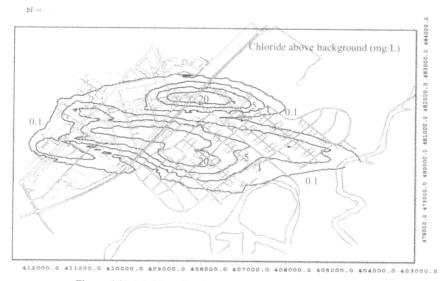

Figure 6.41. Model-predicted ethene and chloride plume contours.

agreement. Based on the simulations, the total masses of PCE and TCE released were estimated to be 1433 kg and 19,250 kg, respectively. The model-predicted chloride data provided an estimate of 19,500 kg in TCE mass that has been degraded over a 40 year period.

A sensitivity analysis was performed to study the effects of flow and transport parameters on the extent of the chlorinated solvent plume. In this particular study

Table 6.19. Calibrated Degradation Constants (day^{-1})

First-Order Rate Constant	Associated Contaminant	Anaerobic Zone 1	Anaerobic Zone 2	Transition Zone	Aerobic Zone
K_P (anaerobic)	PCE	3.2×10^{-4}	4.0×10^{-4}	1.0×10^{-4}	0.0
K_{T1} (anaerobic)	TCE	9.0×10^{-4}	4.5×10^{-4}	1.125×10^{-4}	0.0
K_{T2} (aerobic)	TCE	0.0	0.0	0.4×10^{-5}	1.0×10^{-4}
K_{D1} (anaerobic)	DCE	8.45×10^{-4}	6.5×10^{-4}	1.625×10^{-4}	0.0
K_{D2} (aerobic)	DCE	0.0	0.0	1.6×10^{-3}	4.0×10^{-3}
K_{V1} (anaerobic)	VC	8.0×10^{-3}	4.0×10^{-3}	1.0×10^{-3}	0.0
K_{V2} (aerobic)	VC	0.0	0.0	0.8×10^{-3}	2.0×10^{-3}
K_{E1} (anaerobic)	ETH	2.4×10^{-2}	1.2×10^{-2}	0.3×10^{-2}	0.0
K_{E2} (aerobic)	ETH	0.0	0.0	0.4×10^{-2}	1.0×10^{-2}

Source: Adapted from Clement et al. (2000).

Table 6.20. Estimated Contaminant Source Mass Release Rates (kg/yr)

Source Number	Stress Period #1 PCE	Stress Period #1 TCE	Stress Period #2 PCE	Stress Period #2 TCE	Stress Period #3 PCE	Stress Period #3 TCE	Stress Period #4 PCE	Stress Period #4 TCE
1	1	10	2	10	1	8	1	1
2	1	2	1	2	1	1	1	0
3	1	52	1	1	0	1	0	1
4	1	19	2	19	2	8	2	2
5	25	413	25	165	17	74	0	17
6	1	9	0	1	0	1	0	1
7	1	1	1	1	1	1	0	0
8	0	5	0	5	0	5	0	5
9	10	1	31	517	10	1	1	1
10	0	41	0	41	0	4	0	2
11	2	0	0	413	0	0	0	0
12	0	21	0	21	0	17	0	2
13	0	0	0	0	0	0	0	0

Source: Adapted from Clement et al. (2000).

Table 6.21. Comparison of Plume Mass Estimated Based on Field Measurements and Predicted Based on Model Simulation Results

Species Name	Estimated Mass (kg)	Predicted Mass (kg)
PCE	162	140
TCE	834	843
DCE	692	603
VC	36	68
Ethene	1.8	10
Excess chloride	24271[a]; 18594[b]	15801

[a]If the background value of chloride is 10 mg/L.
[b]If the background value of chloride is 12 mg/L.

Source: Adapted from Clement et al. (2000)

the TCE plume was selected because it was considered to be the largest in terms of its size and mass. The sensitivity analysis was conducted by varying the TCE anaerobic first-order reaction rate (K_{T1}), the source release rate (q_{ST}), and the transmissivity (T) values from the calibrated values, which were considered to be the baseline. Two plume descriptors—the total plume mass and the plume extent—were used to

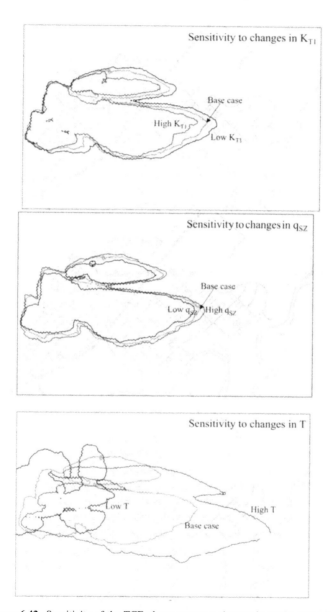

Figure 6.42. Sensitivity of the TCE plume extent to changes in model parameters.

examine the results of the sensitivity study. Results showed that the plume size was most sensitive to transmissivity and least sensitive to source loading rates (Figure 6.42). On the other hand, the total mass of the TCE plume was found to be most sensitive to the degradation rates and least sensitive to the transmissivity values.

REFERENCES

ASTM (American Society for Testing and Materials) (1995). *Standard Guide for Risk-Based Corrective Action Applied at Petroleum Release Sites*, ASTM E-1739-95. ASTM, Philadelphia, PA.

Anderson, M.P. and W.M. Woessner (1992). *Applied Groundwater Modeling: Simulation of Flow and Advective Transport*. Academic Press, San Diego, CA.

Aronson, D. and P.H. Howard (1997). *Anaerobic Biodegradation of Organic Chemicals in Groundwater: A Summary of Field and Laboratory Studies*. American Petroleum Institute, Washington, DC.

Ball, W.P., C. Liu, G. Xia, and D.F. Young (1997). A diffusion-based interpretation of tetrachloroethene and trichloroethene concentration profiles in a groundwater aquitard. *Water Resour. Res.* **33**(12):2741–2757.

Borden, R.C. and P.B. Bedient. (1986). Transport of dissolved hydrocarbons influenced by oxygen-limited biodegradation, 1. Theoretical development. *Water Resour. Res.* **22**(13), 1973–1982.

Clement, T.P. (1997). *RT3D- A Modular Computer Code for Simulating Reactive Multi-Species Transport in 3-Dimensional Groundwater Aquifers*, PNNL-11720. Pacific Northwest National Laboratory, Richland, WA.

Clement, T.P. and C.D. Johnson (1998). *Modeling Natural Attenuation of Chlorinated Solvent Plumes at the Dover Air Force Base Area-6 Site*. Draft Report, Pacific Northwest National Laboratory, Richland, WA.

Clement, T.P. and N.L. Jones (1998). *RT3D Tutorials for GMS Users*, PNNL-11805. Pacific Northwest National Laboratory, Richland, WA.

Clement, T.P., Y. Sun, B.S. Hooker, and J.N. Petersen (1998). Modeling multispecies reactive transport in ground water. *Groundwater Monitoring & Remediation*, 79–92.

Clement, T.P., C.D. Johnson, Y. Sun, G.M. Klecka, and C. Bartlett (2000). Natural attenuation of chlorinated solvent compounds: model development and field-scale application. *J. Contam. Hydrol.* **42**(2-4):113–140.

Doherty, J., L. Brebber, and P. Whyte (1994). *PEST: Model Independent Parameter Estimation*. Watermark Computing, Brisbane, Australia.

Ei, T.A., D.E. Ellis, E.J. Lutz, M.W. Holmes, and G.M. Klecka (1999). *Transect Study of the Intrinsic Bioremediation Test Plot, Dover Air Force Base, Dover, DE*. Remediation Technologies Development Forum, Draft Report.

Environmental Modeling Research Laboratory (2003). Groundwater Modeling System—Tutorials, Volume IV.

Gelhar, L.W., C. Welty, and K.W. Rehfeldt (1992). A critical review of data on field-scale dispersion in aquifers. *Water Resour. Res.* **28**(7):1955–1974.

Goode, D. J. (1990). Particle velocity interpolation in block centered finite difference groundwater flow models. *Water Resour. Res.* **26**(5):925–940.

Gray, W.G. (1984). Comparison of finite difference and finite element methods. In *Fundamental of Transport Phenomena in Porous Media*, J. Bear and M.Y. Corapcioglu (Eds.), NATO ASI Series F, No. 82. Martinus Nijhoff Publishers, The Netherlands, pp. 899–952.

Haggerty, R. and S.M. Gorelick (1994). Design of multiple contaminant remediation: sensitivity to rate-limited mass transfer. *Water Resour. Res.* **30**(2):435–446.

Harbaugh, A.W. and M.D. McDonald (1996). *User's Documents for MODFLOW-96, an Update to the U.S. Geological Survey Modular Finite-Difference Ground-Water Flow Model*. U.S. Geological Survey, Open-File Report 96–485.

Huyakorn, P.S. and G.F. Pinder (1983). *Computational Methods in Subsurface Flow*. Academic Press, San Diego, CA.

Johnson, C.D., R.S. Skeen, D.P. Leigh, T.P. Clement, and Y. Sun (1998). Modeling natural attenuation of chlorinated ethenes at a Navy site using the RT3D code. In *Proceedings of WESTEC 98 Conference*, sponsored by Water Environmental Federation, Orlando, FL, October 3–7.

Kitanidis, P.K. (1997). Comment on "A reassessment of the groundwater inverse problem" by D. McLaughlin and L.R. Townley. *Water Resour. Res.* **33**(9):2199–2202.

Klier, N.J., G.M. Klecka, E.J. Lutz, D.E. Ellis, F.H. Chapelle, and M.E. Witt. (1998). *The Groundwater Geochemistry of Area-6, Dover Air Force Base, Dover DE*. National Technical Information Service, Washington, DC.

Konikow, L.F. and J.D. Bredehoeft (1978). *Computer Model of Two-Dimensional Solute Transport and Dispersion in Ground Water*. U.S. Geological Survey Water-Resources Investigations Book 7, Chap. C2.

Konikow, L.F., D.J. Goode, and G.Z. Hornberger (1996). *A Three-Dimensional Method-of-Characteristics Solute-Transport Model (MOC3D)*. U.S. Geological Survey Open-File Report 96-4267.

Lu, G.P., T.P. Clement, C.M. Zheng, and T.H. Wiedemeier (1999). Natural attenuation of BTEX compounds: model development and field-scale application. *Groundwater* **37**(5):707–717.

McLaughlin, D. and L.R. Townley (1996). A reassessment of the groundwater inverse problem. *Water Resour. Res.* **32**(5):1131–1161.

National Research Council (2001). *Conceptual Models of Flow and Transport in the Fractured Vadose Zone*. National Academy Press, Washington, DC.

Neuman, S.P. (2003). Maximum likelihood Bayesian averaging of uncertain model predictions. *Stochastic Environ. Res. Risk Assess.* **17**(5):291–305.

Neuman, S.P. and P.J. Wierenga (2003). *A Comprehensive Strategy of Hydrogeologic Modeling and Uncertainty Analysis for Nuclear Facilities and Sites*, NUREG/CR-6805. U.S. Nuclear Regulatory Commission, Office of Nuclear Regulatory Research, Washington, DC.

Pankow, J.F. and J.A. Cherry (1996). *Dense Chlorinated Solvents and Other DNAPLs in Groundwater: History, Behavior, and Remediation*. Waterloo Press, Guelph, Ontario.

Poeter, E.P. and M.C. Hill (1999). UCODE, a computer code for universal inverse modeling. *Comput. Geosci.* **25**(4):457–462.

Rifai, H.S., R. Borden, J. Wilson, and C. Ward (1995). Intrinsic bioattenuation for subsurface restoration. In *Intrinsic Bioremediation*, R. Hinchee, J. Wilson, and D.C. Dowey (Eds.). Battelle Press, Columbus, OH, pp. 1–30.

Rifai, H.S. and P.B. Bedient (1990). Comparison of biodegradation kinetics with an instantaneous reaction model for groundwater. *Water Resour. Res.* **26**(4):637–645.

Rifai, H.S., P.B. Bedient, J.T. Wilson, K.M. Miller, and J.M. Armstrong (1988). Biodegradation modeling at aviation fuel spill site. *J. Environ. Eng.* **114**(5):1007–1029.

Rifai, H.S., C.J. Newell, J.R. Gonzales, S. Dendrou, L. Kennedy, and J. Wilson (1997). *BIOPLUME III Natural Attenuation Decision Support System Version 1.0 User's Manual.* Prepared for the U.S. Air Force Center for Environmental Excellence, Brooks Air Force Base, San Antonio, TX.

Smith, G.D. (1985). *Numerical Solution of Partial Differential Equations: Finite Difference Methods*, 3rd ed., Oxford Applied Mathematics and Computing Science Series. Clarendon Press, Oxford, UK.

Sun, Ne-Zheng (1994). *Inverse Problems in Groundwater Modeling.* Kluwer, Dordrecht, The Netherlands.

West, R.J., J.M. Odom, G. Coyle, D.E. Ellis, J.W. Davis, and G.M. Klecka (1999). *Physical and Biological Attenuation of Chlorinated Solvents at Area-6, Dover Air Force Base, Dover, DE.* Remediation Technologies Development Forum, Draft Report.

Wiedemeier, T.H., J.T. Wilson, D.H. Kampbell, R.N. Miller, and J.E. Hansen (1995). *Technical Protocol for Implementing the Intrinsic Remediation with Long-Term Monitoring for Natural Attenuation of Fuel Contamination Dissolved in Groundwater (Volumes I and II).* U.S. Air Force Center for Environmental Excellence, Brooks Air Force Base, San Antonio, TX.

Wiedemeier, T.H., M.A. Swanson, D.E. Moutoux, E.K. Gordon, J.T. Wilson, B.H. Wilson, D.H. Kampbel, J. Hansen, and P. Haas (1996). *Technical Protocol for Evaluating Natural Attenuation of Chlorinated Solvents in Groundwater.* Air Force Center for Environmental Excellence, Technology Transfer Division, Brooks AFB, San Antonio, TX.

Wiedemeier, T.H., H.S. Rifai, C.J. Newell, and J.T. Wilson (1999). *Natural Attenuation of Fuels and Chlorinated Solvents in the Subsurface.* John Wiley & Sons, Hoboken, NJ.

Wu, W., J. Nye, M.K. Jain, and R.F. Hickey (1998). Anaerobic dechlorination of trichloroethylene (TCE) to ethylene using complex organic materials. *Water Res.* **32**(5):1445–1454.

Yeh, W.W.-G. (1986). Review of parameter identification procedures in groundwater hydrology: the inverse problem. *Water Resour. Res.* **22**(1):95–108.

Zhang, D. (2002). *Stochastic Methods for Flow in Porous Media—Coping with Uncertainties.* Academic Press, San Diego, CA.

Zhang, Y.-K. and R. Heathcote (2001). Guidelines for Numerical Modeling in Tier 3 Assessment and Other Corrective Actions, Final Report submitted to Iowa Comprehensive Petroleum Underground Storage Tank Fund Board, AON Risk Services, West Des Moines, IA.

Zheng, C. (1990). *MT3D: A Modular, Three-Dimensional Transport Model for Simulation of Advection, Dispersion and Chemical Reactions of Contaminants in Groundwater Systems.* Report to the U.S. Environmental Protection Agency, Ada, OK.

Zheng, C. and G.D. Bennett (2002). *Applied Contaminant Transport Modeling*, 2nd ed. John Wiley & Sons, Hoboken, NJ.

Zheng, C. and P.P. Wang (1999). *MT3DMS, A Modular Three-Dimensional Multi-species Transport Model for Simulation of Advection, Dispersion and Chemical Reactions of Contaminants in Groundwater Systems; Documentation and User's Guide.* U.S. Army Engineer Research and Development Center Contract Report SERDP-99-1, Vicksburg, MS.

Zienkiewicz, O.C. (1977). *The Finite Element Method*, 3rd ed. McGraw Hill, New York.

7

FIELD AND LABORATORY METHODS TO DETERMINE PARAMETERS FOR MODELING CONTAMINANT FATE AND TRANSPORT IN GROUNDWATER

"We shall not cease from exploration
And the end of all our exploring
Will be to arrive where we started
And know the place for the first time."

—*Four Quartets: Little Darling,* T.S. Elliot

"A monitoring well is just a hole in the ground that lies to you."

—Mike Barcelona

An extensive investment in site characterization is often necessary in preparation for mathematical modeling, planning remediation, or demonstrating the viability of natural attenuation. To obtain site-specific parameters for fate-and-transport models, it is important to characterize the geology, hydrology, and chemistry of the site as well as the budget permits. For an existing contaminant plume, this requires adequate sampling and measurement of many aspects of the site. Important variables that require characterization include contaminant type and its source location, concentration, dimensions, and spatial distribution. Hydrological and geochemical parameters of importance include hydraulic conductivity, specific storage, porosity, bulk density, and organic carbon content of the aquifer. Additionally, site assessment generally requires preparation of lithologic logs, geological cross sections, and contour maps of the groundwater table. From this information, parameters such as hydraulic gradient, retardation coefficients, dispersivity, and biodegradation rate coefficients can be estimated.

Bioremediation and Natural Attenuation: Process Fundamentals and Mathematical Models
By Pedro J. J. Alvarez and Walter A. Illman   Copyright © 2006 John Wiley & Sons, Inc.

This chapter discusses techniques currently in common use for determining the physicochemical parameters that are used in fate-and-transport models. The reader must keep in mind that standard techniques are continuously subject to improvement, and new instruments to measure various parameters are being introduced every year. Therefore, it is important to keep up with new developments in site characterization techniques. In the sections that follow, techniques for measurement of parameters associated with groundwater flow are discussed first, followed by estimation methods for parameters associated with contaminant fate and transport.

7.1. PARAMETERS FOR GROUNDWATER FLOW MODELING

7.1.1. Hydraulic Gradient (dh/dx)

The advective transport of contaminants is controlled by the hydraulic gradient together with hydraulic conductivity and the effective porosity of the aquifer. Determination of hydraulic gradient allows one to obtain the magnitude and direction of groundwater and contaminant transport velocity. For unconfined aquifers, the direction of the hydraulic gradient often follows the topography.

To determine the hydraulic gradient accurately, it is necessary to measure groundwater levels at several locations. Measurements of water levels are taken in monitoring wells or piezometers. Commonly, a water level meter consisting of a permanently marked polyethylene tape embedded with conductors is used to measure the distance from the top of the well casing to the water level (Figure 7.1). The water level is measured when the probe contacts well water, completing the circuit. A buzzer and a light signal the completion of the circuit, indicating the depth to the water level. It is a simple device to collect measurements of water levels, but one main disadvantage of this approach is that a hydrogeologist must physically take a measurement at each sampling point. The water level in this manner is subtracted from the elevation of the measurement point (commonly taken as the top of the well casing) to yield the water level elevation in the well. One can then convert these values to hydraulic head (h) with reference to some arbitrary datum such as sea level.

Alternatively, a pressure transducer can be employed for unattended monitoring of water levels. Data may be stored on a data acquisition system located on the soil surface (Figure 7.2). The data acquisition system controls the time interval at which data is collected from the pressure transducer. Commercially available pressure transducers are easily programmable and allow measurements at selected time intervals. There are now pressure transducers that contain a data logger in a small stainless steel case, which avoids placement of the logger at the soil surface. These new pressure transducers allow for continuous measurement of water levels over an extended period, minimizing costs for site investigations (Figure 7.3).

A pressure transducer measures the pressure (p) above the water column in the well. One can then obtain the hydraulic head by using Eq. [4.11].

Example 7.1. Calculation of Hydraulic Head from Pressure Measurements Made in a Well. Calculate the hydraulic head for a pressure measurement made

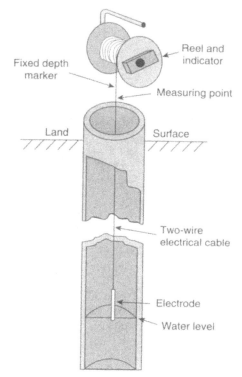

Figure 7.1. Measurement of depth to well water level using an electric tape. (Adapted from Schwartz and Zhang 2003.)

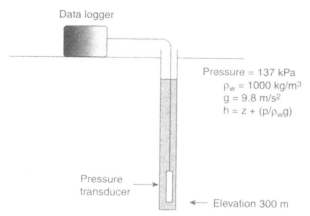

Figure 7.2. Measurement of hydraulic head by means of a pressure transducer and data acquisition system. (Adapted from Schwartz and Zhang 2003.)

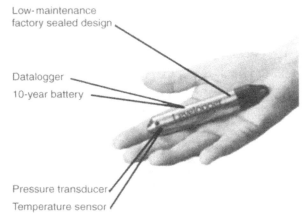

Low-maintenance
factory sealed design

Datalogger
10-year battery

Pressure transducer
Temperature sensor

Figure 7.3. Modern pressure transducers that house dataloggers, allowing for low-cost unattended monitoring of hydraulic head. (From Solinst.)

by a pressure transducer placed in a well at an elevation of 150 m above sea level. The pressure reading is 120 kPa. We note that the density of water (ρ_w) is 1000 kg/m^3 and the gravitational constant (g) is 9.8 m/s^2. Using Eq. [4.11], one can calculate the hydraulic head to be

$$h = z + \frac{p}{\rho g} = 150 \text{ m} + \frac{120 \times 10^3 \text{ Pa}}{(1000 \text{ kg/m}^3)(9.8 \text{ m/s}^2)} = 162.2 \text{ m}$$

The hydraulic gradient is determined by taking measurements of hydraulic head in at least three noncolinear monitoring wells or piezometers within the aquifer of interest at a site. However, more than three measuring locations are desirable because hydraulic gradients can be variable over a short distance within an aquifer. The frequency of water level measurements and the timing of measurements relative to significant recharge events can also affect the inferred hydraulic gradient. As seasonal variations in the hydraulic gradient can have a significant influence on the directions and rates of contaminant transport, quarterly or more frequent hydraulic head measurements should be taken at the same wells over several years. The use of multiple pressure transducers such as the one shown in Figure 7.3 allows one to obtain continuous hydraulic head measurements, which can then be used to monitor the seasonal fluctuations of hydraulic gradient across the field site.

For the *de minimus* three monitor well technique, a trigonometric problem is solved to arrive at the hydraulic gradient magnitude and direction. This technique is improved by placing the wells in a more or less equilateral triangular arrangement. The required data for this method are (1) the spatial locations and distances between the wells and (2) the total head (water level measurements) at each well (Figure 7.4). Groundwater elevation data from the wells can be used to contour equipotential lines and subsequently construct flow nets (Domenico and Schwartz

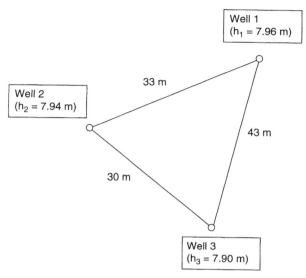

Figure 7.4. Determination of hydraulic gradient and flow direction by triangulation.

1998). The direction of groundwater flow is always from the higher to lower hydraulic heads and in isotropic media is perpendicular to the hydraulic head contour (called the equipotential) lines. The hydraulic gradient can also be calculated using this graphical approach, simply as the difference between the two equipotentials divided by the distance between them.

Example 7.2. Determination of Hydraulic Gradient Based on Trigonometric Approach

Steps in Finding the Hydraulic Gradient and Flow Direction Using Three Wells

1. Identify the well with the intermediate water elevation.
2. Determine the point between the highest and lowest head well that corresponds to the hydraulic head of the intermediate well (Point A). This can be done by linear interpolation.
3. Draw a line between this point (from 2) and the intermediate well and extend this line if necessary. This is an equipotential line for the midlevel hydraulic head.
4. Find a line that is perpendicular to this equipotential line that goes through the well with the lowest hydraulic head. This new line is the flow line and has a flow direction toward the well with the smallest hydraulic head.
5. The hydraulic gradient is equal to the difference between the intermediate and the lowest hydraulic head divided by the length of the flow line.

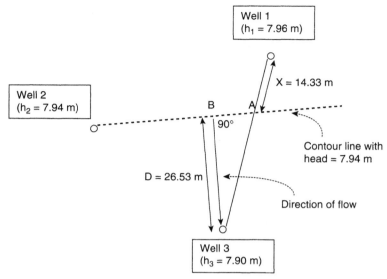

Figure 7.5. Sketch for determining the direction and hydraulic gradient for the example in Figure 7.4.

Sample Calculation Based on Figure 7.4

1. Well 2 has the intermediate hydraulic head ($h_2 = 7.94\,\text{m}$).
2. Determine Point A (i.e., $h = 7.94$ m) between wells 1 and 3, by interpolation (Figure 7.5). The distance between well 1 and well 3 is 43 m. Therefore, Point A is at a distance X from well 1. By linear interpolation, $X = [43(7.96 - 7.94)]/(7.96 - 7.90) = 14.33$ m.
3. Draw the equipotential line connecting Point A and the intermediate hydraulic head well (well 2). This is the dotted line depicted in Figure 7.5.
4. Draw a flow vector that is perpendicular to this equipotential line and points toward the well with the lowest hydraulic head. This vector defines the flow direction.
5. The magnitude of the flow vector can be determined graphically using a map scale. Alternatively, the triangle with vertices at Point A, Point B, and well 3 can be determined using basic trigonometric relationships such as the law of sines and cosines. In this example, the magnitude of the flow vector (Point B to well 3) is denoted by D and it is graphically measured to be 26.53 m. Therefore, the hydraulic gradient is equal to $i = (7.94 - 7.90)/(26.53) = 0.0015$.

A more accurate representation of hydraulic head, and hence hydraulic gradient variation, at a site is obtained with a large array of measuring points. Figure 7.6 is a contour plot of hydraulic head measurements made at the Till Hydrology Research Site (THRS) near Iowa City, Iowa. It shows that the hydraulic head measurements are higher at the southern end of the site relative to the northern portion. From this, one can determine the groundwater flow direction.

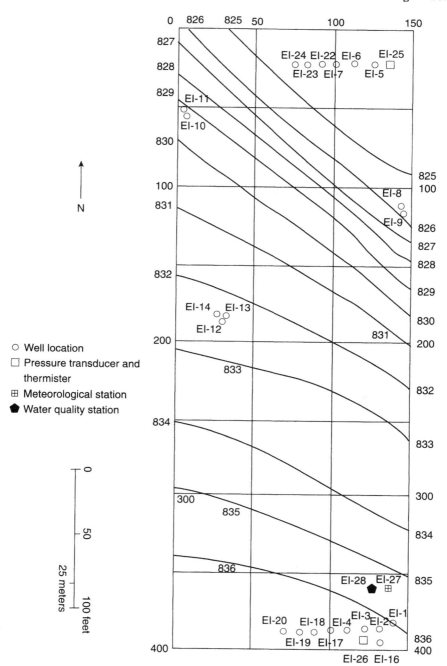

Figure 7.6. Countour plot of hydraulic head from the Till Hydrology Research Site, near Iowa City, IA. The base map indicates the location of meteorological station, observation wells, water quality monitoring station, pressure transducers, and thermistors. (Modified after Bowman 1992.)

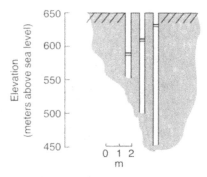

Figure 7.7. Piezometer nest used to determine the vertical hydraulic gradient. (Adapted from Freeze and Cherry 1979.)

It is good practice to extend a monitoring well array beyond the boundaries of a site to obtain a better representation of hydraulic head distribution outside the investigated region. In addition, it is advisable to determine hydraulic head and gradients in three dimensions because at many sites flow can take place vertically. In fact, it is important to remember that groundwater flow takes place in three dimensions. Despite this recognition, measurements of hydraulic head continue with monitoring wells with long screens, which essentially cause depth averaging of hydraulic heads. This depth-averaging process essentially leads to two-dimensional characterization.

Three-dimensional measurements of hydraulic head can be taken using nested piezometers (Figure 7.7) completed at different depths or multilevel measurement and sampling devices (Figure 7.8). Both devices allow for the determination of vertical hydraulic gradients. Measurements of hydraulic head and gradient in three dimensions provide hydrogeologists with a better understanding of the three-dimensional groundwater flow paths and consequently the subsurface distribution of contaminants.

7.1.2. Hydraulic Conductivity (*K*)

Hydraulic conductivity (*K*) is an aquifer property that quantifies the ease of water flow through a porous medium. It has the dimensions of $[LT^{-1}]$ and is a property not only of the porous (or fractured) medium but also of the fluid passing through the medium. The relationship between *K* and permeability (*k*) $[L^2]$ is

$$K = \frac{k\rho g}{\mu} \qquad [7.1]$$

where ρ is fluid density $[ML^{-3}]$ and μ is dynamic viscosity $[ML^{-1}T^{-1}]$. The permeability *k* is only a function of the geologic medium and is a preferred parameter when flow of multiphase fluids is considered. Determination of a representative

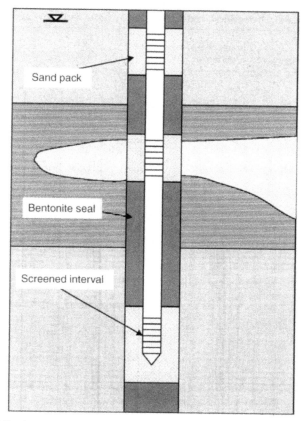

Figure 7.8. Multilevel measurement device to monitor hydraulic head and contaminant concentrations in various units including a sand channel forming a preferential flow path.

hydraulic conductivity value is one of the most important tasks in understanding aquifer hydraulics. As Eq. [4.16] shows, the velocity of groundwater and the transport rate of dissolved contaminants are directly proportional to hydraulic conductivity. Reported values of hydraulic conductivity range over 10 orders of magnitude (Table 7.1). Also, within a given type of material and aquifer, the hydraulic conductivity commonly varies several orders of magnitude. Because of this large variation, good site characterization requires the measurement of hydraulic conductivity at numerous locations at a given site.

There are a number of methods available to determine the hydraulic conductivity of the aquifer material including empirical methods based on correlation with grain sizes, laboratory testing of core samples obtained during drilling, and field aquifer tests. Field aquifer tests are considered to be the most reliable as they are conducted at the investigated site. However, pumping of contaminated groundwater may pose potential problems of waste disposal and may exacerbate the contamination problem by moving the plume to an undesirable location.

Table 7.1. Values of Hydraulic Conductivity for Various Sediment and Rock Types

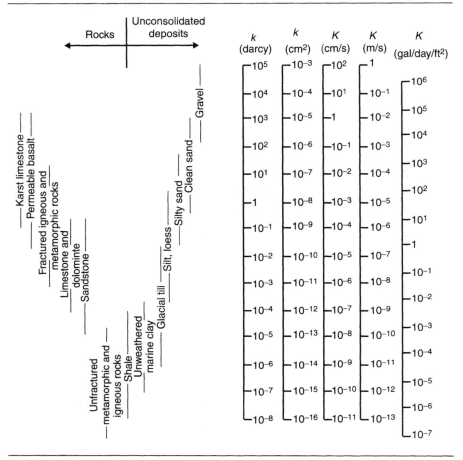

Source: Adapted from Freeze and Cherry (1979).

Laboratory tests can also be performed on samples taken from a site and include a constant head or falling head permeameter test, and particle size analysis. Of the two lab test types, particle size analysis is the least expensive, but also the least accurate, and the representative scale is uncertain, but likely small.

In environmental site assessments, slug tests are widely used because they provide reasonable data reliability at a relatively low cost. Discussion of pumping test techniques and permeameter testing can be found in any standard hydrogeology textbook, and most geotechnical engineering firms have the capability of doing permeameter tests. In the following sections we discuss the often-employed techniques of particle size, slug test analysis, and pumping tests in confined and unconfined aquifers.

Grain Size Analysis

A number of researchers have recognized that there is a relationship between the grain size and hydraulic conductivity and permeability of granular porous media. Such relationships can be useful, particularly when direct measurements of hydraulic conductivity and permeability are sparse. It requires the choice of a representative grain size diameter. Examples of empirical relationships are provided in Table 7.2. Estimating hydraulic conductivity and permeability from grain size analysis works reasonably well when aquifers are composed of well-sorted (fair to poorly

Table 7.2. Empirical Relationships for Estimating Hydraulic Conductivity and Permeability from Grain Size Analysis

Source	Equation	Parameters
Hazen (1911)	$K = Cd_{10}^2$	K = hydraulic conductivity (cm/s) C = constant 100–150 (cm/s)$^{-1}$ for loose sand; see Table 7.3 d_{10} = effective grain size (cm) (10% of particles are finer, 90% coarser)
Harleman et al. (1963)	$k = 6.54 \times 10^{-4} d_{10}^2$	k = permeability (cm^2)
Krumbein and Monk (1943)	$k = 760 d^2 e^{-1.31\sigma}$	k = permeability (darcys) d = geometric mean of grain diameter (mm) σ = log standard deviation of the size distribution
Kozeny (1927)	$k = Cn^3/S^{*2}$	C = dimensionless constant: 0.5, 0.562, and 0.597 for circular, square, and equilateral pore openings, respectively k = permeability (L^2) n = porosity S^* = specific surface–interstitial surface areas of pores per unit bulk volume of the medium
Kozeny–Carman–Bear (1972)	$K = \left(\dfrac{\rho_w g}{\mu}\right) \dfrac{n^3}{(1-n)^2} \left(\dfrac{d_m^2}{180}\right)$	K = hydraulic conductivity ρ_w = fluid density μ = fluid viscosity g = gravitational constant d_m = any representative grain size n = porosity
Shepherd (1989)	$K = a(d_{50})^b$	K = hydraulic conductivity (gal/day/ft^2) d_{50} = median grain diameter (mm) a, b = constants; see Table 7.4

Source: Adapted from Schwartz and Zhang (2003).

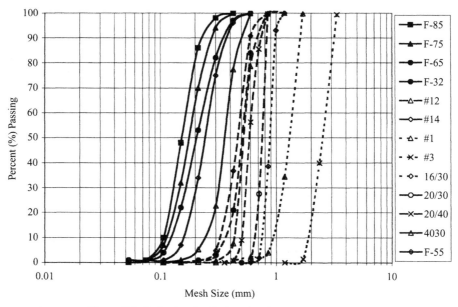

Figure 7.9. Grain size distribution of well-sorted sandy material.

graded), unconsolidated material such as silt, sand, or gravel. These methods are not effective when the aquifer material contains a large fraction of clay size material.

Figure 7.9 shows the grain size distribution of a wide variety of sandy material that is considered to be well sorted. The grain size distribution varies from one sand type to the next. This variation is one important cause for the large variability in hydraulic conductivity.

Hazen Method. The Hazen (1911) method is based on a power law relationship between the effective grain size, d_{10}, and hydraulic conductivity. It takes the form

$$K = C(d_{10})^2 \qquad [7.2]$$

where C is a constant based on the sorting and mean grain size selected from Table 7.3. The d_{10} value is obtained from a grain size gradation curve obtained from sieve

Table 7.3. Values of C for the Hazen Method

Mean Grain Size	Sorting	C (dimensionless)
Very fine to fine sand	Poorly to moderately sorted	40–80
Medium sand	Well sorted	80–120
Coarse sand	Poorly sorted	80–120
Coarse sand	Well sorted	120–150

Source: Fetter (2001).

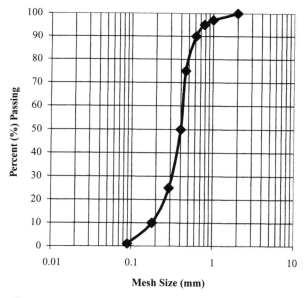

Figure 7.10. Grain size distribution of alluvial aquifer material.

analysis. The d_{10} is defined as the grain size diameter at which 10% by weight of the soil particles are finer and 90% are coarser. This approach is used for materials having grain size diameter ranging from silt (0.1 mm) to gravel (3 mm).

Example 7.3. Determination of Hydraulic Conductivity (K) Using the Hazen Method. Figure 7.10 shows an example particle size distribution curve resulting from a sieve analysis of alluvial aquifer material. From the particle size distribution, we estimate the grain diameter representing 10% of the sample (d_{10}) to be 0.19 mm. This diameter is converted to centimeters and used in Eq. [7.2] to estimate the hydraulic conductivity with units of cm/s for the material:

$$K = (40 \text{ to } 80)(0.019)^2 = 0.013 \text{ to } 0.028 \text{ cm/s, depending on the value of } C$$

where we have taken C to be representative of very fine to fine sand.

Shepherd Method. Another method to estimate hydraulic conductivity from particle size analysis was reported by Shepherd (1989). Similar to the Hazen method, Shepherd's method uses grain size data from sieve analysis, but the median grain diameter, d_{50}, is used instead of d_{10}. For naturally occurring sandy materials, the empirical formula takes the form

$$K = a(d_{50})^b \qquad [7.3]$$

Table 7.4. Values of a and b for the Shepherd Method

Depositional Environment	a	b
Alluvial channel (less mature)	3500	1.65
Beach	12,000	1.75
Dune (more mature)	40,000	1.85
Consolidated	800	1.5

where hydraulic conductivity has units of gallons per day per square foot (gal/day/ft^2), d_{50} has units of mm, and a and b have a range of values depending on texture, as given in Table 7.4. Values of a and b were obtained by regression analysis and represent high correlation coefficients (r^2) for the data sets statistically analyzed by Shepherd. In this method, the environment of deposition and textural maturity (sorting and roundness) of the sediment are related to the hydraulic conductivity. Parameter values are also given for consolidated sediment, but these must be used with caution, as cementation of sediment particles, in conjunction with consolidation, can cause erroneous values of estimated hydraulic conductivity through the use of Eq. [7.3]. The d_{50}, for which this relationship is valid, ranges between silt size particles (<0.1 mm) and gravel (<10 mm).

Example 7.4. Determination of Hydraulic Conductivity (K) Using the Shepherd Method. Using d_{50} from Figure 7.10, and noting the sample is from a buried alluvial channel where a and b are 3500 and 1.65, respectively, Shepherd's formula yields

$$K = 3500(0.4)^{1.65} = 772 \, \text{gpd/ft}^2 = 0.04 \, \text{cm/s}$$

The computed value is similar to the value obtained from the Hazen formula.

Slug and Bail Tests

A slug test permits an estimate of hydraulic conductivity in a piezometer or a monitoring well. It has several important advantages over the traditional pumping test such as its simplicity, significantly lower cost, shorter test duration, and lack of pumped water that can cause disposal problems, particularly when the pumped water is contaminated. However, because the slug test is conducted using a single well, the value of hydraulic conductivity obtained is representative of the geological formation only in the immediate vicinity of the tested well. This local value of hydraulic conductivity may not be representative of a large portion of the aquifer, which may contain higher hydraulic conductivity features that could cause contaminant migration.

Interpretive methods commonly used today derive from the seminal work by Hvorslev (1951) and subsequent theoretical advancements by Cooper et al. (1967) and Bouwer and Rice (1976). Discussions of slug test theory and methodology

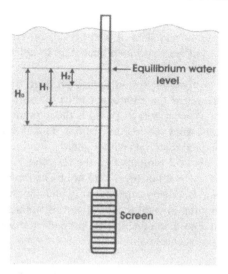

Figure 7.11. Displacement of water level from the equilibrium position in a monitoring well and its recovery during a slug test. The recovery in water level to equilibrium conditions is monitored as a function of time to determine hydraulic conductivity from a slug test. (Adapted from Schwartz and Zhang 2003.)

can also be found in Dawson and Istok (1991), Kruseman and de Ridder (1991), Wiedemeier et al. (1995a,b), and Butler (1998). The slug test interpretive techniques discussed below are limited to unconfined aquifers, which are the shallowest and are most readily contaminated by human activities.

Slug tests involve inducing a sudden displacement of water level in the well from the pretest equilibrium level, then measuring the recovery of the water level in the well (Figure 7.11). The means by which the water level change is initiated commonly involves the use of a solid rod of known volume (the *slug*), although a bailer is sometimes used when initiation involves removing a fixed volume of water from the well (called the *bail test*). When a previously submerged slug is suddenly removed from a well, the test is known as a *rising head test*. For the latter approach, the slug must first be submerged and the water level must come into equilibrium before the test can commence. When a slug is suddenly dropped into the well to a position below the static water level, the test is known as a *falling head test*. Techniques involving pressurized air to induce instantaneous change in the water pressure head have also been developed and are called *pneumatic slug tests* (Butler 1998).

Measurements necessary to accurately interpret slug tests include (1) the well geometry and (2) water level recovery data. To obtain slug test estimates of hydraulic conductivity that are accurate, measurements must be carefully made. For the slugged well, its total radius, its depth, the casing radius, the length of the well screen through which water flows, and the volume of the slug must be accurately known. In wells with partially screened intervals, the sand pack porosity must also be known. For the slug test, the pretest equilibrium water level, the maximum water displacement at test initiation, and the time from initiation to recovery must be measured accurately. Partial recovery data of water levels have also been used to infer hydraulic conductivities but we recommend that data is collected until full recovery in water level is achieved.

Time and recovery are best measured with a pressure transducer placed in the well. The sampling interval of the datalogger must be sufficiently small ($\sim$1 s) at test initiation to capture as accurately as possible the maximum displacement of the water level at early times after the test has begun. This is especially true for slug tests conducted in geological materials with higher hydraulic conductivities, such as sand and gravel. After about 2 minutes into the test, the sampling interval can be lengthened as the recovery in water level may not be as rapid in comparison to the earlier portion of the test. Recovery times for silty and clayey materials can be considerably longer, making the manual measurement of water levels cumbersome. In materials with lower hydraulic conductivities, pressure transducers can be used for unattended monitoring of water levels.

Bouwer and Rice Method. This method is based on the Theim equation governing radial groundwater flow to a well (Bouwer and Rice 1976) and is widely employed in environmental site assessment. A schematic representation of the geometry of a partially penetrating well with a fully submerged screen in an unconfined aquifer is shown in Figure 7.12.

The method of analysis is appropriate for wells with partially or fully penetrating screens, in confined or unconfined aquifers, and, in the latter case, for wells with fully or partially submerged screens. The flow into the well at a particular value of displaced water level, y, can be computed through a modified form of the Theim equation:

$$Q = 2\pi K L_e \frac{y}{\ln(R_e/r_w)} \tag{7.4}$$

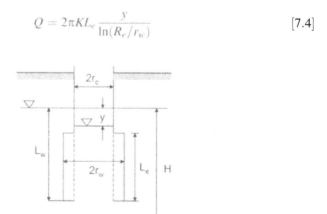

Figure 7.12. Representation of a partially penetrating well with fully submerged screen in an unconfined aquifer. Notations are for well dimensions employed in slug test; analytical methods are described in text.

where Q is the volumetric flow rate into the well $[L^3 T^{-1}]$, L_e is the screen length [L], y is the drawdown induced by a slug [L], R_e is the effective radius over which y is dissipated [L], and r_w is the well radius that includes the gravel pack [L]. Equation [7.4] was derived on the basis of the assumptions that: (1) the aquifer can be treated as homogeneous and isotropic; (2) drawdown of the water table around the well is negligible; (3) flow in the unsaturated zone above the water table is negligible; and (4) well losses are negligible (Bouwer and Rice 1976).

The rate of rise of the water level, dy/dt, in the well can be related to the inflow rate as

$$\frac{dy}{dt} = -\frac{Q}{\pi r_c^2} \qquad [7.5]$$

where r_c is the radius of the well casing [L] and πr_c^2 is the well cross-sectional area $[L^2]$ in which the water level is rising. There is a minus sign on the right-hand side of Eq. [7.5] because as y decreases, t increases. We note that when a filter pack is placed around the well casing with a hydraulic conductivity that is higher than the aquifer material, the filter pack should be included as part of the well. The calculation is based on the total free-water surface area in the well and sand or gravel pack, calculated as $\pi r_c^2 + \pi(r_w^2 - r_c^2)n$ (Bouwer 1989), where r_w is the nominal radius of the well screen [L], n is the drainable porosity $[L^0]$, and $r_w - r_c$ is the thickness of the envelope. The equivalent casing radius (r_{ec}) [L] should be calculated as

$$r_{ec} = \left[(1 - n)r_c^2 + nr_w^2 \right]^{0.5} \qquad [7.6]$$

For example, if the radius of the perforated casing is 10 cm and the casing is surrounded by a 5 cm filter pack with a porosity of 25%, r_{ec} can be computed to be $[(1 - 0.25) \times 10^2 + 0.25 \times 15^2]^{1/2} = 11.5\,\mathrm{cm}$.

To calculate the hydraulic conductivity, we combine Eqs. [7.4] and [7.5], integrate, and solve for K to yield

$$K = \frac{r_c^2 \ln(R_e/r_w)}{2L_e} \frac{1}{t} \left(\ln \frac{y_0}{y_t} \right) \qquad [7.7]$$

where y_0 is the drawdown at $t = 0$ and y_t is the drawdown at t. The above expression is valid when the field data yield a straight line when they are plotted as $\ln y_t$ versus t. The quantity $\ln(R_e/r_w)$ is obtained from an empirical equation that relates the effective radius to the well geometry. For the case when $L_w < H$,

$$\ln \frac{R_e}{r_w} = \left[\frac{1.1}{\ln(L_w/r_w)} + \frac{A + B\ln[(H - L_w)/r_w]}{L_e/r_w} \right]^{-1} \qquad [7.8]$$

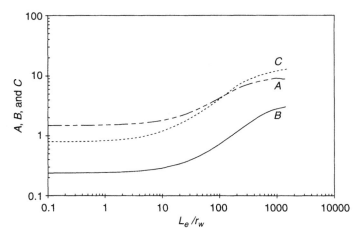

Figure 7.13. Double logarithmic plot of empirically derived curves for estimation of parameters A, B, and C in the Bouwer and Rice method. The curves were obtained by evaluating the polynomial functions fit by Van Rooy (1988).

where A and B are dimensionless coefficients that are function of L_e/r_w (Figure 7.13). For the case when $L_w = H$, Eq. [7.8] should be modified to

$$\ln\frac{R_e}{r_w} = \left[\frac{1.1}{\ln(L_w/r_w)} + \frac{C}{L_e/r_w}\right]^{-1} \qquad [7.9]$$

where C is a dimensionless parameter determined from Figure 7.13. The procedure for computing the hydraulic conductivity is as follows:

1. Plot drawdown y on log scale versus time t on a linear scale using a semilog graph paper.
2. Fit a straight line through the plotted curve and extend the line to $t = 0$.
3. Calculate the quantity $\ln(R_e/r_w)$ using Eq. [7.8] or [7.9] depending on well geometry. In order to do this, find A and B for $L_w < H$ from Figure 7.13. When $L_w = H$, obtain C from Figure 7.13.
4. Record y_0 when $t = 0$ and y for another t on the line. Calculate K using Eq. [7.7].

Bouwer (1989) noted that a sequence of responses to a slug test can sometimes be seen in the data as progressively flattening line segments. This observation was termed the "double straight line effect." The double straight line effect can occur when the formation being tested has a lower hydraulic conductivity than the filter pack placed around the tested well. A straight line should be fit to the second linear segment (B to C on Figure 7.14) for the calculation of hydraulic conductivity.

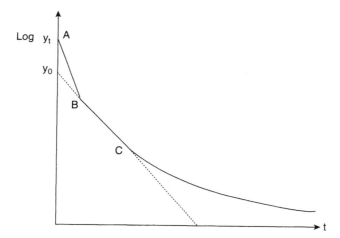

Figure 7.14. Schematic diagram of slug test data revealing a double straight line effect. (Adapted from Bouwer 1989.)

Example 7.5. Determination of Hydraulic Conductivity (K) Using the Bouwer and Rice Method. Use the slug test data reported in Cooper et al. (1967) to calculate the hydraulic conductivity using the Bouwer and Rice method. The well geometry and slug test data are provided in Tables 7.5 and 7.6, respectively.

Step 1. Plotting the drawdown y on log scale versus time t on a linear scale using semilog graph paper reveals a straight line 9 s after the test begins. A straight line is fit to this portion of the data set and the line is extended to $t = 0$. The value of y_0 at $t = 0$ is 0.45 m and $y_t = 0.089$ m at $t = 51$ s as noted from Figure 7.15.

Step 2. Next, the ratio L/r_w is calculated, which is equal to 1290. Since $L_w = H$, Eq. [7.9] is selected to calculate the quantity $\ln(R_e/r_w)$. Then, the dimensionless parameter C is obtained from Figure 7.13 for a value of $L/r_w = 1290$ and is determined to be 12.7.

Step 3. The quantity $\ln(R_e/r_w)$ can then be calculated as

$$\ln\frac{R_e}{r_w} = \left[\frac{1.1}{\ln(9800\,\text{cm}/7.6\,\text{cm})} + \frac{12.7}{9800\,\text{cm}/7.6\,\text{cm}}\right]^{-1} = 6.1$$

Table 7.5. Well Geometry Data for Slug Test Data Obtained from Cooper et al. (1967)

r_w	7.6 cm
R_c	7.6 cm
L_e	9800 cm
L_w	9800 cm
H	9800 cm

Table 7.6. Slug Test Data from Cooper et al. (1967)

Time (s)	y (m)
0	0.560
3	0.457
6	0.392
9	0.345
12	0.308
15	0.280
18	0.252
21	0.224
24	0.205
27	0.187
30	0.168
33	0.149
36	0.140
39	0.131
42	0.112
45	0.108
48	0.093
51	0.089
54	0.082
57	0.075
60	0.071
63	0.065

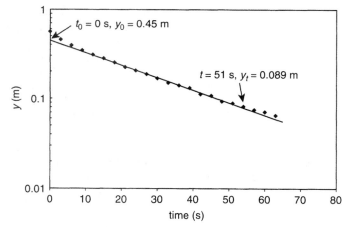

Figure 7.15. A semilogarithmic plot of data from Cooper et al. (1967) with selected points used to determine the hydraulic conductivity with the Bouwer and Rice method. A straight line is fit to the data, providing values for y_0, y_t, and t.

Step 4. The hydraulic conductivity can be calculated using Eq. [7.7] as

$$K = \frac{(7.6 \text{ cm})^2 (6.1)}{2(9800 \text{ cm})} \frac{1}{51 \text{ s}} \left(\ln \frac{45 \text{ cm}}{8.9 \text{ cm}} \right) = 8.8 \times 10^{-4} \text{ cm/s}$$

Pumping Tests

The slug test is useful in obtaining hydraulic conductivity estimates that are representative of geologic media in the immediate vicinity of the screened interval. A site scale estimate of aquifer parameters can be obtained by (1) taking a large number of slug test measurements across the site and taking the average of those values or (2) conducting pumping tests. Here we describe the analysis of pumping tests conducted in confined and unconfined aquifers.

The determination of aquifer characteristics through pumping tests requires an understanding of the hydraulics of groundwater flow into wells. Pumping tests are commonly performed by withdrawing water at a constant (or time-varying) rate from one well and observing the response of water levels within the pumping well and in one or more observation wells. In many cases, water levels decline rapidly during the early stages of a pumping test and slower during the latter stages of the pumping test. The response of the water level, or *drawdown*, is defined as $s(r, t) = H_0 - h(r, t)$, where H_0 is the initial hydraulic head and $h(r, t)$ is the hydraulic head that is the function of both radial distance and time. The manner in which drawdown varies with time in a given well is indicative of hydraulic properties of the aquifer being investigated.

The pressure transient (or wave) created by pumping a well propagates laterally through the aquifer. As time passes, wells located at some distance from the center of the pumped well in the same aquifer will become affected. The magnitude of the drawdown diminishes with radial distance from the pumping well.

The pressure transient moves not only laterally but also upward and downward through the pumped aquifer and layers that confine it. However, the rate of propagation through the confining layers is slower than that in the aquifer. As a result of pressure changes in the confining layers, water may flow out of these layers into the pumped aquifer. When this happens, the pumped aquifer is said to be "leaky." Another form of leakage is encountered when water seeps from another aquifer that is not pumped, through a confining layer into the aquifer from which water is being withdrawn. Both types of leakage—that derived from storage in the confining layers and that derived from storage in an unpumped aquifer—can sometimes be detected from the rate of water level decline in the aquifer being pumped.

Flow to a Well in a Confined Aquifer. We begin by studying the flow to wells in an ideal confined aquifer because it is fundamental to the understanding of aquifer test analysis. In many cases, the confined aquifer is considered to be less susceptible to groundwater contamination in comparison to unconfined aquifers, because they are confined by low-permeability layers called aquitards. In some cases,

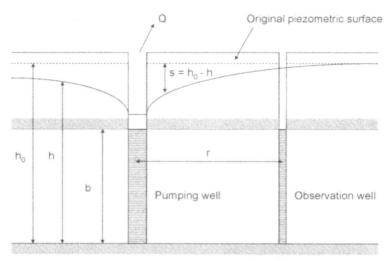

Figure 7.16. Schematic diagram showing a fully penetrating well pumping from a confined aquifer.

these layers could be leaky due to the presence of fractures, allowing contaminants such as DNAPLs to infiltrate into the confined aquifer beneath the confining aquitard.

Consider the flow of groundwater to a well in a confined aquifer of infinite lateral extent, bounded from above and below by impermeable horizontal boundaries. The well discharges at a constant rate, Q. The aquifer is considered to be "uniform," meaning that its hydraulic properties (transmissivity and storativity) are the same at every point in space. In addition, the aquifer is considered to be "isotropic," having the same value of hydraulic conductivity in every direction. Figure 7.16 illustrates a typical pumping test in a confined aquifer, showing a fully penetrating pumping that causes the potentiometric surface to decline. This decline in the potentiometric level (or drawdown) is observed during the test in the fully penetrating observation well located at some distance r and used to obtain aquifer parameters.

The Theis Solution. Let us assume that the pumping well penetrates the entire depth of the aquifer and is open to groundwater flow. When pumping begins under this condition, water converges from every direction toward the pumping well. There is no flow in the vertical direction anywhere in the aquifer because the well is fully penetrating and flow is taking place uniformly though the screened interval. Thus, the partial differential equation (PDE) describing the movement of groundwater in the aquifer takes the form

$$\frac{\partial^2 s}{\partial r^2} + \frac{1}{r}\frac{\partial s}{\partial r} = \frac{S}{T}\frac{\partial s}{\partial t} \qquad [7.10]$$

where s is drawdown or water level decline [L], S is storativity [L^0], T is transmissivity [L^2T^{-1}], r is radial distance (measured in the horizontal plane) from the pumping well [L], and t is time [T]. We see that the governing partial differential equation is written in terms of radial coordinates, reflecting that the drawdown depends only on the radial distance and time.

The initial and boundary conditions take the form

$$s(r,0) = 0 \qquad\qquad [7.11]$$

$$s(\infty, t) = 0 \qquad\qquad [7.12]$$

$$Q = -\lim_{r \to 0} 2\pi r T \frac{\partial s}{\partial r} \qquad\qquad [7.13]$$

where $2\pi r$ is the circumference of the well, and $\partial s/\partial r$ is the hydraulic gradient along the radius. The negative sign in Eq. [7.13] is introduced to compensate for the fact that s decreases as r increases, so that $\partial s/\partial r$ is negative.

Equation [7.11] is an initial condition that states that prior to the beginning of groundwater withdrawal, the drawdown is everywhere zero. Equation [7.12] is an outer boundary condition that states that the drawdown at an infinite distance from the pumping well at all times is zero. Equation [7.13] is an inner boundary condition that governs the flow out of the aquifer into the well that discharges at a constant rate Q. This is exactly Darcy's Law written in radial coordinates. Theis (1935) recognized that the mathematics of the problem could be simplified by treating the well as a line source, that is, setting its radius equal to zero. Therefore, the radius term in Eq. [7.13] has a limit that approaches zero.

The Theis (1935) solution to the boundary value problem consisting of Eqs. [7.10–13] takes the form

$$s(r,t) = \frac{Q}{4\pi T} s_D(t_D) = \frac{Q}{4\pi T} W(u) \qquad\qquad [7.14]$$

where s_D is dimensionless drawdown or $W(u)$ known as the "well function," and $u = 1/(4t_D)$ where t_D is dimensionless time. The dimensionless drawdown can be expressed in one of two ways:

$$s_D = \int_u^\infty \frac{e^{-x}}{x} dx \qquad\qquad [7.15]$$

where the integral in Eq. [7.15] is called the exponential integral in which x is a variable of integration, and t_D is dimensionless time defined as

$$t_D = \frac{Tt}{Sr^2} = \frac{1}{4u} \qquad\qquad [7.16]$$

The exponential integral can also be expressed as a series of the form

$$s_D(t_D) = -0.5772 - \ln u + u - \frac{u^2}{2 \cdot 2!} + \frac{u^3}{3 \cdot 3!} - \cdots \qquad\qquad [7.17]$$

This approximation can be found in handbooks of mathematical functions (e.g., Abramowitz and Stegun 1965). We will later see how this approximation can be

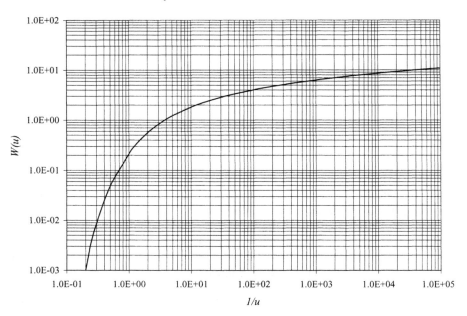

Figure 7.17. The Theis (1935) solution plotted on double logarithmic scale.

useful in analyzing pumping tests using the Jacob–Cooper (Cooper and Jacob 1946) approach.

Theis Type Curve Method. From Eq. [7.17] it can be seen that $W(u)$ is a function of only one parameter, u. When $W(u)$ is plotted versus $1/u$ on double logarithmic paper, one obtains a unique graph known as the Theis type curve (Figure 7.17). It can be used to determine the transmissivity, T, and storativity, S, of the aquifer by matching pumping test data to the type curve, which can be done by hand or through the use of commercially available computer software such as AquiferTest (Waterloo Hydrogeologic, Inc. 2001) and AQTESOLV (Duffield 2000). The procedure for the curve matching by hand consists of the following steps:

Step 1. Prepare the type curve on double logarithmic paper by plotting $W(u)$ on the ordinate, against $1/u$ on the abscissa. Values of $W(u)$ and u can be obtained from Table 7.7.

Step 2. Plot the values of s measured in a given observation well against t, on another sheet of transparent double logarithmic paper having the same scale as the type curve. This is known as the drawdown–time or data plot.

Step 3. Place the drawdown–time plot over the type curve by keeping the horizontal and vertical axes of both sheets parallel to each other.

Step 4. Shift the drawdown–time plot in all directions until it matches the type curve as well as possible. This step is very important as the match determines the parameter estimates.

Table 7.7. Values of the Well Function $W(u)$ for Various u

u	$W(u)$	u	$W(u)$	u	$W(u)$	u	$W(u)$
1×10^{-10}	22.45	7×10^{-8}	15.90	4×10^{-5}	9.55	1×10^{-2}	4.04
2	21.76	8	15.76	5	9.33	2	3.35
3	21.35	9	15.65	6	9.14	3	2.96
4	21.06	1×10^{-7}	15.54	7	8.99	4	2.68
5	20.84	2	14.85	8	8.86	5	2.47
6	20.66	3	14.44	9	8.74	6	2.30
7	20.50	4	14.15	1×10^{-4}	8.63	7	2.15
8	20.37	5	13.93	2	7.94	8	2.03
9	20.25	6	13.75	3	7.53	9	1.92
1×10^{-9}	20.15	7	13.60	4	7.25	1×10^{-1}	1.823
2	19.45	8	13.46	5	7.02	2	1.223
3	19.05	9	13.34	6	6.84	3	0.906
4	18.76	1×10^{-6}	13.24	7	6.69	4	0.702
5	18.54	2	12.55	8	6.55	5	0.560
6	18.35	3	12.14	9	6.44	6	0.454
7	18.20	4	11.85	1×10^{-3}	6.33	7	0.374
8	18.07	5	11.63	2	5.64	8	0.311
9	17.95	6	11.45	3	5.23	9	0.260
1×10^{-8}	17.84	7	11.29	4	4.95	1×10^{-0}	0.219
2	17.15	8	11.16	5	4.73	2	0.049
3	16.74	9	11.04	6	4.54	3	0.013
4	16.46	1×10^{-5}	10.94	7	4.39	4	0.004
5	16.23	2	10.24	8	4.26	5	0.001
6	16.05	3	9.84	9	4.14		

Source: Wenzel (1942).

Step 5. Select an arbitrary "match point" and designate the corresponding values of dimensionless drawdown and time by $W(u)^*$ and $1/u^*$, respectively. It is often convenient to choose a match point corresponding to $W(u)^* = 1/u^* = 1.0$. The value of s and t corresponding to the match point are designated by s^* and t^*, respectively.

Step 6. Compute the transmissivity, T, according to the formula

$$T = \frac{Q}{4\pi s^*} W(u)^* \qquad [7.18]$$

Step 7. Compute the storativity, S, according to the formula

$$S = \frac{4Tt^*u^*}{r^2} \qquad [7.19]$$

Example 7.6. Determination of Transmissivity (T) and Storativity (S) Using the Theis Method. A pumping test is conducted in a confined aquifer with a pumping rate, Q, of $500\,\text{m}^3/\text{day}$ (Table 7.8). The drawdown–time data were collected at an

Table 7.8. Drawdown Data from an Observation Well During a Pumping Test Conducted in a Confined Aquifer

Time (s)	Drawdown (m)
60	0.03
76.2	0.05
96.6	0.09
122.4	0.15
155.4	0.22
197.4	0.31
250.8	0.41
318	0.53
403.2	0.66
511.8	0.80
649.8	0.95
824.4	1.11
1045.8	1.27
1327.2	1.44
1684.2	1.61
2137.2	1.79
2712	1.97
3441.6	2.15
4367.4	2.33
5542.2	2.52
7032.6	2.70
8924.4	2.89
11324.4	3.07
14370	3.26
18235.2	3.45
23139.6	3.64
29363.4	3.83
37261.2	4.02
47283	4.21
60000	4.39

Source: Data from Schwartz and Zhang (2003, p. 274).

observation well that fully penetrates the aquifer 300 m from the pumping well. Use the Theis (1935) type curve matching procedure to estimate the transmissivity and storativity of this aquifer.

Step 1. Prepare the type curve used for matching on a double logarithmic paper.

Step 2. Plot the drawdown versus time data on a double logarithmic paper that is transparent. The scales must be identical to the type curve (Figure 7.18).

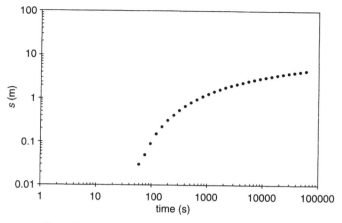

Figure 7.18. Double logarithmic plot of data in Table 7.8.

Step 3. Match the data curve to the type curve by sliding it horizontally and vertically. Once a good match has been obtained, pick an arbitrary match point and the corresponding values for drawdown and time.

Step 4. Compute the transmissivity, T, using Eq. [7.18]:

$$T = \frac{Q}{4\pi s^*} W(u)^* = \frac{(500\,\text{m}^3/\text{day})(1)}{4\pi(0.78\,\text{m})} = 51\,\text{m}^2/\text{day}$$

Step 5. Compute the storativity, S, using Eq. [7.19]:

$$S = \frac{4Tt^*u^*}{r^2} = \frac{4(51\,\text{m}^2/\text{day})\left(22\,\text{min}\,\dfrac{1\,\text{day}}{1440\,\text{min}}\right)(0.1)}{(300\,\text{m})^2} = 3.46 \times 10^{-6}$$

One can also do the curve matching procedure on the computer using a spreadsheet program such as Microsoft Excel. Figure 7.19 is a result of the type curve match obtained in this manner. The procedure is described in Table 7.9.

Semilogarithmic Jacob–Cooper Method. The ratio between transmissivity and storativity is the diffusivity of the aquifer. When pumping tests are conducted in aquifers with large values of diffusivity, one may have a difficult time collecting drawdown data corresponding to the steep early portion of the Theis curve. In such situations, the data cannot be matched to the Theis curve with sufficient accuracy.

An alternative method of analysis that utilizes intermediate to late data has been proposed by Cooper and Jacob (1946). The Jacob–Cooper approach is based on the recognition (Jacob 1950) that when $1/u$ is large (i.e., u is small), Eq. [7.17] can be approximated as

$$s_D(t_D) \approx -0.5772 - \ln u = 2.303 \log_{10}(2.246 t_D) \qquad [7.20]$$

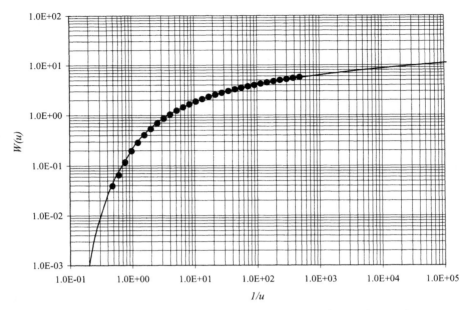

Figure 7.19. Results of matching the data to the Theis type curve. Matching was done on the computer using spreadsheet program MS EXCEL. Matching can also be done manually by superimposing a data curve (Figure 7.18) over the type curve (Figure 7.17).

Table 7.9. Spreadsheet Copied from EXCEL that Allows One to Conduct the Type Curve Matching on the Computer[a]

	A	B	C	D	E	F	G	H
1	**Theis Solution**				**Data**			
2							**8.00E−03**	**1.3**
3	*u*	*1/u*	*w(u)*		time (sec)	*s* (m)	*1/tu*	*W(u)/s*
4	1.00E − 10	1.00E + 10	22.45		60	0.03	*4.80E − 01*	*0.039*
5	2.00E − 10	5.00E + 09	21.76		76.2	0.05	*6.10E − 01*	*0.065*
6	3.00E − 10	3.33E + 09	21.35		96.6	0.09	*7.73E − 01*	*0.117*
7	4.00E − 10	2.50E + 09	21.06		122.4	0.15	*9.79E − 01*	*0.195*
8	5.00E − 10	2.00E + 09	20.84		155.4	0.22	*1.24E + 00*	*0.286*
9	6.00E − 10	1.67E + 09	20.66		197.4	0.31	*1.58E + 00*	*0.403*
10	7.00E − 10	1.43E + 09	20.5		250.8	0.41	*2.01E + 00*	*0.533*
11	8.00E − 10	1.25E + 09	20.37		318	0.53	*2.54E + 00*	*0.689*
12	9.00E − 10	1.11E + 09	20.25		403.2	0.66	*3.23E + 00*	*0.858*
13	1.00E − 09	1.00E + 09	20.15		511.8	0.8	*4.09E + 00*	*1.04*
14	2.00E − 09	5.00E + 08	19.45		649.8	0.95	*5.20E + 00*	*1.235*
15	3.00E − 09	3.33E + 08	19.05		824.4	1.11	*6.60E + 00*	*1.443*
16	4.00E − 09	2.50E + 08	18.76		1045.8	1.27	*8.37E + 00*	*1.651*
17	5.00E − 09	2.00E + 08	18.54		1327.2	1.44	*1.06E + 01*	*1.872*
18	6.00E − 09	1.67E + 08	18.35		1684.2	1.61	*1.35E + 01*	*2.093*

Table 7.9. (*Continued*)

	A	B	C	D	E	F	G	H
19	7.00E − 09	1.43E + 08	18.2		2137.2	1.79	*1.71E + 01*	*2.327*
20	8.00E − 09	1.25E + 08	18.07		2712	1.97	*2.17E + 01*	*2.561*
21	9.00E − 09	1.11E + 08	17.95		3441.6	2.15	*2.75E + 01*	*2.795*
22	1.00E − 08	1.00E + 08	17.84		4367.4	2.33	*3.49E + 01*	*3.029*
23	2.00E − 08	5.00E + 07	17.15		5542.2	2.52	*4.43E + 01*	*3.276*
24	3.00E − 08	3.33E + 07	16.74		7032.6	2.7	*5.63E + 01*	*3.51*
25	4.00E − 08	2.50E + 07	16.46		8924.4	2.89	*7.14E + 01*	*3.757*
26	5.00E − 08	2.00E + 07	16.23		11324.4	3.07	*9.06E + 01*	*3.991*
27	6.00E − 08	1.67E + 07	16.05		14370	3.26	*1.15E + 02*	*4.238*
28	7.00E − 08	1.43E + 07	15.9		18235.2	3.45	*1.46E + 02*	*4.485*
29	8.00E − 08	1.25E + 07	15.76		23139.6	3.64	*1.85E + 02*	*4.732*
30	9.00E − 08	1.11E + 07	15.65		29363.4	3.83	*2.35E + 02*	*4.979*
31	1.00E − 07	1.00E + 07	15.54		37261.2	4.02	*2.98E + 02*	*5.226*
32	2.00E − 07	5.00E + 06	14.85		47283	4.21	*3.78E + 02*	*5.473*
33	3.00E − 07	3.33E + 06	14.44		60000	4.39	*4.80E + 02*	*5.707*
34	4.00E − 07	2.50E + 06	14.15					
35	5.00E − 07	2.00E + 06	13.93					
36	6.00E − 07	1.67E + 06	13.75					
37	7.00E − 07	1.43E + 06	13.6					
38	8.00E − 07	1.25E + 06	13.46					
39	9.00E − 07	1.11E + 06	13.34					
40	1.00E − 06	1.00E + 06	13.24					
41	2.00E − 06	5.00E + 05	12.55					
42	3.00E − 06	3.33E + 05	12.14					
43	4.00E − 06	2.50E + 05	11.85					
44	5.00E − 06	2.00E + 05	11.63					
45	6.00E − 06	1.67E + 05	11.45					
46	7.00E − 06	1.43E + 05	11.29					
47	8.00E − 06	1.25E + 05	11.16					
48	9.00E − 06	1.11E + 05	11.04					
49	1.00E − 05	1.00E + 05	10.94					
50	2.00E − 05	5.00E + 04	10.24					
51	3.00E − 05	3.33E + 04	9.84					
52	4.00E − 05	2.50E + 04	9.55					
53	5.00E − 05	2.00E + 04	9.33					
54	6.00E − 05	1.67E + 04	9.14					
55	7.00E − 05	1.43E + 04	8.99					
56	8.00E − 05	1.25E + 04	8.86					
57	9.00E − 05	1.11E + 04	8.74					
58	1.00E − 04	1.00E + 04	8.63					
59	2.00E − 04	5.00E + 03	7.94					
60	3.00E − 04	3.33E + 03	7.53					
61	4.00E − 04	2.50E + 03	7.25					
62	5.00E − 04	2.00E + 03	7.02					
63	6.00E − 04	1.67E + 03	6.84					

Table 7.9. (*Continued*)

	A	B	C	D	E	F	G	H
64	7.00E − 04	1.43E + 03	6.69					
65	8.00E − 04	1.25E + 03	6.55					
66	9.00E − 04	1.11E + 03	6.44					
67	1.00E − 03	1.00E + 03	6.33					
68	2.00E − 03	5.00E + 02	5.64					
69	3.00E − 03	3.33E + 02	5.23					
70	4.00E − 03	2.50E + 02	4.95					
71	5.00E − 03	2.00E + 02	4.73					
72	6.00E − 03	1.67E + 02	4.54					
73	7.00E − 03	1.43E + 02	4.39					
74	8.00E − 03	1.25E + 02	4.26					
75	9.00E − 03	1.11E + 02	4.14					
76	1.00E − 02	1.00E + 02	4.04					
77	2.00E − 02	5.00E + 01	3.35					
78	3.00E − 02	3.33E + 01	2.96					
79	4.00E − 02	2.50E + 01	2.68					
80	5.00E − 02	2.00E + 01	2.47					
81	6.00E − 02	1.67E + 01	2.3					
82	7.00E − 02	1.43E + 01	2.15					
83	8.00E − 02	1.25E + 01	2.03					
84	9.00E − 02	1.11E + 01	1.92					
85	1.00E − 01	1.00E + 01	1.823					
86	2.00E − 01	5.00E + 00	1.223					
87	3.00E − 01	3.33E + 00	0.906					
88	4.00E − 01	2.50E + 00	0.702					
89	5.00E − 01	2.00E + 00	0.56					
90	6.00E − 01	1.67E + 00	0.454					
91	7.00E − 01	1.43E + 00	0.374					
92	8.00E − 01	1.25E + 00	0.311					
93	9.00E − 01	1.11E + 00	0.26					
94	1.00E + 00	1.00E + 00	0.2119					
95	2.00E + 00	5.00E − 01	0.049					
96	3.00E + 00	3.33E − 01	0.013					
97	4.00E + 00	2.50E − 01	0.004					
98	5.00E + 00	2.00E − 01	0.001					

[a]Columns A and C are values from Table 7.7. Column B is calculated in the spreadsheet. Figure 7.17 is a result of plotting values of $1/u$ and $W(u)$ in columns B and C. The time-drawdown data given in Table 7.8 is listed in Columns E and F. The numbers in italics in Column G are values of $1/tu$. This is calculated by multiplying the values of time given in column E by the number given in cell G2. For example, G4 = E4*G2 and G5 = E5*G2, and so on. The numbers in italics in Column H are values of $W(u)/s$, which is the ratio between the dimensionless drawdown (or well function) and drawdown. These numbers are calculated in an analogous manner for $1/tu$ (i.e., H4 = F4*H2 and H5 = F5*H2), and so on. The user adjusts the bold numbers given in G2 and H2 to move the time-drawdown data over the type curve. The results of matching the time-drawdown data to the Theis curve is shown in Figure 7.19.

Equation [7.20] shows that when t is large or r is small, the dimensionless drawdown varies linearly with the logarithm of the dimensionless time, and s varies linearly with $\log_{10} t$.

The method thus requires plotting the drawdown, s, at a given observation well on semilogarithmic (arithmetic scale for the ordinate and logarithmic scale for the abscissa) paper against the values of time, t. The rest of the procedure consists of the following steps.

A straight line is fit to the drawdown–time data that exhibits straight line behavior. The intersection of this straight line with the time axis corresponding to $s = 0$ is denoted by t_0. The change in s along this line corresponding to a tenfold increase in t (i.e., one logarithmic cycle) is denoted by Δs_{10}.

The transmissivity of the aquifer is determined from

$$T = \frac{2.303 Q}{4 \pi \Delta s_{10}} \tag{7.21}$$

while the storativity, S, is computed using the formula

$$S = 2.246 \frac{T t_0}{r^2} \tag{7.22}$$

Example 7.7. Determination of Transmissivity (T) and Storativity (S) Using the Jacob–Cooper Method. Determine the transmissivity (T) and storativity (S) using the data from Table 7.10. Compare and contrast the values obtained from the Theis (1935) method.

Step 1. Plot the drawdown–time data on semilogarithmic paper (Figure 7.20).

Step 2. Fit a straight line through the straight line portion of the test data.

Step 3. Read off the values of Δs_{10} and t_0 from the graph.

Step 4. Compute the transmissivity, T, using Eq. [7.21]:

$$T = \frac{2.303\, Q}{4 \pi \Delta s} = \frac{2.303(500\,\text{m}^3/\text{day})}{4\pi(1.81\,\text{m})} = 51\,\text{m}^2/\text{day}$$

Step 5. Compute the storativity, S, using Eq. [7.22]:

$$S = 2.246 \frac{T t_0}{r^2} = \frac{2.246(51\,\text{m}^2/\text{day})\left(3.4\,\text{min}\,\dfrac{1\,\text{day}}{1440\,\text{min}}\right)}{(300\,\text{m})^2} = 3.0 \times 10^{-6}$$

The values of transmissivites from the Jacob–Cooper and the Theis methods are identical. However, there is a slight difference in the values of storativities determined from the two approaches. Can you think of the cause of this difference?

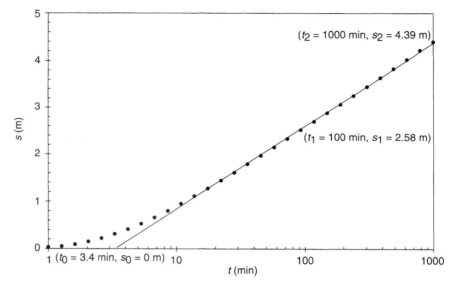

Figure 7.20. Results of the Jacob–Cooper approach.

Flow to a Well in an Unconfined Aquifer. The response of an unconfined aquifer to pumping is considerably different from that of a confined aquifer. The drawdown versus time graph (Figure 7.21) from a pumping test conducted in an unconfined aquifer shows a characteristic *S*-shaped curve. Fitting this data to the Theis type curve yields a poor match (Figure 7.22) for the bulk of the drawdown–time data. At early time, after the commencement of pumping, groundwater is released from storage due to the expansion of water and compression of the aquifer material. During this early period, the unconfined aquifer behaves like a confined aquifer so that early data match the Theis type curve. During intermediate time, groundwater is released from storage through drainage of porewater as the water table drops with pumping. The drawdown behavior during this period differs noticeably from the Theis type curve with the change in drawdown with respect to time being smaller. In fact, it has an appearance as though the drawdown may reach a steady state because of the extra water that is available from the gravity drainage of the pores. The storage property that controls the release of groundwater is the specific yield (S_y) defined as the ratio between the porewater that drains due to gravity and the total volume of the aquifer material. At late time, this delayed gravity response ceases and the response again approaches that of the Theis type curve.

Methods to analyze pumping tests conducted in unconfined aquifers were first introduced by Boulton (1954) and Dagan (1967). These investigators treated the water table as a moving material boundary and assumed that the aquifer is incompressible. This latter assumption restricted the interpretation to intermediate and late data. That is, the model could not account for the early Theis-like response. Kroszynski and Dagan (1975) then incorporated the effect of unsaturated flow above the water table. They treated the water table as merely an isobar ($p = 0$)

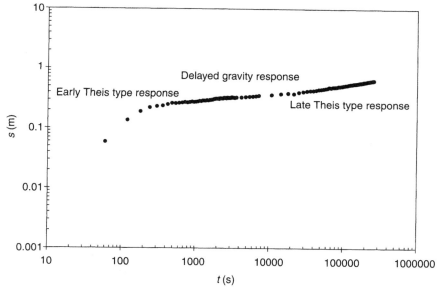

Figure 7.21. Drawdown versus time data from a pumping test conducted in an unconfined aquifer at the Borden Air Force Base, Canada. Note the difference in the drawdown curve compared to Figure 7.18.

and the aquifer was still assumed to be incompressible. This latter assumption again did not allow these investigators to accurately represent the early response except by assuming that well storage is important. A delayed yield model developed by Boulton (1954, 1963) assumed that the Dupuit assumptions held—(1) no seepage face, (2) hydraulic gradient equal to the slope of the water table, and (3) horizontal flow at all times—and that there are two mechanisms of water release: one component that is released instantaneously at time τ and a volume of water that is released according to $\alpha S_y \exp[-\alpha(t - \tau)]$, where $t > \tau$. Here, α is a characteristic constant of the aquifer which lacks a physical meaning. The precise nature in which the water is released by the second mechanism (leakage, gravity, unsaturated flow) is unclear in this model. Other investigators (Prickett 1965; Streltsova 1972; Streltsova and Rushton 1973; Moench 1995) also developed models to analyze flow wells in unconfined aquifer.

Neuman (1972, 1974, 1975) significantly advanced the ability to model the response of an unconfined aquifer through his compressible aquifer model, which considers the delayed aquifer response. In his model, unsaturated flow is not considered and he treats the water table as a moving material boundary. The aquifer is treated to be compressible, at least to the same extent as confined aquifers, but there is a possibility that unconfined aquifers can be more compressible. The elastic storage coefficient S is treated as a constant, and the release of groundwater due to S is instantaneous. S_y is also treated as a constant and the release of water is determined by the rate of water table decline. In contrast to other theories, Neuman's model can treat anisotropy and partial penetration of both the pumping and observation wells.

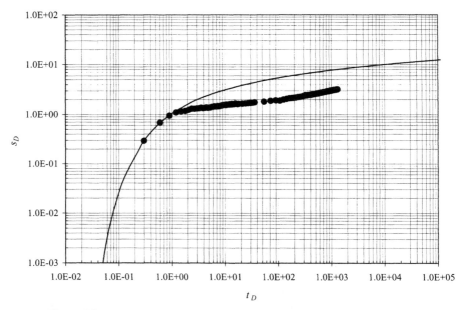

Figure 7.22. A poor match of the drawdown–time data to the Theis (1935) model.

The solution for the case when the pumping and the observation wells fully penetrate the aquifer is given by

$$s(r,t) = \frac{Q}{4\pi T} \int_u^\infty 4y J_0(y\beta^{1/2}) \left[u_0(y) + \sum_{n=1}^\infty u_n(y) \right] dy \qquad [7.23a]$$

where

$$u_0(y) = \frac{\{1 - \exp[-t_s\beta(y^2 - \gamma_0^2)]\}\tanh(\gamma_0)}{\{y^2 + (1 + \sigma)\gamma_0^2 - [(y^2 - \gamma_0^2)^2/\sigma]\}\gamma_0} \qquad [7.23b]$$

$$u_n(y) = \frac{\{1 - \exp[-t_s\beta(y^2 + \gamma_n^2)]\}\tan(\gamma_n)}{\{y^2 - (1 + \sigma)\gamma_n^2 - (y^2 + \gamma_n^2)^2/\sigma\}\gamma_n} \qquad [7.23c]$$

and the terms γ_0 and γ_n are the roots of the equations

$$\sigma\gamma_0\sinh(\gamma_0) - (y^2 - \gamma_0^2)\cosh(\gamma_0) = 0, \qquad \gamma_0^2 < y^2 \qquad [7.23d]$$

$$\sigma\gamma_n\sin(\gamma_n) + (y^2 + \gamma_n^2)\cos(\gamma_n) = 0,$$
$$(2n - 1)(\pi/2) < \gamma_n < n\pi, \quad n \geq 1 \qquad [7.23e]$$

where y is a variable of integration, J_0 is a Bessel function of the first kind with index zero, and Eq. [7.23a–e] are expressed in terms of the dimensionless parameters β, σ, and t_s or t_y.

The curves for the fully penetrating well case are shown in Figures 7.23 and 7.24 with data from the Borden site matched against the curve (see example below). The

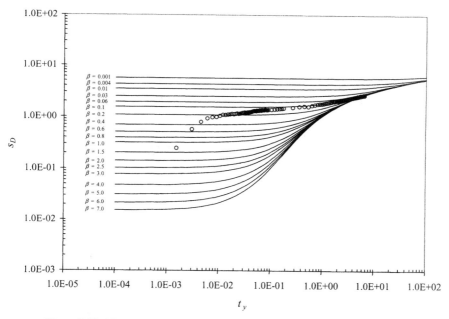

Figure 7.23. Match of the late time data to the type B curve of Neuman (1975).

numerical values for these type curves are listed in Neuman (1975) and can be used to construct one's own type curves using a double logarithmic paper or a spreadsheet program. The curves in Figure 7.23 are called type B curves and correspond to the dimensionless time t_y given as

$$t_y = \frac{Tt}{S_y r^2} \qquad [7.24]$$

where S_y is the specific yield of the aquifer.

The curves in Figure 7.24 are called type A curves and correspond to the dimensionless time t_s given as

$$t_s = \frac{Tt}{S r^2} \qquad [7.25]$$

Type A curves are used to match against early time data and type B curves are used to match against late time data.

Neuman Type Curve Method. Type curve matching is similar in principle to the case for the confined case, but the analysis is more complex because of the larger number of parameters involved in the analysis. We summarize here the steps outlined in Neuman (1975):

Step 1. One first plots the drawdown–time curve at a given observation well on double logarithmic paper.

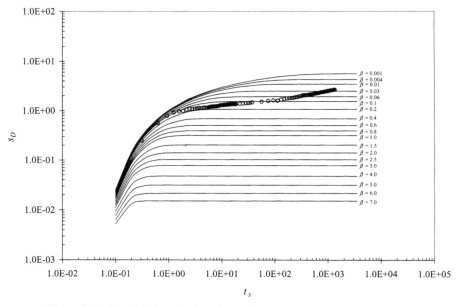

Figure 7.24. Match of the early time data to the type A curve of Neuman (1975).

Step 2. The field data is first superimposed on the type B curves, keeping the vertical and horizontal axes of both graphs parallel to each other and matching as much of the latest drawdown–time data to a particular type curve as one can. The value of β corresponding to this type curve is noted and a match point is chosen anywhere on the overlapping portion of the two sheets of paper. The coordinates of this match point are s^* and $W(u)^*$ along the vertical axis and t^* and t_y^* along the horizontal axis. The transmissivity can then be calculated from

$$T = \frac{Qs_D^*}{4\pi s^*}$$ [7.26]

while the specific yield, S_y, can be computed from

$$S_y = \frac{Tt^*}{r^2 t_y^*}$$ [7.27]

Step 3. Next, one superimposes the drawdown–time curve on the type A curves, keeping the vertical and horizontal axes of both graphs parallel to each other and matching as much of the earliest drawdown–time data to a particular type curve as one can. The value of β corresponding to this type curve must be the same as that obtained earlier from the type B curves. A new match point is selected on the overlapping portion of the two sheets of paper, and its coordinates s^*, s_D^*, t^*, and t_s^* are noted. The transmissivity is again calculated from Eq. [7.26], and its value should be approximately equal to that previously calculated from the late drawdown data. The storage coefficient is obtained from

$$S = \frac{Tt^*}{r^2 t_s^*} \qquad [7.28]$$

Step 4. Having determined the transmissivity of the aquifer, one can calculate the horizontal hydraulic conductivity (K_r):

$$K_r = \frac{T}{b} \qquad [7.29]$$

The degree of anisotropy K_D is obtained from the value of β according to

$$K_D = \frac{\beta b^2}{r^2} \qquad [7.30]$$

By knowing the values of K_r and K_D, one can determine the vertical hydraulic conductivity (K_z) using the relationship

$$K_z = K_D K_r \qquad [7.31]$$

The parameter σ is calculated from

$$\sigma = \frac{S}{S_y} \qquad [7.32]$$

and the specific storage of the aquifer is calculated from

$$S_s = \frac{S}{b} \qquad [7.33]$$

Example 7.8. Determination of Aquifer Parameters of an Unconfined Aquifer Using the Neuman Method. We utilize the Borden test data (Figure 7.21) to illustrate the parameter estimation procedure. In this example, pumping takes place in an unconfined aquifer with a thickness (b) of 5.86 in. at a rate (Q) of 8.02×10^{-4} m^3/s and the data are collected in the observation well at a radical distance (r) 5.13 m away from the pumped well. We then follow the steps below:

Step 1. Prepare the type curve used for matching on a double logarithmic paper.

Step 2. Plot the drawdown versus time data on a double logarithmic paper that is transparent. The scales must be identical to the type curve.

Step 3. Match the late time data to the type B curves by sliding it horizontally and vertically. Once a good match has been obtained, pick an arbitrary match point and read off the corresponding values for drawdown, time, and β.

Step 4. Compute the transmissivity, T, using Eq. [7.26]:

$$T = \frac{Q \, s_D^*}{4\pi \, s^*} = \frac{(8.02 \times 10^{-4} \, \text{m}^3/\text{s})(1)}{4\pi(0.24 \, \text{m})} = 2.68 \times 10^{-4} \, \text{m}^2/\text{s}$$

Step 5. Compute the specific yield, S_y, using Eq. [7.27]:

$$S_y = \frac{T t^*}{r^2 t_Y^*} = \frac{(2.68 \times 10^{-4} \, \text{m}^2/\text{s})(40,000 \, \text{s})}{(5.13 \, \text{m})^2 (1)} = 0.41$$

Step 6. Match the early time data to the type A curves by sliding it horizontally and vertically keeping the value of β constant. Once a good match has been obtained, pick an arbitrary match point and read off the corresponding values for drawdown and time. Obtain another estimate of transmissivity through Eq. [7.26]. This transmissivity value should be close to that computed in Step 4. One can then calculate the storage coefficient, S, using Eq. [7.28]:

$$S = \frac{T t^*}{r^2 t_s^*} = \frac{(2.68 \times 10^{-4} \, \text{m}^2/\text{s})(192.3 \, \text{s})}{(5.13 \, \text{m})^2 (1)} = 1.95 \times 10^{-3}$$

Step 7. Having determined the transmissivity of the aquifer, one can calculate the horizontal hydraulic conductivity (K_r) using Eq. [7.29]:

$$K_r = \frac{T}{b} = \frac{2.68 \times 10^{-4} \, \text{m}^2/\text{s}}{5.86 \, \text{m}} = 4.57 \times 10^{-5} \, \text{m/s}$$

Step 8. The degree of anisotropy K_D is obtained from the value of β according to Eq. [7.30]. Because we know the values of K_r and K_D, one can determine the vertical hydraulic conductivity (K_z) using Eq. [7.31]:

$$K_z = K_D K_r = (0.67)(4.57 \times 10^{-5} \, \text{m/s}) = 3.06 \times 10^{-5} \, \text{m/s}$$

Step 9. The value of the parameter σ can be calculated using Eq. [7.32]:

$$\sigma = \frac{S}{S_y} = \frac{1.95 \times 10^{-3}}{0.41} = 4.76 \times 10^{-3}$$

Step 10. Finally, the value of specific storage, S_s, may be determined through Eq. [7.33]:

$$S_s = \frac{S}{b} = \frac{1.95 \times 10^{-3}}{5.86 \, \text{m}} = 3.34 \times 10^{-4} \, \text{m}^{-1}$$

Table 7.10. Values of Total Porosity, Effective Porosity, and Bulk Density for Common Aquifer Matrix Materials

Aquifer Materials	Bulk Density, ρ_b (g/cm^3)	Total Porosity, ϕ_T	Effective Porosity, ϕ_e
Clay	1–2.4	0.34–0.60	0.01–0.2
Peat	—	—	0.3–0.5
Glacial sediments	1.15–2.1	—	0.05–0.2
Sandy clay	—	—	0.03–0.2
Silt	—	0.34–0.61	0.01–0.3
Loess	0.75–1.60	—	0.15–0.35
Fine sand	1.37–1.81	0.26–0.53	0.1–0.3
Medium sand	1.37–1.81	—	0.15–0.3
Coarse sand	1.37–1.81	0.31–0.46	0.2–0.35
Gravely sand	1.37–1.81	—	0.2–0.35
Fine gravel	1.36–2.19	0.25–0.38	0.2–0.35
Medium gravel	1.36–2.19	—	0.15–0.25
Coarse gravel	1.36–2.19	0.24–0.36	0.1–0.25
Sandstone	1.6–2.68	0.05–0.3	0.1–0.4
Siltstone	—	0.21–0.41	0.01–0.35
Shale	1.54–3.17	0–0.1	—
Limestone	1.74–2.79	0–0.5	0.01–0.2
Granite	2.24–2.46	—	—
Basalt	2–2.7	0.03–0.35	—

Source: Domenico and Schwartz (1990).

7.1.3. Total Porosity (ϕ) and Effective Porosity (ϕ_e)

The total porosity (ϕ) of a porous medium is the volume of voids (V_v) [L^3] divided by the total volume of the medium (V_T) [L^3], making it a dimensionless quantity. Effective porosity (ϕ_e) may be visualized as the connected central passages of a pore network, where fluid flows with negligible resistance from adjacent pore walls and does not stagnate in dead ends. In water-saturated material, total porosity is equivalent to moisture content. Both total and effective porosities are commonly selected from the literature, based on the type of aquifer material (see Table 7.10), but it is always best to obtain in situ estimates.

Total porosity can also be easily obtained by a relatively inexpensive method that can be performed in a geotechnical laboratory. The total porosity test also yields bulk density of the sample. Drilling for site characterization should include taking at least three core samples for porosity determinations.

The effective porosity (ϕ_e) can be obtained using tracer tests run in laboratory columns. For example, an aquifer material core of known volume could be collected and fitted with inlet and outlet valves as a flow-through aquifer column (Anid et al. 1993; Alvarez et al. 1998a). A bromide solution (50 mg/L) can be continuously injected in an up-flow mode with peristaltic or syringe pumps (e.g.,

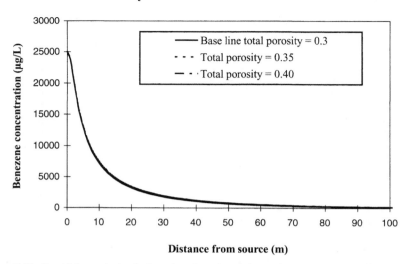

Figure 7.25. Sensitivity analysis of plume length to the total porosity. Figure depicts simulations for steady-state concentration of benzene along plume centerline using different values of the total porosity with Eq. [5.10]. (From Alvarez et al. 1998b.)

at 2 mL/h). Effluent bromide concentrations are then monitored (e.g., sampled every 15 min) until influent concentration is reached. The mean liquid retention time is estimated as the time required for the effluent bromide concentration to reach 50% of the influent concentration. This time is then multiplied by the flow rate to calculate the pore volume through which water can be freely transmitted. The effective porosity (ϕ_e) is then calculated as the ratio of this pore volume to the total volume of the core sample. It should be kept in mind, however, that aquifer materials can have some variability in effective porosity. Therefore, sensitivity analyses should be performed to determine the effect of varying the effective porosity on model simulations.

A sensitivity analysis for total porosity on the plume length shows that simulations are not very sensitive to small errors in porosity (Figure 7.25). Therefore, the cost of site-specific determination of ϕ or ϕ_e may not be justified. However, one should keep in mind that this is the case for aquifers consisting of unconsolidated materials. The porosity of fractured rock aquifers is known to be considerably smaller ($\sim 10^{-5}$–10^{-2}) and highly variable, making site characterization of ϕ or ϕ_e very important.

7.2. PARAMETERS FOR CONTAMINANT TRANSPORT MODELING

7.2.1. Retardation Coefficient (R)

The retardation coefficient describes the extent to which the migration of dissolved contaminants is slowed down by sorption to the aquifer matrix. The degree of retardation depends on both aquifer and contaminant compound properties. The coefficient of retardation is the ratio of the groundwater seepage velocity (also known as the average linear groundwater velocity) (v) to the average velocity of a migrating contaminant (v_c). Assuming the sorption can be described by linear partitioning between the water and the aquifer matrix, the local retardation coefficient is given by Eq. [7.34] (Freeze and Cherry 1979):

$$R = \frac{v}{v_c} = 1 + \frac{\rho_b K_d}{\phi} \qquad [7.34]$$

where ρ_b is the bulk density of the aquifer material $[ML^{-3}]$, K_d is the distribution coefficient (i.e., the linear partitioning coefficient between groundwater and aquifer material) $[L^3 M^{-1}]$, and ϕ is the total porosity $[L^0]$. Field research has shown that the retardation factor is not necessarily constant and can increase with time until some steady state is reached. This trend has been observed in field studies in a sandy aquifer, where R increased over 600 days from about 2 to 6 for PCE and from about 4 to 9 for dichlorobenzene (Roberts et al. 1986). Presumably, this is due to two concentration-dependent phenomena. As contaminant concentrations decrease over time, (1) there is less competition among contaminants and other potential sorbates for sorption sites on the aquifer material, and (2) lower contaminant concentrations favor nonlinear (stronger) adsorption mechanisms onto "hard organic carbon" sites compared to linear absorption mechanisms into more spongelike "soft organic carbon" sites (e.g., humic substances) (Weber et al. 2002).

The retardation of a contaminant relative to the groundwater flow has important implications for intrinsic bioremediation. When retardation is occurring, dissolved oxygen and other nutrients and electron acceptors traveling with groundwater can sweep over the contaminant plume. This enhances the availability of cosubstrates needed for in situ biodegradation of, for example, BTEX. Also, adsorption of a contaminant to the aquifer matrix results in dilution of the dissolved contaminant plume. Although sorption can decrease the bioavailability of hydrophobic contaminants (Lyman et al. 1992), this is generally not a significant factor that limits BTEX degradation kinetics.

As shown in Eq. [7.34], three site-specific parameters, the total porosity (ϕ), the bulk density of the aquifer material (ρ_b), and the distribution coefficient (K_d), are needed to calculate the retardation coefficient. The total porosity (ϕ) was discussed in Section 7.1. The bulk density is defined as the ratio between the mass of the dry solid (M_s) of a sample and the total volume (V_t) of the sample and is related to the particle density (ρ_s) and the total porosity (ϕ) by Eq. [7.35]:

$$\rho_b = \frac{M_s}{V_t} = (1 - \phi_T)\rho_s \qquad [7.35]$$

The distribution coefficient (K_d) can be estimated based on the soil adsorption coefficient for soil organic carbon (K_{oc}) and the fraction of soil organic carbon (f_{oc}) as

$$K_d = f_{oc} K_{oc} \qquad [7.36]$$

where K_{oc} is a chemical-specific partition coefficient between soil organic carbon and the aqueous phase. Larger values of K_{oc} indicate greater affinity of contaminants for the organic carbon fraction of soil. Smaller partitioning coefficients result in lower retardation factors. Typical K_{oc} values for common organic pollutants are listed in Table 7.11. K_{oc} can also be correlated with the octanol–water partitioning coefficient (K_{ow}), which is a measure of compound hydrophobicity (Table 7.11). Examples of such correlations are given in Table 7.12.

Example 7.9. Determination of Retardation Coefficient (R). Calculate the retardation factor for benzene, for a sandy aquifer material with $f_{oc} = 0.1\%$, $\phi = 0.30$, and bulk density $(\rho_b) = 1.86 \, g/cm^3$.

Step 1. Determine the value for K_{oc}: $\log K_{oc} = 1.92$ (Table 7.13). So

$$K_{oc} = 10^{1.92} = 83.2$$

Step 2. Calculate the value for K_d:

$$K_d = K_{oc} f_{oc} = 83.2(0.001) = 0.083$$

Step 3. Calculate the value for R_f:

$$R_f = 1 + [(1.86)(0.083)/0.30] = 1.54$$

Thus, in this aquifer material, benzene would travel 1.54 times slower than groundwater.

If K_{oc} had been calculated using Eq. 3 in Table 7.12 with $\log K_{ow} = 2.13$, the value of K_d would have been

$$\log K_{oc} = 1.00(\log K_{ow}) - 0.21 = 1.92 \quad \text{(same as above)}$$

The value of f_{oc} (which is needed to calculate K_d) can be measured in the lab, using standard techniques such as those described by Nelson and Sommers (1982). An average of three samples can be used to obtain a site-specific value for each parameter (i.e., a total of nine soil samples need to be sent to the laboratory for analysis). Therefore, samples for each parameter must be collected from three different locations considered representative of site conditions. Additionally, samples must

Table 7.11. Compilation of Partitioning Coefficients and Solubilities for Different Organic Compounds

Compound	Log K_{ow}	Log K_{oc}	Water Solubility at 20°C (mg/L)	Equation Number
Halogenated aliphatics				
Bromoform	2.30	2.16	—	7
Carbon tetrachloride	2.64	2.40	800	7
Chloroethane	1.49	1.57	5,740	7
Chloroform	1.97	1.92	8,200	7
Chloromethane	0.95	1.18	6,450	7
Dichlorodifluoromethane	2.16	2.05	280 (25 °C)	7
Dichloromethane	1.26	1.41	20,000	7
Hexachloroethane	3.34	2.81	50	7
Tetrachloroethylene	2.88	2.57	200	7
Trichloroethylene	2.29	2.15	1,100	7
Vinyl chloride	0.60	0.93	90	7
1,1-Dichloroethane	1.80	1.80	400	7
1,1-Dichloroethylene	1.48	1.57	400	7
1,2-Dichloroethane	1.48	1.57	8,000	7
1,2-*trans*-Dichloroethylene	2.09	2.00	600	7
1,1,1-Trichloroethane	2.51	2.30	4,400	7
1,1,2-Trichloroethane	2.07	1.99	4,500	7
Aromatics				
Benzene	2.13	1.92	1,780	3
Benzo[*a*]pyrene	6.06	5.85	0.0038	3
Chlorobenzene	2.84	2.63	500	3
Ethylbenzene	3.34	3.13	152	3
Hexachlorobenzene	6.41	6.20	0.006	3
Naphthalene	3.29	3.08	31	2
Nitrobenzene	1.87	1.66	1,900	2
Pentachlorophenol	5.04	4.83	14	3
Phenol	1.48	1.27	93,000	3
Toluene	2.69	2.48	535	3
1,2-Dichlorobenzene	3.56	3.35	100	3
1,3-Dichlorobenzene	3.56	3.35	123	3
1,4-Dichlorobenzene	3.56	3.35	79	3
1,2,4-Trichlorobenzene	4.28	4.07	30	3
2-Chlorophenol	2.17	1.96	28,500	3
2-Nitrophenol	1.75	1.54	2,100	3
2,4,5,2′,4′,5′-PCB	6.72	6.51	—	3
Pesticides				
Acrolein	0.01	0.86	210,000	6
Alachlor	2.92	2.96	242	1
Atrazine	2.69	2.55	33	4
Dieldrin	3.54	3.33	0.2	3
DDT	6.91	6.70	0.0055	3
Lindane	3.72	3.51	7.52	3
2,4-D	1.78	1.65	900	5

Source: Schnoor (1996).

Table 7.12. Correlations Used to Estimate K_{oc} for Different Classes of Organic Compounds from K_{ow}

	Equation	Number	r^2	Chemical Classes Represented
1.	$\log K_{oc} = 0.544 \log K_{ow} + 1.377$	45	0.74	Wide variety, mostly pesticides
2.	$\log K_{oc} = 0.937 \log K_{ow} - 0.006$	19	0.95	Aromatics, polynuclear aromatics, triazines, and dinitroaniline herbicides
3.	$\log K_{oc} = 1.00 \log K_{ow} - 0.21$	10	1.00	Mostly aromatic or polynuclear aromatics; two chlorinated
4.	$\log K_{oc} = 0.94 \log K_{ow} + 0.02$	9	NA	s-Triazines and dinitroaniline herbicides
5.	$\log K_{oc} = 1.029 \log K_{ow} - 0.18$	13	0.91	Variety of insecticides, herbicides, and fungicides
6.	$\log K_{oc} = 0.524 \log K_{ow} + 0.855$	30	0.84	Substituted phenylureas and alkyl-N-phenylcarbamates
7.	$\log K_{oc} = 0.72 \log K_{ow} + 0.5$	14	0.95	Halogenated hydrocarbons, both aliphatic and aromatic

Source: Schnoor (1996).

be collected at the same depth and in the same stratigraphic unit as that of the maximum soil contamination location. Samples for f_{oc} determination should be collected from an uncontaminated area. Samples for ρ_b and ϕ may be collected from within the plume or from an uncontaminated area.

More natural organic carbon (i.e., higher f_{oc}) generally results in higher adsorption of organic constituents on the aquifer matrix, although the type of soil organic matter present is also important in dictating sorption behavior (Weber et al. 2002). Typical ranges of f_{oc} are given in Table 7.13.

Table 7.13. Representative Values of Total Organic Carbon Content in Various Sediments

Type of Deposit	Texture	Fraction Organic Carbon (f_{oc})
Fluvial-deltaic	Medium sand	0.00053–0.0012
Glaciofluvial	Sands and gravels	0.0004–0.0007
Glacial (lacustine)	Organic silt and peat	0.10–0.25
Back-barrier (marine)	Fine to coarse sand	0.00026–0.007
Eolian	Loess (silt)	0.00058–0.0016
River sediment	Sand, coarse to fine silt	0.0057–0.029

Source: Domenico and Schwartz (1998).

It must be remembered that the retardation factor for a dissolved contaminant characterizes the advective flow of the concentration contour that is 50% of the (constant) source concentration (i.e., $C/C_0 = 0.5$). Because some contaminant molecules migrate further than this contour, the first appearance of a contaminant in low concentrations at some downgradient monitoring well is commonly sooner than one predicts based on the retarded velocity alone.

7.2.2. Hydrodynamic Dispersion Coefficient (D) and Dispersivity (α)

Determination by Means of Theoretical Formula

There are several methods for estimating the apparent longitudinal dispersivity, α_x [L] (see Chapter 4 for definition). The simplest method is the so-called rule of thumb (Gelhar 1993). This rule is based on the observation that the longitudinal dispersivity increases with an overall scale or the travel distance (L) of a contaminant from a source, as shown in Figure 7.26, where a straight line is fitted to the observed values. The slope of this line is 0.1, meaning that the apparent longitudinal dispersivity is one-tenth of L (ASTM 1994) or

$$\alpha_x = 0.1L \qquad [7.37]$$

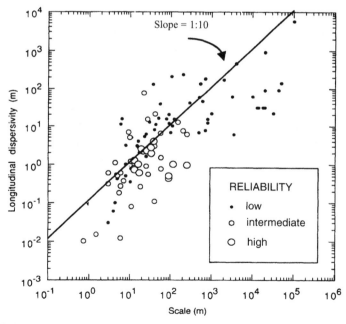

Figure 7.26. The longitudinal dispersivity as a function of an overall scale (revised based on Gelhar et al. 1992).

where L should be the average travel distance of the plume. For a continuous source in a one-dimensional column where dispersion occurs only in the x direction, the average traveled distance of a plume is the distance between the source and the advanced front at which the concentration is half of the source concentration $(C/C_0 = 0.5)$. However, the average travel distance of the plume cannot be easily determined for the more common case when dispersion occurs in three dimensions. Therefore, L in Eq. [7.37] is often represented by the plume length; that is, the downgradient distance between the source and the contour, which is the drinking water standard (e.g., a benzene concentration of 5 µg/L).

Equation [7.37] is slightly different from

$$\alpha_x = 0.1x \qquad [7.38]$$

where x is the downgradient distance along the principal groundwater flow direction. Both formulas incorporate the scale dependence of the longitudinal dispersivity.

Field experiments have shown that one could fit tracer data more accurately by using a constant dispersivity value for all points within the plume at a given time (Dieulin et al. 1981; De Marsily 1986). As the plume spreads, the dispersivity increases with the average traveled distance of all tracer particles rather than with the distance traveled by individual particles. Equation [7.38] predicts that α_x will increase indefinitely with x. However, both theoretical results (Dagan 1989; Zhang and Neuman 1990) and field tracer experiments (Mackay et al. 1986; LeBlanc et al. 1991; Boggs et al. 1992) have shown that α_x reaches an asymptotic (constant) value at large time or large scale in most aquifers. This asymptotic value is usually approached after the plume travels a distance between 10 and 50 m in unconsolidated materials (Dagan 1989; Neuman and Zhang 1990; Zhang and Neuman 1990). In the Borden experiment (Freyberg 1986; Mackay et al. 1986; Sudicky 1986) the longitudinal dispersivity approached a constant value of about 0.5 m after the center of the plume moved 30 m from the injection point. The value of dispersivity stayed at this constant value of 0.5 m even after the plume center moved from 30 m to 100 m from the source in the experiment.

Using universal scaling, Neuman (1990) proposed an improved scale-dependent, empirical method for estimating α_x:

$$\alpha_x = 0.0175L^{1.46} \quad \text{for } L \leq 100 \text{ m} \qquad [7.39a]$$

$$\alpha_x = 0.32L^{0.83} \quad \text{for } L > 100 \text{ m} \qquad [7.39b]$$

This approach generally yields more accurate estimations of α_x compared to Eqs. [7.37] and [7.38].

Other empirical approaches have been proposed that have the potential to yield more accurate estimations of α_x. Some of these methods, however, have extensive data requirements, and the additional cost associated with obtaining these data is not always justified in terms of improved simulation accuracy. Examples of these approaches are given below.

Table 7.14. Typical Values of the Variance of the Natural Logarithm of Hydraulic Conductivity for Various Geological Formations

Medium	Variance of Log Transformed Hydraulic Conductivity, σ_Y^2
Alluvium	0.2–1.5
Fluvial deposits	0.8–4.4
Glacial deposits	0.3–0.6
Limestone	5.25
Sandstone	0.4–4.8

Source: Adapted from Gelhar (1993).

The apparent dispersivity can be related to geostatistical parameters of hydraulic conductivity (Dagan 1989; Domenico and Schwartz 1990):

$$\alpha_x = \sigma_Y^2 \gamma_Y \qquad [7.40]$$

where σ_Y^2 is the variance of log-transformed hydraulic conductivity ($Y = \log K$) and γ_Y is the correlation length of Y. These two parameters can be calculated based on a number of Y values (Domenico and Schwartz 1990). When multiple values of K are not available, one may consider the values of the variance of the natural logarithm of hydraulic conductivity or transmissivity in Table 7.14. For the correlation scale (γ_Y), Gelhar (1993) suggests using one-tenth of the overall scale or one-tenth of the plume length; that is,

$$\gamma_Y = 0.1L \qquad [7.41]$$

In summary, dispersion of a contaminant in a heterogeneous aquifer is a complex process and is mainly caused by groundwater velocity variations at different scales. This variation is, in turn, affected by variations of hydraulic conductivity. Without detailed knowledge of these variations, which is usually the case, accurate calculation of the dispersion coefficients is difficult and one has to rely on rough estimates (e.g., Eqs. [7.37]–[7.39]). It is recommended that Neuman's (1990) method (Eqs. [7.39a] and [7.39b]) should be used when the hydraulic conductivity data is limited. The geostatistical method (Eq. [7.40]) may be used to improve the estimate for the longitudinal dispersivity when multiple measurements of hydraulic conductivity at different points in a site are available.

The transverse dispersivity (α_y) is commonly set to equal 30% of the longitudinal dispersivity, and the vertical dispersivity (α_z) is taken to be 5% of the longitudinal dispersivity (see Chapter 4 for definitions) (ASTM 1994).

Determination by Means of One-Dimensional Column Experiments

When core samples are available from a given site, a column experiment can be conducted to obtain estimates of dispersivities. Figure 7.27 shows a classical

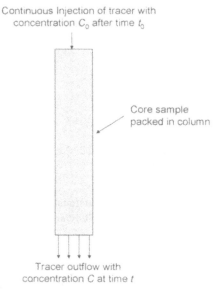

Continuous Injection of tracer with
concentration C_0 after time t_0

Core sample
packed in column

Tracer outflow with
concentration C at time t

Figure 7.27. Classical setup of a one-dimensional column experiment to determine longitudinal dispersivity (α_x).

setup of a one-dimensional column experiment. A tracer with concentration is introduced continuously at time t_0 in a steady flow regime. The tracer outflow is collected during the experiment and analyzed for solute concentration. The concentrations are then plotted against time to obtain a breakthrough curve for the tracer. An example of results from a column experiment conducted by van Genuchten and Wierenga (1986) using chloride as a tracer is shown in Figure 7.28. This data is then analyzed by fitting the breakthrough curve to an analytical solution by Danckwerts (1953) given in terms of relative concentration, $C_e(T) = [C(x, t - C_i]/(C_0 - C_i)$:

$$C_e(T) = \tfrac{1}{2}\mathrm{erfc}\left[(P/4RT)^{1/2}(R - T)\right] \qquad [7.42]$$

where $T = vt/L$ is the number of pore volumes leached through the column, $C(x,t)$ is the measured concentration as a function of distance and time, C_i is the initial concentration, C_0 is the concentration of the applied solution, $P = vL/D_x$ is the column Peclet number, R is the retardation factor, L is the length of the core, v is the pore velocity, t is time, and D_x is the longitudinal hydrodynamic dispersion coefficient. Note that both C_i and C_0 are constants.

The number of pore volumes can be calculated by dividing the amount of water leached through the column (V) by the liquid capacity ($V_0 = A\,\phi L$) of the column, where A is the cross-sectional area and ϕ is porosity (van Genuchten and Wierenga 1986).

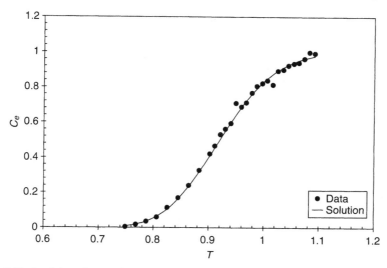

Figure 7.28. Breakthrough curve from a column experiment conducted using bromide as a tracer. The concentration of bromide was obtained by an ion chromatograph.

Estimates of the Peclet number and the retardation factor can be obtained by comparing the experimental data to the analytical solution. One can then vary the values of P and R until a good fit is obtained. Fitting can be done by trial and error by adjusting the fitting parameters until an acceptable fit between the experimental data and the model is obtained or through the aid of software such as the CXTFIT (Toride et al. 1995), which is available from the U.S. Salinity Laboratory. The latter approach relies on a nonlinear least-square parameter optimization method. In this approach the program is used to solve the inverse problem by fitting mathematical solutions of theoretical transport models, based on the convection–dispersion equation (CDE), to experimental data such as the one shown in Figure 7.28.

If the trial and error approach is chosen, the value of the retardation factor can first be obtained by locating the number of pore volumes at which the relative concentration reaches a value of 0.5. When $C_e(T) = 0.5$, the pore volume leached is equal to the retardation factor $(T = R)$ as $\mathrm{erfc}(0) = 1$. The value of the longitudinal hydrodynamic dispersion coefficient may be estimated from the fitted pore volume and the longitudinal dispersivity (α_x) can then be obtained directly through the relation $\alpha_x = D_x/v$. In this particular example, the breakthrough curve is fit to the one-dimensional solute transport model given by Eq. [7.42] to yield $P = 266$, $R = 0.92$, and $\alpha_x = 0.11$ cm.

Details to the experimental procedures, methods of analysis, and analytical solutions to one-dimensional solute transport considering retardation of solutes with various boundary conditions can be found in van Genuchten and Wierenga (1986).

7.2.3. Biodegradation Rate Coefficient (λ)

This is the parameter that describes the rate at which a contaminant is being degraded. As discussed earlier, the degradation rate can usually be approximated by a first-order decay regime with respect to the contaminant concentration (C):

$$\frac{dC}{dt} = -\lambda C \qquad [7.43]$$

There are several different methods to determine the site-specific biodegradation rate coefficient, λ. These methods include mass balances, the technique of Buscheck and Alcantar, normalization of contaminant concentrations to those of a recalcitrant cocontaminant that was present in the initial release, such as trimethylbenzene or tetramethylbenzene for gasoline releases, the use of in situ microcosms, and direct push tests. Chapelle et al. (1996) provide a good case study on the comparison of field and laboratory derived degradation rates. One of the principal findings in this study was the very large sensitivity of degradation rates measured in the laboratory to terminal electron-accepting conditions and the need for experimental conditions to closely mimic field conditions. On the other hand, in situ approaches of obtaining reaction rates are desirable because microcosm studies require the extraction of core samples from the field, which are typically small and may be unrepresentative of field conditions. Core samples can additionally be disturbed or contaminated from sampling, and when core samples are brought into the laboratory, field conditions are difficult to reproduce. Advantages and disadvantages of such approaches are summarized in Table 7.15.

The mass balance approach, described by Chiang et al. (1989), constitutes a rigorous demonstration of natural attenuation. In this method, an extensive monitoring well network is installed to define the complete vertical and horizontal extent of the dissolved hydrocarbon plume. The total mass of a contaminant in the subsurface is monitored over time by interpolation and integration of monitoring well data. The first-order attenuation rate (or the biodegradation rate coefficient) is determined as the percent of contaminant depleted per period of time, usually normalized to a unit of reciprocal days (day^{-1}). The mathematical relationship is stated in Eq. [7.44].

$$\lambda = (-dM/dt)/M = \ln(M_0/M)/\Delta t \qquad [7.44]$$

Here, M_0 is the initial contaminant mass at $t = 0$, M is the contaminant mass remaining at time t, and Δt is the corresponding elapsed time. In using this technique, one has to take into account the net influx rate from the contaminant source and other mechanisms (e.g., sorption, volatilization), which are not part of the actual biodegradation process. This mass balance technique is a direct and quantitative method of demonstrating natural attenuation. However, the cost associated with the extensive monitoring network required for this approach and the relative complexity of the analyses preclude its application at most sites other than research sites (Chiang et al. 1989). Furthermore, in many cases, dissolved gasoline plumes will reach pseudo-steady-state conditions when contaminant concentrations in

Table 7.15. Common Field Methods for Determining the Biodegradation Rate Coefficient

Method	Advantage	Disadvantage
Mass balance	• Direct λ measurement • High accuracy for a pulse release or for shrinking plumes	• Unreliable for continuous source and for steady plumes • Requires extensive amount of time and numerous well clusters screened at different depths • Need to discern other sinks that are not associated with biodegradation process (could overestimate λ)
Buscheck and Alcantar	• Simple analysis • Does not require extensive monitoring	• Applicable only for constant source, steady plume • Need series of wells along the plume centerline • Can overestimate λ since a decrease in contaminant concentration due to horizontal and vertical dispersion is intrinsically attributed to biodegradation
TMB normalization	• Simple analysis • Can be used for steady and transient plumes	• Need series of wells along the plume centerline • TMB tracer may be degraded, which will lead to underestimation of λ
In situ microcosms	• Simple analysis • Can measure rates in different redox zones of the aquifer	• O_2 could be inadvertently introduced when device is installed or when contaminants are added • Sorption processes confound kinetic interpretation and analysis • Does not incorporate potential interactions between degradation, advective, and dispersive processes
Push–pull tests	• Relatively simple field testing procedure • Can be applied to different redox zones of the aquifer • Approach yields local scale degradation rate	• Analysis could be complex • Sampled region can be hard to determine in highly heterogeneous material • When site-wide biodegradation rate is necessary, multiple tests at multiple locations need to be conducted

monitoring wells stabilize (with minor fluctuations) because of the combined effects of contaminant dissolution at the (continuous) source, downgradient transport of the dissolved constituents, and subsequent biodegradation. In this situation, the mass balance approach cannot be used to estimate λ since the mass of dissolved contaminant in the aquifer will be relatively constant. Therefore, this approached works best for pulse releases of dissolved phase contaminants.

Another method that is frequently used to determine λ is the technique of Buscheck and Alcantar (1995). This method is based on an analytical solution for one-dimensional, steady-state contaminant transport that considers advection, longitudinal dispersion, sorption, and first-order biodegradation (Eq. [7.45]).

$$C(x) = C_0 \exp\left\{ \left(\frac{x}{2\alpha_x}\right) \left[1 - \left(1 + \frac{4\lambda\alpha_x}{v_c}\right)^{1/2} \right] \right\} \qquad [7.45]$$

or

$$C(x) = C_0 \left[\exp\left(\frac{-v_c + \sqrt{v_c^2 + 4D\lambda}}{2D}\right) x \right] \qquad [7.46]$$

Buscheck and Alcantar (1995) recognized that contaminant concentrations usually decrease exponentially along the centerline of the plume as the distance from the source increases. This trend is described by Eq. [7.47]:

$$C(x) = C_0 \exp\left(\frac{-kx}{v}\right) \qquad [7.47]$$

where k is the decay coefficient, which incorporates biodegradation, dilution, sorption, and so on, x is the distance from the source, and x/v is the time it takes groundwater to travel a distance x. Buscheck and Alcantar (1995) equated Eqs. [7.46] and [7.47] and solved for λ:

$$\lambda = \left(\frac{v_c}{4\alpha_x}\right) \left\{ \left[1 + 2\alpha_x\left(\frac{k}{v}\right) \right]^2 - 1 \right\} \qquad [7.48]$$

where k/v is the negative of the slope of a line obtained from a log-linear plot of the (centerline) contaminant concentration versus distance downgradient along the flow path, and all other terms are as previously defined. Note that if biodegradation is assumed to occur only in the liquid phase (i.e., adsorption of a contaminant removes it from microbial access), the term v_c in Eq. [7.48] should be replaced by the groundwater seepage velocity, v (ASTM 1998).

To determine λ using this approach also requires knowledge of the longitudinal dispersivity (α_x) and the contaminant (retarded) velocity (v_c). The following example, based on the field data listed in Table 7.16, illustrates how to obtain the slope of the

Table 7.16. Benzene Concentrations from Four Wells Placed Along the Center of Flow

Distance from Source Along Centerline, x (m)	Benzene Concentration, C (µg/l)	ln C
1	4000	8.2940
50	200	5.29831
95	10	2.30258
155	0.2	−1.6094

regression line from the centerline contaminant concentration versus migration distance (i.e., k/v), and how subsequently to calculate λ. This technique is one-dimensional, and may, therefore overestimate the λ value because the loss of contaminant mass from the plume due to transverse dispersion is ignored.

Example 7.10. Determination of Biodegradation Rate Coefficient (λ). The benzene concentrations in Table 7.16 have been measured in a steady plume at four wells placed along the principal flow (x) direction. Seepage velocity is 0.03 m/day, and dispersivity is estimated to be 7.5 m.

Step 1. Plot ln(C) versus x data.

Step 2. Perform a linear regression to obtain the slope (Figure 7.29). The negative of the slope is k/v.

Step 3. If biodegradation is assumed to occur only in the liquid phase, then the biodegradation rate coefficient can be determined from Eq. [7.48]:

$$\lambda = [0.03/(4 \times 7.5)]\{[(1 + 2 \times 7.5(0.0645)]^2 - 1\} = 0.0029\,\text{day}^{-1}$$

This corresponds to a half-life of $\ln(2)/0.0029 = 239$ days.

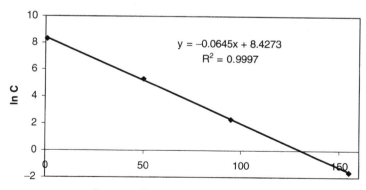

Figure 7.29. Example of linear regression to determine k/v by the Buscheck and Alcantar method. The term k/v is needed to solve Eq. [7.21]. In this example, using the regression routine from Excel, $k/v = 0.0645\,\text{m}^{-1}$.

The method of Buscheck and Alcantar (1995) requires the plume to be at steady state and the monitoring wells to be located along the centerline of the groundwater flow path. In practice, the monitoring wells are rarely ideally located as required by this technique. Furthermore, steady-state conditions and constant sources are not always encountered, which limits the applicability of this approach. It should also be kept in mind that this approach neglects horizontal and vertical dispersion. Thus, it intrinsically assumes that a potential decrease in contaminant concentration due to these processes is due to biodegradation. This can result in an overestimation of λ. Finally, this approach assumes that biodegradation occurs in the liquid but not in the sorbed phase (ASTM 1998), as is commonly the case for organic pollutants. If decay occurs also in the sorbed phase (e.g., radioactive materials), the groundwater velocity (v) should be replaced by the contaminant velocity (v_c) in Eq. [7.48].

Zhang and Heathcote (2003) extended the Buscheck and Alcantar technique for a three-dimensional plume. As shown by Eq. [7.49] a mathematical manipulation is employed to solve the three-dimensional steady-state solution for λ. Providing the values for horizontal transverse and vertical transverse dispersivity at a site are accurate, the λ calculated will be equal to or less than that obtained using the one-dimensional model of Buscheck and Alcantar.

$$C^*(x) = \frac{C(x)}{\mathrm{erf}(\beta_Y)\mathrm{erf}(\beta_Z)} = C_0 \exp\left\{\left(\frac{x}{2\alpha_x}\right)\left[1 - \left(1 + \frac{4\lambda\alpha_x}{v_c}\right)^{1/2}\right]\right\} \qquad [7.49]$$

where

$$\beta_Y = \frac{Y}{4(\alpha_Y x)^{1/2}} \quad \text{and} \quad \beta_Z = \frac{Z}{4(\alpha_Z x)^{1/2}}$$

Y is the width of the contaminant source, and Z is the thickness of that source. Both Y and Z are measured for a plane perpendicular to the flow direction. This method involves correcting the concentrations of interest by dividing by $[\mathrm{erf}(\beta_Y)\,\mathrm{erf}(\beta_Z)]$ to obtain $C^*(x)$. Values for $C^*(x)$ versus x plot as a straight line on a semilogarithmic graph. Therefore, the solution for λ in this method proceeds as in the Buscheck and Alcantar method (Eq. [7.48]).

Chapelle and Bradley (1998) used a similar approach, considering the *natural attenuation capacity (NAC)*. Recall that the solution to the one-dimensional advection–dispersion–degradation equation is given by Eq. [7.50]:

$$C(x) = C_0\left[\exp\left(\frac{-v_c + \sqrt{v_c^2 + 4D\lambda}}{2D}\right)x\right] \qquad [7.50]$$

The slope of C versus x along the centerline is the natural attenuation capacity, mathematically equivalent to

$$NAC = \left(\frac{-v_c + \sqrt{v_c^2 + 4D\lambda}}{2D}\right) \qquad [7.51]$$

Thus, $C(x) = C_0 e^{(NAC)x}$.

Note that NAC increases as v_c decreases or when λ increases.

For petroleum product releases (e.g., BTEX), another commonly used method to determine λ involves normalizing BTEX concentrations to those of a recalcitrant compound that was present in the initial release. This reference compound is assumed to behave as a conservative tracer. Trimethylbenzene (TMB) or tetramethylbenzene is commonly used for this purpose. Similar to the Buscheck and Alcantar approach, this technique also requires that the monitoring wells be placed in the centerline of the plume along the main flow direction. Unlike the Buscheck and Alcantar approach, the TMB normalization approach is applicable to nonsteady plumes.

Briefly, the TMB normalization approach attempts to discern biodegradation from other processes that decrease contaminant concentrations, such as dispersion, dilution from recharge, and sorption. This is accomplished by normalizing the aqueous concentrations of the BTEX compound of interest to the corresponding TMB concentration (Wiedemeier et al. 1995a):

$$C_{\text{corrected}} = C_2 \left(\frac{TMB_1}{TMB_2} \right) \tag{7.52}$$

Here, $C_{\text{corrected}}$ is the normalized concentration of the BTEX compound of interest at (downgradient) point 2, C_2 is the measured BTEX concentration at point 2, TMB_1 is the measured concentration of TMB at point 1, and TMB_2 is the TMB concentration at point 2.

Using Eq. [7.52] and assuming first-order biodegradation kinetics, a relationship between first-order rate coefficient (λ) and the corrected contaminant concentrations at two different points can be obtained:

$$C_{\text{corrected},d} = C_{\text{corrected},u}\, e^{-\lambda t} \tag{7.53}$$

where $C_{\text{corrected}}$ is the TMB-corrected contaminant concentration at the upgradient (u) and downgradient (d) locations, and t is the travel time between the two points. By rearranging Eq. [7.53], λ can be solved for explicitly, which yields a similar expression as Eq. [7.48]:

$$\lambda = \frac{-\left[\ln \left(C_{\text{corrected},d} / C_{\text{corrected},u} \right) \right]}{t} \tag{7.54}$$

The travel time (t) between the two points can be determined from the distance (x) traveled by the contaminant divided by the contaminant velocity (v_c):

$$t = \frac{x}{v_c} \tag{7.55}$$

If data are available for more than two wells located along the centerline of the plume, it is recommended that you do a linear regression of $\ln(C_{\text{corrected}})$ versus

Table 7.17. Benzene and *TMB* Concentrations from Well Samples Along the Center of Flow

Distance, x (m)	Travel Time, t (days)	TMB (μg/L)	Benzene (μg/L)	$(C_{corrected})^a$ (μg/L)	$\ln (C_{corrected})$
0	0	417	5600	5600	8.63052
6.7	50	400	4260	4441	8.39863
33.5	250	380	3000	3292	8.09925

$^a C_{corrected}$ is calculated using Eq. [7.52], where subscript 1 refers to data for $x = 0$.

travel time to determine the biodegradation rate coefficient. This approach is illustrated in the following example.

Example 7.11. Determination of λ by the TMB Normalization Method. The benzene and trimethylbenzene (TMB) concentrations in Table 7.17 have been measured in a steady plume at three wells placed along the principal flow (x) direction. The groundwater velocity is 0.45 m/day. Assume the retardation factor for benzene is 3.36.

Step 1. Determine the contaminant velocity (v_c) according to Eq. [4.16]:

$$v_c = 0.45/3.36 = 0.134\,\text{m/day}$$

Step 2. Determine the travel time (column 2 of Table 7.17) by dividing the travel distance (column 1 of Table 7.17) by the contaminant velocity.

Step 3. Perform a linear regression of $\ln(C_{corrected})$ versus t to obtain the slope. The negative of the slope is λ. In this example (Figure 7.30), $λ = 0.0019$ day^{-1}, which corresponds to a half-life of 1.0 year.

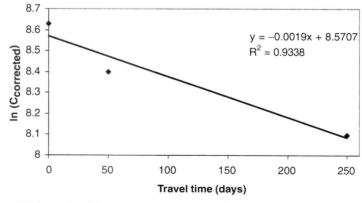

$$y = -0.0019x + 8.5707$$
$$R^2 = 0.9338$$

Figure 7.30. Example of linear regression to determine λ by the TMB normalization method.

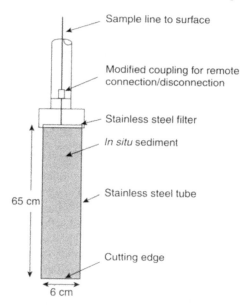

Figure 7.31. In situ microcosm device. (From Gillham et al. 1990a.)

It should be kept in mind that TMB is retarded by sorption to a greater extent than BTEX compounds due to its higher hydrophobicity. In addition, TMB can be biodegraded to some extent (Wiedemeier et al. 1996). Both of these phenomena increase the BTEX/TMB ratio downgradient from the source, which leads to an underestimation of λ. Since lower λ values are conducive to longer plumes, this approach is considered to be conservative.

In situ microcosms (ISMs shown in Figure 7.31) can also be used to determine λ (Gillham et al. 1990a,b; Higgo et al. 1996; Nielsen et al. 1996). ISMs are stainless steel cylinders that isolate about 2 L of the aquifer and can be installed using drilling rigs. The device is equipped with valves that allow for adding the contaminants of interest and for sampling over time from the ground surface. The ISM is open to the ambient aquifer environment at the bottom. Thus, one needs to correct for dilution effects using tracers such as tritium or nonreactive solutes. Since the ISM is hydraulically isolated from the surroundings (aside from capillary exchanges through the bottom) and no replenishment of depleted substrates or electron acceptors will take place during the measurement, the redox conditions should be monitored. Finally, sorption processes may confound the kinetic interpretation of the measurements. Thus, it is suggested that when $K_d < 1$ L/kg, loading of seven pore volumes of spiked groundwater should provide an even distribution of the compound throughout the ISM. This would allow for interpretation of the ISM as a completely mixed batch reactor, and λ to be obtained by fitting the data to Eq. [7.44]. If these conditions are not met, supplementary laboratory sorption studies should be conducted to correct for the disappearance of the added contaminants due to sorption processes.

An alternative method to determine λ is to use well data from the entire site and to use λ as a fitting parameter to a fate-and-transport model, using nonlinear regression techniques. However, the accuracy of this approach is significantly influenced by uncertainty regarding other model parameters.

One method that has recently received increased attention to obtain in situ estimates of biodegradation rates is the push–pull test performed in observation wells or multilevel samplers. Push–pull tests have been utilized to obtained degradation rates for anaerobic transformations of polynuclear aromatics (Borden et al. 1989), BTEX (Istok et al. 1997; Reinhard et al. 1997; Schroth et al. 1998, 2001; Kleikemper et al. 2002; McGuire et al. 2002; Reusser et al. 2002; Pombo et al. 2002), chlorinated solvents (Hageman et al. 2001), and radionuclides (Senko et al. 2002).

The test involves the injection of a mixture of tracers and chemically reactive solutes during the "push" phase and extraction of it during the "pull" phase, during which the mixture is recovered by reversing the flow direction. Figure 7.32a depicts the injection of the test solution in which the tracers are pushed away radially from the screened well. Figure 7.32b depicts the pull phase in which the tracer and reactive solutes are extracted and measured as a function of time. Sometimes a drift or reaction phase is also included between the push and pull sequence (Istok et al. 1997) to allow for the tracers to interact with a larger volume of the aquifer. The injected solution penetrates radially from the screened interval, but in the field these conditions can differ significantly and can depend on a number of factors including the initial saturated thickness, aquifer heterogeneity, length of the well screen, the injection rate, duration of injection, and well construction and development methods (Istok et al. 1997).

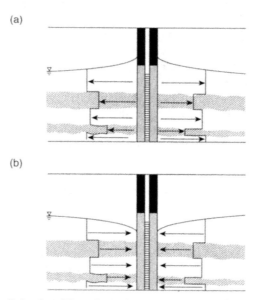

Figure 7.32. Single-well, "push–pull" test for in situ determination of microbial activities: (a) injection phase and (b) extraction phase. (Adapted from Istok et al. 1997.)

The method relies on the use of reactive tracers that are converted to various products by the microbes. Solute concentrations of various geochemical constituents are measured to obtain breakthrough curves of conservative and reactive solutes. This information is then used to compute the quantities of reactant(s) consumed or product(s) formed during a given test and are then used to infer the reaction type and rates in situ. In particular, the conservative tracer is used to evaluate recovery as well as the transport mechanisms (i.e., advection, dispersion, and diffusion), while the breakthrough curves for the reactive solutes and products are used to quantify microbial activity (Istok et al. 1997). For example, to examine whether aerobic processes are occurring and their corresponding rates, one could inject bromide, which acts as a conservative tracer in oxygenated water. If aerobic respiration is occurring, the CO_2 concentrations should increase during the extraction phase. Field tests conducted by Istok et al. (1997) at a formal gasoline bulk terminal located in Oregon showed conclusive results that aerobic respiration was absent at one location (Figure 7.33a) but at another close-by location strong evidence indicated aerobic respiration (Figure 7.33b).

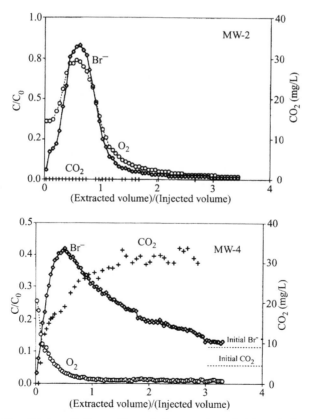

Figure 7.33. Push–pull test results for aerobic respiration experiment in (a) monitoring well MW-2 and (b) monitoring well MW-4. (Adapted from Istok et al. 1997.)

Degradation rates are obtained by using the conservative tracer concentrations to adjust concentrations of the test solution. One then computes the mass balances by integrating the breakthrough curve (e.g., using a spreadsheet) obtained during the extraction phase. The extent of reaction is then used to infer degradation rates. Specifically, the rates are computed from the mass of reactant consumed and the product formed. Rates can also be obtained by fitting the breakthrough curves to available analytical or numerical solutions. A simplified approach for data analysis also has been developed by Haggerty et al. (1998).

Because push–pull tests involve a single well, the method allows for obtaining a larger scale estimate of the biodegradation rate constant that is representative of field conditions in comparison to those obtained from cores. However, in comparison to other methods, such as those by Buscheck and Alacantar (1995), the method yields biodegradation coefficients that are much smaller in scale as representative of local hydrogeological and microbial conditions. Data in Figure 7.33 show that microbial activity can be highly heterogeneous. Therefore, push–pull tests can be used to infer information about the spatial variability of degradation constants (e.g., Schroth et al. 1998).

7.3. DETERMINING IF A PLUME HAS REACHED STEADY STATE

Historical monitoring data collected for the entire plume should be examined to determine if the plume is expanding, steady, or shrinking. This is important for selecting suitable mathematical models to determine biodegradation rate coefficients, and for selecting appropriate remedial actions. If the plume dimensions have not changed significantly over time, and variations in monitoring well concentrations can be attributed to random factors such as water table fluctuations, sampling variability, and analytical uncertainty, the plume can be considered to be at steady state. In essence, this "steady state" is achieved when any additional contamination released from the source is "naturally attenuated" and degraded at the same rate that it is introduced (ASTM 1998). As the source is depleted and the main loading rate of constituents of concern to groundwater decreases, the plume begins to shrink.

To determine the status of a plume, monitoring points or other sampling devices should be located to allow the construction of concentration contour maps for BTEX and other constituents of concern. This monitoring map should include a nondetect or compliance level contour (e.g., the drinking water criterion of 5 ppb for benzene) (ASTM 1998). Based on changes (or lack of changes) over time, the plume can be characterized as shrinking, stable, or expanding.

Alternatively, BTEX concentrations in two or more wells located within the plume and downgradient of the source, and oriented along the main flow direction, can be monitored over time (ASTM 1998). The trends in BTEX concentration will determine if the plume is expanding (i.e., increasing BTEX concentration), stable (i.e., constant BTEX concentration), or shrinking (i.e., decreasing BTEX concentration). One way to visualize such temporal trends is to plot the measured

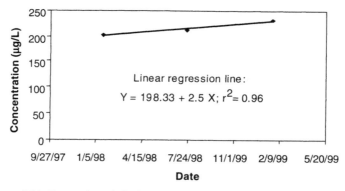

Figure 7.34. Temporal trends for benzene concentrations downgradient from source.

concentrations for a given BTEX compound at a given well as a function of time (Figure 7.34). To analyze temporal trends, a linear regression of concentration versus time data would confirm if concentration is significantly increasing, decreasing, or staying put over time.

The plume should be considered to be at steady state if the slope of the concentration versus time plot for different wells is not statistically discernible from zero. This would require performing Student's *t*-test (see Appendix E for critical values of the Student's *t*-distribution) on the slope of the regression line, with the null hypothesis that the slope equals zero. These statistical tests can be preformed automatically by numerous commercial software packages, including Excel as illustrated below.

Example 7.12. Determining Whether a Plume Has Reached Steady State. The benzene concentrations in Table 7.18 were obtained by sampling a well over time.

Step 1. Plot the concentration versus time data and perform a linear regression. For this example, Excel software was used (Figure 7.34, Table 7.19). Note that concentrations appear to be increasing with time. However, we need to determine if this increase is statistically significant.

Table 7.18. Benzene Concentrations Obtained from a Single Well Over Time

Date	Time (months)	Concentration (μg/L)
1/20/98	0	200
7/21/98	6	210
1/25/99	12	230

Table 7.19. Excel Output of Regression Analysis Results for Data from Table 7.18

Regression Statistics	
Multiple R	0.98
R Square	0.96
Adjusted R Square	0.93
Standard error	4.08
Observations	3

ANOVA	df	SS	MS	F	Significance F
Regression	1	450	450	27	0.12
Residual	1	16.67	16.67		
Total	2	466.67			

	Coefficients	Standard Error	t Statistic	p-Value	Lower 95%	Upper 95%
Intercept	198.33	3.73	53.22	0.01	150.98	245.69
Slope	2.5	0.48	5.20	**0.12**	−3.61	8.61

Step 2. Test the data for slope of concentration versus time equal to zero. Table 7.19 is the Excel output for the regression analysis. The p-value, which represents the attained level of significance for the t-test on the slope, was greater than 0.05 (i.e., $p = 0.12$). Because this slope is not statistically discernible from zero at the 95% significance level, the increase in concentration versus time is not statistically significant. Therefore, this plume meets the criterion to be considered steady.

To better quantify trends in contaminant concentrations, the EPA and the Wisconsin Department of Natural Resources (WDNR) have developed statistical tools for trend analysis. The EPA provides guidance for using tools for statistical analysis entitled "Guidance for Data Quality Assessment," dated July 2000, and the WDNR also provides electronic copies of Mann–Kendall and Mann–Whitney statistical test spreadsheets for the public use, including guidance entitled "Interim Guidance of Natural Attenuation for Petroleum Release Sites," dated October 1999 (WDNR 1999, 2001a,b). To use these spreadsheets, at least four consecutive rounds of data are required. For example, data from each quarter of one year of sampling may be used to analyze trends. Data should be obtained from one or more monitoring wells near the downgradient plume margin, a monitoring well near the source zone, and at least one monitoring well along a flow line between the source zone well and plume margin well (WDNR 2001a,b). Details of the approach are described in Appendix D and an example analysis of a contaminant plume trend is provided in Chapter 9.

We have seen in this chapter various approaches for site characterization necessary in preparation for mathematical modeling (Chapters 5 and 6), designing remediation options for a given problem (Chapter 8), or demonstrating the viability of natural attenuation (Chapter 9). The reader should keep in mind that these techniques are continuously subject to improvement, and new instruments to measure various parameters are being introduced every year. Therefore, it is important to keep up with new developments in site characterization techniques. Also, it is very important to keep in mind that parameters are just parameters. Recognizing whether a certain parameter makes sense or not in a particular field situation and its judicious use are some of the most important attributes of a well-trained contaminant hydrogeologist or environmental engineer. In the next chapter, we discuss various active and passive remediation technologies and case studies of their application.

REFERENCES

Abramowitz, M. and I.A. Stegun (1965). *Handbook of Mathematical Functions with Formulas, Graphs, and Mathematical Tables*. Dover Publications, New York.

ASTM (1994). *1994 Annual Book of ASTM Standards: Emergency Standard Guide for Risk-Based Corrective Action Applied at Petroleum Release Sites (Designation: ES 38-94)*. American Society for Testing and Materials, West Conshohocken, PA, pp. 1–42.

ASTM (1998). *1998 Annual Book of ASTM Standards: Standard Guide for Remediation of Ground Water by Natural Attenuation at Petroleum Release Sites (Designation: E 1943-98)*. American Society for Testing and Materials, West Conshohocken, PA, pp. 875–917.

Alvarez, P.J.J., L.C. Cronkhite, and C.S. Hunt (1998a). Use of benzoate to establish reactive buffer zones for enhanced attenuation of BTX migration. *Environ. Sci. Technol.* **32**:509–515.

Alvarez, P.J., Y.K. Zhang, and N. Lovanh (1998b). The Evaluation of the IA Tier-2 Model Based on Benzene Plume Dimensions and a Literature Search. Submitted to the Iowa Underground Storage Tank Board.

Anid, P.J., P.J.J. Alvarez, and T.M. Vogel (1993). Biodegradation of monoaromatic hydrocarbons in aquifer columns amended with hydrogen peroxide and nitrate. *Water Res.* **27**:685–691.

Boggs, J.M. and others (1992). Field study of dispersion in a heterogeneous aquifer. 1. Overview and site description. *Water Resour. Res.* **28**(12):3281–3291.

Borden, R.C., M.D. Lee, J.M. Thomas, P.B. Bedient, and C.H. Ward (1989). In situ measurement and numerical simulation of oxygen limited biotransformation. *Ground Water Monitoring Rev.* **9**(1):83–91.

Boulton, N.S. (1954). The drawdown of the water-table under non-steady conditions near a pumped well in an unconfined formation. *Proc. Inst. Civil Eng.* **3**:564–579.

Boulton, N.S. (1963). Analysis of data from nonequilibrium pumping tests allowing for delayed yield from storage. *Proc. Inst. Civil Eng.* **26**:469–492.

Bouwer, H. (1989). The Bouwer and Rice slug test—an update. *Ground Water* **27**(3):304–309.

Bouwer, H. and R.C. Rice (1976). A slug test for determining hydraulic conductivity of unconfined aquifers with completely or partially penetrating wells. *Water Resour. Res.* **12**(3):423–428.

Bowman, P.R. (1992). *Hydrologic Data for a Study of Pre-Illinoian Glacial Till in Linn County, Iowa, Water Year 1991.* USGS Open-File Report 92–500, Iowa City, IA.

Buscheck, T.E. and C.M. Alcantar (1995). Regression techniques and analytical solutions to demonstrate intrinsic bioremediation. In *Proceeding of the 1995 Battelle International Symposium on In Situ and On-Site* (April 1995) **3**(1):109–116.

Butler, J.J. (1998). *The Design, Performance, and Analysis of Slug Tests.* Lewis Publishers, Chelsea, MI.

Chapelle F.H and P.M. Bradley (1998). Selecting remediation goals by assessing the natural attenuation capacity of groundwater systems. *Bioremediation J.* **2**(3&4):227–238.

Chapelle, F.H., P.M. Bradley, D.R. Lovley, and D.A. Vroblesky (1996). Measuring rates of biodegradation in a contaminated aquifer using field and laboratory methods. *Ground Water* **34**:691–698.

Chiang, C.Y., J.P. Salanitro, E.Y. Chai, J.D. Colthart, and C.L. Klein (1989). Aerobic biodegradation of benzene, toluene, and xylene in a sandy aquifer—data analysis and computer modeling. *Ground Water* **27**:823–834.

Cooper, H.H. Jr. and C.E. Jacob (1946). A generalized graphical method for evaluating formation constants and summarizing well field history. *EOS Trans.* **27**:526–534.

Cooper, H.H., J.D. Bredehoeft, and I.S. Papadopulos (1967). Response of a finite-diameter well to an instantaneous charge of water. *Water Resour. Res.* **3**(1):263–269.

Dagan, G. (1967). A method of determining the permeability and effective porosity of unconfined anisotropic aquifers. *Water Resour. Res.* **3**:1059–1071.

Dagan, G. (1989). *Flow and Transport in Porous Formation.* Springer-Verlag, New York.

Danckwerts, P.V. (1953). Continuous flow systems. *Chem. Eng. Sci.* **2**:1–13.

Dawson, K.J. and J.D. Istok (1991). *Aquifer Testing—Design and Analysis of Pumping and Slug Tests.* Lewis Publishers, Chelsea, MI.

De Marsily, G. (1986). *Quantitative Hydrogeology: Groundwater Hydrology for Engineers.* Academic Press, Orlando, FL.

Dieulin, A., G. Matheron, G. de Marsily, and B. Beaudoin (1981). Dependence of an "equivalent dispersion coefficient" for transport in porous media. In *Proceedings of Euromech 143, Delft,* A. Verruijt, and F.B.J. Barends (Eds.). Balkema, Rotterdam, The Netherlands, pp. 199–202.

Domenico, P.A. and F.W. Schwartz (1990). *Physical and Chemical Hydrogeology.* John Wiley & Sons, Hoboken, NJ.

Domenico, P.A. and F.W. Schwartz (1998). *Physical and Chemical Hydrogeology,* 2nd ed. John Wiley & Sons, Hoboken, NJ.

Duffield, G.M. (2000). AQTESOLV. HydroSOLVE, Inc.

Fetter, C.W. (2001). *Applied Hydrogeology,* 4th ed. Prentice Hall, Upper Saddle River, NJ.

Freeze, R.A. and J.A. Cherry (1979). *Groundwater.* Prentice Hall, Upper Saddle River, NJ.

Freyberg, D.L. (1986). A natural gradient experiment on solute transport in a sand aquifer, 2. Spatial moments and the advection and dispersion of nonreactive tracers. *Water Resour. Res.* **22**(13):2031–2046.

Gelhar, L.W. (1993). *Stochastic Subsurface Hydrology.* Prentice Hall, Englewood Cliffs, NJ.

Gelhar, L.W., C. Welty, and K.R. Rehfeldt (1992). A critical review of data on field-scale dispersion in aquifers. *Water Resour. Res.* **28**(7):1955–1974.

Gillham, R.W., M.J.L. Robin, and C.J. Ptacek (1990a). A device for in situ determination of geochemical transport parameters. 1. Retardation. *Ground Water* **28**(5):666–672.

Gillham, R.W., R.C. Starr, and D.J. Miller (1990b). A device for in situ determination of geochemical transport parameters. 2. Biochemical reactions. *Ground Water* **28**(6):858–862.

Hageman, K.J., J.D. Istok, J.A. Field, T.E. Buscheck, and L. Semprini (2001). In situ anaerobic transformation of trichlorofluoroethene in trichloroethene-contaminated groundwater. *Environ. Sci. Technol.* **35**(9):1729–1735.

Haggerty, R., M.H. Schroth, and J.D. Istok (1998). Simplified method of "push–pull" test data analysis for determining in situ reaction rate coefficients. *Ground Water* **36**(2):314–324.

Hazen, A. (1911). Discussion: dams on sand foundations. *Trans. Am. Soc. Civil Eng.* **73**:199.

Higgo, J.J.W., P.H. Nielsen, M.P. Bannon, I. Harrison, and T.H. Christensen (1996). Effect of geochemical conditions on fate of organic compounds in groundwater. *Environ. Geol.* **27**:335–346.

Hvorslev, J. (1951). *Time Lag and Soil Permeability in Ground-Water Observations*, Bulletin No. 36. Waterways Experiment Station, U.S. Army Corps of Engineers.

Istok, J.D., M.D. Humphrey, M.H. Schroth, M.R. Hyman, and K.T. O'Reilly (1997). Single well, "push–pull" test for in situ determination of microbial activities. *Ground Water* **35**(4):619–631.

Jacob, C.E. (1950). Flow of groundwater. In *Engineering Hydraulics*, H. Rouse (Ed.). John Wiley & Sons, Hoboken, NJ.

Kleikemper, J., M.H. Schroth, W.V. Sigler, M. Schmucki, S.M. Bernasconi, and J. Zeyer (2002). Activity and diversity of sulfate-reducing bacteria in a petroleum hydrocarbon-contaminated aquifer. *Appl. Environ. Microbiol.* 1516–1523.

Kroszynski, U.I. and G. Dagan (1975). Well pumping in unconfined aquifers: the influence of the unsaturated zone. *Water Resour. Res.* **11**(3):479–490.

Kruseman, G.P. and N.A. de Ridder (1991). Analysis and Evaluation of Pumping Test Data. International Institute for Land Reclamation and Improvement, The Netherlands.

LeBlanc, D.R., S.P. Garabedian, K.M. Hess, L.W. Gelhar, R.D. Quadri, K.G. Stollenwerk, and W.W. Wood (1991). Large-scale natural gradient tracer test in sand and gravel, Cape Cod, Massachusetts, 1. Experimental design and observed tracer movement. *Water Resour. Res.* **27**(5):895–910.

Lyman, W.J., P.J. Reidy, and B. Levy (1992). *Mobility and Degradation of Organic Contaminants in Subsurface Environments*. C.K. Smoley, Chelsea, MI.

Mackay, D.M., D.L. Freyberg, and P.V. Roberts (1986). A natural gradient experiment on solute transport in a sand aquifer, 1. Approach and overview of plume movement. *Water Resour. Res.* **22**(13):2017–2029.

McGuire, J.T., D.T. Long, M.J. Klug, S.K. Haack, and D.W. Hyndman (2002). Evaluating behavior of oxygen, nitrate, and sulfate during recharge and quantifying reduction rates in a contaminated aquifer. *Environ. Sci. Technol.* **36**:2693–2700.

Moench, A. (1995). Combining the Neuman and Boulton models for flow to a well in an unconfined aquifer. *Ground Water* **33**:378–384.

National Research Council (1994). *Alternatives for Groundwater Cleanup*. Report of the National Research Council Committee on Groundwater Cleanup Alternatives. National Academy Press, Washington, DC.

Nelson, D.W. and L.E. Sommers (1982). Total carbon, organic carbon, and organic matter. In *Methods of Soil Analysis Part 2, Chemical and Microbiological Properties of Soil*, 2nd ed. Soil Science Society of America, Madison, WI.

Neuman, S.P. (1972). Theory of flow in unconfined aquifers considering delayed response of the water table. *Water Resour. Res.* **8**(4):1031–1045.

Neuman, S.P. (1974). Effect of partial penetration on flow in unconfined aquifers considering delayed gravity response. *Water Resour. Res.* **10**(2):303–312.

Neuman, S.P. (1975). Analysis of pumping test data from anisotropic unconfined aquifers considering delayed gravity response. *Water Resour. Res.* **11**(2):329–342.

Neuman, S.P. (1990). Universal scaling of hydraulic conductivities and dispersivities in geological media. *Water Resour. Res.* **26**(8):1749–1758.

Neuman, S.P. and Y.-K. Zhang (1990). A quasi-linear theory of non-Fickian and Fickian subsurface dispersion: 1. Theoretical analysis with application to isotropic media. *Water Resour. Res.* **26**(5):887–902.

Nielsen, P.H., P.L. Bjerg, P. Nielsen, P. Smith, and T.H. Christensen (1996). In situ and laboratory determined first-order rate constants of specific organic compounds in an aerobic aquifer. *Environ. Sci. Technol.* **30**:31–37.

Pombo, S.A., O. Pelz, M.H. Schroth, and J. Zeyer (2002). Anaerobic transformations:BTEX field-scale ^{13}C-labeling of phospholipid fatty acids (PLFA) and dissolved inorganic carbon:tracing acetate assimilation and mineralization in a petroleum hydrocarbon-contaminated aquifer. *FEMS Microbiol. Ecol.* **41**:259–267.

Prickett, T.A. (1965). Type curve solution to aquifer tests under water table conditions. *Ground Water* **3**(3):5–14.

Reinhard, M., S. Shang, P.K. Kitanidis, E. Orwin, G.D. Hopkins, and C.A. Lebron (1997). In situ BTEX biotransformation under enhanced nitrate- and sulfate-reducing conditions. *Environ. Sci. Technol.* **31**(1):28–36.

Reusser, D.E., J.D. Istok, H.R. Beller, and J.A. Field (2002). In situ transformation of deuterated toluene and xylene to benzylsuccinic acid analogues in BTEX-contaminated aquifers. *Environ. Sci. Technol.* **36**(19):4127–4134.

Roberts, P.V., M.N. Goltz, and M.M. Douglas (1986). A natural gradient experiment on solute transport in a sand aquifer. 3. Retardation estimates and mass balances for organic solutes. *Water Resour. Res.* **22**(13):2047–2058.

Schnoor, J.L. (1996). *Environmental Modeling: Fate and Transport of Pollutants in Water, Air, and Soil.*, John Wiley & Sons, Hoboken, NJ.

Schroth, M.H., J.D. Istok, G.T. Conner, M.R. Hyman, R. Haggerty, and K.T. O'Reilly (1998). Spatial variability in in situ aerobic respiration and denitrification rates in a petroleum-contaminated aquifer. *Ground Water* **36**(6):924–937.

Schroth, M.H., J. Kleikemper, C. Bolliger, S.M. Bernasconi, and J. Zeyer (2001). In situ assessment of microbial sulfate reduction in a petroleum-contaminated aquifer using push–pull tests and stable sulfur isotope analyses. *J. Contam. Hydrol.* **51**:179–195.

Schwartz, F.W. and H. Zhang (2003). *Fundamentals of Ground Water.* John Wiley & Sons, Hoboken, NJ.

Senko, J.M., J.D. Istok, J.M. Suflita, and R. Lee (2002). In-situ evidence for uranium immobilization and remobilization. *Environ. Sci. Technol.* **35**:1491–1496.

Shepherd, R.G. (1989). Correlations of permeability and grain size. *Ground Water* **27**(5):633–638.

Streltsova, T.D. (1972). Unsteady radial flow in an unconfined aquifer. *Water Resour. Res.* **8**(4):1059–1066.

Streltsova, T.D. and K. Rushton (1973). Water table drawdown due to a pumped well. *Water Resour. Res.* **9**:236–242.

Sudicky, E.A. (1986). A natural gradient experiment on solute transport in a sand aquifer: spatial variability of hydraulic conductivity and its role in the dispersion process. *Water Resour. Res.* **22**(13):2069–2082.

Theis, C.V. (1935). The relationship between the lowering of the piezometric surface and the rate and duration of discharge of a well using ground-water storage. *Trans. AGU* **16**:519.

Toride, N., F.J. Leij, and M. Th. van Genuchten (1995). *The CXTFIT Code for Estimating Transport Parameters from Laboratory or Field Tracer Experiments, Version 2.0*, Research Report No. 137. U.S. Salinity Laboratory, USDA, ARS, Riverside, CA.

van Genuchten, M.Th. and P.J. Wierenga (1986). Solute dispersion coefficients and retardation factors. In *Methods of Soil Analysis, Part 1—Physical and Mineralogical Methods*, 2nd ed., A. Klute (Ed.). American Society of Agronomy, Madison, WI.

Van Rooy, D.A. (1988). A note on the computerized interpretation of slug test data. *Inst. Hydrodyn. Hydraulic Eng. Prog. Rep. 66*, Tech. Univ., Denmark, p. 47.

Waterloo Hydrogeologic, Inc. (2001). *AquiferTest Version 3.0*.

Weber, W.J., S.H. Kim, and M.D. Johnson (2002). Distributed reactivity model for sorption by soils and sediments. 15. High-concentration co-contaminant effects on phenanthrene sorption and desorption. *Environ. Sci. Technol.* **36**(16):3625–3634.

Wenzel, L.K. (1942). *Methods for Determining Permeability of Water-Bearing Materials with Special Reference to Discharging Well Methods*. USGS Water-Supply Paper 887.

Wiedemeier, T.H., J.T. Wilson, D.H. Kampbell, R.N. Miller, and J.E. Hansen (1995a). *Technical Protocol for Implementing the Intrinsic Remediation with Long-Term Monitoring Option for Natural Attenuation of Dissolved-Phase Fuel Contamination in Groundwater*. U.S. Air Force Center for Environmental Excellence, Brooks Air Force Base, San Antonio, TX.

Wiedemeier, T.H., M.A. Swanson, J.T. Wilson, D.H. Kampbell, R.N. Miller, and J.E. Hansen (1995b). Patterns of intrinsic bioremediation of BTEX at two U.S. Air Force bases. In *Intrinsic Bioremediation*, R.E. Hinchee, J.T. Wilson, and D.C. Downey (Eds.). Battelle Press, Columbus, OH.

Wiedemeier, T.H., M.A. Swanson, J.T. Wilson, D.H. Kampbell, R.N. Miller, and J.E. Hansen. (1996). Approximation of biodegradation rate constants for monoaromatic hydrocarbons (BTEX) in ground water. *Ground Water Monitoring and Remediation* **16**:186–194.

WDNR (Wisconsin Department of Natural Resources) (1999). *Interim Guidance on Natural Attenuation for Petroleum Releases*. PUB-RR-614.

WDNR (Wisconsin Department of Natural Resources) (2001a). Mann–Kendall Statistical Test. Form 4400-215. Available at www.dnr.state.wi.us/org/aw/rr/archives/pub_index.html#15.

Wisconsin Department of Natural Resources (2001b). Mann–Whitney Statistical Test. Form 4400-216. Available at www.dnr.state.wi.us/org/aw/rr/archives/pub_index.html#15.

Zhang, Y.-K. and S.P. Neuman (1990). A quasi-linear theory of non-Fickian and Fickian subsurface dispersion, 2. Application to anisotropic media and the Borden site. *Water Resour. Res.* **26**(5):903–913.

Zhang, Y.-K. and R.C. Heathcote (2003). An improved method for estimation of biodegradation rate with field data. *Ground Water Monitoring and Remediation*, **23**(3):112–116.

8

BIOREMEDIATION
TECHNOLOGIES

"Nature, to be commanded, must be obeyed."

—*Novum Organum,* Sir Francis Bacon

8.1. INTRODUCTION

Modern civilization has become heavily dependent on a wide variety of hazardous substances that are used extensively in industrial, agricultural, service, and domestic activities. Without proper destruction, recycling or disposal, hazardous wastes become a major source of toxic pollution and a threat to our lives and the natural environment. A paradigm shift toward pollution prevention, supported by several international agreements, has curtailed the spread of hazardous wastes on a global basis. Most notably, the adoption of the precautionary principle in the late 1980s and early 1990s has been a major driving force for pollution prevention and industrial ecology. This principle states that "*when an activity raises threats of harm to human health or the environment, precautionary measures should be taken even if some cause-and-effect relations are not fully established scientifically*" (1998 Wingspread Statement). At the same time, two-thirds of the Earth's surface—the oceans—were closed to waste management through global bans on dumping industrial and radioactive waste and incineration at sea, and the Basel Convention of 1994 banned toxic waste exports from highly industrialized countries to less developed ones. Nevertheless, environmental contamination by hazardous chemicals continues to be an unintended consequence of human activities, often resulting in an unacceptable risk to public or environmental health. Consequently, numerous physical, chemical, and biological approaches have been developed to clean up contaminated sites. Among these, bioremediation is playing an increasingly important role as a practical, robust, and cost-effective approach to manage groundwater

---

Bioremediation and Natural Attenuation: Process Fundamentals and Mathematical Models
By Pedro J. J. Alvarez and Walter A. Illman   Copyright © 2006 John Wiley & Sons, Inc.

and soil contamination by toxic organic chemicals (National Research Council 1993, 1994, 1997a,b, 1999, 2000).

The use of bioremediation extends over a range of environmental media and contaminants, including air, water, soils, sediments, and industrial waste sludge. The range of contaminants is also diverse, including both organic (mainly) and inorganic compounds that are biotransformed into immobile forms. However, this chapter will devote special attention to bioremediation technologies that target aquifer contamination by petroleum hydrocarbons and chlorinated solvents. As discussed in Chapter 1, these are the two most commonly encountered classes of toxic organic pollutants. Indeed, environmental contamination is common during petroleum exploration, refinement, transportation, and storage operations. For example, gasoline releases from leaking underground storage tanks are widespread, with over 440,000 releases confirmed in the United States as of 2003 (U.S. Environmental Protection Agency 2003). Chlorinated compounds are also common environmental pollutants; occupying 12 out of the top 20 places in the 2003 CERCLA Priority List of organic hazardous substances frequently encountered at Superfund sites (Table 8.1). It should be kept in mind, however, that the technologies and case

Table 8.1. Organic Hazardous Substances Most Commonly Found at Facilities on the National Priorities List[a] (NPL)

NPL Rank	Substance Name
1	*Vinyl chloride*
2	*Polychlorinated biphenyls*
3	**Benzene**
4	**Polycyclic aromatic hydrocarbons**
5	**benzo[*a*]pyrene**
6	**benzo[*b*]fluoranthene**
7	*Chloroform*
8	*DDT*
9	*Aroclor 1254*
10	*Aroclor 1260*
11	**Dibenzo[*a,h*]anthracene**
12	*Trichloroethylene*
13	*Dieldrin*
14	*Chlordane*
15	*DDE*
16	Hexachlorobutadiene
17	**Coal tar creosote**
18	*DDD*
19	Benzidine
20	*Aldrin*

Source: Agency for Toxic Substances and Disease Registry (http: //www.atsdr.cdc.gov/clist.html).

[a] Hydrocarbons (bold) and chlorinated compounds (italic) dominate the list.

studies that will be presented in this chapter might also be extrapolated to other priority pollutants. Specifically, aerobic approaches to bioremediate hydrocarbon contamination might be applicable to other organic pollutants that are thermodynamically susceptible to be oxidized, whereas anaerobic bioremediation approaches for highly chlorinated solvents are likely to be effective for the removal of redox-sensitive pollutants that are prone to be reduced.

8.2. CONSIDERATIONS FOR SITE REMEDIATION

There are many physical and/or chemical processes and technologies that can be applied to partially, or completely, achieve site remediation. In cases where partial contaminant removal is attained, bioremediation can subsequently be used as an "in situ polishing step" to remove residual contamination. When complete physical removal of contamination is achieved, ex situ bioremediation may be employed to treat the contaminant mass removed. In some cases, physical and/or chemical processes are used in conjunction with bioremediation. Thus, site remediation can involve a combination of the following strategies:

- Physical removal of the pollutant or the highly contaminated media by means of extraction, excavation, thermal desorption, and soil washing.
- Chemical treatment, such as the addition of strong oxidants, to degrade high concentrations of organic chemicals near the source zone, and chemical reduction through the addition of strong reducing agents to stimulate the reductive precipitation of some heavy metals or the dechlorination of chlorinated solvents.
- Pump-and-treat systems using aboveground reactors that are based on numerous water or wastewater treatment processes such as electrodeposition, nanofiltration, microfiltration, activated carbon adsorption, volatilization in air strippers, and reverse osmosis.
- Biological treatment, including bioremediation and phytoremediation.
- Monitored natural attenuation.

These processes can be arbitrarily grouped into two categories: physicochemical and biological. In this section, we will survey physicochemical treatment processes that are frequently used in support of bioremediation, followed by a brief description of common biological treatment approaches.

8.2.1. Physicochemical Treatment

Physical and chemical processes often target the removal of the source of contamination in soils and aquifers to decrease the mass of contamination before bioremediation is employed. For example, excavation of highly contaminated soil near the

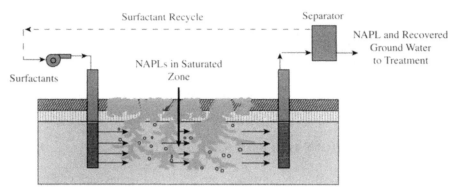

Figure 8.1. Surfactant or cosolvent application for flushing. (From Hughes et al. 2002.)

spill zone or removal of leaking underground storage tanks is a common first step in site remediation. The excavated material can be treated in situ by incineration or in slurry bioreactors (discussed later in this chapter) or transported to treatment and disposal facilities.

Physical extraction (i.e., pumping) of the free product, which is often present as a nonaqueous phase liquid (NAPL), is also important for source control. Certain processes such as surfactant and cosolvent flushing target the removal of light nonaqueous phase liquids (LNAPLs), such as gasoline, and dense non-aqueous phase liquids (DNAPLs), such as chlorinated solvents (Figure 8.1). Hydrophobic pollutants that are strongly sorbed to soil and aquifer sediments can also be mobilized for enhanced removal through extraction wells by the addition of surfactants or organic cosolvents (e.g., ethanol) that promote con-taminant dissolution. In temperate or cold regions, steam can also be injected to promote desorption of hydrophobic contaminants for enhanced physical removal (e.g., by volatilization) and aboveground treatment (Figure 8.2). Although these approaches only transfer the pollutants from one phase to another, treatment can subsequently be provided aboveground using a wide variety of methods.

Organic and inorganic contaminants may also be treated chemically to cause immobilization, mobilization for extraction, or detoxification. For example, the source zone, which may have contaminants at high concentrations that are in-hibitory to microorganisms, may be treated by the injection of strong oxidizer such as Fenton's reagent (a mixture of hydrogen peroxide and ferrous iron that generates hydroxyl radicals) (Figure 8.3). This promotes degradation of the contaminants with the production of by-products that facilitate the participation of microorganisms in the cleanup process during a subsequent polishing step. Contaminants that are thermodynamically susceptible to undergo reduction (e.g., highly chlorinated compounds and some heavy metals and radionuclides) could also be treated chemically through the addition of reducing agents, such as

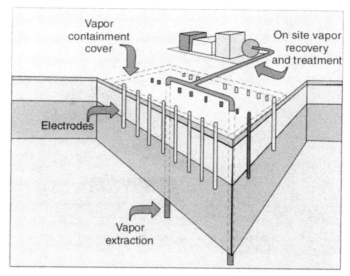

Figure 8.2. Thermal desorption system. (Adapted from www.sandia.gov/Subsurface/factshts/ert/teves.pdf.)

metabisulfide or iron nanoparticles. Permeable reactive barriers (PRBs) made with granular iron can also be installed perpendicular to the flow direction of the plume to intercept the transport of contaminants from the source and reduce contaminant flux from the source or treat the dissolved or aqueous phase contamination in the

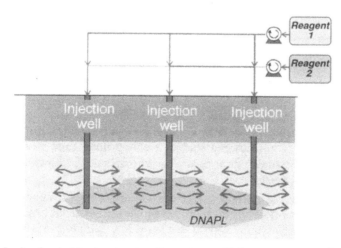

Figure 8.3. In situ chemical treatment system. Reagents are added to promote contaminant oxidation or reduction. (Adapted from enviro.nfesc.navy.mil/erb/restoration/technologies/remed/phys_chem/phc-43. asp.)

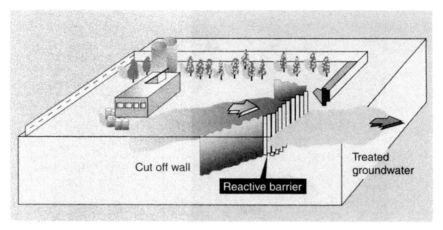

Figure 8.4. Permeable reactive iron barrier. (From Taisei Corporation web site (http://www.taisei.co.jp/english/technology/5b_permeable.html).)

plume (Figure 8.4). In such cases, natural attenuation might be an effective approach to remove pollutants or by-products that break through the PRB.

An in-depth discussion of such physical and chemical processes is beyond the scope of this chapter. However, Table 8.2 provides a summary of physicochemical processes that can be used in support of bioremediation.

8.2.2. Biological Treatment

Microbial degradation of organic pollutants is one of the most important mechanisms responsible for the assimilative capacity of aquifers, and bioremediation technologies constitute a range of natural and engineered process configurations. For example, biological treatment can be accomplished through:

1. Extraction and aboveground treatment of contaminated groundwater in conventional bioreactors such as those used to treat municipal and industrial wastewaters (e.g., activated sludge, trickling filters, and rotating biological contactors), including prior separation of suspended solids through screening or settling as needed.

2. Biostimulation of indigenous microorganisms through manipulation of environmental conditions in situ, by adjusting the pH and redox potential or adding selected nutrients or substrates that provide a competitive advantage to and stimulate specific degraders.

3. Bioaugmentation, or injection of specialized microorganisms into the contaminated zone, for cases when the target pollutant is recalcitrant and the indigenous concentration of specific degraders is negligible.

Table 8.2. Common Physicochemical Processes and Applications Used Prior to (or in Conjunction with) Bioremediation

Treatment Technology	Region of Action[a]	Process Description
Excavation	S, I, P	Removal of contaminated soil with earth-moving equipment followed by transport to appropriate location.
Dredging	S, I, P	Removal of contaminated sediments (and associated water) with a dredge followed by transport to appropriate location.
Groundwater extraction	P	Removal of groundwater using extraction wells and pumps. Not capable of mobilizing trapped globules of DNAPL contamination.
Soil washing	NA	Separation of soil or sediment particles after excavation or dredging based on particle size to retain fraction with the highest contamination levels (typically fine particles).
Thermal extraction	NA	Vaporization of contaminants from soil or sediment particles using heat after excavation or dredging.
Free-product recovery	S	Use of groundwater extraction wells to collect and remove LNAPL in pure form (may lower water table).
Surfactant flooding	S	Acceleration of dissolution or mobilization of DNAPL using surfactant solutions, injection wells, and extraction wells.
Cosolvent washing	S	Acceleration of dissolution of DNAPL using cosolvent solutions (e.g., alcohol solutions), injection wells, and extraction wells.
Heating (steam injection, electric resistive heating, radiofrequency heating)	S	Heating source zone to increase mobilization, volatilization, and stripping of organic contaminants.
Wet oxidation	S	Injection of a strong oxidizing agent such as permanganate that rapidly reacts with target contaminants.
Redox manipulation	I,P	Injection of redox active reagents such as dithionate to modify reactivity of mineral surfaces toward target contaminants.
Reactive barrier	I, P	System installed perpendicular to groundwater flow that contains materials such as elemental iron or microorganisms capable of chemically reacting with target pollutants.
Acid leaching	NA	Used with excavated or dredged materials to remove heavy metals by reducing pH and increasing metal solubility and mobility.

[a]Region is defined as the source (S), intercepting the transport from the source to the plume (I), the freely dissolved plume (P), and not applicable (NA) since it is carried out ex situ.

Source: Hughes et al. (2002).

8.3. OVERVIEW OF BIOREMEDIATION STRATEGIES

Bioremediation is a managed or spontaneous process in which biological, especially microbiological (bacteria and fungi), catalysis acts on pollutants, thereby remedying or eliminating environmental contamination present in water, wastewater, sludge, soil, aquifer material, or gas streams. The common approach to in situ bioremediation is to engineer the environment to overcome limitations to natural degradation processes. For example, fertilizers and oxygen can be injected into hydrocarbon-contaminated aquifers to overcome the insufficient supply of nutrients and electron acceptors. This approach is called *active* or *engineered* bioremediation. In some cases, however, natural conditions at contaminated sites meet all the essential environmental factors so that bioremediation can occur without human intervention to stimulate microbial activity. This process is called *natural* or *intrinsic* bioremediation and differs from no-action alternatives in that it requires thorough documentation of the role of microorganisms in eliminating the target contaminants. This is accomplished via tests and monitoring at field sites or onsite derived samples of soil, sediment, or water to ensure that the natural attenuation process continues to provide adequate risk protection.

Although bioremediation can be considerably less costly than traditional pump-and-treat (for contaminated groundwater) or excavation and incineration approaches (for contaminated soil), it can be scientifically more intense. Successful implementation often requires the integration of microbial ecology, contaminant hydrogeology, geochemistry, and process kinetics and reaction engineering, to name a few supporting disciplines. Therefore, the delineation of the applicability and limitations of bioremediation to clean up complex sites and the optimization of desirable microbial activities often require the participation of multidisciplinary teams.

Various international congresses have documented a wide variety of case studies where bioremediation has been successfully implemented. These are summarized in *Proceedings of the Battelle In Situ and On Site Bioremediation Symposia* (Alleman and Leeson 1991, 1993, 1995, 1997, 1999), *Proceedings of the Battelle International Conference on Remediation of Chlorinated and Recalcitrant Compounds* (Wickramanayake and Hinchee 1998; Wickramanayake et al. 2000), various publications of the Federal Remediation Technologies Roundtable 1997, 2000a,b, 2001a,b), and numerous publications by the U.S. Environmental Protection Agency (2000, 2001; Grindstaff 1998).

The rising popularity of bioremediation over the past decade is due primarily to its relatively low cost and its capacity to destroy a wide variety of organic pollutants, rather than transferring the target pollutants from one phase to another (as is the case in air stripping and activated carbon adsorption). Thus, bioremediation minimizes risks to environmental and public health while minimizing potential liability associated with hazardous waste transportation and disposal.

Microorganisms can also be induced to oxidize or reduce heavy metals (e.g., chromium) and radionuclides (e.g., uranium) to alter the valence of such inorganic pollutants for enhanced immobilization. For example, microbial U(VI) reduction could be an attractive alternative for in situ remediation of uranium-contaminated groundwater. In this case, extracellularly produced U(IV) precipitates as uraninite (UO_2), a highly immobile mineral (Gorby and Lovley 1992). Many bacteria have been identified that immobilize U(VI) by utilizing it as electron acceptor during respiration, including metal-reducing species belonging to the genera *Geobacter* and *Shewanella* as well as *Pseudomonas fluorescens*, *Desulfovibrio desulfuricans*, and *Deinococcus radiodurans* (Finneran et al. 2002; Liu et al. 2002; Anderson et al. 2003). However, the availability of appropriate electron donors can limit U(VI) bioreduction in situ. Common additional challenges to in situ bioreduction of U(VI) include the presence of heavy metals (e.g., Al and Ni) at inhibitory levels, acidity associated with the use of nitric acid in the processing of uranium waste, and the preferential reduction of competing electron acceptors such as nitrate. Previous work suggests that U(VI) bioreduction strategies should focus on nitrate removal as a preliminary step, not only because nitrate competes for reducing equivalents that would otherwise be available to immobilize U(VI), but also because NO_3^--dependent microbial U(IV) oxidation may reverse U(VI) reduction and decrease the stability of U(IV) precipitates (Elias et al. 2003; Istok et al. 2004).

The above discussion illustrates that the factors that control the efficacy of bioremediation can be very complex and exhibit considerable variation in relative importance from one site to another. In many cases, it is difficult to distinguish the contribution of physical, chemical, and biodegradation processes to contaminant removal. Albeit, the list of contaminants that could be removed by bioremediation is rapidly expanding, because the concept of recalcitrance is changing with increasing evidence that even persistent xenobiotics (e.g., 1,4-dioxane or methyl-*tert*-butyl ether [MTBE]) can be rapidly biodegraded under the right conditions (Spain 1997). As discussed in Chapter 3, recalcitrance does not imply indefinite resistance, and it may be due to suboptimal environmental conditions (e.g., extreme pH or temperature, unfavorable redox potential, or presence of toxic compounds). Therefore, one of the first steps in bioremediation engineering is to conduct a through site investigation and identify the factors that limit natural degradation processes at a given site. This information is then used to design approaches to overcome such limitations and enhance specific biocatalytic activities in situ.

There is considerable variability from site to site in how bioremediation is applied. Yet, all engineered bioremediation systems share a common feature— the stimulation and maintenance of microbial metabolism. However, some strategies may be applicable to one site but not to another, depending on the contamination scenario and the local hydrogeochemical conditions. Therefore, successful implementation of bioremediation may require integration of a broad range of sciences (including biochemistry, microbiology, mechanics, and geosciences) and engineering principles (including mathematical modeling, systems analysis, and reactor design).

8.3.1. Aerobic Versus Anaerobic Bioremediation

The oxidation state of target contaminants is the single most important factor determining whether the bioremediation process should be aerobic or anaerobic. Aerobic bioremediation is thermodynamically most favorable for the cleanup of reduced pollutants such as hydrocarbons. Such pollutants degrade faster under aerobic conditions than under anaerobic conditions, and oxygen availability is a common rate-limiting factor. In these cases, the pollutants serve as electron donors and carbon source to support microbial growth. On the other hand, highly chlorinated compounds such as perchloroethylene (PCE, $CH_2=CCl_2$) and hexachlorobenzene (HCB, C_6Cl_6) are already oxidized (by chlorine) and their aerobic degradation is not as feasible thermodynamically. Such compounds degrade faster anaerobically by reductive dechlorination. In such cases, the pollutants do not serve as carbon sources for growth but as electron acceptors in cometabolic or respiratory processes (i.e., dehalorespiration) as discussed in Chapter 3, and the availability of suitable electron donors can be rate limiting.

8.3.2. Biostimulation Versus Bioaugmentation

Based on the source of microorganisms used, there are two main bioremediation approaches:

- *Biostimulation*, which involves the addition of nutrients, electron acceptors (or electron donors), and sometimes auxiliary substrates to stimulate the growth and activity of specific indigenous microbial populations (Table 8.3).
- *Bioaugmentation*, which refers to the addition of exogenous, specialized microorganisms with enhanced capabilities to degrade the target pollutant.

Biostimulation is commonly selected for the cleanup of hydrocarbon-contaminated sites, often through the addition of oxygen and macronutrients such as nitrogen and phosphorus. Bioaugmentation is often unnecessary in such cases because indigenous bacteria that degrade hydrocarbons under aerobic conditions are ubiquitous in nature. Indeed, hydrocarbons have a natural pyrolitic origin and have been in contact with microorganisms throughout evolutionary periods of time, exerting selective pressure for the development of a wide variety of degradation pathways.

Table 8.3. Substances Commonly Injected into the Saturated Subsurface to Enhance In Situ Biodegradation

Electron Acceptors (Aerobic Processes)	Cometabolism (Aerobic Processes)	Electron Donors (Anaerobic Processes)
Oxygen	Butane	Hydrogen
Air	Methane	Lactate
Pure oxygen	Phenol	Molasses
Hydrogen peroxide	Propane	Polylactate esters, a hydrogen-releasing compound
Magnesium peroxide and other solid, oxygen-releasing compounds	Toluene	Vegetable oils

Furthermore, exogenous organisms that degrade hydrocarbons may not be as well adapted physiologically to site-specific conditions as the indigenous degraders, and their addition might result in marginal benefits. Bioaugmentation is most often selected for the cleanup of recalcitrant contaminants such as MTBE and PCE and will be discussed later in this chapter.

8.3.3. Ex Situ Versus In Situ Bioremediation

Based on the location where the bioremediation process occurs, bioremediation processes can be classified as follows:

- *Ex situ*, which involves the use of aboveground bioreactors to treat contaminated soil (in slurry bioreactors) or groundwater (in conventional suspended or attached growth bioreactors) that has been extracted from the contaminated site.
- *In situ*, which refers to bioremediation processes that occur below the ground surface, where the contaminated zone becomes the bioreactor.

In situ bioremediation often relies on manipulations of environmental conditions in the contaminated zone to overcome specific limitations to natural degradation processes and increase the rate and extent of pollutant degradation. This can be accomplished through biostimulation and/or bioaugmentation.

The principal advantages of in situ relative to ex situ bioremediation are:

- No remediation wastes are produced because treatment occurs below ground. This eliminates transportation costs and disposal liability.
- Minimum land and environmental disturbance, which is important if contamination migrates below existing buildings.
- Can treat relatively large volumes of contaminated soil and groundwater and can attack hard-to-withdraw hydrophobic pollutants that may be located at depths that preclude excavation.
- Although site characterization and monitoring requirements are higher, treatment costs are generally lower on a per volume basis.
- It is an environmentally sound process for which it is easy to get public acceptance.
- Unlike pump-and-treat approaches, it does not dewater the aquifer.

The principal disadvantages are:

- In situ treatment is generally slower than ex situ treatment due to inefficiencies in distributing stimulatory substrates throughout the contaminated zone and difficulties in overcoming mass transfer limitations through mixing or surfactant addition, which may result in comparatively slow degradation of hydrophobic pollutants.

- It is difficult to implement in highly stratified soils that hinder the vertical distribution of injected air or other gases through the contaminated zone.
- A major limitation is the inability to effectively treat mixtures of organic pollutants with heavy metals. However, innovative electrokinetic approaches in combination with in situ bioremediation show promise for the treatment of organometallic mixtures.

In situ bioremediation could be implemented as an aerobic or anaerobic treatment process. Aerobic treatment requires the addition of oxygen, which can be injected into the contaminated zone (along with nutrients) using air spargers (especially for shallow contamination) or hydrogen peroxide dissolved in the nutrient solution (Figure 8.5). Oxygen can also be released into the groundwater from (solid) oxygen-releasing compounds that are buried the contaminated zone. Anaerobic processes can also be stimulated. In this case, organic substrates or hydrogen gas can be injected as electron donors (e.g., acetate, lactate, methanol, or hydrogen gas) in concentrations that exceed the biochemical oxygen demand to stimulate anaerobiosis. Nutrients are typically injected also to satisfy the physiological requirements of anaerobes. For both aerobic and anaerobic biostimulation, the location, depth, and number of injection wells depend on the hydrogeologic characteristics of the sites and the contamination profile.

Ex situ bioremediation is generally selected for the treatment of highly contaminated material that has been extracted from the source zone, or when hydrophobic pollutants (such as PAHs and PCBs) are not effectively removed in situ due to low bioavailability. Ex situ treatment may also be selected when rapid removal of the contamination is required.

The general approach is to excavate contaminated soil and treat it in slurry or solid phase. Earth excavation and movement can account for a significant fraction

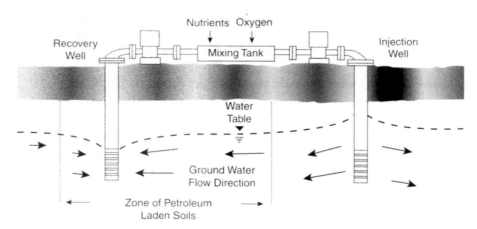

Figure 8.5. In situ aerobic biostimulation system. Electron acceptor (oxygen) may be replaced with electron donor (lactate, molasses, vegetable oil) for anaerobic biostimulation. (Adapted from Raymond 1974; Hughes et al. 2002.)

of the cost, often exceeding 25% of the total. Soil handling can be a complicated task when the soil is contaminated with radioactive, flammable, or explosive wastes. For slurry treatment, water (about 60–80% by weight) and nutrients are added with the contaminated soil to a batch reactor to obtain a slurry solution. Other amendments can include seed bacteria or fungi, acids or bases to adjust the pH, and surfactants to enhance the dissolution of hydrophobic pollutants. Mixing, aeration, and temperature are typically controlled in slurry bioreactors to optimize microbial activity and enhance solubilization of organic pollutants. However, this represents additional costs associated with energy and waste material and water handling. After treatment is completed, the slurry is dewatered and disposed of. Solid phase bioremediation is typically a less expensive but slower process due to decreased ability to control the process. Solid phase treatment is often accomplished through land farming or similar spreading and conditioning of the contaminated soil in special facilities that are lined with impermeable fabrics to prevent the migration of contaminated leachate to the underlying soil and groundwater. Moisture and nutrients (and sometimes microorganisms) can be provided by sprinklers or sprayers, whereas oxygen is provided by mixing and tilling, or through air pumps that can be installed at the bottom of the contaminated soil pile. To facilitate aeration, straw or other bulking agents may be added.

In the case of hydrocarbon-contaminated soils, the lighter constituents are typically removed by evaporation, and the remaining aliphatic and aromatic compounds are degraded by a wide variety of microorganisms (e.g., *Pseudomonas, Corynebacterium*, and *Mycobacterium*). Oxygen availability is a common rate-limiting factor, and its addition can result in removal efficiencies of about 80% within six months, with accompanying thousandfold to millionfold increases in the concentration of hydrocarbon degraders. Some hydrocarbons, such as branched aliphatics, cyclic, and polynuclear aromatic compounds, may persist for longer periods, especially if they reach anaerobic pockets. Optimum conditions for hydrocarbon biodegradation are given in Table 8.4.

Table 8.4. Optimum Environmental Conditions for Hydrocarbon Degradation in Soil

Environmental Factor	Optimum Levels
Nutrients	Ratio of C:N:P $= 120{:}10{:}1$
Temperature	25–35 °C (mesophilic)
Moisture	50–85% of the soil field capacity
Redox potential	Greater than $+50$ mV
pH	5.5–8.5
Microbial community	Low concentration of predatory protozoa; high number and diversity of hydrocarbon degraders
Contamination levels	Sufficiently low to preclude toxicity ($<$30,000 mg/kg total petroleum hydrocarbons [TPHs]) but sufficiently high to induce the required enzymes

8.3.4. Microbial Versus Plant-Based Bioremediation

There are two principal biological approaches to treat the contamination in situ: traditional bioremediation systems that rely on microbial metabolism for site cleanup (extensively discussed in this chapter) and phytoremediation, which relies on vegetation (briefly introduced).

Phytoremediation is a biological treatment process that utilizes natural processes harbored in (or stimulated by) plants to enhance degradation and removal of contaminants in contaminated soil or groundwater. Broadly, phytoremediation can be cost effective for (1) large sites with shallow residual levels of contamination by organic, nutrient, or metal pollutants, where contamination does not pose an imminent danger and only "polishing treatment" is required; and (2) where vegetation is used as a final cap and closure of the site (Schnoor et al. 1995; Schnoor 1997).

Advantages of using phytoremediation include cost effectiveness, aesthetic advantages, and long-term applicability (Table 8.5). Furthermore, the use of phytoremediation as a secondary or polishing in situ treatment step minimizes land disturbance and eliminates transportation and liability costs associated with offsite treatment and disposal. Increasing public and regulatory acceptance are likely to extend the use of phytoremediation beyond current applications.

Phytoremediation utilizes physical, chemical, and biological processes to remove, degrade, transform, or stabilize contaminants within soil and groundwater (McCutcheon and Schnoor 2003). Hydraulic control, uptake, transformation, volatilization, and rhizodegradation are important processes used during phytoremediation (Figure 8.6). These mechanisms and their applicability are discussed below and are summarized in Table 8.6.

Phytoextraction. This approach refers to the extraction and translocation of heavy metals (e.g., Cd, Ni, Hg, Se, and radionuclides) from shallow contaminated soil to plant tissue. Certain plants known as metal hyperaccumulators are known to extract and selectively absorb large quantities of heavy metals, resulting in their accumulation in plant tissue at greater concentrations than in the contaminated soil. Commonly used hyperaccumulators include sunflower (*Helianthus agnus*), Indian mustard (*Brassica juncea*), crucifers (*Thlaspi caerulescens, T. elegans*), violets (*Viola calaminaria*), serpentines (*Alyssum bertolonii*), corn, nettles, and dandelion. The normal procedure is to plant selected species in the contaminated area, allow

Table 8.5. Advantages and Disadvantages of Phytoremediation with Respect to Traditional (Microbial-Based) Bioremediation

Advantages	Disadvantages
Relatively low cost	Longer remediation times
Easily implemented and maintained	Climate dependent
Several mechanisms for removal	Effects to food web might be unknown
Aesthetically pleasing	Ultimate contaminant fates might be unknown
Harvestable plant material	Limited to shallow, residual contamination

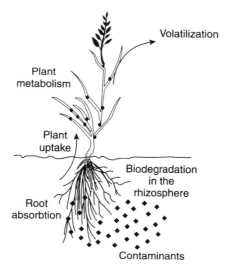

Volatilization

Plant
metabolism

Plant
uptake

Biodegradation
in the
rhizosphere

Root
absorbtion

Contaminants

Figure 8.6. Common phytoremediation mechanisms. (Adapted from Phytokinetics Inc.)

the plants to grow, and then harvest and incinerate the plant tissue. Although the heavy metals are not destroyed, this approach results in considerable reduction in heavy metal mobility and volume of contaminated media. Depending on the type and concentration of the heavy metals, extraction from the plant ashes for recycling purposes might be feasible.

Rhizofiltration. This application refers to the use of aquatic plants in wetlands or hydroponic reactors. The submerged roots of such plants act as filters for the adsorption of a wide variety of contaminants. When the sorption capacity of the submerged roots is saturated, the plants are removed and replaced.

Phytodegradation. Moderately hydrophobic (log $K_{ow} = 1.0$–3.0) organic pollutants can be removed from soil and groundwater through direct plant uptake. In

Table 8.6. General Application of Different Phytoremediation Approaches

Application	Medium[a]	Contaminants
Phytoextraction	S, WW	Metals
Rhizofiltration	WW	Metals, organics
Phytodegradation	S, GW, WW	Organics
Hydraulic control	GW, WW	Metals, organics
Phytovolatization	GW, WW	Volatile organics
Rhizoremediation	S, WW	Organics
Phytostabilization	S	Metals, organics

[a]S = soil, GW = groundwater, WW = wastewater discharges from municipal or industrial treatment plants, used for irrigation.

general, uptake of hydrophilic compounds (log $K_{ow} < 1$) is poor due to their low affinity for root membranes; whereas uptake efficiency is hindered for hydrophobic compounds such as PAHs (log $K_{ow} > 4$), which strongly sorb to soil and are, therefore, not bioavailable. Uptake efficiency depends on the transpiration rate and varies with plant species, age, health, and physicochemical properties of the root zone. Transpiration rate also varies dramatically and depends on the plant type, leaf area, nutrients, soil moisture, temperature, wind conditions, and relative humidity. Once the organic xenobiotic enters the plant system, it is partitioned to different plant parts through translocation. Unlike microbial species that metabolize organic contaminants to carbon dioxide and water, plants use detoxification mechanisms that transform parent chemicals to nonphytotoxic metabolites. The detoxification mechanism within plants is often described using the "green liver" concept. Once a contaminant enters the plant, any number of reactions within the following series may occur:

- Phase I—Conversion
- Phase II—Conjugation
- Phase III—Compartmentation

Conversion reactions include oxidations, reductions, or hydrolysis that the plant uses to begin detoxification. Conjugation reactions chemically link the phase I products to glutathione, sugars, or amino acids and, thus, the plant alters the solubility and toxicity of the contaminant. Once conjugated, xenobiotics can be removed as waste or compartmentalized. During compartmentation, chemicals are conjugated and segregated into vacuoles or bound to the cell wall material (hemicellulose or lignin). Phase III conjugates are often described as "bound residues" because chemical extraction methods do not recover the original contaminants.

Whereas many priority pollutants can be transformed within plants, some contaminants such as RDX may accumulate in leaves (Thomson et al. 1998). This is of concern because leaves could fall to the ground (potentially reintroducing the contaminant to the environment) or be eaten by animals (potentially impacting food webs).

Hydraulic Control. Phytoremediation applications can be designed to capture contaminated groundwater plumes to prevent off-site migration and/or decrease downward migration of contaminants. Trees and grasses act as solar "pumps," removing water from soils and aquifers through transpiration. The key to forming a successful barrier against plume migration is for trees to be rooted into a shallow water table aquifer. Phreatophytes, deep-rooted plants including hybrid poplars (e.g., *Populus deltoides* × *nigra*), and willows (e.g., *Salix alba* × *matsudana*) are most often used for hydraulic control. When planted densely (more than 600 trees per acre), poplars and willows usually reach optimum working conditions after 3–4 years during canopy closure when almost all the direct sunlight is intercepted. This application of phytoremediation requires that the bottom of the aquifer be confined by materials of low hydraulic conductivity such as clay, shale, or rock (hydraulic conductivity

$<10^{-6}$ cm/s) and does not "leak" water vertically down to another unit. However, plume capture is not limited to shallow aquifers, as poplar trees planted in well casings have been used to tap water tables at a depth of 10 m (Gatliff 1994).

Phytovolatilization. The natural ability of a plant to volatilize a contaminant that has been taken up through its roots can be exploited as a natural air-stripping pump system. Phytovolatilization is most applicable to those contaminants that are treated by conventional air-stripping, that is, contaminants with a Henry's constant $K_H > 10$ atm-m^3 water/(m^3 air), such as BTEX, TCE, vinyl chloride, and carbon tetrachloride. Chemicals with $K_H < 10$ atm-m^3 water/(m^3 air) such as phenol and PCP are not suitable for the air-stripping mechanism because of their relatively low volatility. Volatile pollutants diffuse from the plant into the atmosphere through open stomata in leaves. Radial diffusion through stem tissues has also been reported. For example, methyl-*tert*-butyl ether (MTBE) can escape through leaves, stems, and the bark to the atmosphere. Once released into the atmosphere, compounds with double bonds such as TCE and perchloroethylene (PCE) could be rapidly oxidized in the atmosphere by hydroxyl radicals. However, under certain circumstances (e.g., poor air circulation), phytovolatilization may not provide a terminal solution. For example, MTBE is long-lived in the atmosphere and can pose a risk to shallow groundwater during precipitation. In such cases, simple mass balance models can be utilized to determine if phytovolatilization poses a significant risk to humans and/or the environment. Nevertheless, the rate of release of VOCs from plant tissues is generally small relative to other emissions (Aitchinson et al. 2000). Thus, phytovolatilization is a potentially viable remediation strategy for many volatile organic chemicals.

Rhizoremediation. This application refers to bioremediation in the root zone. Microbial degradation in the rhizosphere might be the most significant mechanism for removal of hydrophobic compounds such as PAHs and PCBs. The strong sorption of such compounds to soils decreases their bioavailability for plant uptake and phytotransformation but increases their retention in the root zone, which facilitates the participation of microorganisms in the cleanup process. The rhizosphere of most plants promotes a wealth of microorganisms that can contribute significantly to the degradation of petroleum hydrocarbons during phytoremediation. Thus, though a plant may not directly act on these contaminants, a plant can influence the microbial community within its root zone to a great extent. Deposition of plant-derived carbon sources through root exudation and/or root turnover provides rhizosphere bacteria with numerous organic substrates. Rhizodeposition can account for release of 7–27% of the total carbon fixed during plant photosynthesis and varies between plants. The availability of simple organic carbon sources that can be used for growth promotes rhizosphere microbial populations, which have been reported to be fourfold to 100-fold greater than that observed in surrounding bulk soils. Selection of competent microorganisms during phytoremediation has been hypothesized. Miya and Firestone (2000) observed greater percentages of phenanthrene-degrading bacteria in rhizosphere soil than bulk soils and suggested the rhizosphere selected for PAH degraders. Siciliano et al. (2003) observed a higher frequency of catabolic genes in tall fescue rhizosphere than in bulk soil, suggesting that gene transfer or another mechanism of selection exists in the rhizosphere. Induction of microbial aromatic degradation has also been hypothesized due to the deposition of phenolic compounds that are structurally analogous to known inducers of enzymes responsible for degradation of aromatic contaminants. Gilbert and Crowley (1997) demonstrated induction of polychlorinated biphenyl (PCB) degradation in *Arthrobacter* sp. strain B1B using spearmint products and identified *l*-carvone as the compound responsible. Induction of PAH-degrading enzymes by plant root products has not been demonstrated in the literature (Kamath et al. 2004). However, co-oxidation of high molecular weight (HMW) PAH within the rhizosphere appears to be an important mechanism for phytoremediation. Generally, HMW PAHs do not serve as carbon and energy sources for microbial populations during degradation. The use of plants as a method to "inject" growth substrates to contaminated soil could overcome this limitation to degradation.

Phytostabilization. This application aims to prevent the dispersion of contaminated sediments and soil by using plants (mainly grasses) to minimize erosion by wind or rain action.

8.3.5. Dealing with Heavy Metals

Out of the 106 known elements, 84 are categorized as heavy metals (based on having a specific gravity greater than 4.0). Thus, the presence of heavy metals—from both natural and anthropogenic sources—is common at contaminated sites. Many heavy metals are priority pollutants that can hinder bioremediation performance when present at concentrations that inhibit microbial activities (typically on the order of 1 mg/L in groundwater or 2500 mg/kg in soil). Unlike organic pollutants, heavy metals cannot be destroyed by natural chemical, physical, or biological processes and tend to be persistent in the environment.

The bacteriostatic and bactericidal properties of heavy metals represent a major challenge to bioremediation. Heavy metal toxicity depends not only on the concentration and time of exposure, but also on the metal speciation, which is governed by the local pH and oxidation reduction potential, and soil characteristics. Most heavy metals are more mobile and more bioavailable under acidic (low pH) conditions, but their fate depends on intrinsic physical and chemical properties, the associated waste matrix, and the soil. Leachate and percolation of metals from the soil surface occurs when the metal retention capacity of the soil is overloaded, or when metals are solubilized (e.g., by oxidation–reduction reactions or by low pH). Other transport mechanisms include cotransport with dust and soils due to erosion caused by wind and surface water runoff. The extent of vertical contamination will depend on both the soil solution and the surface chemistry of the aquifer materials.

Common heavy metals that can be toxic to humans and inhibitory to bacteria are discussed below in alphabetical order (Federal Remediation Technologies Roundtable 2002):

- *Arsenic.* Arsenic (As) can be found as arsenate (As(V), AsO_4^{3-}) or as arsenite (As(III), AsO_2^-). Whereas arsenite is the more toxic form, arsenate is more common in soils. The behavior of arsenate in soil is similar to that of phosphate, which may reflect a high similarity in chemical properties between P and As (e.g., both belong to group VI of the periodic table). Arsenate sorbs strongly to soil and thus is relatively immobile. Sorption may be enhanced by iron (Fe) and other metals that form insoluble complexes with arsenate, such as aluminum (Al) and calcium (Ca). Under anaerobic conditions, arsenate may be reduced to arsenite, which is four to ten times more soluble and more prone to leaching.

- *Cadmium.* Cadmium (Cd) is commonly found in small quantities associated with zinc, copper, and lead ores and is used for metal plating and coating operations for transportation equipment, machinery and baking enamels, photography, and television phosphors. Cd is also used in nickel–cadmium and solar batteries, in pigments, as a stabilizer in plastics and synthetic products, as alloys, and in many types of solder (e.g., silver solder). Cadmium oxide and sulfide are relatively insoluble, especially under alkaline conditions, while the chloride and sulfate salts are soluble. Cadmium can be desorbed and mobilized under acidic conditions (pH <6). Immobilization of cadmium by sorption can be enhanced by the addition of anions such as humate or tartrate. Cd associated with carbonate minerals, precipitated as stable solid compounds, or coprecipitated with hydrous iron oxides is less likely to be mobilized by resuspension of sediments or biological activity, whereas Cd adsorbed to mineral surfaces (e.g., clay) or organic materials is more readily bioaccumulated or released in the dissolved state when sediments are disturbed (e.g., during flooding events).

- *Chromium.* Chromium (Cr) is commonly found in three forms: trivalent Cr(III) (Cr^{3+}) and hexavalent Cr(VI) [$(Cr_2O_7)^{2-}$ and $(CrO_4)^{2-}$]. Hexavalent chromium is broadly used in industry (e.g., tanneries) and wood preservatives and predominates as a chromate ion $(CrO_4)^{2-}$ above pH 6 or as dichromate ion $(Cr_2O_7)^{2-}$ below pH 6. Both forms of Cr(VI) ions are more toxic than Cr(III) ions, with dichromate presenting a greater health hazard. Because of its anionic nature, Cr(VI) binds mainly with positively charged soil surfaces. Sorption increases with decreasing pH due to a higher degree of protonation (and higher density of positive charges) at the soil surface. Iron and aluminum oxides also adsorb chromate ions at acidic or neutral pH. Chromium(III) precipitates in soil as hydroxy compounds at pH > 4.5 and is thus relatively immobile in soil, although Cr(III) can form complexes with soluble organic ligands that increase its mobility. Cr(VI) is commonly reduced in soil to Cr(III) by reduced organic matter, Fe(II) minerals, or microorganisms that donate the electrons.

- *Copper.* Copper (Cu) sorbs strongly to most soil constituents through ion exchange, although it has a high affinity for soluble organic ligands that may greatly increase its mobility. Cu is highly toxic to aquatic organisms.

- *Lead.* Lead (Pb) is a most common contaminant at battery recycling sites and exists in three oxidation states: 0, 2 + (II), and 4 + (IV). Pb tends to accumulate in top soil, usually within 3–5 cm of the surface. Pb concentrations decrease with depth. Insoluble lead sulfide is typically immobile in soil as long as reducing conditions are maintained. Lead can also be biomethylated, forming volatile tetramethyl and tetraethyl lead. The Pb sorption capacity of soil increases with pH, cation exchange capacity, organic carbon content, redox potential, and phosphate concentrations. Pb exhibits a high degree of sorption on clay-rich soil. Although only a small percent of the total Pb is leachable, surface runoff can transport soil particles containing lead. Groundwater, which is typically low in suspended solids, does not normally create a major pathway for Pb migration. Pb can be dissolved and mobilized also at low pH or high pH values, such as those induced by solidification/stabilization treatment.

- *Mercury.* Mercury (Hg) is very toxic and highly mobile in the environment. Volatile Hg forms (e.g., metallic mercury and highly toxic dimethylmercury, which is produced biologically usually under sulfate-reducing conditions) evaporate to the atmosphere, whereas solid forms sorb to soil. Hg exists primarily in the mercuric (II) and mercurous (I) valence states as a number of complexes with varying water solubilities. Sorption to soils and sediments is one of the most important processes for Hg removal from solution, and this process is usually more effective with increasing pH. Hg can also be coprecipitated with sulfides and complexed with natural organic matter (e.g., humic materials). Inorganic mercury sorbed to soils is not readily desorbed, which makes freshwater and marine sediments important repositories for inorganic Hg.

- *Selenium.* Selenium (Se) constitutes about 0.09 ppm of the earth's crust and is commonly found in the sulfide ores of heavy metals. Se is the most enriched element in coal, being present as an organoselenium compound, a chelated species, or an adsorbed element. Se is also used in many industrial processes such as the production of glass, pigments, rubber, metal alloys, textiles, petroleum, medical therapeutic agents, photographic emulsions, catalysts, and drugs. Selenium occurs in several valence states: II (hydrogen selenide, sodium selenide, dimethyl selenium, trimethyl selenium, and selenoamino acids such as selenomethionine); 0 (elemental selenium); IV (selenium dioxide, selenious acid, and sodium selenite); and VI (selenic acid and sodium selenate). Toxicity of selenium varies with valence state and water solubility of the compound in which it occurs, with the most oxidized forms at higher pH being more bioavailable (including for plant uptake) and more toxic. In acidic or neutral soils, Se tends to remain relatively insoluble and less bioavailable. Se volatilizes from soils when biomethylated by microorganisms to volatile compounds such as dimethyl selenide.

- *Zinc.* Zinc (Zn) is commonly found in soil and sediments polluted with iron (Fe) and manganese (Mn) oxides. Although clay carbonates and hydrous oxides adsorb Zn, rainfall mobilizes it from soil because zinc compounds are highly soluble. As with all cationic metals, Zn adsorption increases with pH. Zn also hydrolyzes at pH > 7.7, resulting in hydrolyzed species that sorb strongly to soil surfaces.

In summary, the toxicity and mobility of heavy metals depend primarily on their speciation, which is significantly influenced by soil pH, redox conditions, and surface chemistry. All of these are environmental factors that can be optimized by manipulating microbial activities to reduce the risk posed by heavy metals in aquifer environments. For example, indigenous microorganisms can be induced to alter the valence state (and thus the solubility) of heavy metals through redox reactions discussed earlier in Chapter 3. Specific strategies may include:

- Reducing heavy metals that precipitate in their reduced state, such as Cr(VI), to decrease their toxicity and facilitate their immobilization, which decreases their bioavailability. Reduction could be achieved through direct (anaerobic) microbial activity or indirect natural reductants produced by microorganisms (e.g., Fe(II)-bearing solids such as green rust, which are produced by Fe(III)-reducing bacteria, and sulfide produced by sulfate-reducing bacteria). For example, sulfate and electron donors can be injected to stimulate microbial sulfide production for subsequent precipitation of metal sulfides.

- Oxidizing heavy metals that precipitate in their oxidized state, such as As(III), for the same purpose. This would require providing suitable conditions for the proliferation and functioning of aerobic bacteria that catalyze this immobilization process.

- Stimulating the opposite redox reactions described above to enhance the solubility of the heavy metals and facilitate their extraction for subsequent (aboveground) treatment and recovery, perhaps by phytoextraction.

- Stimulating anaerobic conditions that are conducive to biomethylation for removal of some heavy metals such as Hg and Se by volatilization, provided that the volatile emissions are controlled and do not pose a threat to human or environmental health.

Of course, microorganisms are likely to be ineffectual when the heavy metal concentrations are relatively high, due to toxicity. In such cases, the inhibitory effects of heavy metals could be alleviated to some extent (prior to bioremediation) by the addition of acids or bases to manipulate the pH, and reducing or oxidizing agents to change the redox potential.

8.4. GENERAL IMPLEMENTATION APPROACH FOR IN SITU BIOREMEDIATION

The use of bioremediation extends over a range of environmental media, including air, water, soils, sediments, and industrial waste sludges. The range of contaminants is also diverse, including both organic (mainly) and inorganic compounds that are biotransformed into immobile forms. Such a broad range of applications requires a wide variety of implementation approaches that will be discussed later in this chapter. Despite fundamental differences among these bioremediation options, all alternatives generally follow a general approach consisting of the following five stages:

1. Site investigation—to diagnose problem prior to selecting an appropriate remediation strategy.
2. Physical measures to prevent further spreading of the contamination (e.g., free-product recovery) and contain plume migration if needed (e.g., install interdiction wells for hydraulic control to prevent off-site migration of contaminated groundwater).
3. Evaluation of bioremediation potential (e.g., extrapolate experience from similar sites and/or conduct treatability studies in the lab).
4. Select and design an appropriate system (e.g., aerobic versus anaerobic, biostimulation versus bioaugmentation, type of nutrient and electron acceptor or donor delivery system).
5. Implementation and operation of the selected system, using an adequate monitoring system for performance evaluation.

8.4.1. Site Investigation

Prior to selecting and designing any type of remediation system, including bioremediation, the site must be carefully and adequately characterized to obtain detailed information on the local geology, stratigraphy, hydrology, geochemistry, type and distribution of contamination, and microbiology. Critical information required includes:

- The nature of the geologic deposits, including the presence of fractures, thickness and extent of aquifers, location of confining units, texture, and fraction of organic carbon (f_{oc}), which influences the sorption potential as discussed in Chapter 2.
- The hydraulic conductivity, porosity, and variability of these parameters in the water-transmitting units (including heterogeneity and fluctuations in water table that cause smearing).
- The direction and slope of the naturally occurring hydraulic gradient (including seasonal variations), and the presence of potential preferential flow pathways along the backfill of utility lines.
- As much information as possible on the location, type, and mass of the contamination source, and the extent and distribution of contaminant plumes, including:
  - Plume delineation (recognizing that existing buildings may preclude the installation of monitoring wells needed to bracket the plume)

- Concentration range, aerial extent, and vertical distribution (including smearing potential)
- Physicochemical characteristics of pollutant (e.g., floater versus sinker, hydrophobic versus hydrophilic)

- Intrinsic supply and consumption rates of geochemical species affecting biodegradation:

  - Macronutrients, N, P
  - Dissolved oxygen
  - Fe(II) and other reduced minerals that could precipitate and cause clogging if oxygen is added
  - Oxidation capacity (Eq. [2.19]) and electron acceptors (e.g., nitrate, sulfate, Fe(III))
  - Reduction capacity of aquifer solids
  - pH and buffering capacity

- Basic information on microbiological activity (presence and concentration of desirable phenotypes), which could be obtained thorough the use of molecular probes or microcosm studies.

An ideal site for in situ bioremediation has the following general features:

- The contaminant is readily biodegradable.
- Hydraulic conductivity is relatively homogeneous, isotropic, and large ($>10^{-3}$ cm/s).
- Contaminant concentrations are not excessive (e.g., heavy metals <2000 mg/kg, TPH $<25,000$ mg/kg; NAPL concentrations <10 g/kg) to prevent a reduction in aquifer permeability and microbial access to the biodegradable contaminants, and preclude toxicity to microorganisms.
- The contamination is relatively shallow in order to minimize costs of drilling and sampling.
- For aerobic systems, dissolved [Fe(II)] <20 mg/L to prevent clogging due to iron oxide precipitation.

The feasibility of in situ bioremediation depends primarily of the following three factors:

1. The hydraulic conductivity, K, which determines the ability to distribute nutrients and electron acceptors throughout the contaminated zone. Ideally, K should exceed 10^{-3} cm/s.
2. The sorption capacity of the aquifer materials, which influences the retardation factor (R, Eq. [4.49]) and potential bioavailability. Bioremediation is generally easier to implement for cases in which $R \leq 3.0$.

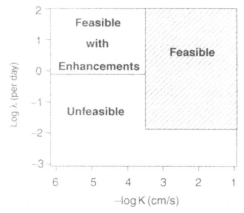

Figure 8.7. Feasibility of aerobic in situ bioremediation as a function of hydraulic conductivity (K) and the achievable biodegradation rate coefficient (λ).

3. The potential biodegradation rate (i.e., under biostimulated conditions), which should ideally result in a half-life no greater than 1 week (i.e., $\lambda > 0.1 \, \text{day}^{-1}$)

Figure 8.7 depicts the feasibility of in situ aerobic bioremediation as a function of hydraulic conductivity ($-\log K$) and biodegradation rate coefficient ($\log \lambda$). The hydraulic conductivity is generally the most important parameter determining the feasibility of biostimulation (Figure 8.8), although it should be recognized that values of K are typically spatially averaged. Often, there is insufficient appreciation of the heterogeneity of K. Even if a high value of K is determined at a given site, localized areas may have very low K values and contaminants trapped there will not be readily accessed by stimulatory materials added to the subsurface (see Table 8.3).

8.4.2. Physical Measures to Prevent Spreading of the Contamination

There are various physical measures that can be used to prevent spreading and off-site migration of the contamination, including:

- Source removal, which involves digging out leaking underground storage tanks and/or excavating highly contaminated soil, which are then treated onsite (e.g., by incineration or in slurry bioreactors) and/or disposed of in hazardous waste landfills.

- Installation of an interdiction well field for hydraulic control of plume migration. Several wells can be installed across the plume width (perpendicular to the main flow direction) to intercept and extract contaminated groundwater. The wells can be installed either downgradient of the plume to form a hydraulic barrier that captures the front of the plume, or upgradient near the source zone to reverse the gradient and hinder plume migration. Care

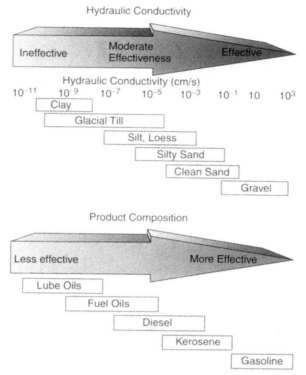

Figure 8.8. Screening for potential effectiveness of bioremediation of hydrocarbon-contaminated sites as a function of hydraulic conductivity of the aquifer and the composition of the hydrocarbon product released. (Adapted from http://www.epa.gov/swerust1/pubs/tum_ch10.pdf.)

must be taken to ensure that well installation does not cause vertical smearing of the pollution and does not facilitate vertical migration of contaminated groundwater into other water-bearing layers. Particular attention should be paid when installing wells in DNAPL source zones, since DNAPLs are denser than water and can readily migrate along the wells to greater depths.

- Free-product recovery, which typically removes between 50% and 70% of the total mass of contaminants in the case of fuel spills. One common approach is to use a dual-pump system (Figure 8.9). A deep well pumps groundwater to create a cone of depression on the water table, and the "floating" immiscible phase that accumulates in this cone of depression is pumped out by the shallow well, which transfers the LNAPL through an oil/water separator.

Another alternative to remove LNAPLs is bioslurping, which is also known as dual-phase extraction. As illustrated in Figure 8.10, this approach is based on vacuum-enhanced free-product recovery to extract the LNAPL from the capillary fringe and the water table. This system withdraws groundwater, free product, and soil vapors during the same process, using a single pump or multiple pumps

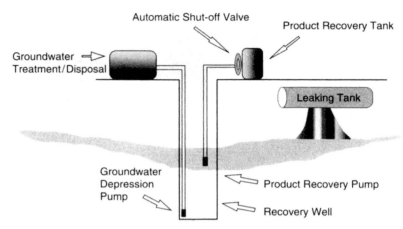

Figure 8.9. Dual-pump well system to extract free product with water table depression.

(Figure 8.11). The vadose zone soils are also aerated during soil gas vapor extraction. Therefore, this system promotes air circulation and aerobic biodegradation. The extracted groundwater is then passed aboveground through oil/water separators to remove the free product and is treated (when required) and discharged. The free product that is recovered can sometimes be recycled. The extracted soil gas vapor may also be treated (when required) and discharged.

A common phenomenon associated with pump-and-treat and extraction systems is that an asymptotic behavior is typically observed with respect to the rates of recovery of free product and groundwater, as well as a reduction in vapor

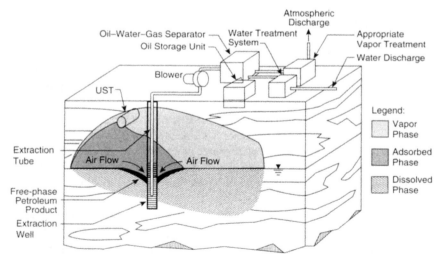

Figure 8.10. Dual-phase extraction system (a.k.a. bioslurping) to remove free product and fuel vapors near the water table while promoting air circulation and aerobic bioremediation. (From U.S. EPA 1985.)

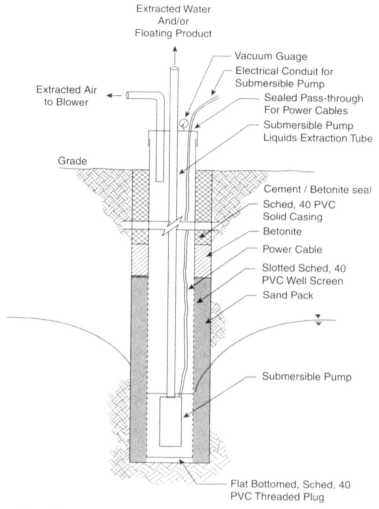

Figure 8.11. Multipump bioslurping extraction well. (From U.S. EPA 1985.)

concentration and the overall rate of mass removed (Figure 8.12). In such cases, it is common practice to shut down the system after about six months of asymptotic behavior, provided that the increase in dissolved volatile organic compound (VOC) concentration (due to desorption/deissolution) after shutdown remains relatively small.

8.4.3. Choice of Nutrient and Stimulatory Material Delivery System

There are many variations of in situ bioremediation systems, based on the way in which stimulatory materials are added (e.g., liquid, gas, or solid delivery systems)

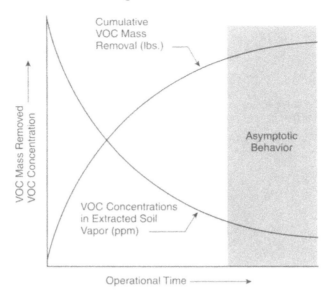

Figure 8.12. Asymptotic behavior of pumping systems. (From U.S. EPA 1985.)

(Hughes et al. 2002). The first in situ bioremediation system, which was patented by Raymond (1974), employed a liquid delivery system (see Figure 8.5). Such systems are widely used to treat hydrocarbon-impacted aquifers and commonly consist of injection and recovery wells for hydraulic control of the groundwater (which can be recirculated) and aboveground equipment to introduce nutrients and oxygen (e.g., hydrogen peroxide, which decomposes into water and oxygen: $2H_2O_2 \rightarrow O_2 + 2H_2O$). Delivery of dissolved anaerobic electron acceptors such as nitrate (Thomas et al. 1997) and sulfate (Anderson and Lovley 2000), or soluble electron donors such as lactate or acetate to stimulate microbial dechlorination of chlorinated solvents (Alleman and Leeson 1999 5(2); Wickramanayake et al. 2000 C2-4) is also possible. Whereas liquid delivery systems can achieve good hydraulic control, one potential concern with aerobic systems is clogging in areas near the injection well due to excessive microbial growth or precipitation of metal oxides.

Solid phase delivery systems provide for a long-term source of electron acceptors or electron donors that passively dissolve in the contaminated media (Koenigsberg and Norris 1999). Examples of solid phase amendments include Oxygen Releasing Compound (ORC®), such as magnesium peroxide (MnO_2), which is used to stimulate aerobic processes; and Hydrogen Releasing Compound (HRC®), such as polylactate, a polymer that is hydrolyzed in groundwater with the subsequent fermentation of by-products that yields hydrogen gas to stimulate reductive dechlorination. Granular (zero-valent) iron in permeable reactive barriers (see Figure 8.4) can also act both as an abiotic reductant and as a hydrogen-releasing compound during anaerobic corrosion (i.e., $Fe^0 + 2H_2O \rightarrow Fe^{2+} + 2OH^- + H_2$) (Till et al. 1998; Oh et al. 2001; Gandhi et al. 2002).

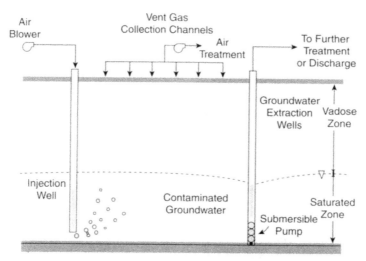

Figure 8.13. Gas phase delivery system for aerobic (with air) or anaerobic (with hydrogen) delivery system. (From Federal Remediation Technologies Roundtable 2002.)

One major advantage of solid phase delivery is negligible energy requirements that result in lower operation costs. Major concerns include depletion of the stimulatory material and the potential for contaminated water to bypass the biostimulated zone, due to lack of hydraulic control. Another potential concern associated with any amendments might be changes in pH and oxidation–reduction potential that induce changes in water quality (e.g., dissolution of metals under anaerobic, low pH conditions, increase in volatile fatty acids production from HRC.)

Gas phase delivery systems can inject air or pure oxygen to stimulate aerobic bioremediation, or pure hydrogen to stimulate anaerobic bioremediation (Figure 8.13) (Wilson and Ward 1986; Alleman and Leeson 1993 2(5), 1995 3(6), 1999 5(3); Hinchee 1994a,b; Van Eyk 1997; Newell et al. 2000; Groundwater Services, Inc. 2001). One potential advantage of these systems is enhanced stripping of volatile compounds, which can be captured by vacuum pumps and subsequently treated aboveground, although volatilization could be a concern if the vapors are not captured and spread out. Higher energy requirements and safety concerns in the case of hydrogen are potential disadvantages. The latter concern can be addressed by the use of hydrogen detectors, ventilation equipment, or both.

Examples of liquid, solid, and gas phase delivery systems will be provided in the next sections as we discuss the bioremediation of two common types of contamination problems: petroleum product releases and chlorinated solvent spills.

8.5. BIOSTIMULATION SYSTEMS TO TREAT FUEL SPILLS

Groundwater contamination by petroleum product releases is a common occurrence. In such cases, the contaminants of greatest concern are generally the monoaromatic hydrocarbons, benzene, toluene, ethylbenzene, and xylenes (BTEX), which are relatively mobile and are toxic to the central nervous system. Benzene is also a known human carcinogen, and its presence often drives the need for site remediation. Early approaches to remove BTEX from contaminated aquifers involved pumping the contaminated groundwater for aboveground treatment with activated carbon or air strippers. Nevertheless, BTEX compounds are moderately hydrophobic and tend to sorb to the aquifer material. This makes them difficult to withdraw by pumping and serves as a slow-release mechanism for sustained groundwater contamination. Pump-and-treat technologies alone can result in prohibitively long time periods for the removal of the residual contamination and are often economically unfeasible (Environmental Engineering Research Council 1990). In addition, treatment with activated carbon or air strippers merely transfers the contaminants from one phase to another, rather than transforming them into less harmful compounds. In situ bioremediation, which relies on indigenous microorganisms to degrade the target compounds within the aquifer, is receiving increasing attention due to its potential cost-effectiveness.

The ubiquity of microorganisms capable of degrading BTEX and other hydrocarbons associated with fuel spills is well established. Thus, bioremediation of fuel spills often involves stimulating microbial activities by supplying nutrients and electron acceptors (mainly O_2 and sometimes NO_3^-), with success often limited by the ability to distribute the stimulating materials throughout the contaminated zone. Biostimulation works best for relatively homogeneous aquifers of high permeability ($>10^{-5}$ m/s) after the free product has been removed. The most common biostimulation approaches include:

- *Bioventing.* This approach is used to stimulate aerobic degradation processes above the water table by the action of vacuum pumps that pull air through the unsaturated zone (Van Eyk 1997). Bioventing is often used with infiltration galleries that deliver water (with nutrients) to prevent desiccation in the unsaturated zone.

- *Water Circulation Systems.* This method is based on extracting contaminated groundwater for aboveground treatment and its reinjection into the ground with stimulatory amendments (e.g., H_2O_2 as oxygen source, and nutrients) (Ward et al. 1989; McCarty et al. 1998). Clogging near injection well screens and infiltration galleries can occur due to bacterial growth and mineral precipitation. In general, pulsing nutrients results in less clogging than continuous delivery. Occasional pulsing of Cl_2 or H_2O_2 to control biofouling can also prevent clogging.

- *Air Sparging.* This engineered system involves injection of compressed air into the contaminated subsurface to deliver oxygen and strip the BTEX into a vapor-capture system. Air sparging can be a relatively effective and inexpensive BTEX bioremediation approach, but it is not effective when low-permeability soil traps or diverts the airflow (Hinchee 1994a,b; Holbrook et al. 1998; Salanitro et al. 2000).

- *Biobarriers.* This term refers to biologically active zones that are placed in the path of (preferably shallow and narrow) BTEX plumes, often incorporating air spargers (i.e., air curtains) or oxygen-releasing compounds to enhance oxidative biodegradation processes (Salanitro et al. 2000). Hydraulic or physical controls on groundwater movement may be required to ensure that BTEX pass through the barrier.

The following sections provide a more detailed description of these biostimulation alternatives, including some case studies. It should be kept in mind that although emphasis has been placed on BTEX bioremediation, these techniques can be applied to remove a wide variety of organic hazardous wastes that are susceptible to aerobic biodegradation.

8.5.1. Bioventing

Bioventing is an in situ remediation technology that uses indigenous microorganisms to biodegrade organic constituents adsorbed to soils in the unsaturated zone. The activity of the indigenous bacteria is enhanced by inducing air (or oxygen) flow into the unsaturated zone, commonly by using vacuum pumps that pull air through the contaminated soil (Figure 8.14). If necessary, nutrients are added. This is often accomplished by using infiltration galleries that percolate water (with nutrients) to prevent desiccation. Ideally, the soil moisture should be 40–85% of the water holding capacity. Care must be taken in that water infiltration does not saturate the porous medium, which would hinder soil permeability to gas flow.

Bioventing is a medium- to long-term remediation technology. Cleanup ranges from 6 months to 5 years with a typical cost of $10–50/yd^3. Air circulation through the contaminated area can be induced by vertical or horizontal wells (Figure 8.14b). The latter is preferred for shallow groundwater conditions and for larger areal coverage. Bioventing is best suited for sites with groundwater depths of 10 ft or greater and is not appropriate for sites with groundwater tables located less than 3 ft below the surface (Table 8.7). In such cases, groundwater upwelling can occur within bioventing wells under vacuum pressures, potentially occluding screens and reducing or eliminating vacuum-induced soil vapor flow.

The principal advantages and disadvantages of bioventing are essentially the same as for bioslurping.

Advantages

- Uses readily available equipment that is easy to install.
- Creates minimal disturbance to other site operations.
- Can be used to clean up inaccessible areas (e.g., under a building).
- Requires short treatment times: usually 6 months to 2 years under optimal conditions.

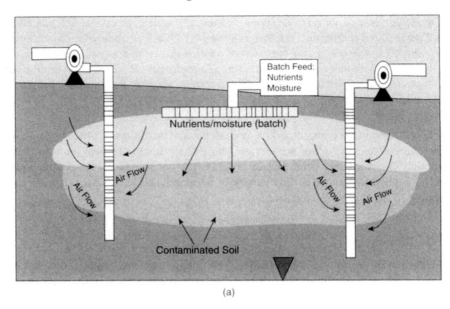

(a)

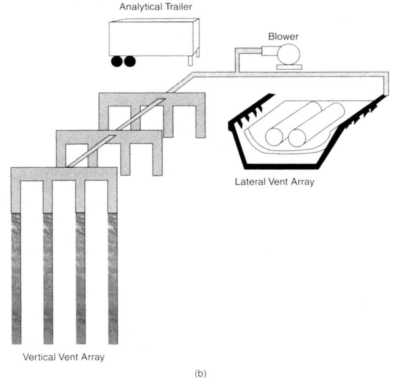

(b)

Figure 8.14. (a) Bioventing stimulates the aerobic degradation of pollutants in the unsaturated zone. (b) Bioventing system utilizing vertical and horizontal extraction well layouts to promote air circulation through contaminated soil. (From U.S. EPA 1985.)

Table 8.7. Effect of Depth to Water Table on Effectiveness of Bioventing

Depth to Water Table	Effectiveness
>10 ft	Effective
3–10 ft	Need special controls (e.g., horizontal wells or pumping)
<3 ft	Not effective

- Is easily combinable with other technologies (e.g., air sparging, groundwater extraction).
- May not require costly offgas treatment.

Disadvantages

- High contaminant concentrations may initially be toxic to microorganisms.
- Not applicable for soils with low permeability (e.g., high clay content).
- Cannot always achieve very low cleanup standards.
- Permits generally required for nutrient injection wells (if used).

Bioventing resembles soil vapor extraction (SVE) since both apply a vacuum to the vadose zone to force the movement of air through the contaminated area. However, these techniques have different purposes and also differ on their principal removal mechanisms as discussed below.

SVE

The main purpose of soil vapor extraction is to remove contaminants through volatilization. Common characteristics of SVE include:

- It is generally used for contamination at shallow depths (<30 ft).
- The radius of influence of extraction wells typically ranges from 15 to 30 ft.
- Vacuums of 0.1–0.2 atm (75–150 mm Hg) are commonly applied.
- Extraction rates are typically 1–6 m^3/min (0.6–3.5 cfs).
- Works only for "very" volatile contaminants.
- Moisture in the vadose zone will likely be evaporated under these conditions (which desiccates microorganisms and prevents their significant participation in the cleanup process).

Bioventing

This approach uses lower air flow rates because a major purpose is to stimulate biodegradation in the vadose zone.

- The purpose of the vacuum is to provide oxygen for aerobic biodegradation.
- The soil must be sufficiently permeable to gas flow.

- The spacing of vacuum points, although similar to those given above for SVE, are dependent on the soil's permeability and the applied vacuum.

- If nutrients are needed (which is fairly common), it is possible to add them in gaseous form (e.g., NH_3 or N_2O as N sources or triethylphosphate as P source).

- Methane, propane, or butane can also be added to stimulate the aerobic cometabolism of chlorinated aliphatic compounds such as TCE and DCE.

The target contaminants must be sufficiently volatile (but not too volatile) for bioventing to be most effective. Otherwise (e.g., if the dimensionless Henry's constant (H/RT) exceeds 0.1), removal will be due to volatilization rather than biodegradation. Hinchee (1994b) compiled Figure 8.15 to depict the limits of feasibility as a function of the physico chemical properties of various contaminants.

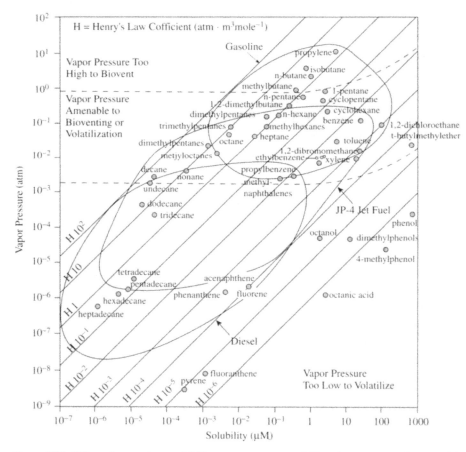

Figure 8.15. Effect of contaminant solubility, vapor pressure, and Henry's constant on the potential effectiveness of bioventing. (From Hinchee 1994b.)

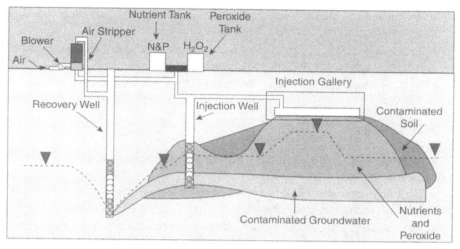

Figure 8.16. Water recirculation system for groundwater bioremediation.

8.5.2. Water Recirculation Systems

This alternative represents the first engineered bioremediation approach that was developed. Contaminated groundwater is extracted and treated aboveground using physicochemical processes (e.g., air stripping, activated carbon) or bioreactors. The treated groundwater (or a portion of it) is amended with nutrients and reinjected into the aquifer to biostimulate the contaminated zone (Figure 8.16).

This system was pioneered by Richard Raymond (1974; Raymond et al. 1986) for the bioremediation of hydrocarbon contamination in the saturated zone, and it is often referred to as the Raymond process. Because this is the original bioremediation system, in situ bioremediation is commonly associated as a recirculation system, although this alternative is not currently applied as frequently for hydrocarbon bioremediation as monitored natural attenuation or engineered options such as bioventing, air sparging, or land farming/composting.

Water recirculation systems are subject to the following advantages and disadvantages.

Advantages

- This system generally remediates contaminants that are adsorbed onto or trapped within the geologic materials of which the aquifer is composed along with contaminants dissolved in the groundwater.
- This application uses equipment that is widely available and easy to install.
- This process creates minimal disruption and/or disturbance to on-going site activities.
- Time required for subsurface remediation may be shorter than other approaches (e.g., pump and treat).

- This process generally is recognized as being less costly than other remedial options (e.g., pump and treat, excavation).
- Recirculation systems can be combined with other technologies (e.g., bioventing, soil vapor extraction) to enhance site remediation.
- In many cases, this technique does not produce waste products that must be disposed of.

Disadvantages

- Injection wells and/or infiltration galleries may become plugged by microbial growth or mineral precipitation.
- High concentrations (TPH > 50,000 ppm) of low-solubility constituents may be toxic and/or not bioavailable.
- Bioremediation with this system is difficult to implement in low-permeability aquifers ($<10^{-4}$ cm/s).
- Reinjection wells or infiltration galleries may require permits or may be prohibited. Some states require a permit for air injection.
- This process may require continuous monitoring and maintenance.
- Remediation may only occur in the more permeable layer or channels within the aquifer.

Water recirculation systems are perhaps the most versatile among biostimulation options because they attempt to handle contamination above and below the water table (Figure 8.16) and can, in theory, attack dissolved, sorbed, and NAPL contamination. Another aspect of their versatility is that, if stimulation of anaerobic conditions is desired (e.g., to induce microbial reductive dechlorination of chlorinated solvents), the electron acceptor delivery system may be modified to introduce electron donors instead, such as lactate, molasses, and vegetable oils. Specialized microorganisms (i.e., bioaugmentation) can also be added to the subsurface with these delivery systems. However, the recovery of all the circulating water may require aboveground treatment prior to reinjection, depending on regional environmental regulations. The time required for treatment with water recirculation systems can vary between 6 months and 5 years, and the cost can range from $1 to $25 per pound of contaminant removed.

There are many alternatives to introduce molecular oxygen to the contaminated subsurface (Table 8.8). A key feature of liquid delivery systems is the ability to achieve hydraulic control and the ability to rapidly dissolve amendments in aboveground delivery systems. Among the possible disadvantages of this approach include the potential to clog areas around the injection wells with precipitates and microorganisms, relatively high capital and operating costs, and the limited solubility of certain amendments (in particular, oxygen). Solid phase (case study 5) and gas phase delivery systems (case studies 2 and 5) were developed to overcome some of the limitations inherent to liquid delivery systems (Koenigsberg et al.

Table 8.8. Alternative Sources of Oxygen for Biostimulation of the Contaminated Zone

1. Air spargers:	Can dissolve only 8–12 mg/L of O_2 at 1 atm pressure. This approach may require continuous sparging at many wells.
2. O_2 spargers:	Can dissolve five times more O_2 than with air (O_2 makes up 21% of air, or about one-fifth). However, pure O_2 is more costly and poses an explosion hazard.
3. O_2 generator:	Cryogenically separates O_2 from air. This is generally more economical than O_2 sparging.
4. H_2O_2:	Hydrogen peroxide can be injected to produce oxygen as follows: $$2\,H_2O_2 \rightarrow 2\,H_2O + O_2$$ However, uncontrolled decomposition can occur if catalase enzymes (derived from microbial or plant sources) are abundant. Note that H_2O_2 can be toxic at concentrations of 3%. In fact, H_2O_2 is commonly used for sterilization in medical applications.
5. O_3:	Ozone can be used as a source of oxygen, $2\,O_3 \rightarrow 3\,O_2$, although it is relatively expensive.
6. ORCs:	Several solid substances, known as oxygen-releasing compounds (ORCs) can release oxygen upon hydrolysis. These include calcium peroxide (CaO_2), magnesium peroxide (MgO_2), sodium percarbonate ($Na_2CO_3 \bullet 1.5H_2O_2$), and urea hydrogen peroxide (urea-H_2O_2) powder mixed with concrete briquettes. An example reaction is $$2\,CaO_2 + 2H_2O \rightarrow 2Ca(OH)_2 + O_2$$

1999), but these are not without their own constraints (e.g., ORCs and HRCs have a finite release capacity).

Potential oxygen-releasing substances include magnesium peroxide (MgO_2), calcium peroxide (CaO_2), sodium percarbonate ($Na_2CO_3 \bullet 1.5H_2O_2$), and urea hydrogen peroxide (urea-H_2O_2) (Borden et al. 1997). These oxygen-releasing compounds are commonly powdered substances that can also be incorporated into a cement matrix to obtain an ORC concrete. This concrete can be prepared as briquettes to avoid significant reductions in aquifer permeability. The total available oxygen is about 5.2% by weight in mixture with concrete at 20% ORC (Bianchi-Mosquera et al. 1994).

Laboratory experiments have shown that urea peroxide and sodium percarbonate are most likely not acceptable for use in long-term projects due to a rapid depletion of oxygen release capabilities. The relatively high solubility of these compounds would require encapsulation for sustaining oxygen release capabilities. In contrast, a cement mix of 14% MgO_2 or 14% CaO_2 cement can release oxygen for hundreds of days (Bianchi-Mosquera et al. 1994). One major advantage of commercial CaO_2 (prepared at a mass purity of 60–80%) over commercial MgO_2 (prepared at a mass purity of 15–25%) is that the former generally delivers three to four times more oxygen (White et al. 1998). In addition, CaO_2 is less expensive and is easily

Table 8.9. System Monitoring Recommendations

Phase	Frequency	What to Monitor	Where to Monitor
Start-up (1–2 weeks)	At least daily	• Extraction volume • Injection volume	• Extraction or discharge wells and manifolds
	Every 2–3 days	• Electron acceptor concentration • Groundwater levels	• Monitoring wells
Remedial	Weekly	• Extraction and injection flow rates • Concentration of electron acceptor, ammonia, phosphate, nitrate, pH, and conductivity	• Extraction or discharge wells and manifolds • Monitoring wells
	Monthly or quarterly	• Concentration of contaminants in groundwater and soil	• Extraction, injection, and monitoring wells • Soil borings

produced in the field by heating lime with hydrogen peroxide (Vol'nov 1966). The rate of oxygen release can be controlled by the size of the concrete briquettes containing the ORCs, with smaller briquettes generally resulting in higher release rates (Borden et al. 1997). Field experiments found that ORCs are able to maintain elevated dissolved oxygen concentrations, typically between 6 and 19 mg/L (Bianchi-Mosquera et al. 1994).

In all cases, oxygen reacts with dissolved, reduced metals (e.g., Fe^{2+}) that may precipitate as oxides. These precipitation reactions may clog well screens and reduce the permeability of the aquifer (Hinchee and Downey 1988). Excessive bacterial growth can also contribute to this clogging problem (Cunningham et al. 1991). However, this problem can be prevented by pulsing the biostimulatory materials rather than introducing them continuously. In some cases, it may be necessary to occasionally flush the wells with a chlorine or hydrogen peroxide solution to control biofouling.

Process performance and evaluation requires an appropriate monitoring system. Table 8.9 summarizes monitoring recommendations for water circulation systems. A successful example of a water recirculation system is presented below in case study 1.

Case Study 1. Liquid Delivery of Electron Acceptor for In Situ Bioremediation of Hydrocarbons at Traverse City, MI
(*Source*: Hughes et al. 2002; based on Ward et al. 1989)

Indigenous aquifer microorganisms in the plume of an aviation fuel spill were stimulated by injection of an oxygen source, nitrogen, phosphorus, and trace elements to degrade hydrocarbons. The contaminated aquifer was pretreated with pure oxygen and inorganic nutrients, followed by treatment with hydrogen peroxide (H_2O_2) and inorganic nutrients. After eight months of treatment, reductions in benzene, toluene, ethylbenzene, and xylene (BTEX) concentrations were observed between the injection point and the 15 m (50 ft) wells.

This pilot-scale demonstration of in situ bioremediation was modeled after the liquid delivery system developed by Raymond (see Figure 8.5). The use of hydrogen peroxide (H_2O_2) as an oxygen source was also assessed since H_2O_2 provides increased dissolved oxygen (DO) concentrations, but can also be toxic to the microflora and cause reactions that have the potential to plug the aquifer.

Site Description

Approximately 38,000 L (10,000 gal) of aviation fuel contaminated the subsurface and entered the groundwater at a Coast Guard Air Station in Traverse City, Michigan, from a 1969 spill. The aquifer consisted of highly permeable sand and gravel. The major groundwater contaminants were benzene, toluene, and the isomers of xylene. The contaminant plume traveled in a northeasterly direction from the source at the hanger/administration building, then north across the property line into a residential area, and entered the East Bay (Traverse Bay) of Lake Michigan (Figure 8.17).

Field Design/Operation

The field demonstration was conducted in a 9 × 31 m (30 × 100 ft) plot upgradient of and in the contaminant plume (Figure 8.18). The well system consisted of nine

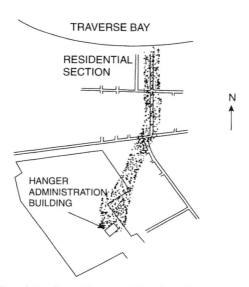

Figure 8.17. Aviation fuel spill and resulting plume, Traverse City, Michigan.

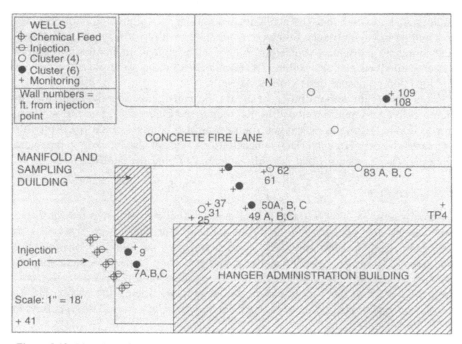

Figure 8.18. Plan view of in situ liquid delivery field demonstration area, Traverse City, Michigan.

4 in. monitoring wells with 3.7 m (12 ft) well screens and twelve cluster wells having four, five, or six screens at different depths at distances of approximately 2.1, 9.4, and 15.2 m (7, 31, and 50 ft) from the injection point. Cluster wells with four screens (Cluster(4)) were sampled at depths of 5, 5.5, 5.9, and 6.4 m (16.5, 18.0, 19.5, and 21 ft) below ground surface (BGS). Cluster wells with five or six screens (Cluster (6)) were sampled at 4.3, 5, 5.5, 5.9, 6.4, and 6.9 m (14.0, 16.5, 18.0, 19.5, 21.0, and 22.5 ft) BGS.

The inorganic nutrient feed contained ammonium chloride (NH_4Cl), potassium diphosphate (KH_2PO_4), and sodium monophosphate (Na_2HPO_4) to produce chloride, phosphorus, and ammonia nitrogen concentrations of 250, 75, and 100 mg/L, respectively, in the feed. It was pumped from the mixing tank to the chemical feed injection wells. Pure oxygen or hydrogen peroxide (H_2O_2) was added to the chemical feed after leaving the mixing tank. Water sparged with pure oxygen, resulting in an average DO concentration of 40 mg/L, was injected into the 4.3 and 5.9 m (14 and 19 ft) BGS depths through five injection wells at a rate of 3 L/min (11 gal/min). After three months of operation, or Julian day 150 (time from start of project or time of injection plus 60 days), when an increase in DO concentration was observed in well 7-B-2 (second screening depth of well 7-B), H_2O_2 replaced pure oxygen as the oxygen source, and the concentration was increased gradually. For one week, 50 mg/L of H_2O_2 was supplied to yield 25 mg/L DO. Then it was increased to 100 mg/L for one week. Next, the H_2O_2 concentration was increased to 250 mg/L for nine weeks and, finally, was increased to 500 mg/L for nine weeks (1 mg/L H_2O_2 yields 0.5 mg/L O_2).

Results

As oxygen and inorganic nutrients were consumed, an overall decrease in BTEX concentration was observed. No increase in DO concentration was observed until hydrocarbons were depleted. Decreased BTEX and increased DO concentrations were observed at the 7 ft well within 150 days, at the 31 ft well within 250 days, and at the 50 ft well within 300 days (Figure 8.19). The change in oxygen source from pure oxygen to H_2O_2 occurred at Julian day 150, or after three months of injection. Biodegradation of BTEX contaminants continued while aquifer permeability was not affected, unlike the plugging observed during H_2O_2 addition at Eglin AFB in Florida due to iron oxide precipitation (Hinchee and Downey 1988). Microbial numbers, as determined on nutrient and hydrocarbon degrading

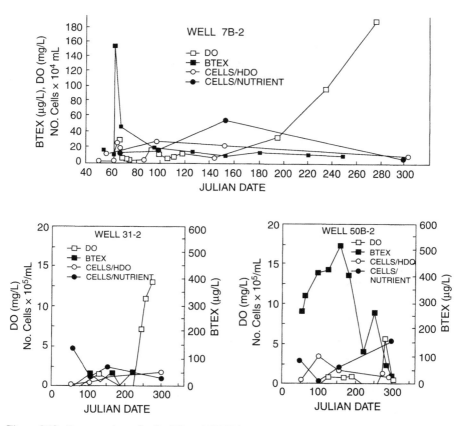

Figure 8.19. Concentrations of cells, DO, and BTEX in samples from wells 2.1, 9.4, and 15.2 m (7, 31, and 50 ft) from injection wells, Traverse City, Michigan.

organism (HDO) agars, increased over time in all downgradient wells until Julian day 300. No detectable viable microorganisms were found in well 7-B 5 m (16 ft) from the injection wells, probably due to H_2O_2 toxicity, at this last sampling period.

8.5.3. Air Sparging

This biostimulation approach consists of the injection of compressed air directly into the contaminated zone through a series of injection wells. Air sparging stimulates aerobic biodegradation at the same time that it strips volatile organic compounds (VOCs) into the unsaturated zone, which can subsequently be removed by a vapor-capture system (Figure 8.20). Thus, this system is also known as in situ air stripping or in situ volatilization. Air sparging is most often used together with soil vapor extraction (SVE), but it can also be used with other remedial technologies. When air sparging is combined with SVE, the SVE system creates a negative pressure in the unsaturated zone through a series of extraction wells to control the vapor plume migration. This combined system is called AS/SVE.

The effectiveness of air sparging depends primarily on two factors. First are the partitioning characteristics of the contaminants that can be present in the vapor, dissolved, and sorbed phases. In particular, vapor/dissolved phase partitioning is a significant factor in determining the rate at which dissolved constituents can be transferred to the vapor phase. Second, the permeability of the soil determines the rate at which air can be injected into the saturated zone. This, in turn, determines the mass transfer rate of the contaminants from the dissolved to the vapor phase.

As is the case with other biostimulation approaches, air sparging has some relative advantages and disadvantages:

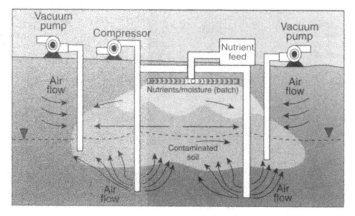

Figure 8.20. Air sparging system to stimulate the aerobic bioremediation of BTEX.

Advantages

- The equipment need for air sparging is readily available and easily installed.
- This system can be installed with minimal disturbance to other site operations.
- The time required for cleanup is relatively short: usually less than 1–3 years under optimal conditions.
- At about $20–50/ton of saturated soil, air sparging is less costly than aboveground treatment systems.
- Requires no removal, treatment, storage, or discharge considerations for groundwater.
- Air sparging can significantly enhance VOC removal by SVE.

Disadvantages

- This system should not be used if free product exists. Any free product must be removed prior to air sparging, which can create groundwater mounding and cause the free product to spread out.
- Confined aquifers cannot be treated by air sparging because injected air would be trapped by the saturated confining layer and could not escape to the unsaturated zone.
- Air sparging should be avoided if there are nearby basements, sewers, or other subsurface confined spaces present at the site. Potentially dangerous VOC concentrations could accumulate in such confined spaces unless a vapor extraction system is used to control vapor migration.
- Stratified soils may cause air sparging to be ineffective due to preferential channeling of air flow.
- There is generally a lack of sufficient field and laboratory data to support design considerations.
- There is a potential for inducing migration of volatile contaminants if these are not captured by a vacuum pump system in the unsaturated zone.
- Successful implementation requires detailed pilot testing and monitoring to ensure adequate vapor control to limit VOC migration.

Air sparging is frequently used at U.S. Air Force bases to treat jet fuel releases, and it generally is a very cost-effective approach provided low-permeability soils (which trap or divert airflow) do not prevail at the contaminated site. Cleanup by air sparging involves a medium- to long-term duration, typically requiring 6 months to 2 years to remediate a site. Typical costs range from $150,000 to $350,000 per acre.

The predominant removal mechanism by air sparging systems is a function of the contaminant volatility (Figure 8.21). Compounds of low volatility such as No. 6 fuel oil, waste oil, and diesel fuel are removed primarily by aerobic biodegradation, whereas more than 50% of the removal of volatile compounds such as gasoline, mineral spirits, and jet fuel is due to volatilization. Figure 8.22 summarizes the

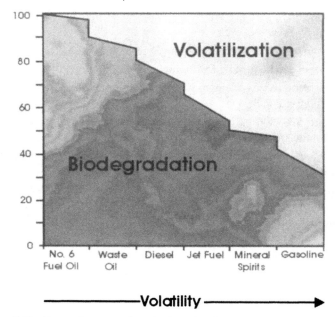

Figure 8.21. Air sparging removal mechanisms as a function of contaminant volatility.

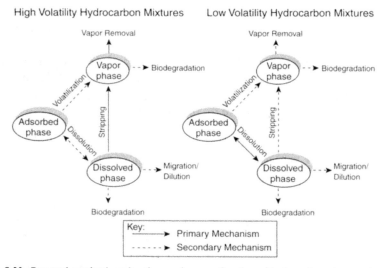

Figure 8.22. Removal mechanisms by air sparging as a function of hydrocarbon phase and volatility. (Adapted from Hinchee 1994a.)

primary and secondary removal mechanisms by air sparging for hydrocarbons in dissolved, sorbed, or vapor phase as a function of the volatility of the mixture.

Based on interphase contaminant transfer, air sparging can have both beneficial and detrimental effects, which are summarized below.

- *Enhanced oxygenation.* Injection of air replenishes O_2 used in biological and chemical reactions. This is beneficial because it stimulates contaminant biodegradation.

- *Enhanced dissolution.* Volatile and semivolatile compounds are present in both the dissolved and sorbed phases. With air injection, mixing enhances volatilization where there are gas–water interfaces. Along with enhanced biological activity, this causes a decrease in aqueous concentrations, which drives desorption of sorbed components and NAPL dissolution. Whereas enhanced NAPL dissolution leads to faster cleanup, this could be a concern if there are no controls to protect nearby downgradient receptors that could be impacted by higher contaminant flux.

- *Desorption with enhanced biodegradation* is beneficial; however, enhanced volatilization and dissolution may be detrimental due to potential spreading of the contaminant, unless the plume is captured or contained (or, alternatively, if the enhanced biological activity reduces plume dimensions) and the off-gases collected.

- *Volatilization.* In the vadose zone, sorbed contaminants will be carried off in the gas stream. This is beneficial if the mass is significant and if the off-gases are collected and treated.

- *Groundwater stripping.* If sufficient air is provided, stripping of VOCs from the groundwater is possible. This may be beneficial or detrimental, as described above.

- *Physical displacement.* If air flow rates are very high, water can be rapidly displaced. If the displaced water is highly contaminated, it will spread the contamination (especially if LNAPL is present). Aquifers tend to be very heterogeneous and in many cases we do not know (nor can we accurately predict) the direction of groundwater flow.

Based on these observations, successful application of air sparging is subject to the following constraints:

- *Type of contaminant.* The process works best for contaminants that are both volatile and biodegradable, such as BTEX. If the contaminants are soluble and/or nonvolatile, they must be biodegradable.

- *Heterogeneity.* Homogeneous aquifers are more amenable to air sparging than heterogeneous formations that offer preferential pathways to the air flow, leaving impermeable areas relatively inaccessible to the added air.

- *Permeability and hydraulic conductivity (K).* K must be sufficiently high to allow air movement in both vertical and horizontal directions. If the ratio of horizontal to vertical K is low ($<2:1$), air sparging can be effective even if the aquifer material has low permeability ($K < 10^{-5}$ cm/s). However, if the ratio of horizontal to vertical K is high ($>3:1$), the aquifer material must be relatively permeable ($K > 10^{-4}$ cm/s) for air sparging to be effective. Otherwise, hydraulic fracturing may be used to enhance K.

- There is both a minimum and maximum saturated zone thickness for air sparging to be effective. It needs to be at least 4 ft to ensure confinement of the air, and less than 30 ft to ensure control and predictability of air flow.

Some lithologic heterogeneities can inhibit the efficiency of air sparging. For example, impervious barriers such as aquicludes or other confining layers inhibit the flow of air in the vertical direction, which increases the horizontal migration of contaminant vapors (Figure 8.23). Similarly, horizontal channeling of air flow can occur if there is a high-permeability layer (Figure 8.24), which contributes to the spreading of contaminant vapors and to lower treatment efficiency.

There are two common air sparging options: in situ aquifer sparging (IAS) and in-well aeration (IWA) (Figure 8.25). A major difference between these options is the air flow rate. For IAS, air injection flow rates are typically 0.05–0.2 m^3/min (0.03–0.12 cfs). For IWA, the rates are much higher, about 10–20 cfs. In general, the radius of influence of IAS wells is only a few meters because of channeling of the air and its tendency to rise quickly through the water column. The pressures and energy required to achieve gas injection at significant depth usually restrict applications to near-surface contaminated sites.

The air injection wells should be screened below the lowest point of contamination, regardless of whether a vertical or horizontal well arrangement is used

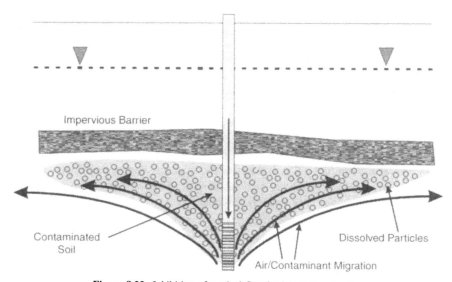

Figure 8.23. Inhibition of vertical flow by impervious barrier.

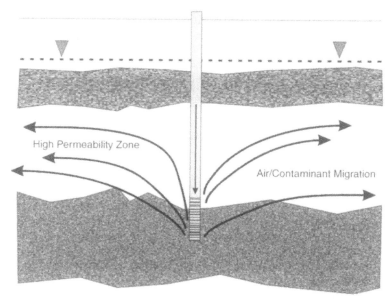

Figure 8.24. Channeled air flow through highly permeable zone.

(Figure 8.26). The required driving pressure for air sparging can be estimated as follows:

$$P_{air} = \rho_w\, gH \qquad [8.1]$$

where P_{air} = air pressure required to empty the well (Pa)

ρ_w = density of water (about 1 kg/m^3)

g = gravitational constant, 9.81 m/s^2

H = distance from top of the screened section of the well to the water table

Some researchers have been working on *anaerobic biosparging* by injecting H_2 gas into aquifers at rates <1 cfs to stimulate halorespiration and cometabolic reductive dechlorination of chlorinated aliphatic hydrocarbons (case study 5). In this case, H_2 detectors, ventilation equipment, or both need to be installed for safety concerns. An example of oxygen sparging with bioaugmentation to form a biobarrier for the interception and treatment of an MTBE plume is presented below in case study 2.

Case Study 2. Air Sparging and Bioaugmentation to Remediate MTBE at Port Hueneme, CA
(*Source*: Hughes et al. 2002; based on Salanitro et al. 2000)

A field pilot in situ bioaugmentation and biostimulation test for the remediation of methyl tertiary butyl ether (MTBE) was conducted at the Strategic Environmental Research and Development Program (SERDP)-sponsored U.S. Navy National

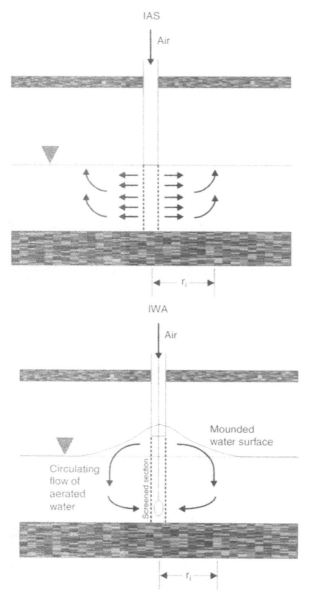

Figure 8.25. In situ air sparging (IAS) and in well aeration (IWA) systems.

Environmental Test Site at Port Hueneme, California. Reduction in MTBE concentrations was observed when oxygen gas was injected (sparged) into the subsurface and when the oxygen sparge was combined with inoculation of MC-100, a MTBE-degrading bacterial consortium previously referred to as BC-4. MTBE degradation began much more rapidly (within 30 days) in the bioaugmented plot than in the

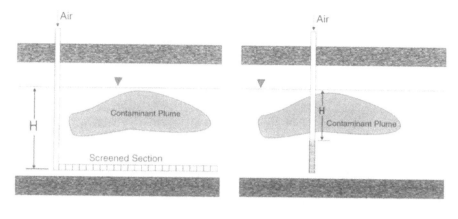

Figure 8.26. Horizontal and vertical air injection wells need to be screened below the plume.

O_2-only plot (186–231 days). Additionally, decreases in MTBE and its metabolite, *tert*-butyl alcohol (TBA), concentrations were greater overall in the bioaugmented plot (reduced to ≤0.001–0.01 mg/L MTBE in bioaugmented compared to 0.01–0.1 mg/L in the O_2-only plot). A small reduction in MTBE concentration was observed in the control plot, attributed to natural attenuation.

Process

Biological degradation of MTBE occurs in aerobic environments if sufficient numbers of MTBE-degrading microorganisms are present. The slow growth rate and low numbers of MTBE degraders in the subsurface may limit the natural attenuation of MTBE in groundwater, however. In addition, low O_2 transfer rates limit the growth of the indigenous degraders. In some cases, natural ether degraders may not be present at a site. Therefore, in situ biodegradation of MTBE may be enhanced by supplying O_2 to the aquifer (biostimulation) and/or through the inoculation of specialized MTBE-degrading microorganisms (bioaugmentation) combined with increased oxygenation. The effectiveness of creating a biobarrier inoculated with a specialized MTBE-degrading consortium of bacteria, MC-100, and maintaining oxygenated conditions through oxygen sparging for removal of MTBE from groundwater was compared to treatment by oxygen sparging only and no treatment (natural attenuation control).

Site Description

Several thousand gallons of leaded gasoline was spilled on the Port Hueneme, CA, Construction Battalion Center Naval Base in 1984–1985. The source zone was approximately 100 m wide by 300 m long. A 150 m (490 ft) wide MTBE plume was produced that traveled more than 1300 m (4300 ft) downgradient by 1999

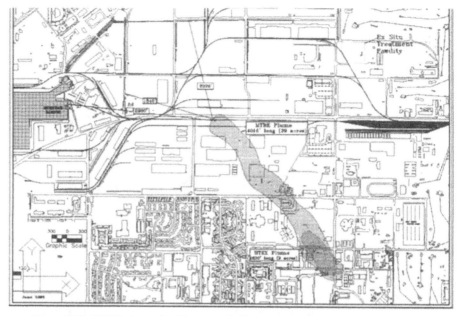

Figure 8.27. MTBE plume, Port Hueneme, California (source at southeast edge of plume).

(Figure 8.27). Initial MTBE and dissolved oxygen (DO) concentrations ranged from 2–9 mg/L and ≤1 mg/L, respectively, in the area of the test plots. Groundwater flowed in a southwesterly direction in the uppermost aquifer at a maximum flow velocity of 90 m/yr (295 ft/yr). The aquifer was composed of an upper clayey–silt unit from 0 to 3 m (0 to 10 ft) below ground surface (BGS), a fine- to medium-grained sand unit from 3 to 6 m (10 to 20 ft) BGS, and a clay layer beginning at about 6 m (20 ft) BGS. The water table was generally 2 m (6.6 ft) BGS.

Field Design/Operation

The test site was located approximately 450 m (1475 ft) downgradient of the gasoline-containing source zone in a plume area containing only MTBE and consisted of three experimental plots (Figure 8.28). One of the plots received O_2 injection only, one was bioaugmented with MC-100 as well as biostimulated with O_2 injection, and the third received no treatment. The third plot served as the natural attenuation control. Each 6 m × 15 m (20 ft × 45 ft) plot was separated by 3 m (10 ft) and was aligned with the initial estimated groundwater flow direction. A total of 188 monitoring wells were placed 1.5 and 3.0 m (5 and 10 ft) upgradient and 1.5, 4.5, and 10.5 m (5, 15, and 34.5 ft) downgradient of the treatment zones. Ninety-six O_2 gas delivery (sparge) wells were installed in the O_2-only and O_2 + MC-100 plots; however, many were not needed. A pulse injection of O_2 gas (1700 L (60 ft^3)) was delivered to each plot four to eight times each day for

(a)

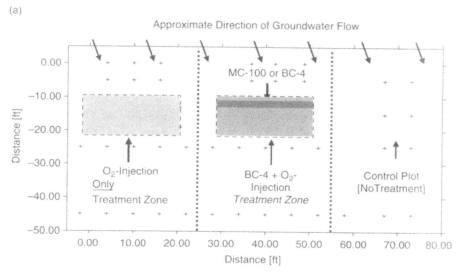

+ --monitoring wells (shallow and deep)

(b)

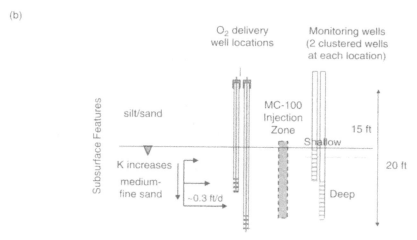

Figure 8.28. (a) Plan view of bioaugmentation field test layout, Port Hueneme, California. (b) Vertical positioning in field test layout, Port Hueneme, California.

seven weeks before bioaugmentation to ensure adequate oxygenation while minimizing contaminant loss through volatilization.

Results

The dissolved oxygen (DO) concentration in the O_2-only and $O_2 + MC$-100 plots increased from an initial concentration of ≤ 1 mg/L to 5–20 mg/L within a few weeks of the initiation of O_2 injection. MTBE concentrations decreased from an

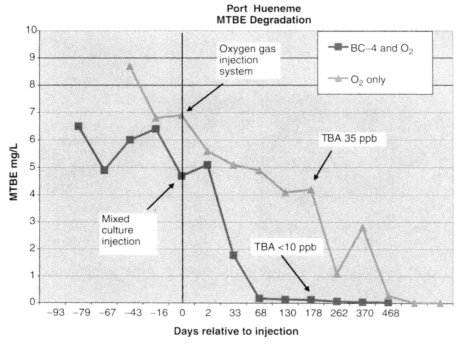

Figure 8.29. MTBE degradation, Port Hueneme, California.

initial concentration of 2–9 mg/L to 0.01–0.1 mg/L after a lag period of 186–261 days in the O_2-only plot (Figure 8.29). Growth of indigenous ethene-degrading bacterial populations was apparently stimulated by the introduction of O_2. MTBE concentrations decreased more rapidly (30 day lag period) and to a greater extent ($\leq$0.001–0.01 mg/L) in the O_2 + MC-100 plot. The concentration of TBA also decreased in the bioaugmented plot from 0.02–0.25 mg/L to $\leq$0.01 mg/L. TBA is a metabolite of MTBE biodegradation and was more slowly degraded by indigenous microorganisms in the oxygenated environment.

8.5.4. Biobarriers

Biobarriers represent a new technology that is emerging as an alternative to traditional impermeable barriers (e.g., slurry walls, grout curtains, and sheet pile cutoff walls), which are containment tools designed to prevent or control groundwater flow into, through, or from a certain location. Such impermeable barriers keep fresh groundwater from contacting a contaminated aquifer zone or contaminated groundwater from moving into pristine areas. In contrast, biobarriers are permeable and rely on the creation of a biologically active zone perpendicular to the path of the plume to intercept and degrade contaminants that flow through it (Figure 8.30). Thus, biobarriers are not intended as a conventional bioremediation approach that

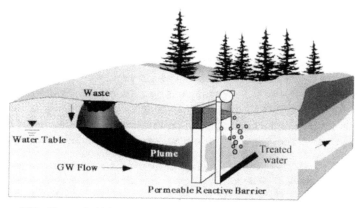

Figure 8.30. Air curtain biobarrier to intercept and treat a dissolved hydrocarbon plume.

seeks to remove the mass of the contaminant within the contaminated zone, but rather as a management strategy to intercept and treat the dissolved plume. Biobarriers are best implemented to manage plumes where source remediation is not feasible due to technical, operational, or economic constraints. Other names for biobarriers include biowalls, trench biosparge, microbial fences, bubble curtains, air curtains, bioscreens, and sparge curtains.

The general goal of biobarriers is to establish and maintain a biological treament system with a high density of competent microorganisms under a controlled process that protects specific degraders and prevents their loss under environmental stress conditions. One simple and common approach to establish a biobarrier is to dig a trench and fill it with crushed stone, which serves as a medium for biofilm growth. Biostimulation is accomplished by the addition of nutrients and electron acceptors. Oxygen can be introduced by air sparging to form an air curtain (Figure 8.30), by oxygen sparging (case study 2), or by burying oxygen-releasing compounds. Nitrate-releasing briquettes could also be used to introduce nitrate as an alternative electron acceptor. As discussed in case study 2, specialized microorganisms that degrade recalcitrant compounds could also be injected into the biobarrier to bioaugment sites with negligible concentrations of indigenous degraders. Alternatively, the barrier could be biostimulated by also adding electron donors to produce thick biofilms that plug up pore spaces, to create a vertical, impermeable biobarrier.

Anaerobic biobarriers could also be established to intercept and remove reducible pollutants such as trichloroethylene, hexavalent chromium, sulfate, nitrate, nitroaromatic explosives, and some radionuclides (Fruchter et al. 1996). Such pollutants tend to persist in aerobic systems with a low reduction capacity. Therefore, barriers containing reducing equivalents (e.g., organic material) have been used to intercept and stimulate the anaerobic biodegradation of such redox-sensitive contaminants. Two applications of organic barriers that have proved successful in field applications include (1) sulfate reduction and iron precipitation in the treatment of acid mine drainage (Benner et al. 1997), and (2) stimulation of heterotrophic denitrification to remove nitrate (Robertson and Cherry 1995; Schipper and Vojvodič-Vukovič, 1998). Potential barrier materials providing organic carbon include peat, sawdust, leaf compost, and seeds (Kao and Borden 1997; Schipper and Vojvodič-Vukovič 1998; Waybrant et al. 1998).

Hydraulic or physical controls on groundwater movement (e.g., funnel and gate) may be required to ensure that contaminants pass through the biobarrier. Conditions that limit the applicability of biobarriers include a close proximity of the plume to site boundaries or receptors (which may require a more aggressive and faster remediation approach), and large depth and width of the contaminant plume (which

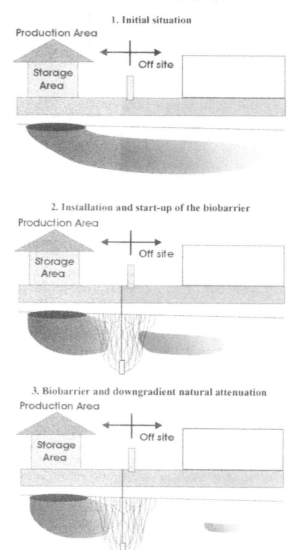

Figure 8.31. Effect of biobarrier and downgradient natural attenuation on a dissolved plume.

would result in prohibitively expensive installation costs). Note that biobarriers could be installed at sites even after the plume has migrated beyond the property boundary, provided that the contaminants are amenable to natural attenuation (Figure 8.31).

8.6. BIOSTIMULATION SYSTEMS TO TREAT CHLORINATED SOLVENT SPILLS

Chlorinated solvents are common priority pollutants that pose a threat to human and environmental health (see Table 8.1). Unlike hydrocarbons, most of these compounds have a synthetic origin and have not been in contact with microorganisms through evolutionary periods of time. As a result, chlorinated solvents are not as frequently metabolized by indigenous microorganisms as hydrocarbons, which are more labile in the environment. Nevertheless, several biotransformation mechanisms have been identified that could be exploited for site remediation. Common chlorinated solvents and their known biodegradation mechanisms were discussed in Chapter 3 and are summarized in Table 8.10.

Table 8.10. Common Chlorinated Solvents that Contaminate Aquifers and Degradation Pathways Known to Exist[a]

| Chlorinated Solvent | Formula | Used as Primary Substrate | | Anaerobic Acceptor | Transformed Cometabolically | |
		Aerobic Donor	Anaerobic Donor		Aerobic	Anaerobic
Methanes						
Carbon tetrachloride	CCl_4					•
Chloroform	$CHCl_3$				•	•
Dichloromethane	CH_2Cl_2	•	•		•	•
Chloromethane	CH_3Cl	•			•	•
Ethanes						
1,1,1-Trichloroethane	CH_3CCl_3				•	•
1,1,2-Trichloroethane	$CH_2ClCHCl_2$				•	•
1,1-Dichloroethane	CH_3CHCl_2				•	•
1,2-Dichloroethane	CH_2ClCH_2Cl	•	•		•	•
Chloroethane	CH_3CH_2Cl	•			•	•
Ethenes						
Tetrachloroethene	$CCl_2=CCl_2$			•		•
Trichloroethene	$CHCl=CCl_2$			•	•	•
cis-1,2-Dichloroethene	$CHCl=CHCl$		•	•	•	•
trans-1,2-Dichloroethene	$CHCl=CHCl$		•		•	•
1,1-Dichloroethene	$CH_2=CCl_2$				•	•
Vinyl chloride	$CH_2=CHCl$	•	•	•	•	•

[a]Indicated by black dot (•).

Source: Adapted from Rittmann and McCarty (2001).

The main biotransformation pathways for chlorinated ethenes are:

1. *Aerobic Oxidation.* Here, the pollutant serves as the primary substrate for growth. Oxygen (O_2) serves as the electron acceptor and is supplied by air sparging, bioventing, H_2O_2, or an oxygen-releasing compound as discussed in Section 8.5. Since chlorinated compounds are volatile, some volatilization losses may occur with air sparging or bioventing. This process is essentially identical to the biostimulation strategies used to bioremediate petroleum compounds (e.g., BTEX), although aerobic metabolism is limited to the less chlorinated compounds such as chloromethane, dichloromethane, chloroethane, 1,2-dichloroethane, and vinyl chloride.

2. *Aerobic Cometabolism.* In addition to providing oxygen and nutrients, this approach requires that an electron donor also be added. Numerous chlorinated aliphatics are amenable to aerobic cometabolism (Table 8.10). In general, the fewer the number of Cl atoms, the better the cometabolic process will work. As discussed in Chapter 3, toluene, methane, propane, butane, and phenol have been used as primary substrates to support such cometabolic transformations.

3. *Anaerobic Oxidation.* In this mechanism, the chlorinated organic serves as an electron donor for growth. Only a few chlorinated aliphatics are amenable to this treatment (i.e., dichloromethane; 1,2-dichloroethane; *cis-* and *trans-*dichloroethene, and vinyl chloride). Nitrate and sulfate can serve as electron acceptors in such cases, and dichloromethane can also be fermented. Nevertheless, degradation rates are relatively slow and this process has not yet been demonstrated or exploited for site remediation.

4. *Anaerobic Reductive Dechlorination.* In this process, the compound serves as an electron acceptor. All chlorinated aliphatics are susceptible to anaerobic, *cometabolic*, reductive dechlorination. This requires a suitable electron donor, and it works mainly under sulfate-reducing or methanogenic conditions. An exception is carbon tetrachloride, which can also be dechlorinated under denitrifying conditions. This is being shown at full-scale in Michigan using acetate as an electron donor, nitrate as an electron acceptor, and bioaugmented with a *Pseudomonas* sp. isolated by Dr. Craig Criddle and colleagues (Dybas et al. 1998).

Much current research activity is focused on *dehalorespiration*, where PCE, TCE, DCE, and VC serve as terminal electron acceptors in support of growth. This mechanism requires a source of H_2 as electron donor (See case study 3). A "slow-release" source of H_2 generally works best to avoid reaching high H_2 concentrations that provide a competitive advantage to hydrogenotrophic methanogens that are not as efficient dechlorinators as dehalorespirers such as *Dehalococcoides ethenogenes*.

Other electron donors being tested to stimulate microbial dechlorination include propionate, butyrate, biomass, oleate, molasses, vegetable oil, and a proprietary hydrogen-releasing compound (HRC), which is a polylactate ester that hydrolyzes slowly in water. The EPA recently tested a sugar derivative (cyclodextrin) that complexes TCE and enhances its solubilization and biodegradation.

Case Study 3. Hydrogen Biosparging for Enhanced Reductive Dechlorination at Cape Canaveral, FL
(*Source*: Hughes et al. 2002; based on Groundwater Services 2001; Newell et al. 2000)

An 18 month, low-volume pulsed hydrogen biosparging pilot test was conducted at Launch Complex 15, Cape Canaveral Air Station, Florida, to evaluate the effectiveness of hydrogen biosparge technology for bioremediation of chlorinated solvents. Hydrogen was successfully distributed in the test zone and was rapidly removed by biodegradation processes. When compared to the nitrogen and natural attenuation controls (17–18% reduction for each), significant reductions in chlorinated solvent concentrations occurred. A 94% reduction was observed in wells close to hydrogen sparge points, and a 45% reduction was observed in wells approximately 4.6 m (15 ft) downgradient.

Process

Direct addition of the electron donor, hydrogen, without addition of fermentation substrates or carbon sources to enhance reductive dechlorination is an in situ bioremediation technology that is under development. A hydrogen gas mixture was injected (sparged) into the contaminated aquifer in low volume pulses. The hydrogen was used as an electron donor in the biologically mediated dechlorination of the chlorinated solvents.

Site Description

Launch Complex 15 is one of the rocket launching facilities located along the easternmost edge of the Cape Canaveral Air Station, Florida, adjoining the Atlantic Ocean. The uppermost water-bearing unit, a sandy aquifer 1.8–2.1 m (6–7 ft) and extending to at least 21.3 m (70 ft) BGS, was contaminated with chlorinated solvents, including tetrachloroethene (PCE), trichloroethene (TCE), *cis*-1,2-dichloroethene (*cis*-DCE), and vinyl chloride (VC). The groundwater had a low dissolved oxygen (DO) concentration (<0.4 mg/L), conducive to reductive dechlorination. Groundwater flow was to the northeast.

Field Design/Operation

A 14.5 m × 27.5 m (48 ft × 90 ft) test zone was established 4.6–6 m (15–20 ft) below the water table (Figure 8.32). The biosparging system had four sparge wells, 20 single-level monitoring wells, and six multilevel saturated zone sample wells installed immediately adjacent to the sparge wells. Three hydrogen sparge and 12 monitoring wells were located within the treatment zone, while the fourth sparge well and four monitoring wells were used as nitrogen sparge controls to determine loss through volatilization. The final four monitoring wells were placed outside the sparge area to serve as natural attenuation controls. The rows of monitoring wells were spaced approximately 9.1 m (30 ft) apart, which corresponded to

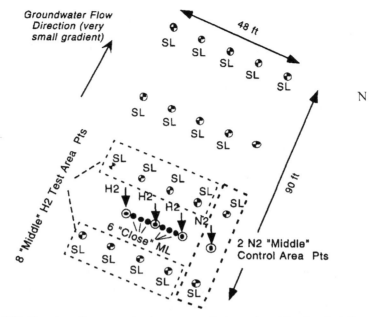

Figure 8.32. Plan view of test zone showing sparge wells (arrows), multilevel wells (ML), and single-level monitoring wells (SL), Cape Canaveral, Florida. "Close" monitoring wells are 0.9–1.8 m (3–6 ft) from sparge wells, and "Middle" wells are approximately 4.6 m (15 ft) downgradient from sparge wells. "Pts" are sample points or wells.

a groundwater travel distance at approximately 3 month intervals. The pilot tests began in February 1999 and were completed in August 2000.

A gas mixture of 49% hydrogen, 49% helium, and 2% sulfur hexafluoride (SF_6) (used as a tracer) was used to produce an expected hydrogen equilibrium concentration of approximately 0.8 mg/L. Initially, the gas mixture was sparged in a burst of 1 min duration one time per day (approximately 5–7 standard ft^3 per well) at a depth between 6.9 and 7.6 m (22.5 and 25 ft). Injection pressures were approximately 20 psig (15 scfm) to overcome the hydrostatic pressure in the sparge wells, which was approximately 9 psig. Later, one large pulse (50–60 ft^3 per well) once a week was used, delivering an average of 7–8 standard ft^3 per well. No fouling was observed. These larger weekly pulses delivered hydrogen to the treatment zone more effectively than the smaller daily pulses.

Results

Distribution of hydrogen and tracer gases was achieved throughout the injection zone by short-term sparging (Figure 8.33). The larger once-a-week pulses used later in the test were apparently more successful than the smaller once-a-day pulses used initially for delivery of the hydrogen to the treatment zone.

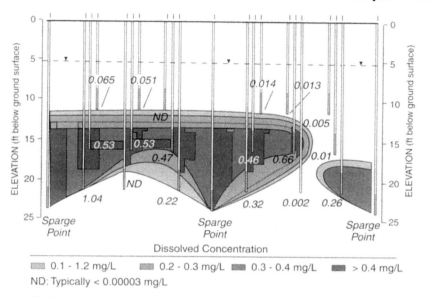

Figure 8.33. Vertical profile of dissolved H_2 concentration 4 days after weekly sparge event, 18 months after startup, Cape Canaveral, Florida. Distance between sparge wells is 3.7 m (12 ft).

When compared to the tracer gases, helium and SF_6, hydrogen was preferentially removed from the system, indicating that the hydrogen was being used in biodegradation processes. A 94% reduction in concentration of total chlorinated ethenes (TCE, *cis*-DCE, and VC) was observed in monitoring wells close to the sparge points (1–2 m or 3–6 ft) and a 45% reduction was observed in the wells approximately 4.5 m (15 ft) from the sparge points. Reductions of 17–18% were observed in both the nitrogen and natural attenuation controls (Table 8.11). TCE was preferentially removed compared to *cis*-DCE or VC, which would be expected in an

Table 8.11. Reduction in Total Chlorinated Ethenes Concentration, Cape Canaveral, Florida

Group of Wells	Total Chlorinated Ethenes (TCE, *cis*-DCE, VC) (mg/L)		% Change over 18 Months
	Baseline	After 18 Months	
H_2 zone 3–6 ft	291	16	−94%
H_2 zone 15 ft	294	151	−45%
N_2 control 15 ft	74	62	−17%
NA[a] control 20 ft	207	165	−18%

[a]NA, natural attenuation.

Table 8.12. Change in Concentration of TCE, *cis*-DCE, VC, Ethene, and Methane, Cape Canaveral, Florida

		Concentration (mg/L)			
Contaminant	Group of Wells	Baseline	After 4 Months	After 18 Months	% Change over 18 Months
TCE	H$_2$ zone - close	14.1	0.5	0.15	−99 %
	H$_2$ zone - 15 ft	6.7	1.8	1.37	−80%
	N$_2$ control - 15 ft	<0.55	<0.1	<0.25	ND[a]
	NA control - close	21	27	20	−5%
cis-DCE	H$_2$ zone - close	237.1	88.2	13.16	−94%
	H$_2$ zone - 15 ft	244.9	183.7	101.21	−59%
	N$_2$ control - 15 ft	21.0	27.0	20.0	−5%
	NA control - close	169.4	158.7	124.8	−26%
VC	H$_2$ zone - close	39.5	21.1	2.49	−94%
	H$_2$ zone - 15 ft	42.1	48.0	48.23	+15%
	N$_2$ control - 15 ft	21.0	29.0	17.0	−19%
	NA control - close	21.4	20.7	23.64	+10%
Ethene	H$_2$ zone - close	2.9	1.5	0.37	−42%
	H$_2$ zone - 15 ft	5.7	5.9	5.75	+1%
	N$_2$ control - 15 ft	7.3	5.2	0.88	−88%
	NA control - close	1.0	1.0	0.50	−50%
Methane	H$_2$ zone - close	0.5	0.1	0.54	+8%
	H$_2$ zone - 15 ft	2.2	1.3	1.25	−43%
	N$_2$ control - 15 ft	3.7	1.7	0.49	−87%
	NA control - close	0.2	0.2	0.07	−65%

[a]ND, nondetect.

anaerobic, reductive dechlorination environment. The reductive dechlorination of TCE produced *cis*-DCE as a daughter product, which in turn was reduced to produce VC. The increase in VC concentration in the 4.6 m (15 ft) wells under hydrogen sparge and natural attenuation conditions was indicative of bioremediation of TCE. Reduction in chlorinated solvent concentrations exceeded the reductions in methane and ethene concentrations, indicating that volatilization alone could not account for the removal of the chlorinated solvents (Table 8.12). Apparently, microbial competition from methanogens was not an important process since no excess methane production was observed.

8.7. BIOAUGMENTATION

Bioaugmentation is the addition of specialized microbial cultures, which are typically grown separately under well-defined conditions, to perform a specific remediation task in a given environment (in situ or in a bioreactor). This approach has been utilized in agriculture since the 1800s (e.g., addition of nitrogen-fixing *Rhizobium* spp. to legume roots (Gentry et al. 2004)). In bioremediation, bioaugmentation is often used to enhance the degradation of recalcitrant contaminants that cannot be mineralized by the indigenous consortium even under optimum conditions. The introduced microorganism augments the indigenous population's degradation capacity, hence the term bioaugmentation.

Bioaugmentation is based on the ecological principle that indigenous organisms are not exploiting contaminant biodegradation as an ecological niche (Vogel and Walter 2002). Thus, an inoculum has a good chance to compete if the contaminant is present and the niche is presently unoccupied. The benefits of bioaugmentation have been demonstrated in field trials for a wide variety of recalcitrant contaminants, including MTBE (Salanitro et al. 2000), carbon tetrachloride (Dybas et al. 1998), and TCE (Duba et al. 1996; Ellis et al. 2000; Major et al. 2001). Whereas a competent indigenous consortium could develop in the long run at some contaminated sites (resulting in the eventual degradation of some persistent organic pollutants), bioaugmentation results in shorter acclimation periods and faster degradation, often with less objectionable by-products. Furthermore, if a rapid response is important, relying on small numbers of indigenous microorganisms may not be appropriate. In such cases, bioaugmentation can enhance the reliability and rate of the cleanup process.

Bioaugmentation has received increasing interest over the past decade, and the Green Pages (www.eco-web.com) lists 263 bioaugmentation companies. However, the success of bioaugmentation is partly determined by the ability to assess whether the added microbes are responsible for increased removal and degradation of the target contaminant, which requires comparison with nonbioaugmented controls.

In wastewater treatment, bioaugmentation of activated sludge systems with specialized bacterial strains can improve several aspects of the treatment processes, such as improved flocculation and degradation of recalcitrant compounds. This is relatively easy to accomplish because the added microorganisms can readily be mixed in the reactor and reaction conditions can be manipulated to enhance their survival and performance. Bioaugmentation of aquifers is a more challenging task. Critical issues include survival of added strains, their distribution throughout the contaminated zone (which is hindered by filtering through the porous medium), and low concentration of nutrients and target contaminants that serve as substrates to the added microorganisms. Fungi, which are larger than bacteria, are restricted to reactor or surface soil applications, whereas bacteria are more adaptable to subsurface or surface applications. In addition to the challenge of delivering the inoculant to the desired subsurface location, the survival of exogenous microorganisms may be hindered by abiotic and biological stress. These include fluctuating or extreme temperatures, pH, water activity, low nutrient levels, toxic pollutant concentrations, and competition with indigenous microorganisms (Gentry et al. 2004). Also, some of the bacteria used for bioaugmentation (e.g., *Burkholderia cepacia*) could, under certain conditions, exhibit pathogenic properties and cause infection.

Different types of microorganisms can be used for bioaugmentation. These are usually bacteria, although lignolytic fungi can also be used to treat contaminated soil ex situ. Common inocula injected into aquifers include:

1. *Enrichment cultures.* These generally are a collection of indigenous organisms that have been highly enriched on the contaminant(s) of interest. The seed culture to be enriched can be originally obtained from contaminated soil or wastewater treatment plants.

2. *Pure cultures specific for a given contaminant.* For example, dehalorespirers such as *Dehalococcoides ethenogenes* or *Desulfomonile tiedjei* could be added to enhance the removal of tetrachloroethene or chlorobenzoate, respectively (Table 8.13).

3. *Commercially available cultures.* These are typically freeze-dried microorganisms that are sold by numerous companies as either pure or mixed cultures.

4. *Genetically modified microorganisms* (GMOs). GMOs represent a research frontier with broad implications, although their use for in situ bioremediation has not occurred yet. The potential benefits of using GMOs that integrate enhanced biodegradation and survival capabilities are significant and extend beyond improved contaminant removal efficiency and lower operation and maintenance costs. Nevertheless, whereas GMOs have been extensively used in agriculture, little research has been conducted to assess their long-term life cycle impacts, including the consequences of increased genetic drift across species on biodiversity and biological community structure. This gives rise to much speculation and polarization regarding the consequences of in vitro genetic manipulation, which represents a significant political barrier to the use of GMOs in bioremediation.

There are several options for introducing microorganisms into the contaminated zone. The principal delivery methods are:

1. *Liquid injection.* High microbial concentrations, typically 10^9–10^{12} cells/mL, are mixed with the nutrient solution that is introduced using liquid delivery pumps.

2. *Injection after immobilization* by mixing with a variety of carriers, such as agarose, guar, and mineral oil. The microorganisms can also be introduced after immobilization on carriers such as foam (e.g., polyurethane mixed with soil), clays (some may be injected, others mixed with soil), wood chips (then mixed with soil), granular activated carbon, and lava slag. Presterilization of the carrier generally increases the inoculant's shell life.

3. *Encapsulation and then injection.* Microorganisms can be encapsulated in microbeads (e.g., agar, alginate, or polyurethane) or in porous ceramics (e.g., isolite) to enhance their survival. Some of these products can be injected as "hydraulic fracturing" fluids to remediate more relatively impermeable formations.

4. *Biocassettes.* These are modular units with support media (oyster shell support matrix) that are inoculated with the organism of interest.

The encapsulation technologies shelter the organisms in a nontoxic matrix (protecting them from high pollutant concentrations) where nutrients and gases can diffuse. Encapsulation may also facilitate the creation of microsites with a unique microbial community that synergistically interacts to degrade various pollutants.

To enhance microbial transport following bioaugmentation, researchers have tested various techniques, including the use of adhesion-deficient bacteria, starved bacteria, ultramicrobacteria, and surfactants (for reviews, see Gentry et al. 2004). These techniques hold significant potential but must address specific challenges such as surfactant toxicity or the loss of biodegradation capabilities (e.g., plasmid curing) due to starvation or mutation.

Following bioaugmentation, process monitoring is implemented to ensure that the added microorganisms thrive and perform their intended function. This can be accomplished by using molecular probes that target the added strains and by determining whether contaminant removal rates increase and expected metabolites appear. Table 8.13 summarizes several projects where bioaugmentation has been successfully implemented. A detailed example is provided in case study 4.

Case Study 4. Bioaugmentation for In Situ Reductive Dechlorination at Dover Air Force Base, DE
(*Source*: Hughes et al. 2002; based on Ellis et al. 2000)

A microbial enrichment culture from the Department of Energy's (DOE's) Pinellis site in Largo, Florida, capable of dechlorinating TCE to ethene, was used in a bioaugmentation pilot study for the bioremediation of a TCE-contaminated aquifer at Dover Air Force Base (AFB), Delaware. TCE and *cis*-DCE concentrations in the groundwater averaged approximately 4800 and 1200 ppb, respectively. The indigenous microflora was capable of dechlorinating TCE to DCE but could not proceed past DCE. Two sequential injections of the Pinellis enrichment culture, a total of 351 L, were introduced into the test cell, a hydraulically controlled system. VC and ethene appeared in monitoring wells after an initial lag period of 90 days following the first injection. Conversion of TCE to DCE and ethene occurred by day 518. The pilot was operated for 577 days.

Process

For reductive dechlorination to occur, a readily biodegradable organic compound was provided in the form of lactate to exhaust all the oxygen present (halorespiring bacteria are obligate anaerobes) and to produce fermentative biogenic H_2. Hydrogen was used as an electron donor in the biologically mediated dechlorination of the chlorinated solvents. Liquid delivery of lactate, molasses, or vegetable oils is one of the most common applications of engineered in situ bioremediation to remediate sites contaminated with chlorinated solvents. At some sites, like Dover AFB, the indigenous microorganisms are not capable of dechlorinating PCE or

Table 8.13. Selected Bioaugmentation Case Studies

Author(s)	Year	Inoculum	Scale	Results and Conclusions
Duba et al.	1996	*Methylosinus trichosporium* OB3b	Field-scale	Bacteria (1800 L) were injected into groundwater (425 ppb TCE) through a single well. Approximately 50% of the bacteria attached to sediment, forming a fixed-bed reactor, which biodegraded 40% of the mass of TCE (20 g) in a 40 day period. TCE concentrations dropped to regulatory limits for only 2 days.
Imamura et al.	1997	Strain JM1	Pilot-scale	A lysimeter was filled with sandy soil and contaminated with TCE. Strain JM1, which cometabolically transforms TCE, was injected into the lysimeter and growth of JM1 and degradation of TCE were monitored. Growth of strain JM1 was hindered by low oxygen and TCE concentrations in the soil.
Bourquin et al.	1997	*Burkholderia cepacia* PR1$_{301}$	Field-scale	First injection of a cometabolic TCE-degrading bacterium reduced TCE to a nondetect level in 24 h and maintained for 4 days. Bacterial concentration was too high and plugged the soil formation. Second stepwise injections at lower microbial concentrations resulted in complete degradation when cell concentrations reached 10^8/mL.
Dybas et al.	1998	*Pseudomonas stutzeri* KC	Field-scale	Bacteria (1500 L) were injected into a carbon tetrachloride (CT)-contaminated aquifer. CT levels decreased ~65% through aerobic and cometabolic transformations. Final sediment analysis indicated a 60–88% removal and persistence of *P. stutzeri* KC.
Steffan et al.	1999	*Burkholderia cepacia* ENV 435	Field-scale	*B. cepacia* aerobically and cometabolically transforms chlorinated ethenes except PCE. Bacteria (550 L) were injected with oxygen into an aquifer contaminated with chlorinated ethenes at concentrations of 1000–2500 µg/L. Strain traveled through the test plot at similar linear velocities to bromide tracer. Total mass of TCE, *cis*-DCE, and vinyl chloride was reduced by as much as 78% within 2 days.
El Fantroussi et al.	1999	*Desulfomonile tiedjei*	Bioreactor (pilot-scale)	*D. tiedjei* was injected into soil bioreactors contaminated with 3-chlorobenzoate (2 mM). Complete dechlorination was observed. First demonstration of applicability and limitations of pilot-scale soil bioaugmentation with a pure anaerobic dechlorinating strain of bacterium.

Reference	Year	Culture	Scale	Description
Ellis et al.	2000	Mixed enrichment culture (DOE's Pinellas culture)		Aquifer contaminated with TCE (4800 µg/L) and *cis*-DCE (1200 µg/L) was bioaugmented with a microbial enrichment culture capable of dechlorinating TCE to ethene. Natural attenuation was observed at the site previous to bioaugmentation with an accumulation of *cis*-DCE. Within 90 days, vinyl chloride and ethene appeared in monitoring wells. By day 509, TCE and *cis*-DCE were fully dechlorinated to ethene.
Salanitro et al.	2000	Mixed consortium (MC-100)		MC-100 is a microbial consortium capable of aerobically degrading MTBE. Within 30 days of bioaugmentation with MC-100 and oxygen, and through 261 days, MTBE was degraded to nondetect levels. A by-product, *tert*-butyl alcohol, was also degraded to nondetect levels.
Hensse et al.	2001	Mixed culture	Bioreactor (field-scale)	Bioreactors anaerobically dechlorinated PCE to approximately 60% ethene. Effluent was allowed to infiltrate into a contaminated aquifer along with a carbon source. Natural attenuation of PCE to *cis*-DCE occurred. Complete dechlorination (ethene production) was observed after 4 weeks in monitoring wells 3.5 m and after 8 weeks in monitoring wells 7 m from infiltration point.
Lendvay et al.	2001	*Dehalococcoides* and *Desulfuromonas*	Field-scale	Natural attenuation of PCE was occurring with TCE, *cis*-DCE, and vinyl chloride observed in the plume. Lactate (0.1 mM) and nutrients were added prior to bioaugmentation and continued throughout study. Approximately 210 L of culture was added. Complete dechlorination to ethene was observed in less than 50 days.
Major et al.	2001	Mixed culture KB-1 (*Dehalococcoides ethenogenes*)	Field-scale	Site was contaminated with PCE and lesser amounts of TCE and *cis*-DCE. A closed loop recirculation test plot was bioaugmented with KB-1 mixed culture containing *Dehalococcoides ethenogenes*. Molecular analysis prior to bioaugmentation showed no *D. ethenogenes*. Trace amounts of vinyl chloride were present after 16 days, and ethene was detected after 52 days. By day 142, ethene was the dominant product. Molecular analysis showed *D. ethenogenes* was present throughout the test plot.

Source: Hughes et al. (2002).

TCE completely to ethene. In these cases, DCE and VC concentrations increase as these metabolic by-products accumulate. To achieve complete dechlorination, the site may be augmented with a microbial culture capable of degrading the contaminants along with substrate and/or nutrient injection.

Site Description

The alluvial aquifer underlying Dover AFB was contaminated with TCE and DCE. The aquifer was composed of glacial-fluvial deposits of sand, silt, and clay and was contaminated with chlorinated solvents. Before the pilot test began, the concentrations of PCE, TCE, DCE, 1,2-dichloroethane (DCA), and VC averaged 46, 7500, 1200, 96, and 34 ppb, respectively. Methane was also present at an average concentration of 80 ppb, and ethene was present in trace amounts.

Field Design/Operation

The pilot system, 12 m × 18 m (39.4 ft × 59.1 ft) was constructed 15 m (49.2 ft) BGS in an alluvial (glacial deposits of sand, silt, and clay lenses) aquifer to stimulate indigenous anaerobic biodegradation of TCE through the injection of nutrients and substrate, primarily sodium lactate (Figure 8.34). Three extraction wells, 6 m (19.7 ft) apart, and three injection wells, 18 m (59.1 ft) upgradient of the extraction wells, hydraulically controlled the system and created a recirculation cell to ensure sufficient residence time and adequate substrate concentration for reductive dechlorination of the chlorinated contaminants (Figure 8.35).

In the first phase of the pilot study, nutrients and substrate were pulse-fed on a 7 day cycle at a total pumping rate of 11.6 L/min, or 3.91 L/min/well, from the deep zone of the aquifer. The residence time for groundwater introduced into the injection wells was approximately 60 days. The initial substrate (lactate) feed rate was 100 mg/L as lactate (94.5 L of 60% sodium lactate, 198.5 L of water, and 100 g of yeast extract). The substrate was injected for 3.75 days, followed by circulation of unamended groundwater for 0.5 day. For the next 2.75 days, nutrients (5 mg/L ammonia and 5.5 mg/L phosphate) were injected, followed by another 0.5 day of

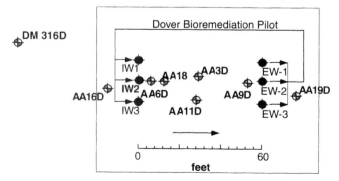

Figure 8.34. Pilot-test system for nutrient and substrate addition and bioaugmentation, Dover AFB, Delaware. (IW, injection well; EW, extraction well; AA, monitoring well within test plot; DM, monitoring well upgradient of test plot.)

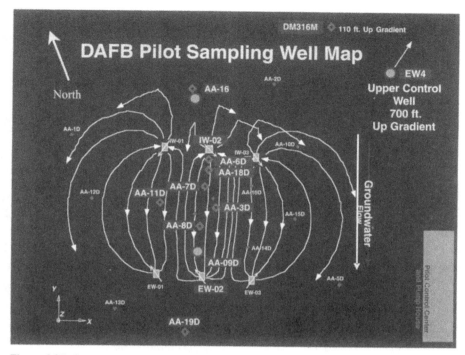

Figure 8.35. Groundwater recirculation cell established in pilot-test plot, Dover Air Force Base, Delaware. (IW, injection well; EW, extraction well; AA, monitoring well within test plot; DM, monitoring well upgradient of test plot.)

groundwater circulation. The cycle was repeated throughout the operation of the pilot system.

In the second phase of the pilot study, an enrichment microbial culture capable of reductive dechlorination of TCE to ethene was also injected into the groundwater after it became apparent that the indigenous microflora was capable of degrading TCE to DCE but not beyond. The culture was obtained from the DOE's Pinellis site in Largo, Florida. Bacterial injection, or bioaugmentation, was performed in the central injection well, well 2. The outer wells, wells 1 and 3, were used to control the flow and groundwater recirculation. Approximately 180 L of bacterial culture was injected after 269 days of lactate/nutrient injection. On day 284 of the lactate/nutrient injection, an additional 171 L of culture was injected. Groundwater recirculation rate was maintained at 1.51 L/min during injection of the microorganisms. Lactate feed concentration was returned to 200 mg/L as lactate one day following bioaugmentation.

Results

The initial substrate feed concentration of about 100 mg/L as lactate was found to be insufficient to dechlorinate all of the TCE to *cis*-DCE by day 144. The feed

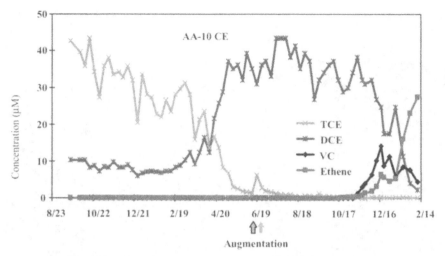

Figure 8.36. Reductive dechlorination of TCE and accumulation of *cis*-DCE prior to bioaugmentation, Dover AFB, Delaware.

concentration was increased to approximately 130 mg/L as lactate. An increase in *cis*-DCE concentration and a decrease in TCE concentration were then observed in monitoring wells closest to the injection wells. By day 291, a nearly stoichiometric conversion of TCE to *cis*-DCE was observed; however, no dechlorination of *cis*-DCE was observed. The indigenous microflora appeared to be unable to dechlorinate beyond *cis*-DCE, which began to accumulate prior to bioaugmentation (Figure 8.36).

Bioaugmentation was conducted on days 269 and 284. The appearance of VC in a monitoring well downgradient from the injection point 91 days after bioaugmentation (day 360) was the first indication that bioaugmentation was effective (Figure 8.37). Ethene was detected in this well, indicative of complete dechlorination, shortly afterward. Both VC and ethene were then observed in the other monitoring wells and continually increased while *cis*-DCE concentrations decreased throughout the pilot area. By day 479, influent TCE and *cis*-DCE were stoichiometrically reduced to levels below EPA maximum contaminant levels (MCLs), without the production of toxic by-products. Approximately 75–80% of the original molar concentrations of TCE and *cis*-DCE were recovered as ethene.

8.8. EX SITU (ABOVEGROUND) BIOREMEDIATION SYSTEM

Aboveground, or ex situ, bioremediation involves the use of biological treatment processes that are similar to those used for municipal and industrial wastewater treatment. Suspended-growth (e.g., activated sludge) and biofilm reactors (e.g., trickling filters or rotating biological contactors) can be used to treat contaminated

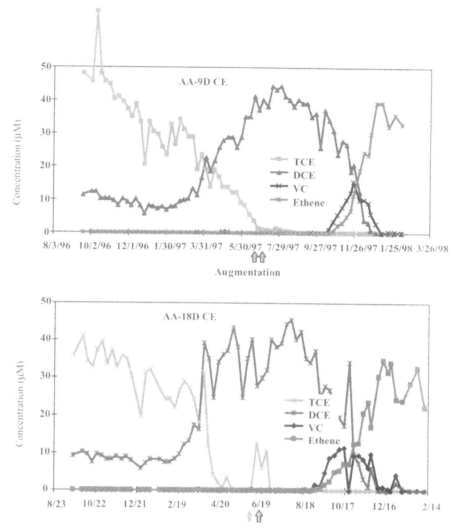

Figure 8.37. Reductive dechlorination of TCE prior to bioaugmentation and of *cis*-DCE and VC to produce ethene following bioaugmentation in two monitoring wells (AA-9D and AA-18D).

groundwater as components of so-called pump-and-treat systems, whereas slurry (batch) reactors, composting, and land farming can be used to treat contaminated soils. The term biopiles is sometimes used to describe composting and land farming operations. Contaminated gas streams extracted from the unsaturated zone by SVE systems can also be treated ex situ in biofiltration reactors. The most common ex situ bioremediation systems are slurry reactors, composting, and land farming. These are discussed below.

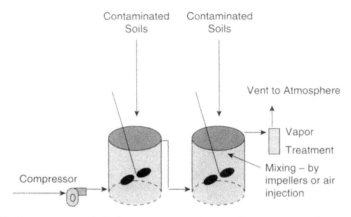

Figure 8.38. Slurry reactor typical of the reactors used in series in case study 5. (From Hughes et al. 2002.)

8.8.1. Slurry Reactors

These systems typically contain about 5–40% solids that are homogenized and mixed with water to form a slurry, which is vigorously mixed in batch reactors (Figure 8.38). Slurry reactors are typically high-energy systems where the high mixing intensity promotes aerobic conditions and rapid mass transfer of contaminants from the solid phase to the liquid phase (Manning et al. 1996; Zhang et al. 2000). Enhancements such as nutrients and surfactants (to emulsify NAPLs) can also be added. The high mixing intensity generally results in excellent contact between contaminants, amendments, and microbes. A wide variety of contaminants can be treated effectively with this method. Advantages and disadvantages withrespect to other ex situ methods for treating contaminated soils and sediments include:

Advantages

- Greater and more uniform process control.
- Enhanced solubilization of organic compounds.
- Physical breaking of soil/sludge particles (size reduction).
- Increased microbe–contaminant contact.
- Improved distribution of nutrients, electron donors and acceptors, and amendments.
- Faster biodegradation rates than in land farming or composting.

Disadvantages

- Increased energy costs associated with reactors, mixing, and aeration.
- Increased material handling.

- Increased complexity (need liquid and solids separation capability).
- Need to dewater solids.

Case Study 5. Ex Situ Biodegradation of Dinitrotoluene at Chattanooga, TN and Baraboo, WI
(Source: Hughes et al. 2002; based on Zhang et al. 2000)

Aerobic slurry reactors were used to treat soils contaminated with 2,4-dinitrotoluene (2,4-DNT) and 2,6-dinitrotoluene (2,6-DNT) from Volunteer Army Ammunition Plant (VAAP) at Chattanooga, Tennessee, and Badger Army Ammunition Plant (BAAP) at Baraboo, Wisconsin. Concentrations of 2,4-DNT and 2,6-DNT were 19,000 and 1380 mg/kg in VAAP and 8900 and 480 mg/kg in BAAP soils. A soil slurry (70 L) created by soil homogenization and washing was fed into an Eimco bioreactor operated in a draw-and-fill mode. Bioaugmentation with DNT-degrading mixed bacterial cultures was required for mineralization. Complete degradation of 2,4-DNT in both soils was achieved within two days. High concentrations of 2,4-DNT inhibited 2,6-DNT degradation. Complete degradation of both contaminants was achieved by operating two bioreactors in series. The 2,4-DNT was degraded in the first reactor, and the 2,6-DNT in the second.

Process

Under anaerobic conditions, the aryl nitro groups of DNT may undergo cometabolic reduction. The contaminant undergoes transformation, but not mineralization.

Under aerobic conditions, 2,4-DNT and 2,6-DNT are carbon and energy substrates for microbial degradation. The aromatic ring of the DNT is mineralized during oxidative metabolism, and the nitro groups are released as nitrate.

Field Design/Operation

Soils were initially homogenized by repeated sieving and tumbling followed by use of a sample splitter. Soil washing in an upward jet of warm (60 °C) water separated the DNT-contaminated fines from the clean sand. The resulting soil slurry was used as feed into the bioreactors, and the sand was discarded.

The Eimco Biolift slurry reactors (75 L) were operated in the draw-and-fill mode. The reactor contents were drained, and 7 L of slurry was reserved for reinoculation. Soil slurry (60 L), phosphate buffer (NaH_2PO_4 and Na_2HPO_4), and the recycled inoculum were added and brought to 70 L total volume with tap water. Nominal solids loading rates, or the mass of soil introduced into the bioreactor (kg) for a given cycle divided by the reactor volume (70 L), of 20% and 30% were used for the VAAP soil and 40% was used for the BAAP soil.

Bioaugmentation with DNT-degrading bacteria was required. A mixed culture of the 2,4-DNT-degrading strain *Burkholderia cepacia* JS872 and the 2,6-DNT-degrading strains *B. cepacia* JS850 and *Hydrogenophaga paleronii* JS863 was added to the bioreactors.

To achieve both 2,4-DNT and 2,6-DNT mineralization, two reactors were operated in series. The first reactor was charged with slurry, and its contents were transferred to the second reactor after attaining the desired 2,4-DNT degradation. Initially, the second reactor was filled with. a 50% dilution of the effluent from the first reactor and inoculated with a 2,6-DNT-degrading culture. After establishing 2,6-DNT degradation, the volume of effluent from the first reactor was increased gradually until undiluted effluent was being used as the feed for the second reactor. Each cycle received supplemental nutrients, including nitrogen and potassium.

Results

Initial 2,4-DNT concentrations were up to 9750 micromolar (μM) in VAAP soil and up to 10,840 μM in BAAP soil. Essentially complete degradation of 2,4-DNT was achieved in approximately two days for both soils when the bioreactors were operated in a draw-and-fill mode. However, degradation of 2,6-DNT was slow and incomplete (about 40%) although initial concentrations were approximately ten times less than the 2,4-DNT concentrations. Low level 2,6-DNT degradation activity could only be maintained by repeated bioaugmentation.

Degradation of 2,4-DNT was nearly complete for up to the 20% nominal solids loading rate for VAAP soil. At a loading rate of 30%, complete degradation was not observed. A nominal solids loading rate of up to 40% with BAAP soil achieved almost complete 2,4-DNT degradation. Degradation of 2,6-DNT was negligible at greater than a 5% loading rate.

To determine the effect of feeding strategies, reactor performance was evaluated for a continuous feed at a flow rate of 5–8 mL/min. During continuous feed of BAAP soil, 2,4-DNT accumulated in the reactor to a concentration of 770 μM and then decreased to below the detection limit. Complete degradation was established within two days. Approximately 90% of the 2,6-DNT was also degraded during this time period under continuous feed operation.

High 2,4-DNT concentrations (500–1000 μM) inhibited 2,6-DNT degradation in draw-and-fill operation. A sequential reactor system was developed to provide rapid and sustainable 2,6-DNT degradation in the slurry reactors. The first reactor was operated to degrade 2,4-DNT, and undiluted effluent was fed to the second reactor for complete 2,6-DNT degradation. Degradation of 2,6-DNT was sustained over a concentration range of 50–275 μM in BAAP soil and 100–300 μM in VAAP soil.

8.8.2. Composting

Composting is an aerobic process for treating contaminated solids, with a moisture content typically adjusted to 50–60% (40–50% solids). Contaminated soils or sediments are mixed with uncontaminated amendments (bulking agents), such as manure, straw, wood chips, or grasses, to create biologically active decomposing environments (Figure 8.39). Although compost piles are exposed to the atmosphere, the interior is often anaerobic due to the oxygen demand of the contaminants and amendments. Thus, air is drawn through the compost (by vacuum, although aerated piles have been used to enhance the drainage) to supply O_2 and remove evaporated water. Compost piles are subjected to intermittent mixing using specially manufac-

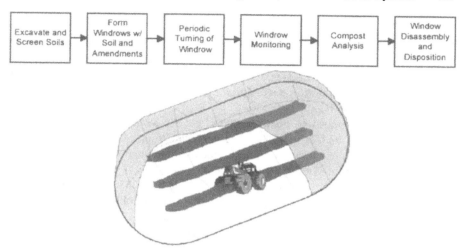

Figure 8.39. Windrow composting, typical of the system used in case study 6. (From Hughes et al. 2002.)

tured equipment that is capable of turning the pile over onto itself. Temperatures can increase to 60–70 °C due to the exothermic nature of biodegradation, and mixing, aeration, and moisture addition help dissipate excess heat that could be inhibitory to biodegradation.

Composting is appropriate to treat soils contaminated with munitions wastes that contain 2,4,6-trinitrotoluene (TNT) and other contaminants that are susceptible to cometabolic reduction (Spain et al. 2000). Composting of soils contaminated with polycyclic aromatic hydrocarbon (PAH) is also common (sometimes referred to as "biopiles"). In these systems, contaminant degradation is accomplished by nonspecific oxidation by lignolytic fungi (e.g., *Phaenerochaeta chrysosporium*, *Phaenerochaeta sordida*, and *Trametes hirsuta*) and nonspecific oxygenase enzymes that also participate in the decomposition of complex organic matter (humification). An example of composting is presented below in case study 6.

Case Study 6. Ex Situ Composting of Soils Contaminated with Explosives, Umatilla Army Depot Activity, Hermiston, OR (*Source*: Hughes et al. 2002; based on TRW Inc. 1997)

The first full-scale application of windrow composting for the biodegradation of explosives in contaminated soils was conducted at Umatilla Army Depot Activity (UMDA), Hermiston, Oregon. A total of 10,969 yd^3 of contaminated soils were treated in 14 windrows. The average concentration of 2,4,6-trinitrotoluene (TNT) was reduced from 190 to <30 ppm, and the concentration of the other target compound, hexahydro-1,3,5-trinitro-1,3,5-triazine (RDX), was reduced from 227 to <30 ppm.

Process

During composting, many organic explosives undergo cometabolic biodegradation. They are fortuitously biodegraded by microorganisms, usually at elevated temperatures, as the microorganisms degrade other carbon substrates. Heat is produced by the microorganisms during degradation, and typical compost temperatures are in the range of 55–65 °C.

For windrow composting, contaminated soils are excavated and screened to remove large rocks and debris. The soils are then transported to a composting pad within a temporary structure that provides containment and protection from weather. Bulking agents (alfalfa, straw, manure, agricultural wastes, wood chips, etc.), which also serve as carbon substrates, are layered with the contaminated soils into long piles, called windrows. The windrow is periodically mixed using a commercially available windrow turning machine. Moisture, pH, and temperature are monitored to maintain optimum conditions for microbial degradation.

Site Description

From approximately 1955 to 1965, the Umatilla Army Depot Activity (UMDA) operated a munitions washout facility in Hermiston, Oregon. Hot water and steam were used to remove explosives from munitions casings. The contaminated washwater was disposed into two settling lagoons on site. The soils and groundwater underlying the lagoons were contaminated with explosive compounds, primarily TNT, RDX, octahydro-1,3,5,7-tetranitro-1,3,5,7-tetrazocine (HMX), and 2,4,6-trinitrophenylmethylnitramine (tetryl). TNT and RDX soil concentrations ranged from 100 to 2000 ppm, and HMX concentrations ranged from <1 to 100 ppm.

Contamination was present in the upper 2 m (6 ft) of soil, which was predominantly composed of fine sand and loamy fine sand. The pH of the soil was relatively uniform (from 7.6 to 8.3) and typical of mineral soils in arid regions. The moisture content ranged from 3.5% to 16.7%, with a mean value of 7.2%. The moisture content was higher for the silt lenses than for the sand. The total organic content of the soil ranged from 0.8% to 7.3%, with a mean value of 2.6%. The total organic content corresponded with the level of explosives contamination.

Field Design/Operation

The windrows for composting were constructed within a 60 m × 27.5 m (200 ft × 90 ft) structure. Thirteen windrows with approximately 210 m^3 (810 yd^3) of contaminated soils each and one windrow with 110 m^3 (439 yd^3) were constructed individually within the structure (batch processing) with approximately 30% soil loading. The majority of the material in the windrows was bulking agent and consisted of cow manure (21%), alfalfa (18%), sawdust (18%), potato waste (10%), and hen manure (3%). A total of 2800 m^3 (10,969 yd^3) of explosives-contaminated soils were treated. The windrows were turned once every 24 hr for the first 5 days, followed by less frequent turning on subsequent days (three times a week). The moisture content was maintained at 30–35%. The batches

Table 8.14. Results of Windrow Composting Bioremediation, Umatilla Army Depot Activity, Hermiston, Oregon

Contaminant	Initial Concentration (ppm)	Final Concentration (ppm)
TNT	190	<30
RDX	227	<30

required approximately 22 days for cleanup goals (<30 ppm) of TNT and RDX to be achieved. Composting was initiated July 1995 and completed September 1996.

Results

Cleanup goals (<30 ppm) for TNT and RDX were achieved in approximately 22 days of composting; from 190 to <30 ppm for TNT and from 227 to <30 ppm for RDX (Table 8.14). Each windrow was sampled, and only two of the almost 300 samples did not meet cleanup goals after the initial treatment phase.

8.8.3. Land Farming

This method is perhaps the most common ex situ treatment process, and it involves mixing of contaminated materials into the top layer of topsoil (Figure 8.40). This is often accomplished by using heavy equipment to periodically till the soil.

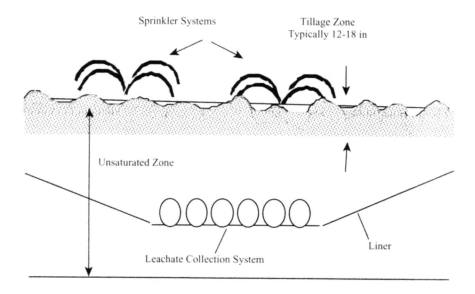

Figure 8.40. Land farming system to treat contaminated soil.

Impermeable liners may have to be installed to avoid leaching of contaminants to the underlying soil and groundwater. Amendments such as nutrients and bulking agents can also be added as needed together with moisture until contaminant concentrations reach acceptable levels and the treated soil is ready for safe disposal. Land farming is mainly applicable to treat hydrocarbon-contaminated solids, where oxygen availability often limits biodegradation rates. In these cases, the combination of mixing and exposure to the atmosphere allow for adequate aeration. An example of land farming is provided below in case study 7.

Case Study 7. Ex Situ Land Farming of Petroleum Hydrocarbons at Wichita Falls, TX
(*Source*: **Hughes et al. 2002; based on Loehr et al. 2000)**

On-site land treatment was evaluated as a technology to remediate petroleum refinery waste pond contents and associated contaminated soils. The study was conducted in two phases over 140 weeks in a pilot land treatment unit (PLTU). The concentration of specific hydrocarbons, as measured by 16 target polycyclic aromatic hydrocarbons (PAHs) and BTEX compounds, was reduced. Total reduction in PAHs was 98%, and total BTEX reduction was 46–98%. The toxicity of the pond material was reduced, as measured by Microtox and earthworm studies, and soil extracts were found to be nontoxic to rat liver cell cultures after land treatment. Vertical migration of the organic and inorganic contaminants was not observed in the PLTU soils, and no contaminants were transported below the PLTU treatment zone. While hydrocarbons remained in the soil following treatment, little relative soil toxicity or chemical mobility remained, and an acceptable treatment end point was achieved.

Process

Biodegradation of petroleum hydrocarbons occurs more rapidly in aerobic environments and is often limited by insufficient concentration of oxygen and possibly inorganic nutrients, primarily nitrogen and phosphorus. Introduction of the limiting electron acceptor and inorganic nutrients stimulates the indigenous microflora to increase hydrocarbon degradation. Tilling of the contaminated soil in a land treatment unit increases mixing of contaminated soils, oxygen, and nutrients. Remediation performance is determined by (1) loss of specific organic chemicals, (2) reduction in toxicity, and (3) reduction in mobility of the contaminants remaining.

Site Description

A 4 acre waste pond was constructed at the Conoco Wichita Falls, Texas, Product Terminal for the disposal of refinery wastes and other materials. Contaminants included approximately 24,000 m^3 (31,000 yd^3) of straight-chain and polyaromatic hydrocarbon (PAHs) wastes and 46,000 m^3 (60,000 yd^3) of contaminated soil and rubble.

The first continuous aquifer was approximately 3.7 m (12 ft) BGS. A shallow, noncontinuous aquifer hydraulically connected to the waste pond was perched about 1 m (3 ft) BGS. It contained slightly elevated levels of hydrocarbon and lead contaminants. No contamination was found in the first or second continuous aquifers, separated by a dense clay layer, located approximately 3.7 m (12 ft) and 4.5–9 m (15–30 ft) BGS.

The soils at the site consisted of a top layer of silty clays and clayey silts about 3–4.5 m (10–15 ft) thick in which the PLTU was built. This silty clay had a cation exchange capacity of 18.5 meq/100 g, a total organic carbon (TOC) content of 1%, a moisture field capacity of 30%, and a pH of 7.8. Beneath this was a dense clay layer about 1.5–2.4 m (5–8 ft) thick, with silty sand beneath that layer. Rock was found about 12 m (39 ft) BGS.

Field Design/Operation

The PLTU was operated in two phases. During the first phase, two separate 15 m by 15 m (50 by 50 ft) land treatment plots were designed, constructed, and operated at an initial waste loading of 5% oil-in-soil for 70 weeks. During the second phase, the two plots plus additional area were combined into one large PLTU of about 1020 m^2 (11,000 ft^2), which treated an initial waste loading of 10% oil-in-soil for 73 weeks.

No liner was needed for the PLTU because of the silty clay nature of the soil. The PLTU was surrounded by a 1 m (3 ft) berm. The plow zone, or the zone of incorporation (ZOI) of the contents of the waste pond and the soil in the PLTU, was approximately 25 cm (10 in.) in depth. Before any contaminated materials were placed in the PLTU, fertilizer was applied to the plots and tilled into the base soil layer to supply nitrogen, phosphorus, sulfur, potassium, and trace elements for microbial degradation.

The contents of the waste pond and surrounding contaminated materials were applied to the plots. Periodic tilling, usually twice a week using a tractor and roto-tiller, mixed the pond contents into the soil of the PLTU and ensured oxygen was available in the ZOI for aerobic biodegradation of the petroleum hydrocarbons. Water was provided only if the soil was dry throughout the soil volume as determined by soil cores.

Results

The average total oil and grease concentration decreased from 22,250 to 2040 mg/kg, a 91% reduction, during the first phase of treatment (Table 8.15). Total petroleum hydrocarbons (TPHs) concentrations also decreased. Total polycyclic aromatic hydrocarbons (PAHs) concentrations decreased from 70.1 mg/kg to less than 5 mg/kg at week 40 and to below detection limits by the end of the first phase (week 67).

During the second phase of land treatment, the initial oil and grease and TPH concentrations were much higher. Within 73 weeks, the initial oil and grease concentration in the ZOI had been reduced by 41%, from 109,600 to 64,300 mg/kg. TPH was reduced by 45%, and 15 of the PAHs analyzed for were reduced to

Table 8.15. Contaminant Concentration Reductions During Land Farming, Wichita Falls, Texas

	Phase 1 Concentration (mg/kg wet weight)		Phase 2 Concentration (mg/kg wet weight)	
Contaminant	Initial	Final	Initial	Final
Oil and grease	22,250	2,040	109,600	64,300
Total petroleum hydrocarbons	22,000	<5,900	84,800	46,800
Polycyclic aromatic hydrdocarbons	70.1	ND[a]	597	ND[a] (except pyrene 133 to 34.5)

[a]ND, not detect.

nondetect levels from an initial total PAH concentration of 597 mg/kg. Only pyrene was not removed to below detection limits within the 73 week period.

Three toxicological assays were performed on soils or soil extracts over time. Microtox and rat liver cell assay evaluated the relative toxicity of the mobile hydrocarbon fraction that could be extracted from the soil, while the earthworm survival assay determined the relative toxicity of the soil itself. Nontoxic conditions were achieved by week 21–50 in the first phase. Toxicity measurements, from Microtox only, during the second phase showed a general decline from a highly toxic EC_{50} value of 6.7 to 68.1 for the aqueous soil extract (Table 8.16). There was a rapid initial decrease in toxicity during the first five weeks of treatment, with a continued gradual decrease through week 73. EC_{50} is a measure of the effective concentration of the extract that caused a 50% decrease in light output of the bioluminescent Microtox organisms. Higher EC_{50} values correspond to lower toxicities.

There was little or no mobility of hydrocarbons or metals to the aqueous phase during the first phase of treatment. During the second phase, slightly higher concentrations of metals, particularly chromium, and PAHs were observed in soil extracts but not in lysimeter results, indicative of a potential mobility for metals

Table 8.16. Microtox Toxicity of Phase II PLTU Aqueous Soil Extracts over Time, Wichita Falls, Texas

Length of PLTU Operation (weeks)	Effective Concentration or Relative Toxicity (EC_{50})
1	6.7
5	39.0
18	20.3
24	27.5
32	63.4
49	49.0
57	59.0
73	68.1

and PAHs. The higher waste loading and higher residual concentration of organic contaminants and metals in the second phase may have caused the difference in the results between the first and second phases.

8.9. TREATMENT TRAINS

Most contaminated sites exhibit a high degree of complexity in hydrogeologic characteristics and in the type and distribution of contaminants that are present, which may require a very complex remediation procedure. In such cases, no single system or unit is likely to be appropriate to address all of the physical, chemical, and biological constraints that influence the removal of different contaminants that may be present. Many single technologies might provide 90–95% of the mass removal from the source zone but would fail to produce a concurrent reduction in contaminant concentration in groundwater to meet cleanup regulatory standards. Consequently, several unit operations may have to be combined in series (i.e., treatment trains) or in parallel to meet cleanup standards in a cost-effective manner. This typically involves aggressive treatment approaches for the source zone, followed by more passive approaches for plume management.

The balance between performance, complexity, and cost controls the success of treatment trains. Strategies that are cost-effective, easier to implement, less complex, and easier to maintain would be more widely accepted. For example, most remediation projects begin by removing the source, followed by the installation of pumping systems that remove free product floating on the water table. Barriers may also be constructed to slow an advancing plume or to reduce the amount of water requiring treatment. Biostimulation may be feasible in diluted areas of the plume. In some circumstances, a site may reach final restoration goals without human intervention, due to natural biological and chemical processes. In such cases, an adequate monitoring program would be required to ensure that natural attenuation processes continue to effectively remove the contamination. In all cases, an understanding of the hydrogeologic and geochemical characteristics of the site is required to design a system that will adequately integrate the selected remediation techniques, and enhance their collective reliability and restoration cost-effectiveness. An example of a treatment train is given below in case study 8.

Case Study 8. A Treatment Train: Permeable Iron Barrier Followed by Biosparge at Naval Air Station Alameda, CA
(*Source*: Hughes et al. 2002; based on Fiorenza et al. 2000)

A granular (zero-valent) iron reactive barrier in series with a biosparge zone effectively treated a high-concentration mixed petroleum and chlorinated hydrocarbons plume to near MCL levels. At high influent concentrations, the granular iron was responsible for the abiotic degradation of more than 91% of the chlorinated solvents. At lower influent concentrations, almost complete degradation (>99%) was observed. The effectiveness of the biosparge treatment was unable to be assessed because of the sorption of toluene and other hydrocarbons to the granular iron (retardation); however, aerobic degradation of *cis*-DCE and VC that broke through the iron wall was supported in the biosparge zone, even though the primary removal mechanism was volatilization.

Process

Under highly reducing conditions and in the presence of metallic surfaces that release electrons during corrosion, certain dissolved chlorinated hydrocarbons, like tetrachloroethene (PCE) and trichloroethene (TCE), in groundwater will degrade to nontoxic compounds such as ethene, ethane, and chloride (Scherer et al. 2000). In this abiotic reductive dechlorination, the metal (iron in particular) lowers the redox potential and serves as an electron donor in the reaction.

Aerobic biodegradation of petroleum hydrocarbons, dichloroethene (DCE), and vinyl chloride (VC) can be promoted with the addition of oxygen in a biosparge system. Oxygen is added to increase the dissolved oxygen (DO) concentrations to approximately 8 mg/L with air sparge and approximately 20 mg/L with pure oxygen sparge to provide the electron acceptor necessary for the degradative reactions of indigenous aerobic bacteria attached to the growth support material.

Passive and semipassive in situ anaerobic–aerobic treatment sequences were assessed for the treatment of a mixed petroleum and chlorinated hydrocarbons plume at Naval Air Station (NAS) Alameda, near Oakland, California. A permeable barrier containing granular iron was used for the reductive dechlorination of solvents, followed by biosparging with oxygen gas for biodegradation of the petroleum hydrocarbons. Carbon dioxide was added to reduce the high pH levels (8–11), produced in the granular iron zone, to a pH range of 7–8. These technologies were installed in a Funnel-and-Gate^{TM} (University of Waterloo) system.

Site Description

Site 1 was located in the northwestern part of NAS Alameda, on the northwestern tip of Alameda Island. Starting in the 1940s cleaning solvents and waste petroleum hydrocarbons were disposed of in unlined waste pits, creating an apparently narrow plume of chlorinated solvents (*cis*-DCE, VC, and lesser concentrations of TCE) and petroleum hydrocarbons (toluene and benzene) that discharged into San Francisco Bay (Figure 8.41).

The field demonstration site at site 1 was located on artificial fill placed on top of natural bay mud estuarine deposits of silt and clay. The shallow aquifer, composed of the sandy artificial fill material, extended from the ground surface to a depth of approximately 6 m (20 ft). Depth to groundwater was 1.2–2.1 m (4–7 ft) BGS. Groundwater within this unit was unconfined, with a water table that experienced seasonal fluctuations. The bay mud unit was approximately 4.6–6 m (15–50 ft) thick and acted as a confining unit.

In the area of the field demonstration, the groundwater contained up to 210 mg/L of *cis*-DCE, 26 mg/L of VC, and <1 mg/L of other solvents such as TCE, 1,1-DCE, and DCE. Moderate concentrations of BTEX were detected. Toluene concentrations ranged from nondetect to 7 mg/L, benzene up to 0.3 mg/L, and concentrations of

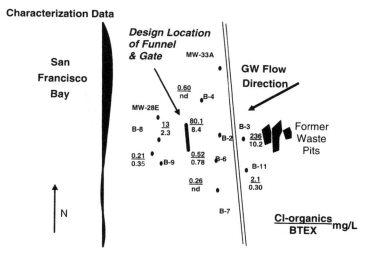

Characterization Data

Figure 8.41. Location of Funnel-and-Gate, Naval Air Station (NAS) Alameda, California.

ethylbenzene and xylenes were <1 mg/L. The highest concentrations of both chlorinated and petroleum hydrocarbons occurred within the upper portion of the aquifer.

Field Design/Operation

At the Alameda site, the treatment sequence consisted of a permeable iron barrier
followed by biosparge using a Funnel-and-Gate system (Figure 8.42). The artificial
fill was excavated to the top of the bay mud or confining layer, and a concrete pad
approximately 0.6 m (2 ft) thick was installed. Both the abiotic reactive barrier and
the aerobic biosparge systems were contained in a gate structure, constructed on the

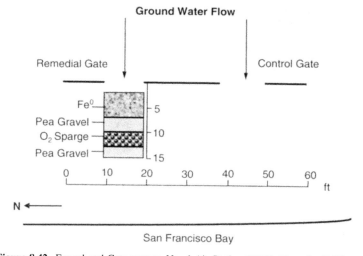

Figure 8.42. Funnel-and-Gate system, Naval Air Station (NAS) Alameda, California.

concrete pad to prevent settling, approximately 3 m (10 ft) wide (cross-gradient direction) and 6 m (20 ft) deep. Sheet piling funnels 9 m (30 ft) long were installed to a depth of 10 m (33 ft), 3.5 m (11.5 ft) below the top of the bay mud aquitard, at the upgradient end of the gate for the capture of an approximately 7.6 m (25 ft) wide section of the contaminant plume. Groundwater flowing through the gate encountered 0.6 m (2 ft) of a 5% by weight mixture of sand and granular iron, 1.5 m (5 ft) of 100% granular iron, and 0.9 m (3 ft) of pea gravel in the iron wall. This porous media system was followed by the biosparge zone, which consisted of a 0.6 m (2 ft) open-water zone partially filled with plastic Yeager™ Tripak balls. The balls served as support for the microbial biomass. Finally, the groundwater flowed through a final 0.6 m (2 ft) of pea gravel, which served as the final zone for compliance monitoring. The downgradient side of the gate was sealed with sheet piling and a constant pumping rate was maintained to establish an average groundwater flow velocity of about 6 cm/day (2.4 in./day). The maximum flow rate that the system could process was determined by pumping. A second gate was left undisturbed to serve as a control for assessment of natural attenuation.

Oxygen sparging for 20 min six times a day at a delivery pressure of 20 psi established aerobic conditions in the bio-ball-packed biosparge zone. Carbon dioxide sparging intervals of once per month for 10 min at a delivery pressure of 20 psi were sufficient to neutralize the elevated pH of the groundwater that had passed through the granular iron.

Results

Based on high percent removals, the overall performance of the iron wall–biosparge treatment train was determined to be excellent. The percent removal was calculated based on the differences between averaged contaminant concentrations entering the remedial gate and averaged concentrations leaving the gate (Table 8.17). Remedia-

Table 8.17. Percent Removal of Contaminants During the Treatment Sequence, NAS Alameda, California

Contaminant	% Removal 16 June 1997	% Removal 31 July 1997	% Removal 30 September 1997
TCE	100	NA[a]	100
1,1-DCE	NA[a]	NA[a]	100
cis-DCE	95.7	98.5	99.8
trans-DCE	NA[a]	NA[a]	100
VC	96.7	97.6	99.0
Benzene	NA[a]	NA[a]	100
Toluene	100	NA[a]	99.7
Ethylbenzenes	NA[a]	NA[a]	100
Xylenes (O)	NA[b]	NA[b]	100
Xylenes (P and M)	NA[b]	NA[b]	100

[a]NA: % removal could not be calculated because both concentrations were <MDL.
[b]NA: only total xylene was analyzed, but % removal could not be calculated because both concentrations were <MDL.

Table 8.18. Percent Removal of Chlorinated Contaminants in Granular Iron Wall, NAS Alameda, California

Contaminant	Influent Mass (mg)	Effluent Mass (mg)	Mass Removed (%)
TCE	107,500	278	99.7
1,1-DCE	3595	137	96.1
cis-DCE	1,347,000	116,500	91.3
trans-DCE	2997	66	97.8
VC	571,900	37,180	93.5

tion of cis-DCE and VC by the iron wall was incomplete (Table 8.18), probably because localized high influent concentrations resulted in breakthrough. However, most of the cis-DCE and VC that passed through the iron wall was aerobically biodegraded or volatilized in the biosparge zone (Table 8.19). Therefore, even with breakthrough, the total percent removal of cis-DCE was 99.8% and 99% for VC.

An accurate assessment of the performance of the biosparge on removal of BTEX could not be inferred because the contaminant concentrations in the biosparge zone were at or below their limits of quantification. Although the results in Table 8.17 suggest that there was excellent removal (100%) of BTEX compounds in the iron wall, most of the mass was likely removed by sorption within the iron wall, not by degradation. The field demonstration was not of sufficient length to allow the BTEX compounds to break through the iron wall, and therefore, the effectiveness of the biosparge on BTEX removal could not be determined in this case. However, the aerobic bioremediation of BTEX contaminants is well documented.

8.10. MONITORED NATURAL ATTENUATION

In many contaminated sites, natural conditions meet all the essential environmental and nutritional requirements for biodegradation, and bioremediation proceeds without human intervention. This process is called natural attenuation and differs from no-action alternatives in that it requires thorough documentation and

Table 8.19. Estimates of Percent Removal of cis-DCE and VC by Biodegradation and Volatilization Within Biosparge Zone, NAS Alameda, California

Contaminant/Process	Influent Mass (mg)	Effluent Mass (mg)	Mass Removed (%)
cis-DCE			
Biodegradation	142	58.05	59
Volatilization	142	93.57	34
Actual total removal	142	47.91	66
VC			
Biodegradation	283	216.20	24
Volatilization	283	108.88	62
Actual total removal	283	97.31	66

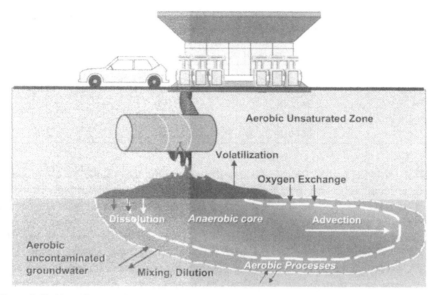

Figure 8.43. Natural attenuation mechanisms that reduce the mass and concentration of hydrocarbons in a gasoline-contaminated aquifer.

monitoring of the role of microorganisms and other attenuation processes in eliminating the target contaminants. Sometimes, natural attenuation is refereed to as intrinsic bioremediation.

Natural attenuation can be defined as *the combination of natural biological, chemical, and physical processes that act without human intervention to reduce the mass, toxicity, mobility, volume, or concentration of the contaminants (e.g., intrinsic bioremediation, dispersion, dilution, sorption, and volatilization).* These processes are illustrated in Figure 8.43 for the natural attenuation of a gasoline leak from an underground storage tank.

Monitored natural attenuation (MNA) is often the most cost-effective approach to manage groundwater contamination by benzene and other hydrocarbons. For example, a study of 42 Air Force sites contaminated with benzene (with plume areas ranging from 0.3 to 60 acres) concluded that the average cost of MNA was $126,000, including site characterization, laboratory analysis, data analysis, mathematical modeling, and reporting (Parsons Engineering science 1999). Although the average cleanup time for these 42 sites with MNA would be about 30 years, more aggressive cleanup technologies that would have reduced the cleanup time to 15 years would incur an average cost of $816,000 (i.e., 6.5 times higher). The duration of MNA projects depends on site-specific conditions and regulatory requirements and can range from as low as 2 to 10 years until desired degradation levels are achieved. Typical cost ranges from $50,000 to $250,000 per acre. MNA case studies were presented in Chapter 6.

MNA is receiving increasing attention due to its potential cost-effectiveness. In particular, MNA experienced significant growth during the 1990s, from laboratory

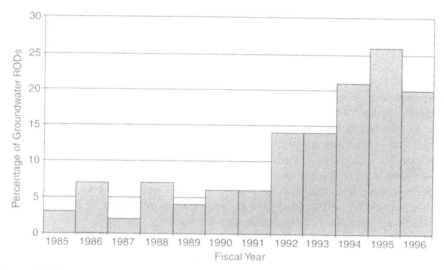

Figure 8.44. Use of natural attenuation in the cleanup of contaminated groundwater at Superfund sites, 1985–1996, shown as a percentage of the total number of remedies for contaminated groundwater selected in the indicated year. *Note*: RODs are records of decision (regulatory documents specifying remedies for Superfund sites). (Adapted from National Research Council 2000.)

research projects to a widely used approach to manage contaminated groundwater (Figure 8.44). MNA is highly applicable to manage groundwater pollution by petroleum product releases from leaking underground storage tanks, and it has been already selected to treat about 50% of sites contaminated with BTEX compounds (Figure 8.45). MNA may also be effective to manage plumes originating from

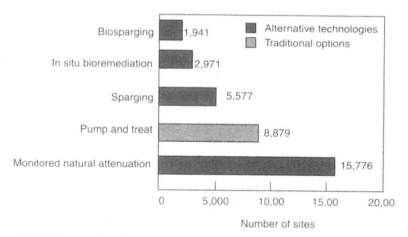

Figure 8.45. Methods used to clean up groundwater contamination from leaking underground storage tanks as of 1997. (Adapted from Tullis et al. 1998.)

municipal and industrial landfills, manufacturing facilities, and refineries. MNA is typically used in conjunction with active remediation measures (e.g., source control) or as a follow-up to such measures and is most appropriate for sites experiencing low exceedences of cleanup levels. In general, active remedial processes should be stopped when treatment is yielding marginal benefits (if any), or when treatment is no longer faster than MNA. The success of MNA as a remedial option depends on the following: (1) adequate site characterization, (2) a long-term monitoring plan consistent with the level of knowledge regarding subsurface conditions at the site, (3) evaluation and (if needed) control of the contaminant source, and (4) a reasonable time frame to achieve the remedial objectives. Whether a time frame is reasonable is a site-specific decision, generally meaning not excessive compared to other remedies. Reasonableness depends on current and future uses of the aquifer, public acceptance of extended time for remediation, reliability of monitoring and institutional controls, and regional resource issues.

Relative to more aggressive bioremediation alternatives such as biostimulation and bioaugmentation, MNA has the following advantages and disadvantages.

Advantages

- Lower overall remediation costs.
- Generation of lesser volume of remediation wastes.
- Reduced potential for cross-media transfer of contamination.
- Reduced risk of human exposure to contaminated media.
- Less intrusion.

Disadvantages

- Longer time to achieve remediation objectives.
- Site characterization may be more complex and costly.
- Toxicity of some transformation products may exceed that of the parent compound (e.g., reductive dechlorination of dichloroethylene to vinyl chloride).
- Long-term monitoring is required.
- Institutional controls (deed restrictions, well-drilling prohibitions, etc.) and more extensive education and outreach efforts may be needed.
- There is a potential for remobilization of previously stabilized metals and radionuclides.

Selection of MNA as a remediation technology does not guarantee its approval by the appropriate regulatory agency. To get permission to implement MNA over more aggressive (and more costly) approaches, it is important to keep in mind the following:

- The burden of proof that MNA is an appropriate selection for a given site is on the proponent, not on the regulator, and evidence of its effectiveness should emphasize proof of biodegradation.
- MNA is not a default technology or presumptive remedy. One should not assume that it will work without site-specific data demonstrating its effectiveness (see Chapter 9).
- Remediation by MNA is not complete until the goals of the regulatory agency have been reached to their satisfaction.
- Although MNA tends to be more economical, it can take much longer to achieve cleanup, and it may appear as if officials are walking away from contaminated sites. Thus, early public involvement is critical to minimize such controversy.

A plume size comparison for different pollutants shows that the median length of BTEX plumes is relatively small compared to that of chlorinated solvent plumes (Figure 8.46), due mainly to the effect of natural attenuation. BTEX, which have a natural pyrolytic origin, are often easier to biodegrade (and thus are attenuated to a greater extent) than chlorinated solvents that have a synthetic origin.

8.10.1. Considerations for MNA Selection

One of the most important factors influencing the suitability of MNA is whether the plume is expanding, steady, or shrinking. Historical monitoring well data collected for the entire plume should be examined to determine if the plume dimensions have not changed significantly over time. If variations in contaminant concentrations are

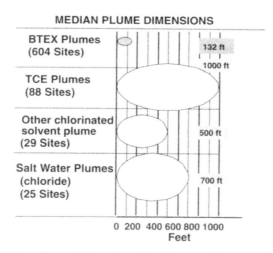

Figure 8.46. Natural attenuation mechanisms that reduce the mass and concentration of hydrocarbons in a gasoline-contaminated aquifer. (Adapted from Newell and Connor 1998.)

not due to random factors such as water table fluctuations, sampling variability, and analytical uncertainty, the plume can be considered to be at steady state. This "steady state" is achieved when any additional contamination released from the source is "naturally attenuated" and degraded at the same rate that it is introduced (ASTM 1998). As the source is depleted and the contaminant loading rate to groundwater decreases, the plume begins to shrink until it disappears, completing its life cycle.

To determine the status of a plume, monitoring points or other sampling devices should be located to allow the construction of concentration contour maps for contaminants and other constituents of concern. This monitoring map should include a nondetect or compliance level contour (e.g., the drinking water criterion of 5 ppb for benzene or TCE) (ASTM 1998). Based on changes (or lack of changes) over time, the plume can be characterized as shrinking, stable, or expanding. Alternatively, contaminant concentration(s) in two or more wells located within the plume and downgradient of the source, and oriented along the main flow direction, can be monitored over time (ASTM 1998). The trends in contaminant concentrations (and, more importantly, total mass) will determine if the plume is expanding (i.e., increasing dissolved mass and concentrations), stable (i.e., constant mass and concentrations), or shrinking (i.e., decreasing mass and concentrations).

Other questions that should be considered prior to selection of MNA include:

- Are the target contaminants known to be degraded or immobilized by natural processes?
- Do the transformation products pose a lower risk?
- Are redox conditions at the site conducive to natural attenuation? (For example, for chlorinated solvents, are native or anthropogenic electron donors present in sufficient concentration to drive dechlorination?)
- Have contamination sources been adequately controlled?
- Are drinking water supplies or sensitive ecosystems likely to be impacted if one selects MNA?
- How will existing and proposed active remediation measures affect the MNA component of the remedy?
- Is the estimated time frame of remediation "reasonable"? (What are the current or projected demands for the affected aquifer in this time period?)
- Are there reliable vehicles for implementing institutional controls (e.g., zoning ordinances) and can a local institution monitor and enforce these controls?

An understanding of the following factors is also important to support the selection and ensure the success of MNA:

1. *Understand how the plume was formed in the first place*

   - Understand the three-dimensional distribution of the original source of contamination, and the movement of contaminated groundwater and vapors.

- Must be able to explain the existing contamination pattern based on what is known about the original source material and site hydrogeology.

2. *Understand rate of transport and rate of attenuation*

- What is the variation in groundwater flow velocity and direction?
- What is the seepage velocity of the lithology that actually carries the plume? Are there preferential conduits present?
- What is the mass flux of the contaminants? Is it decreasing along the flow path?
- What is the relative importance in quantifying hydraulic conductivity, gradient, dispersivity, and rate of biodegradation at this site (sensitivity and uncertainty analyses)?
- What is the confidence in the methods used to measure them? Is the resolution of the monitoring well data defined and documented?
- Will the current rate of attenuation be maintained? Is there a sufficient supply of electron donors and/or acceptors?

3. *Understand the persistence of the contaminant mass*

- Evaluate effectiveness of source control. Is there a new plume forming from a previously unidentified source? Is there a NAPL moving downgradient?
- How fast is the plume expanding or shrinking?
- How fast will other remedies approach cleanup goals?

MNA is an established remedial option for only a few types of contaminants (e.g., BTEX, alcohols, ketones, esters) and should be accepted as a formal remediation option only when the processes that control contaminant immobilization and/or destruction are well understood at a given site and demonstrated to be sustainable (National Research Council 2000). Table 8.20 summarizes the rankings of an NRC committee for the likelihood of natural attenuation succeeding in remediating different classes of contaminants. The level of understanding of the underlying attenuation mechanisms for different compounds is also depicted. This table should not to be taken as a recipe for which chemicals should be remediated by MNA nor as a prescription for a mechanism that will control the attenuation process. It should also be kept in mind that knowledge changes rapidly in the environmental sciences, and some contaminants not rated as having high natural attenuation potential could achieve this status in the future.

MNA has been successfully demonstrated in more than 15,000 BTEX-contaminated sites (MacDonald 2000) but there is a tendency to overprescribe it due to its relatively low cost. The current trend is to use MNA in conjunction with *risk-based corrective action* (RBCA), which is a site cleanup philosophy that is gaining increasing importance worldwide due to resource allocation problems (e.g., limited time, funds, regulatory oversight, and qualified professionals). Specifically, RBCA is a decision-making process to focus resources on contaminated sites that pose the

Table 8.20. Likelihood of Success of Natural Attenuation

Chemical Class	Dominant Attenuation Processes	Current Level of Understanding[a]	Likelihood of Success Given Current Level of Understanding[b]
	Organic		
Hydrocarbons			
BTEX	Biotransformation	High	High
Gasoline, fuel oil	Biotransformation	Moderate	Moderate
Nonvolatile aliphatic compounds	Biotransformation, immobilization	Moderate	Low
Polycyclic aromatic hydrocarbons	Biotransformation, immobilization	Moderate	Low
Creosote	Biotransformation, immobilization	Moderate	Low
Oxygenated Hydrocarbons			
Low-molecular-weight alcohols, ketones, esters	Biotransformation	High	High
MTBE	Biotransformation	Moderate	Low
Halogenated Aliphatics			
Tetrachloroethene, trichloroethene, carbon tetrachloride	Biotransformation	Moderate	Low
Trichloroethane	Biotransformation, abiotic transformation	Moderate	Low
Methylene chloride	Biotransformation	High	High
Vinyl chloride	Biotransformation	Moderate	Low
Dichloroethylene	Biotransformation	Moderate	Low
Halogenated Aromatics			
Highly chlorinated PCBs, tetrachlorodibenzofuran, pentachlorophenol, multichlorinated benzenes	Biotransformation, immobilization	Moderate	Low
Less chlorinated PCBs, dioxins	Biotransformation	Moderate	Low
Monochlorobenzene	Biotransformation	Moderate	Moderate
Explosives			
TNT, RDX	Biotransformation, abiotic transformation, immobilization	Moderate	Low

Table 8.20. (*Continued*)

Chemical Class	Dominant Attenuation Processes	Current Level of Understanding[a]	Likelihood of Success Given Current Level of Understanding[b]
		Inorganic	
Metals			
Ni	Immobilization	Moderate	Moderate
Cu, Zn	Immobilization	Moderate	Moderate
Cd	Immobilization	Moderate	Low
Pb	Immobilization	Moderate	Moderate
Cr	Biotransformation, immobilization	Moderate	Low to moderate
Hg	Biotransformation, immobilization	Moderate	Low
Nonmetals			
As	Biotransformation, immobilization	Moderate	Low
Se	Biotransformation, immobilization	Moderate	Low
Oxyanions			
Nitrate	Biotransformation	High	Moderate
Perchlorate	Biotransformation	Moderate	Low
Radionuclides			
^{60}Co	Immobilization	Moderate	Moderate
^{137}Cs	Immobilization	Moderate	Moderate
^{3}H	Decay	High	Moderate
^{90}Sr	Immobilization	High	Moderate
^{99}Tc	Biotransformation, immobilization	Low	Low
^{238,239,240}Pu	Immobilization	Moderate	Low
^{235,238}U	Biotransformation, immobilization	Moderate	Low

[a]Levels of understanding: "High" means that there is good scientific understanding of the process involved, and field evidence confirms attenuation processes can protect human health and the environment; "moderate" means that studies confirm the dominant attenuation process occurs but the process is not well understood scientifically; "low" means that scientific knowledge is inadequate to judge if and when the dominant process will occur and whether it will be protective.

[b]Likelihood of success relates to the probability that, at any given site, natural attenuation of a given contaminant is likely to be protective of human health and the environment. "High" means scientific knowledge and field evidence are sufficient to expect that natural attenuation will protect human health and the environment at more than 75% of contaminated sites. "Moderate" means natural attenuation can be expected to be protective at about half of the sites. "Low" means natural attenuation is expected to be protective at less than 25% of contaminated sites. A "low" rating can also result from a poor level of scientific understanding.

Source: National Research Council (2000).

greatest risk, to ensure that site assessment activities are focused on collecting only that information that is necessary to make RBCA decisions, and to ensure that the selected remedial option is the most economically favorable one to achieve the negotiated degree of exposure and risk reduction (ASTM 1994). Thus, following RBCA procedures (Chapter 1), decisions can be made about site-specific cleanup levels that result in an acceptance risk at the point where potential receptors (including sensitive ecosystems) are located. For example, if potential receptors are beyond the reach of the plume, aggressive remediation steps are generally not taken and site cleanup relies primarily on MNA.

There is a critical need to demonstrate whether MNA is actually working in a particular site at a rate that is sufficient to protect public health and the environment. Such a demonstration, also referred to as "performance assessment," involves integrated performance and risk assessment using state-of-the-art characterization—geochemical, microbial, forensic analysis, and computer modeling techniques (discussed in Chapter 9). Thus, performance assement requires a staff with broad experience in computer modeling, engineering, microbiology, and the earth sciences. A meaningful MNA performance assessment can only be achieved when the remediation objectives have been clearly defined, including the stipulation of the required cleanup levels, establishment of acceptable cleanup time frames, and the determination of points of compliance. Once these objectives are defined, the evaluation of natural attenuation as a sole remediation option or a component of a comprehensive plan that includes active remediation and source control can be done.

During the past decade a large number of protocols have been developed to provide guidance on evaluating and assessing natural attenuation. Table 8.21 lists 14 protocols reviewed by the Committee on Intrinsic Remediation (National Research Council 2000). These protocols were developed primarily by federal and state agencies but also by corporations, professional and industrial associations, and public/private consortia. A comprehensive protocol should cover the broad subject areas of community concerns, scientific and technical issues, and implementation issues. The committee's main conclusion was that the existing body of MNA protocols is limited in several important areas including: (1) the lack of guidance on certain contaminants (e.g., polycyclic aromatic hydrocarbons, polychlorinated biphenyls, explosives, and other classes of persistent organic compounds); (2) little or no discussion of when and how to involve the public in site decisions and when and how to implement institutional controls; (3) discussion of when and how to implement contingency plans in case natural attenuation does not work also is inadequate in many of the protocols; (4) protocols provide insufficient guidance on when engineered methods to remove or contain sources of contamination benefit natural attenuation and when they interfere with it; (5) inadequate guidance on how to conduct long-term monitoring to ensure that natural attenuation remains protective of public health and the environment; (6) lack of description of necessary training for implementation; (7) use of scoring systems that may yield a numerical value but could lead to an erroneous conclusion about whether natural attenuation will succeed or not, due to the complexity of the processes involved and the tendency of the scoring system to oversimplify them; (8) lack of sufficient guidance on which

**Table 8.21. Natural Attenuation Protocols Reviewed in the
National ResearchCouncil Study**

Federal Agencies

Environmental Protection Agency (EPA)

- *Use of Monitored Natural Attenuation at Superfund, RCRA Corrective Action, and Underground Storage Tank Sites*, Final OSWER Directive (OSWER Directive Number 9200 4-17P), April 21, 1999, EPA Office of Solid Waste and Emergency Response.
- *Technical Protocol for Evaluating Natural Attenuation of Chlorinated Solvents in Ground Water*, Todd H. Wiedemeier, M.A. Swanson, D.E. Moutoux, E.K. Gordon, J.T. Wilson, B.H. Wilson, D.H. Kampbell, P.E. Hass, R.N. Miller, J.E. Hansen, and F.H. Chapelle, EPA/600/R-98/128, September 1998, EPA Office of Research and Development.
- *Draft Region 4 Suggested Practices for Evaluation of a Site for Natural Attenuation (Biological Degradation) of Chlorinated Solvents*, Version 3.0, November 1997, EPA Region 4.

Department of Energy

- *Site Screening and Technical Guidance for Monitored Natural Attenuation at DOE Sites*, P.V. Brady, B.P. Spalding, K.M. Krupka, R.D. Waters, P. Zhang, D.J. Borns, and W.D. Brady, Draft, August 30, 1998, Sandia National Laboratory.

Air Force

- *Technical Protocol for Implementing Intrinsic Remediation with Long-Term Monitoring for Natural Attenuation of Fuel Contamination in Groundwater*, T. Wiedemeier, J.T. Wilson, D.H. Kampbell, R.N. Miller, and J.E. Hanson, Volumes I and II, November 11, 1995, Air Force Center for Environmental Excellence, Technology Transfer Division, Brooks AFB, San Antonia, TX.
- *Technical Protocol for Evaluating Natural Attenuation of Chlorinated Solvents in Groundwater*, T. Wiedemeier, M.W. Swanson, D.E. Moutoux, E.K. Gordon, J.T. Wilson, B.H. Wilson, D.H. Kampbell, J.E. Hansen, P.E. Hass, and F.H. Chapelle, Draft-Revision 2, July 1997, Air Force Center for Environmental Excellence, Technology Transfer Division, Brooks AFB, San Antonio, TX.

Navy

- *Technical Guidelines for Evaluating Monitored Natural Attenuation at Naval and Marine Corps Facilities*, T.H. Wiedemeier. and F.H. Chapelle, Draft-Revision 2, March 1998.

State Agencies

Minnesota Pollution Control Agency

- *Draft Guidelines—Natural Attenuation of Chlorinated Solvents in Ground Water*, Working Draft, December 12, 1997, Minnesota Pollution Control Agency, Site Response Section.

New Jersey

- *New Jersey Administrative Code 7: 26E—Technical Requirements for Site Remediation, and Classification Exception Areas: Final Guidance 4-17-1995.*

Table 8.21. (*Continued*)

Corporations

Chevron

- *Protocol for Monitoring Intrinsic Bioremediation in Groundwater*, T. Buscheck and K. O'Reilly, March 1995, Chevron Research and Technology Company, Health, Environment, and Safety Group.
- *Protocol for Monitoring Natural Attenuation of Chlorinated Solvents in Groundwater*, T. Buscheck and K. O'Reilly, February 1997, Chevron Research and Technology Company, Health, Environment, and Safety Group.

Professional and Industrial Associations

American Society of Testing Materials

- *Standard Guide for Remediation of Ground Water by Natural Attenuation at Petroleum Release Sites*, Draft, February 4, 1997.

American Petroleum Institute

- *Methods for Measuring Indicators of Intrinsic Bioremediation: Guidance Manual*, American Petroleum Institute, Health and Environmental Sciences Department, Publication Number 4658, November 1997, API Publishing Services, Washington, DC.

Public/Private Consortia

Remediation Technologies Development Forum

- *Natural Attenuation of Chlorinated Solvents in Groundwater: Principles and Practices*, Industrial Members of the Bioremediation of Chlorinated Solvents Consortium of the Remediation Technologies Forum, Version 3.0, August 1997.

---

protocols are appropriate for use in various regulatory programs; and (9) the existing protocols, on the most part, have not been technically reviewed (National Research Council 2000).

The National Research Council (2000) recommended that the EPA take the lead in developing national consensus guidelines for protocols on natural attenuation. These protocols should provide a directive providing a framework for utilizing natural attenuation as a remedy for remediation.

8.11. POLITICAL AND SCIENTIFIC CHALLENGES FOR A BROADER IMPLEMENTATION OF BIOREMEDIATION AND NATURAL ATTENUATION

A National Research Council (1993) Committee that evaluated bioremediation concluded that:

> Bioremediation is clouded by controversy over what it does and how well it works, partly because it relies on microorganisms, which cannot be seen, and partly because it has become attractive for "snake-oil salesmen" who claim to be able to solve all

types of contamination problems. As long as the controversy remains, the full potential of this technology cannot be realized.

This reflects that bioremediation is not universally understood or trusted by those who must approve it. Whereas significant advances have been made toward understanding the biochemical and genetic basis for biodegradation, bioremediation (which involves exploiting these processes at the field scale) is still an underutilized technology that has not reached pedagogical maturity. Currently, bioremediation is utilized primarily for fuel contamination from leaking underground storage tanks and is practiced more as an art of trial and error than as an empirical science. To take full advantage of the potential of bioremediation and MNA, practitioners, researchers, and teachers need to communicate better its capabilities and limitations, and answer question such as:

- Which attenuation processes are responsible for the loss of contaminants at a given site?
- Why are some processes occurring at some sites but not others?
- How are different microbes degrading different contaminants, and who is doing what?
- How fast is biodegradation proceeding, and can we control it and make it go faster?
- When can we meet cleanup standards in a cost-effective manner?
- Can we reliably predict that what we want to happen will happen?

To answer such questions, it is important to adopt an appropriate research philosophy, the selection of which will be addressed in the next section.

8.11.1. Epistemology of Bioremediation

Epistemology (Greek, *espisteme* = knowledge) is the theory of the methods and basis used to acquire knowledge, including the possibility and opportunity to advance fundamental understanding, sphere of action, and the philosophy of the scientific disciplines that we rely upon. There are two contrasting philosophies to conduct scientific research in the search to obtain new knowledge:

1. *Reductionism.* This philosophy is based on system analysis through separation of its components, to eliminate complexity and enhance data interpretation. Reductionism is based on the premise that knowledge of a system can be improved by studying its components, and that an idea can be understood better if we understand its concepts separately. Reductionism is used increasingly in bioremediation research to investigate biotransformation and mass transfer mechanisms.

2. *Holism.* This approach is based on the concept that the totality of a system is greater than the sum of its parts, due to potential synergistic and antagonistic

interactions among system components that enhance or hinder bioremediation (e.g., microorganisms interacting with mineral surfaces to enhance degradation rates, or relying on dilution of inhibitory contaminant concentrations to lower levels that permit microbial participation in the cleanup process).

8.11.2. Epistemology's Uncertainty Principle

Figure 8.47 depicts critical differences and outcomes of relying on reductionism or holism as bioremediation research philosophies. Both approaches have their advantages and disadvantages. As we unfold each layer of reductionism (moving left to right on Figure 8.47), we simplify the system, enhance experimental control, make it easier to test hypotheses, and facilitate the interpretation of the results. However, reductionism may augment lab artifacts and hinder the relevance of the information that we obtain with regard to site cleanup.

These contrasting research philosophies are, to some extent, scale dependent. For example, microbiologists and geologists are professionals who often contribute to the development of bioremediation. Yet, they often work at different scales and rely on different disciplines with diverging philosophical tendencies. Specifically, many microbiologists work at the in vitro and in vivo scales (right side of Figure 8.47) and rely on reductionist disciplines such as genetics, molecular biology, and biochemistry to advance knowledge about biodegradation. Microbiologists are often experts at taking apart the components of a cell and learning how each works. Since microbes are difficult to observe in their natural habitats, the notion that advancement of knowledge must rely on hypothesis-driven experimentation is deeply rooted in microbiologists. In contrast, geology has traditionally been an observational science. Many geologists conduct research at larger scales (left side of Figure 8.47) and rely more on field-scale observation than on experimentation

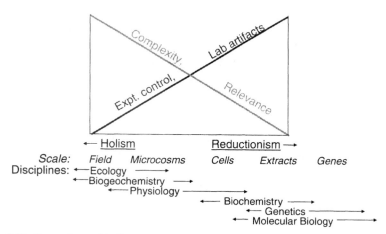

Figure 8.47. Comparison of holism versus reductionism as bioremediation research philosophies. (Adapted from Madsen 1998.)

because the enormous temporal and spatial scale of many of the earth's phenomena cannot be reproduced in the lab. For example, the most important unifying concept in geology is perhaps the plate tectonics model. This model, which was proposed by Wegener and Holmes, explains continental drift, oceanic trenches, geosynclines, and other phenomena was entirely based on observational evidence. Experimental evidence, indicating that rocks were insufficiently deformable for continents to move at all, largely muddied the issues involved (Chappelle 2001).

This comparison between reductionist and holistic disciplines suggests the possibility of taking advantage of the positive aspects of both approaches to enhance the application of bioremediation. However, to do so may require consideration of the following axioms.

- Quantitative extrapolation from reductionist lab experiments to complex field systems should be avoided. For example, rates measured in the lab are often one order of magnitude faster than in the field, where nutrient and mass transfer limitations are greater and temperatures lower.
- It is recommended to rely more on holistic disciplines (e.g., microbial ecology and biogeochemistry) and iterate between the field and the lab, and between basic and applied research.
- Collaborations in multidisciplinary teams should be encouraged, keeping in mind that some of the most intellectually stimulating and technologically productive research lies at the interstices between disciplines. However, researchers should be careful not to try to be all encompassing if greater breadth comes at the cost of depth. Rather, it is recommended to build on strength and think in systems while acting in one's own discipline. This is analogous to thinking globally while acting locally.
- Finally, it is important to balance the scientific drive to understand nature with the engineering mission to solve problems. Thus, it may be opportune to quote *Aurea Mediocridad* from San Ignacio de Loyola, founder of the Jesuit order. This quote literally translates to "the gold is in the middle" and can be interpreted to mean "the virtue lies in the equilibrium." It is also relevant to highlight Albert Einstein's philosophy: *science should be humane in spirit, wise in its uses, and moral in purpose.*

In conclusion, indigenous microorganisms can destroy a wide variety of common groundwater contaminants, which often makes bioremediation approaches technically sound and economically feasible. Numerous techniques have also been developed to overcome limitations to natural degradation processes through bioaugmentation. However, there has been a recent paradigm shift in site remediation toward monitored natural attenuation. This alternative can be more economical, but care must be taken not to overprescribe it. Furthermore, MNA requires much longer operation to achieve cleanup, and it may give the impression that officials are walking away from contaminated sites. Thus, early public involvement is critical to minimize such controversy.

It is important to keep in mind that neither bioremediation nor natural attenuation is a panacea that is universally applicable, and that some sites may require the application of physicochemical approaches as the principal removal mechanism of the contaminant mass, especially near the source zone. Nevertheless, bioremediation is quickly achieving pedagogical maturity and has earned an important place in the menu of alternatives from which we select solutions to our environmental pollution problems. Methods to assess the performance of bioremediation and natural attenuation are discussed in the next chapter.

REFERENCES

Aitchison E., S. Kelley, J.L. Schnoor, and P.J.J. Alvarez (2000). Phytoremediation of 1,4-dioxane by hybrid poplar trees. *Water Environ. Res.* **72**:313–321.

Alleman, B.C. and A. Leeson (Eds.) (1991). *First International In Situ and On-Site Bioreclamation Symposium*, April, San Diego, CA. Battelle Press, Columbus, OH.

    1(1) *On-Site Bioreclamation: Processes for Xenobiotic and Hydrocarbon Treatment.*

    1(2) *In Situ Bioreclamation: Applications and Investigations for Hydrocarbon and Contaminated Site Bioremediation.*

Alleman, B.C. and A. Leeson (Eds.) (1993). *Second International In Situ and On-Site Bioreclamation Symposium*, April, San Diego, CA. Battelle Press, Columbus, OH.

    2(1) *Bioremediation of Chlorinated and Polycyclic Aromatic Hydrocarbon Compounds.*

    2(2) *Hydrocarbon Remediation.*

    2(3) *Applied Biotechnology for Site Remediation.*

    2(4) *Emerging Technology for Bioremediation of Metals.*

    2(5) *Air Sparging for Site Bioremediation.*

Alleman, B.C. and A. Leeson (Eds.) (1995). *Third International In Situ and On-Site Bioreclamation Symposium*, April, San Diego, CA. Battelle Press, Columbus, OH.

    3(1) *Intrinsic Bioremediation.*

    3(2) *In Situ Aeration: Air Sparging, Bioventing, and Related Remediation Processes.*

    3(3) *Bioaugmentation for Site Remediation.*

    3(4) *Bioremediation of Chlorinated Solvents.*

    3(5) *Monitoring and Verification of Bioremediation.*

    3(6) *Applied Bioremediation of Petroleum Hydrocarbons.*

    3(7) *Bioremediation of Recalcitrant Organics.*

    3(8) *Microbial Processes for Bioremediation.*

    3(9) *Biological Unit Processes for Hazardous Waste Treatment.*

    3(10) *Bioremediation of Inorganics.*

Alleman, B.C. and A. Leeson (Eds.) (1997). *Fourth International In Situ and On-Site Bioremediation Symposium*, April, New Orleans, LA. Battelle Press, Columbus, OH.

    4(1) *In Situ and On-Site Bioremediation: Volume 1.*

    4(2) *In Situ and On-Site Bioremediation: Volume 2.*

    4(3) *In Situ and On-Site Bioremediation: Volume 3.*

    4(4) *In Situ and On-Site Bioremediation: Volume 4.*

    4(5) *In Situ and On-Site Bioremediation: Volume 5.*

Alleman, B.C. and A. Leeson (Eds.) (1999). *Fifth International In Situ and On-Site Bioremediation Symposium*, April, San Diego, CA. Battelle Press, Columbus, OH.

5(1)   *Natural Attenuation of Chlorinated Solvents, Petroleum Hydrocarbons, and Other Organic Compounds.*

5(2)   *Engineered Approaches for In Situ Bioremediation of Chlorinated Solvent Contamination.*

5(3)   *In Situ Bioremediation of Petroleum Hydrocarbon and Other Organic Compounds.*

5(4)   *Bioremediation of Metals and Inorganic Compounds.*

5(5)   *Bioreactor and Ex Situ Biological Treatment Technologies.*

5(6)   *Phytoremediation and Innovative Strategies for Specialized Remedial Applications.*

5(7)   *Bioremediation of Nitroaromatic and Haloaromatic Compounds.*

5(8)   *Bioremediation Technologies for Polycyclic Aromatic Hydrocarbon Compounds.*

Anderson, R.T. and D.R. Lovley (2000). Anaerobic bioremediation of benzene under sulfate-reducing conditions in a petroleum-contaminated aquifer. *Environ. Sci. Technol.* **34**:2261–2266.

Anderson, R.T., H.A. Vrionis, I. Ortiz-Bernad, C.T. Resch, P.E. Long, R. Dayvault, K. Karp, S. Marutzky, D.R. Metzler, A. Peacock, D.C. White, M. Lowe, and D.R. Lovley (2003). Stimulating the in situ activity of *Geobacter* species to remove uranium from the groundwater of a uranium-contaminated aquifer. *Appl. Environ. Microbiol.* **69**(10):5884–5891.

ASTM (1994). *1994 Annual Book of ASTM Standards: Emergency Standard Guide for Risk-Based Corrective Action Applied at Petroleum Release Sites* (Designation: ES 38-94). American Society for Testing and Materials, West Conshohocken, PA.

ASTM (1998). *1998 Annual Book of ASTM Standards: Standard Guide for Remediation of Ground Water by Natural Attenuation at Petroleum Release Sites* (Designation: E 1943-98). American Society for Testing and Materials, West Conshohocken, PA, pp. 875–917.

Benner, S.G., D.W. Blowes, and C.J. Ptacek (1997). A full-scale reactive wall for prevention of acid mine drainage. *Ground Water Monitoring and Remediation* **Fall**:99–107.

Bianchi-Mosquera, G.C., R.M. Allen-King, and D.M. Mackay (1994). Enhanced degradation of dissolved benzene and toluene using a solid oxygen-releasing compound. *Ground Water Monitoring and Remediation* **Winter**:120–128.

Borden, R.C., R.T. Goin, and C.-M. Kao (1997). Control of BTEX migration using a biologically enhanced permeable barrier. *Groundwater Monitoring and Remediation* **Winter**:70–80.

Bourquin, A.W., D.C. Mosteller, R.L. Olsen, M.J. Smith, and R.F. Reardon (1997). Aerobic bioremediation of TCE-contaminated groundwater: bioaugmentation with *Burkholderia cepacia* PR1. In: *Fourth International In Situ and On-Site Bioremediation Symposium*, New Orleans, LA, Battelle Press, Columbus, OH.

Chapelle, F.H. (2001). *Ground-Water Microbiology and Geochemistry*, 2nd ed. John Wiley & Sons, Hoboken, NJ.

Cunningham, A.B., W.G. Characklis, F. Abedeen, and D. Crawford (1991). Influence of biofilm accumulation on porous-media hydrodynamics. *Environ. Sci. Technol.* **25**:1305–1311.

Duba, A.G., K.J. Jackson, M.C. Jovanovich, R.B. Knapp, and R.T. Taylor (1996). TCE remediation using in-situ, resting-state bioaugmentation. *Environ. Sci. Technol.* **30**:1982–1989.

Dybas, M.J., M. Barcelona, S. Bezborodnikov, S. Davies, L. Forney, H. Heuer, O. Kawka, T. Mayotte, L. Sepulveda-Torres, K. Smalla, M. Sneathen, J. Tiedje, T. Voice, D.C. Wiggert, M.E. Witt, and C.S. Criddle (1998). Pilot-scale evaluation of bioaugmentation for in-situ remediation of a carbon tetrachloride-contaminated aquifer. *Environ. Sci. Technol.* **32**:3598–3611.

Elias, D.A., L.R. Krumholz, D. Wong, P.E. Long, and J.M. Suflita (2003). Characterization of microbial activities and U reduction in a shallow aquifer contaminated by uranium mill tailings. *Microbial Ecol.* **46**:83–91.

El Fantroussi, S., M. Belkacemi, E.M. Top, J. Mahillon, H. Naveau, and S.N. Agathos (1999). Bioaugmentation of a soil bioreactor designed for pilot-scale anaerobic bioremediation studies. *Environ. Sci. Technol.* **33**:2992–3001.

Ellis, D.E., E.J. Lutz, J.M. Odom, J. Buchanan, J. Ronald, C.L. Bartlett, M.D. Lee, M.R. Harkness, and K.A. Deweerd (2000). Bioaugmentation for accelerated in situ anaerobic bioremediation. *Environ. Sci. Technol.* **34**:2254–2260.

Environmental Engineering Research Council (1990). Groundwater protection and reclamation. *J. Environ. Eng. Div. A Soc. Civ. Eng.* **116**:654–662.

Federal Remediation Technologies Roundtable (1997). *Remediation Case Studied: Bioremediation and Vitrification, Volume 5.* EPA 542-R-97-008.

Federal Remediation Technologies Roundtable (2000a). *FRTR Cost and Performance: Remediation Case Studies and Related Information.* EPA 542-C-00-001.

Federal Remediation Technologies Roundtable (2000b). *Abstracts of Remediation Case Studies, Volume 4.* EPA 542-R-00-006.

Federal Remediation Technologies Roundtable (2001a). *FRTR Cost and Performance Remediation Case Studies and Related Information.* EPA 542-C-01-003.

Federal Remediation Technologies Roundtable (2001b). *Abstracts of Remediation Case Studies, Volume 5.* EPA 542-R-01-008.

Federal Remediation Technologies Roundtable (2002). *Remediation Technologies Screening Matrix and Reference Guide*, 4th ed. Office of Management and Budget, Paperwork Reduction Project (0704-0188), Washington, DC.

Finneran, K.T., M.E. Housewright, and D. Lovley (2002). Multiple influences of nitrate on uranium solubility during bioremediation of uranium-contaminated subsurface sediments. *Environ. Microbiol.* **4**:510–516.

Fruchter, J.S., J.E. Amonette, C.R. Cole, Y.A. Gorby, M.D. Humphrey, J.D. Istok, K.B. Olsen, F.A. Spane, J.E. Szecsody, S.S. Teel, V.R. Vermeul, M.D. Williams, and S.B. Yabusaki (1996). *In Situ Redox Manipulation Field Injection Test Report—Hanford 100 H Area*, PNNL-11372/UC-602. Pacific Northwest National Laboratory, Richland, WA.

Gatliff, E.G. (1994). *Remediation* **8**:343–352.

Gandhi, S., B.-T. Oh, J.L. Schnoor, and P.J.J. Alvarez (2002). Degradation of TCE, Cr(VI), sulfate, and nitrate mixtures by granular iron in flow-through columns under different microbial conditions. *Water Res.* **36**(8):1973–1982.

Gentry, T.J., C. Rensing, and I.L. Pepper (2004). New approaches for bioaugmentation as a remediation technology. *Crit. Rev. Environ. Sci. Technol.* **34**(5):447–494.

Gilbert, S. and D.E. Crowley (1997). Plant compounds that induce polychlorinated biphenyl biodegration by *Arthrobacter* sp. strain BIB. *Appl. Environ. Microbiol.* **63**:1933.

Gorby, Y.A. and D.R. Lovley (1992). Enzymatic uranium precipitation. *Environ. Sci. Technol.* **30**:205–207.

Grindstaff, M. (1998). *Bioremediation of Chlorinated Solvent Contaminated Groundwater.* U.S. EPA Technology Innovation Office, Washington, DC.

Groundwater Services, Inc. (2001). Hydrogen Biosparging to Stimulate and Sustain In Situ Dechlorination Activity, Final Report, Launch Complex 15, Cape Canaveral Air Station, Florida. GSI Job No. G-2050. Air Force Center for Environmental Excellence (AFCEE), Contract No. F41624-97-C-8020, CDRL No. A006.

Hensse, M.J.C., A.W. van der Werf, S. Keuning, C. Hubach, R. Blokzijl, E. van Keulen, B. Alblas, C. Haasnoot, H. Boender, and E. Meijerink (2001). Engineered full scale bioremediation of chlorinated ethylenes. In *Sixth International In Situ and On-Site Bioremediation Symposium*, San Diego, CA. Battelle Press, Columbus, OH.

Hinchee, R.E. (Ed.) (1994a). *Air Sparging for Site Remediation.* Lewis Publishers, Boca Raton, FL.

Hinchee, R.E. (1994b). Bioventing of petroleum hydrocarbons. In *Handbook of Bioremediation*, R.D. Norris, R.E. Hinchee, et al. (Eds.). Lewis Publishers, Boca Raton, FL.

Hinchee, R.E. and D.C. Downey (1988). The role of hydrogen peroxide in enhanced bioreclamation. In *Proceedings, NWWA/API Conference on Petroleum Hydrocarbons and Organic Chemicals in Ground Water: Prevention, Detection and Restoration*, November 9–11, Houston, TX, pp. 931–948.

Holbrook, T.B., D.H. Bass, P.M. Boersma, J.J. Eisenbeis, N.J. Hutzler, and E.P. Roberts (1998). *Innovative Site Remediation Technology—Design & Application: Vapor Extraction and Air Sparging*, Vol. 7, W.C. Anderson (Ed.). American Academy of Environmental Engineers, Washington, DC.

Hughes, J.B., K.L. Duston, and C.H. Ward (2002). *Engineered Bioremediation. Technology.* Evaluation Report TE-02-03. Ground-Water Remediation Technologies Analysis Center, Pittsburgh, PA.

Imamura, T., S. Kozaki, A. Kuriyama, M. Kawaguchi, Y. Touge, T. Yano, E. Sugawa, Y. Kawabata, H. Iwasa, A. Watanabe, M. Iio, and Y. Senshu (1997). Inducer-free microbe for TCE degradation and feasibility study in bioaugmentation. In *Fourth International In Situ and On-Site Bioremediation Symposium*, New Orleans, LA. Battelle Press, Columbus, OH.

Istok, J.D., J.M. Senko, L.R. Krumholz, D. Watson, M.A. Bogle, A. Peacock, Y.J. Chang, and D.C. White (2004). In situ bioreduction of technetium and uranium in a nitrate-contaminated aquifer. *Environ. Sci. Technol.* **38**:468–475.

Kamath, R., J.L. Schnoor, and P.J.J. Alvarez (2004). Effect of root-derived substrates on the expression of *nah-lux* genes in *Pseudomonas fluorescens* HK44. Implications for PAH biodegradation in the rhizosphere. *Environ. Sci. Technol.* **38**:1740–1745.

Kao, C.-M. and R.C. Borden (1997). Enhanced TEX biodegradation in nutrient briquet–peat barrier system. *J. Environ. Eng.* **123**(1):18–24.

Kawabata, Y., H. Iwasa, A. Watanabe, M. Iio, and Y. Senshu (1997). Inducer-free microbe for TCE degradation and feasibility study in bioaugmentation. In *Fourth International In Situ and On-Site Bioremediation Symposium*, New Orleans, LA. Battelle Press, Columbus, OH.

Koenigsberg, S.S. and R.D. Norris (Eds.) (1999). *Accelerated Bioremediation Using Slow Release Compounds—Selected Battelle Conference Papers: 1993–1999*. Regenesis Bioremediation Products, San Clemente, CA.

Lendvay, J., P. Adriaens, M. Barcelona, C.L. Major, Jr., J. Tiedje, M. Dollhopf, F. Loffler, B. Fathepure, E. Petrovskis, M. Gebhard, G. Daniels, R. Hickey, R. Heine, and J. Shi (2001). Preventing contaminant discharge to surface waters: plume control with bioaugmentation. In *Sixth Internation In Situ and On-Site Bioremediation Symposium*, San Diego, CA. Battelle Press, Columbus, OH.

Liu, C.X., Y.A. Gorby, J.M. Zachara, J.K. Fredrickson, and C.F. Brown (2002). Reduction kinetics of Fe(III), Co(III), U(VI), Cr(VI), and Tc(VII) in cultures of dissimilatory metal-reducing bacteria *Biotechnol. Bioeng.* **80**:637–649.

MacDonald, J. (2000). Evaluating natural attenuation for groundwater cleanup. *Environ. Sci. Technol.* **34**:346A–353A.

Madsen, E.L. (1998). Epistemology of environmental microbiology. *Environ. Sci. Technol.* **32**(4):429–439.

Major, D.W., M.L. McMaster, E.E. Cox, B.J. Lee, E.E. Gentry, E. Hendrickson, E. Edwards, and S. Dworatzek (2001). Successful field demonstration of bioaugmentation to degrade PCE and TCE to ethene. In *Sixth International In Situ and On-Site Remediation Symposium*, San Diego, CA. Battelle Press, Columbus, OH.

Manning, J.F. Jr., R. Boopathy, and E.R. Breyfogle (1996). Field Demonstration of Slurry Reactor Biotreatment of Explosives-Contaminated Soils. Report No. SFIM-AEC-ET-CR-96178. U.S. Army Environmental Center.

McCarty, P.L., M.N. Goltz, G.D. Hopkins, M.E. Dolan, J.P. Allan, B.T. Kawakami, and T.J. Carrothers (1998). Full-scale evaluation of in situ cometabolic degradation of trichloroethylene in groundwater through toluene injection. *Environ. Sci. Technol.* **32**:88–100.

McCutcheon, S. and J.L. Schnoor (Eds.) (2003). *Phytoremediation. Transformnation and Control of Contaminants*. John Wiley & Sons, Hoboken, NJ.

Miya, R.K. and M.K. Firestone (2000). Enhanced phenanthrene biodegradation in soil by slender oat root exudates and root debris. *J. Environ. Quality* **29**:584.

National Research Council (1993). *In Situ Bioremediation: When Does It Work?* National Academy Press, Washington, DC.

National Research Council (1994). *Alternatives for Ground Water Cleanup*. National Academy Press, Washington, DC.

National Research Council (1997a). *Barrier Technologies for Environmental Management*. National Academy Press, Washington, DC.

National Research Council (1997b). *Innovations in Ground Water and Soil Cleanup: From Concept to Commercialization*. National Academy Press, Washington, DC.

National Research Council (1999). *Groundwater & Soil Cleanup: Improving Management of Persistent Contaminants*. National Academy Press, Washington, DC.

National Research Council (2000). *Natural Attenuation for Groundwater Remediation*. National Academy Press, Washington, DC.

Newell, C.J. and J.A. Connor (1998). Characteristics of dissolved petroleum hydrocarbon plumes: results from four studies. In *Proceedings of the 1998 Petroleum Hydrocarbons and Organic Chemicals in Ground Water: Prevention, Detection and Remediation. Conference and Exposition*, Houston, Texas, pp. 51–59.

Newell, C.J., P.E. Haas, J.B. Hughes, and T.A. Khan (2000). Results from two direct hydrogen delivery field tests for enhanced dechlorination. In *Bioremediation and Phytoremediation of Chlorinated and Recalcitrant Compounds*, G.B. Wickramamayake, A.R. Gavaskar, B.C. Alleman, and V.S. Magar (Eds.). Battelle Press, Columbus, OH pp. 31–37.

Oh, B.-T., C.L. Just, and P.J.J. Alvarez (2001). Hexahydro-1,3,5-trinitro-1,3,5-triazine (RDX) mineralization by zero-valent iron and mixed anaerobic cultures. *Environ. Sci. Technol.* **35**(21):4341–4346.

Parsons Engineering Science, Inc. (1999). Natural Attenuation of Fuel Hydrocarbons Performance and Cost Results from Multiple Air Force Demonstration Sites Technology Demonstration—Technical Summary Report. Prepared for Air Force Center for Environmental Excellence Technology Transfer Division, Brooks Air Force Base, Texas.

Raymond, R.L. (1974). Reclamation of hydrocarbon contaminated groundwater. U.S. Patent. 3,846,290.

Raymond, R.L., R.A. Brown, R.D. Norris, and E. T. O'Neill (1986). Stimulation of biooxidation processes in subterranean formation. U.S. Patent. 4,588,505.

Rittmann B.E. and P.L McCarty (2001). *Environmental Biotechnology: Principles and Applications*. McGraw Hill, New York.

Robertson, W.D. and J.A. Cherry (1995). In situ denitrification of septic-system nitrate using reactive porous media barriers: field trials. *Ground Water* **33**(1):99–111.

Salanitro, J.P., P.C. Johnson, G.E. Spinnler, P.M. Maner, H.L. Wisniewski, and C. Bruce (2000). Field-scale demonstration of enhanced MTBE bioremediation through aquifer bioaugmentation and oxygenation. *Environ. Sci. Technol.* **34**:4152–4162.

Scherer, M.M., S. Richter, R.L. Valentine, and P.J.J. Alvarez (2000). Chemistry and microbiology of permeable reactive barriers for in situ groundwater cleanup. *Crit. Rev. Environ. Sci. Technol.* **30**:363–411.

Schipper, L. and M. Vojvodič-Vukovič (1998). Nitrate removal from groundwater using a denitrification wall amended with sawdust: field trial. *J. Environ. Qual.* **27**(3): 664–668.

Schnoor, J.L. (1997). Phytoremediation, Technology Evaluation Report TE-98-01. Groundwater Remediation Technologies Analysis Center (GWRTAC), Pittsburgh, PA.

Schnoor J.L., L.A. Licht, S.C. McCutcheon, N.L. Wolfe, and L.H. Carriera (1995). Phytoremediation of organic and nutrient contaminants. *Environ. Sci. Technol.* **29**:318A–323A.

Siciliano, S.D., J.J. Germida, K. Banks, and C.W. Greer (2003). Changes in microbial community composition and function during a polyaromatic hydrocarbon phytoremediation field trial. *Appl. Environ. Microbiol.* **69**:483.

Spain, J. (1997). Synthetic chemicals with potential for natural attenuation. *Bioremediation. J.* **1**:1–9.

Steffan, R.J., K.L. Sperry, M.T. Walsh, S. Vainberg, and C.W. Condee (1999). Field-scale evaluation of in situ bioaugmentation for remediation of chlorinated solvents in groundwater. *Environ. Sci. Technol.* **33**:2771–2781.

Thomas, J.M., C.L. Bruce, V.R. Gordy, K.L. Duston, S.R. Hutchins, J.L. Sinclair, and C.H. Ward (1997). Assessment of the microbial potential for nitrate-enhanced bioremediation of a JP-4 fuel contaminated aquifer. *J. Indus. Microbiol.* **18**:152–160.

Thompson, P.L., L.A. Ramer, and J.L. Schnoor (1998). Uptake and transformation of TNT by hybrid poplar trees. *Environ. Sci. Technol*, **32**(7): pp. 975–980.

Till, B.A., L.J. Weathers, and P.J.J. Alvarez (1998). Fe(0)-supported autotrophic denitrification. *Environ. Sci. Technol.* **32**(5):634–639.

TRW Inc. (1997). Windrow Composting to Treat Explosives-Contaminated Soils at Umatilla Army Depot Activity (UMDA), Hermiston, Oregon (full-Scale Remediation). Report No. SFIM-AEC-ET-CR-96184. U.S. Army Environmental Center, Aberdeen Proving Ground, MD. aec.army.mil

Tullis, D., P.E. Prevost, and P. Kostecki (1998). Study points to new trends in use of alternative technologies at LUST sites. *Soil Groundwater Cleanup* **7**:12–17.

U.S. Environmental Protection Agency (1985). How to Evaluate Alternative Cleanup Technologies for Underground Storage Tank Sites: A Guide for Corrective Action Plan Reviewers. EPA 510-B-94-003 and EPA 510-B-95-007.

U.S. Environmental Protection Agency (2000). Engineered Approaches to In Situ Bioremediation of Chlorinated Solvents: Fundamentals and Field Applications. EPA-542-R-00-008.

U.S. Environmental Protection Agency (2001). Remediation Technology Cost Compendium—Year 2000. EPA-542-R-01-009.

Van Eyk, J. (1997). *Petroleum Bioventing*. A.A. Balkema, Rotterdam.

Vogel, T.M. and M.V. Walter (2002). Bioaugmentation. In *Manual of Environmental Microbiology*, 2nd ed., C.J. Hurst, R.L. Crawford, G.R. Knudsen, M.J. McIrney, and L.D. Stezenback (Eds.). ASM Press, Washington, DC.

Vol'nov, I.I. (1966). *Peroxides, Superoxides, and Ozonides of Alkali and Alkaline Earth Metals*. Plenum Press, New York.

Ward, C.H., J.M. Thomas, S. Fiorenza, H.S. Rifai, P.B. Bedient, J.T. Wilson, and R.L. Raymond (1989). In situ bioremediation of subsurface material and ground water contaminated with aviation fuel: Traverse City, Michigan. In *Proceedings of 1989 A&WMA/ EPA International Symposium on Hazardous Waste Treatment: Biosystems for Pollution Control*, Cincinnati, OH.

Waybrant, K.R., D.W. Blowes, and C.J. Ptacek (1998). Selection of reactive mixtures for use in permeable reactive walls for treatment of mine drainage. *Environ. Sci. Technol.* **32**(13):1972–1979.

White, D.M., R.L. Irvine, and C.W. Woodlard (1998). The use of solid peroxides to stimulate growth of aerobic microbes in tundra. *J. Hazard. Mater.* **57**:71–78.

Wickramanayake, G.B. and R.E. Hinchee (1998). *The First International Conference on Remediation of Chlorinated and Recalcitrant Compounds*, Monterey, CA, May 18–21. Battelle Press, Columbus, OH.

   C1-1 *Risk, Resource, and Regulatory Issues: Remediation of Chlorinated and Recalcitrant Compounds.*

   C1-2 *Nonaqueous-Phase Liquids: Remediation of Chlorinated and Recalcitrant Compounds.*

   C1-3 *Natural Attenuation: Remediation of Chlorinated and Recalcitrant Compounds.*

   C1-4 *Bioremediation and Phytoremediation: Remediation of Chlorinated and Recalcitrant Compounds.*

   C1-5 *Physical, Chemical, and Thermal Technologies: Remediation of Chlorinated and Recalcitrant Compounds.*

   C1-6 *Designing and Applying Treatment Technologies: Remediation of Chlorinated and Recalcitrant Compounds.*

Wickramanayake, G.B., A.R. Gavaskar, M.E. Kelley, and K.W. Nehring (2000). *The Second International Conference on Remediation of Chlorinated and Recalcitrant Compounds*, Monterey, CA, May 22–25. Battelle Press, Columbus, OH.

C2-1 *Risk, Regulatory, and Monitoring Considerations: Remediation of Chlorinated and Recalcitrant Compounds.*

C2-2 *Treating Dense Nonaqueous-Phase Liquids (DNAPLs): Remediation of Chlorinated and Recalcitrant Compounds.*

C2-3 *Natural Attenuation Considerations and Case Studies: Remediation of Chlorinated and Recalcitrant Compounds.*

C2-4 *Bioremediation and Phytoremediation of Chlorinated and Recalcitrant Compounds.*

C2-5 *Physical and Thermal Technologies: Remediation of Chlorinated and Recalcitrant Compounds.*

C2-6 *Chemical Oxidation and Reactive Barriers: Remediation of Chlorinated and Recalcitrant Compounds.*

C2-7 *Case Studies in the Remediation of Chlorinated and Recalcitrant Compounds.*

Zhang, C., J.B. Hughes, S.F. Nishino, and J.C. Spain (2000). Slurry-phase biological treatment of 2,4-dinitrotoluene and 2,6-dinitrotoluene: role of bioaugmentation and effects of high dinitrotoluene concentration. *Environ. Sci. Technol.* **34**:2810–2816.

9

PERFORMANCE ASSESSMENT AND DEMONSTRATION OF BIOREMEDIATION AND NATURAL ATTENUATION

"In the end, it will be OK. If it is not OK, it is not the end."

—*Anonymous*

9.1. INTRODUCTION TO ENVIRONMENTAL FORENSIC ANALYSIS

Environmental forensics is the systematic investigation of a contaminated site or an event that has impacted the environment (Stout et al. 1998). The purpose of such investigations can range from liability assessment to evaluating the performance of a given remedial technology. In this chapter we introduce forensic techniques to quantify the efficacy of bioremediation and demonstrate natural attenuation, including the characterization of relevant biogeochemical processes. Both traditional assessment techniques recommended by regulatory agencies as well as new approaches that are rapidly moving from the laboratory to the field will be discussed, including the use of stable isotope, chemical fingerprinting, and molecular techniques (Table 9.1).

Monitoring requirements and site characterization efforts necessary to demonstrate the efficacy of natural attenuation are generally more comprehensive than those needed to evaluate other active remediation options. In the United States, the Environmental Protection Agency (EPA) directive (and the ASTM standard developed for the remediation industry) requires essentially the same lines of evidence to support bioremediation by natural attenuation. These lines of evidence are as follows (EPA 1999):

Bioremediation and Natural Attenuation: Process Fundamentals and Mathematical Models
By Pedro J. J. Alvarez and Walter A. Illman

Table 9.1. Comparison of Established and Emerging Techniques to Assess Natural Attenuation

Method of Investigation	Applicability	Comments
Graphical and statistical analyses, analytical and numerical models	Established	Determine whether plume is stabilizing, increasing, or decreasing in size.
Geochemical parameters	Established	Observe contaminant trends, daughter product formation, and presence of geochemical footprints to indicate biogeochemical process responsible for biodegradation.
Push–pull tests	Established	Single-well tracer tests used to evaluate in situ degradation rates and metabolite formation.
Microcosm studies	Established	Observe presence of microorganisms, protozoan predators, cellular biomarkers, nucleic acids and phospholipids; determination of cellular contents of ribosomes, intracellular energy reserves, nutritional status; measure cell growth and uptake of physiological substrates, respiratory activity from the reduction of dyes; determination of novel organisms from DNA sequencing; compute relative abundances of each phylogenetic group; observe unique intermediary metabolites indicative of biodegradation.
Stable isotope analysis	Emerging but in greater use	Demonstrate biodegradation of organic compounds; quantify microbial activities in situ; quantify relative extent of biodegradation between zones of contaminant plume.
Chemical fingerprinting	Emerging but in greater use	Identification of fuel types, determination of contaminant sources that will assist in the mapping of subsurface contamination and backward in time contaminant transport simulations.
Molecular techniques	Emerging	Used to increase understanding of biochemical reactions and their mechanisms; techniques allow one to obtain answers to questions such as: Why are microorganisms there? What are microorganisms doing? When are the microorganisms active? It also likely will provide a comprehensive mechanistic understanding of processes that regulate the genes that encode the proteins that actually degrade the contaminants.

1. *Primary Lines of Evidence.* "Historical groundwater and/or soil chemistry data that demonstrate a clear and meaningful trend of decreasing contaminant mass and/or concentration over time at appropriate monitoring or sampling locations. In the case of a groundwater plume, decreasing concentration should not be solely the result of plume migration. In the case of inorganic contaminants, the primary attenuating mechanism should also be understood." The primary lines of evidence are considered to be *direct* and can be used also to determine whether the plume is stable, receding, or growing.

2. *Secondary Lines of Evidence.* "Hydrogeologic and geochemical data that can be used to demonstrate indirectly the type(s) of natural attenuation processes active at the site, and the rate at which such processes will reduce contaminant concentrations to required levels." Secondary lines of evidence are often used to demonstrate specific natural attenuation *mechanisms.* For example, biodegradation can be demonstrated using geochemical indicators such as stoichiometric consumption of electron acceptors, detection of intermediate metabolites, and increased ratio of nondegradable to degradable components. Other geochemical parameters such as dissolved oxygen, nitrate, iron(II), sulfate, methane, and carbon dioxide can provide information on the degradation mechanisms and their rates. The secondary lines of evidence provide *indirect* evidence for natural attenuation that is taking place at the site and its rate.

3. *Tertiary Lines of Evidence.* "Data from field or microcosm studies (conducted in or with actual contaminated site media) which directly demonstrate the occurrence of a particular natural attenuation process at the site and its ability to degrade the contaminants of concern (typically used to demonstrate biological degradation processes only)." The tertiary evidence is considered to be *direct* evidence because it involves the use of field or microcosm studies that demonstrate directly the biodegradation processes and mechanisms that are destructing the contaminants of concern. Currently, microbiological studies are conducted using samples taken from the field and incubated at conditions similar to the field. From this, rates of contaminant destruction or immobilization are inferred. These studies can provide estimates of potential metabolic activity. However, they do not provide detailed insight into microorganisms and the biogeochemical reactions that are causing bioremediation. This could, however, change in the near future, through the concurrent use of novel techniques such as stable isotope analysis, chemical fingerprinting, and molecular techniques to identify specific organisms and mechanisms responsible for biodegradation.

Collection of both primary and secondary lines of evidence at a given site is recommended to support a decision to use monitored natural attenuation (EPA 1999). However, if the overseeing regulatory agency determines that there are sufficient historical data to characterize the site, then secondary lines of evidence may not be necessary. The quantity and type of information that need to be collected at a given site are determined on a site-to-site basis. Site-specific factors such as the size and nature of the problem, the potential risk to receptors, the type of contaminants involved, and the complexity in hydrogeologic conditions drive the quantity and type of information that are collected. The tertiary lines of evidence may become necessary when the first two lines of evidence are inadequate or inconclusive.

9.2. LINES OF EVIDENCE OF INTRINSIC BIOREMEDIATION

9.2.1. Direct Evidence of Contaminant Mass Loss in the Field

To demonstrate a clear and meaningful trend of decreasing contaminant mass and/or concentration over time at the field scale, data from several monitoring or sampling locations must be collected over several time periods. These data analyses must show that decreasing groundwater concentrations are not solely due to plume migration, dispersion, and dilution. In the case of inorganic contaminants, the primary attenuating mechanism should also be understood. An important aspect in such contaminant hydrogeology analyses is to ensure that the contaminant concentrations are measured in representative samples (Keith 1986). A thorough investigation should obtain multiple samples at a given point in space and time and conduct a thorough error analysis to calculate confidence intervals of collected data. This aspect is beyond the scope of this book, but there are excellent resources that discuss statistical and error analyses in greater detail (e.g., Gilbert 1987; Berthouex and Brown 1994; Gibbons and Coleman 2001; see also Example 9.1). We recommend the reader to peruse these references and incorporate error analysis as part of the laboratory and field investigation.

The first step in documenting the direct evidence of contaminant mass loss in the field is to develop an initial site conceptual model (Chapter 6). At this stage, at which minimal data is available, the focus is on formulating a basic conceptual model that utilizes techniques presented in Chapter 7 to determine the hydrogeological conditions at the site, including the direction of groundwater flow, background concentrations, and the location of the contaminant plume. Once an initial conceptual model is formulated for the site, groundwater monitoring strategies may be developed based on the preliminary information. As discussed in Chapter 6, the formulation of a conceptual model is an iterative process with increased availability of information allowing for the updating of the model. The differences between a conceptual model and a site analytical or numerical model are also discussed in Chapter 6. Briefly, a conceptual model is an evolving hypothesis identifying the important features, processes, and events controlling fluid flow and contaminant transport of consequence at a specific field site in the context of a recognized problem (National Research Council 2001). A conceptual model may also be a mental construct or hypothesis accompanied by verbal, pictorial, diagrammatic, and/or tabular interpretations and representations of site conditions as well as corresponding flow and transport dynamics (Neuman and Wierenga 2003). The main difference between a site conceptual model and analytical or numerical model is that the latter is a quantitative representation of the former.

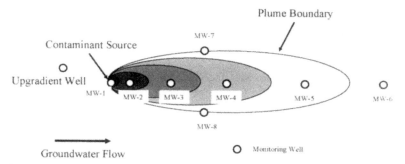

Figure 9.1. Recommended groundwater monitoring network. (Adapted from McAllister and Chiang 1994.)

Once a preliminary site conceptual model is formulated, decisions need to be made to allocate resources for additional characterization efforts. The goal here is to document contaminant mass loss; hence, detection of changes in contaminant concentration both spatially and temporally will depend heavily on the proper location and construction of monitoring wells. The location of monitoring wells is important in all remediation efforts, but in the case of natural attenuation, this takes on greater importance because contaminant migration is controlled mainly by advective transport provided by ambient groundwater flow. The installation of monitoring wells to adequately quantify contamination concentrations is very important in determining the physical, chemical, and biological contributions to mass destruction. The location, number, and other pertinent data on monitoring wells to evaluate natural attenuation are site-specific problems as the network design will be determined by plume size, site complexity, source strength, groundwater–surface water interactions, distance to receptors, and the confidence limits parties involved wish to place on the data (Azadpur-Keeley et al. 2001). One suggested monitoring network design by McAllister and Chiang (1994) is shown in Figure 9.1. It requires the monitoring of at least two well transects throughout the plume, one in the direction of groundwater flow and another in the transverse direction to flow. At least one well should be placed upgradient of the source to obtain background concentrations as well as beyond the downgradient edge of the plume to bracket the region of influence.

Using a monitoring network such as the one shown in Figure 9.1, different groundwater monitoring strategies are employed at various stages of remediation:

- *Site characterization* to describe disposition of contamination and forecast its future behavior. Also used to set site-specific cleanup levels in accordance with the risk-based corrective action (RBCA) paradigm.
- *Validation monitoring* to determine whether predictions of site characterization are accurate (demonstrate performance according to expectations, ensure that there is no impact on downgradient receptors, detect any new releases that could impact effectiveness, detect changes in environmental conditions, etc.)

- *Long-term or performance monitoring* to ensure that plume behavior does not change. Required for as long as contamination remains above cleanup goals, and beyond.

It is important to recognize that the network suggested in Figure 9.1 is for a generic plume. We discussed in Chapter 4 that the subsurface is generally heterogeneous, and this is the rule rather than the exception, so plumes will take different shapes. Hence, the monitoring network will need to be customized for a given site. There are several issues that one must also consider in placing wells. For example, the longitudinal dispersivity of plumes is typically 10 times larger than the transverse dispersivity. This causes the plume to be elongated like a cigar; therefore, a sparse network of wells could potentially cause the plume to bypass a monitoring well network (see Chapter 4 for further details). Also, buried channels and backfill for infrastructure such as sewers, water mains, and cables may offer higher hydraulic conductivity and serve as preferential channels for contaminants to travel, causing the plume to miss the monitoring well network completely. These factors remind us to be vigilant and fully utilize the available hydrogeological information to construct a monitoring well network that will serve the interests of all parties.

The EPA (1999) stresses that performance monitoring is of even greater importance for monitored natural attenuation than for other types of remediation action due to the potential for ongoing contamination, potential for longer time frames to accomplish remediation objectives, and uncertainties associated with using MNA (Chapter 8). Once a sampling network is developed, there is a need to specify the sampling frequency, location, and sample types and measurements. In addition, all monitoring programs should be designed to accomplish the following goals (EPA 1999):

- "Demonstrate that natural attenuation is occurring according to expectations."
- "Detect changes in environmental conditions (e.g., hydrogeologic, geochemical, microbiological, or other changes) that may reduce the efficacy of any of the natural attenuation processes. Identify any potentially toxic and/or mobile transformation products."
- "Verify that the plume(s) is not expanding (either downgradient, laterally or vertically)."
- "Verify no unacceptable impact to downgradient receptors."
- "Detect new releases of contaminants to the environment that could impact the effectiveness of the natural attenuation remedy."
- "Demonstrate the efficacy of institutional controls that were put in place to protect potential receptors."
- "Verify attainment of remediation objectives."

Sampling frequency should be designed so that changes in conditions listed above can be detected. "Triggers" should be established to signal unacceptable performance (e.g., increasing concentrations indicative of new or renewed release, detection in sentinel wells outside the original plume boundary, concentrations not decreasing at a sufficiently rapid rate to meet remediation objective). The EPA (1999) also suggests that flexibility in sampling frequency should be built-in to the performance monitoring program. For example, when remediation is progressing well, the sampling frequency could be decreased; otherwise, the sampling frequency may have to be increased. The length of performance monitoring differs from one site to the next and hinges on whether site remediation objectives have been achieved or not. At some sites, long-term monitoring may be required to ensure that the contaminants do not pose a threat to humans and the environment.

9.2.2. Documentation of Contaminant Mass Loss

To analyze and document contaminant mass loss at a given site, we advocate the hierarchical approach proposed by the National Research Council (2000). There are four levels of data analysis that one can perform at a given site with the complexity of analysis increasing as we go down the list:

1. Graphical and statistical analyses of contaminant concentration trends.
2. Mass budgeting and mass flux analysis to track contaminant mass.
3. Analytical modeling of solute transport.
4. Numerical modeling of flow and solute transport.

There are a number of factors that determine the suitability of one approach over another at a given site. Factors such as the characteristics of the contaminants, site hydrogeology, and the anticipated risk associated with the spreading of the contamination are all considered to be important in determining the approach. Table 9.2 provides guidance in choosing the level of data analysis appropriate for a site under investigation. In general, simple analyses such as graphical and statistical analyses of data are sufficient for sites with uniform and simple hydrogeological conditions and when the contaminant concentrations are low. If the contaminants readily attenuate (Table 8.20), then a simple analysis is considered in most cases to be adequate. On the other hand, more detailed and complex analyses are required when the contaminant concentrations are high, when the contaminants are not readily biodegradable, and/or when the site hydrogeology is considered to be complex. The simple methods of analyses should also be performed when a suite of complex analyses are required at a given site. In such cases, a simple analysis should be conducted at the initial stages of the project, and as more data become available, more complex analyses should be performed.

9.2.3. Graphical and Statistical Analyses of Contaminant Concentration Trends

A plume stability analysis can be conducted to determine whether a particular plume is shrinking, stable, or expanding. There are several techniques that are

Table 9.2. Recommended Levels of Natural Attenuation Data Analysis for Different Contaminants and Site Conditions

Hydrogeology	Contaminant Characteristics				
	Biodegrades Under Most Conditions (e.g., BTEX)	Immobile Under Most Conditions (e.g., Pb)	Biodegrades Under Limited Conditions (e.g., Chlorinated Ethenes)	Immobile Under Some Conditions (e.g., Cr)	Mobile and Degrades or Decays Slowly (e.g., Tritium, MTBE)
Simple flow, uniform geochemistry, and low concentrations	Graphical and statistical analyses	Graphical and statistical analyses	Mass budgeting and mass flux analysis	Analytical solute transport model	Mass budgeting and mass flux analysis with numerical solute transport model
Simple flow, small-scale physical or chemical heterogeneity, and medium–high concentrations	Mass budgeting and mass flux analysis	Analytical solute transport model	Mass budgeting and mass flux analysis with analytical solute transport model	Analytical solute transport model	Numerical flow and solute transport models
Strongly transient flow, large-scale physical or chemical heterogeneity or high concentrations	Mass budgeting and mass flux analysis or analytical solute transport model	Mass budgeting and mass flux analysis or analytical solute transport model	Numerical flow and solute transport models	Numerical flow and solute transport models	Numerical flow and solute transport models

Notes: BTEX = benzene, toluene, ethyl benzene, and xylene; MTBE = methyl *tert*-butyl-ether. The table lists the recommended levels of natural attenuation data analysis for different contaminants and site conditions. In the site descriptions given along the left-hand side, the recommended data analysis strategy applies when all of the conditions are satisfied unless the term "or" is used. Data completeness and consistency are to be evaluated in all cases. All techniques listed in higher rows of the same column are to be applied, along with the methods in the applicable row. Where mixed concentrations are present, the most thorough analysis recommended for any single contaminant should be applied to the entire site.

Source: Adapted from National Research Council (2000).

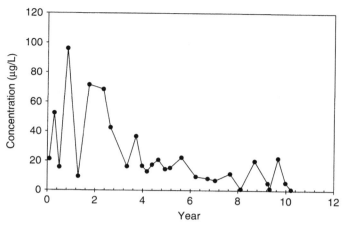

Figure 9.2. Concentration versus time for well GW-12A, manufacturing plant site, California. (Data from Buscheck et al. 1997.)

available, but here we briefly describe the ones that are most commonly used. The first approach utilizes concentration trends of contaminants obtained from monitoring wells over several time periods. Figure 9.2 shows an example of such a time series plot of TCE obtained from a well at a manufacturing plant site in California. It reveals that, at this particular location, contaminant concentrations are decreasing with time, which is evidence that contaminant mass loss is occurring. The detection and estimation of trends can be complicated because of problems associated with changes in procedures of sample collection or analytical techniques, seasonality, correlated data, and corrections for groundwater flow (Gilbert 1987).

Contaminant concentration can also be obtained and plotted in the direction of contaminant transport. For example, concentrations of contaminants can be obtained in the source zone and at a number of locations downgradient to the source. Such plots can be used to assess whether a plume is shrinking, stable, or expanding (see Section 7.3).

Other graphical methods include the preparation of contour maps of contaminant concentrations obtained through synoptic sampling. That is, concentrations are determined spatially on a given date and these data are contoured using various interpolation schemes to examine the spatial distribution of contaminant concentrations. Figure 9.3 shows contour plots of bromide, lithium, and molybdate at several transport times during a natural gradient tracer test conducted at a Cape Cod site in Massachusetts (LeBlanc et al. 1991). These data depict that lithium and molybdate are retarded relative to bromide, which is a conservative tracer. An important feature to notice about these plumes is that they are highly elongated and this may cause the plumes to miss monitoring wells if the spacing between the wells is large. Well spacing design is dependent on site characteristics and varies from one site to another, but typically ranges from several meters to several tens of meters.

Contour plots can also be produced over multiple dates to obtain the spatiotemporal coverage of the plume. This can provide additional insight into the

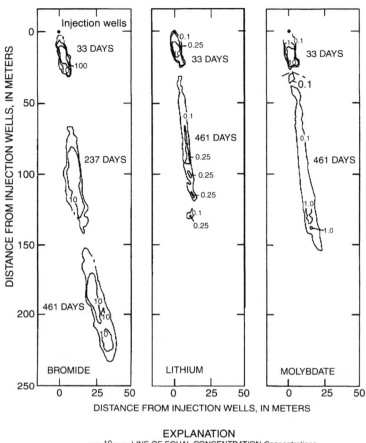

Figure 9.3. Contour plots of bromide, lithium, and molybdate at times during the natural gradient experiment conducted at a Cape Cod site in Massachusetts. (Adapted from LeBlanc et al. 1991.)

evolution of plume status over time. Users can employ commercial software (e.g., Tecplot (http://www.tecplot.com) and Surfer (http://www.goldensoftware.com/)) to build contour plots on the computer and load sequential images to produce animations of the evolving plume.

Contouring is a commonly used technique to study the spatial distribution of environmental as well as other variables. However, depending on the interpolation schemes used and the sampling grid, the resulting contour plot may be physically unrealistic. It is recommended that the investigator try out different interpolation schemes when constructing contour plots and see which ones make the best physical sense. It is also important to know that the investigator should not rely solely on contouring programs. Physical insight and common sense must accompany pictorial renditions of contaminant plumes.

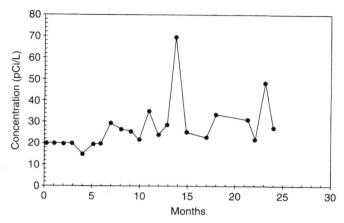

Figure 9.4. ^{238}U concentrations in groundwater in well E at the former St. Louis Airport storage site for January 1981 through January 1983. (Adapted from Gilbert 1987; original data from Clark and Berven 1984.)

As natural attenuation progresses, a decline in contaminant concentration should be observed, while an increase in daughter product concentration occurs. Evaluation of contaminant trends can determine whether the plume is decreasing, stable, or increasing in size. Therefore, establishing that there is a trend and that contaminant concentrations are indeed decreasing with time can be a powerful message to the stakeholders that natural attenuation is taking place.

We have discussed earlier that, sometimes, it is difficult to spot a trend in a time series plot of contaminant concentrations at a sampling point. Graphical analysis may provide a semiquantitative indication of plume trend. In many cases, such trends are not clearly visible and a trend may be a product of the analyst's perceptions. For example, Figure 9.4 is a time-series plot of ^{238}U concentrations obtained from a well at the former St. Louis Airport storage site for January 1981 through January 1983. Examination of the figure shows that a clear trend is not visible. In such a case, statistical analysis of trends in concentrations of contaminants and other substances is required to quantitatively show spatial and temporal trends in concentrations. The statistical analysis of this data to show whether there is a trend or not is presented in Example 9.1 and the reader is also referred to Chapter 7 as well as Appendix D for additional details.

Example 9.1. Statistical Analyses (Adapted from National Research Council 2000). The examples presented below show how statistical analyses can be used to analyze field data and to evaluate whether a trend in contaminant concentrations is apparent or not.

How representative are the measured contaminant concentrations?

Consider a site where five measurements of contaminant concentrations are available: 13.55, 6.39, 13.81, 11.20, and 13.88. The mean of the five measurements is

$\bar{x} = 11.77$, and the population standard deviation is $\sigma = 3.20$ (McBean and Rovers 1998). The question of how good the calculated mean value is based on the available measurements (i.e., how close $\bar{x}$ is to the true mean concentration) can be provided in terms of a "confidence interval," given by

$$CI = \bar{x} \pm \frac{t\sigma}{\sqrt{n}} = 11.77 \pm \frac{(2.70)(3.20)}{\sqrt{5}} = 11.77 \pm 3.99$$

The above calculation suggests that the mean value is likely between $11.77 - 3.99$ and $11.77 + 3.99$. The t-value obtained from Student's t-distribution can be obtained from any textbook on statistics or from computer packages (see also Appendix E). The confidence interval in the above example is relatively large but this can be narrowed by increasing the number of samples n.

Has a trend in contaminant concentration developed over time?

Figure 9.4 shows a time series of uranium-238 concentrations obtained from groundwater samples from the former St. Louis Airport storage site during the time period 1981–1983 (Clark and Berven 1984). Examination of the plot shows that there may be an upward trend in the concentration but this is unclear. Statistical techniques can be used to determine whether there is an increasing trend or not. A Mann–Kendall test (see Appendix D for additional details) for determining trends was used by Gilbert (1987) to analyze the data set. The test compares changes in signs between values collected at each sampling time and all others collected at a later time. The Mann–Kendall test is done by testing the null hypothesis of no trend to the alternative hypothesis that there is an upward trend. Gilbert (1987) showed that there is a 95% probability of a true upward trend through this statistical test.

9.2.4. Mass Budgeting and Mass Flux Analysis to Track Contaminant Mass

Mass Budgeting

Simple mass budgeting should probably be done at a given site to determine the mass of contaminants present and their degradation products. It is also very useful to conduct such a first-order analysis before one spends a considerable amount of time developing analytical and numerical models. It involves determining the mass balance of geochemical constituents of interest at a given site. That is, the procedure involves evaluating whether the rate at which geochemical footprint compounds are being produced is commensurate with the rate at which the contaminant is destroyed or immobilized (National Research Council 2000). The computed relative rates reveal the relative importance of natural attenuation processes occurring at the site. It is important to remember that mass balance models may provide an overall mass balance but will not yield spatial and temporal distributions of contaminants and their degradation products. They also do not predict the kinetics (speed) of a reaction (National Research Council 2000).

Consider a situation in which one-dimensional advection dominates contaminant transport under steady-state conditions. Further consider that inputs and outputs of contaminants and their metabolites are at steady state (not changing at a specific point over time). Under these conditions, the net mass-per-time reaction rate of a material within the domain is

$$\gamma = R(XYZ) = q\left(C_{\text{up}} - C_{\text{down}}\right)(YZ) \qquad [9.1]$$

where γ is the reaction rate in mass per time $[MT^{-1}]$, R is the concentration decrease rate $[ML^{-3}T^{-1}]$, XYZ is the cubic element for the budget calculation $[L^3]$, q is the Darcy velocity of groundwater flow $[LT^{-1}]$; C_{up} and C_{down} are the upstream and downstream concentrations of the material $[ML^{-3}]$, YZ is the area perpendicular to the flow direction $[L^2]$, and X is the length of the domain along the flow direction $[L]$. One important component of mass budgeting is to have all the rates on a common mass-per-time basis, $R(XYZ)$. Through the use of input and output rates to estimate $R(XYZ)$ for some contaminant, their electron acceptors, inorganic carbon, alkalinity, or any other constituent, their stoichiometric ratios can be computed and compared to what ought to occur if natural attenuation is acting (National Reseach Council 2000). The reaction rates can also be used to estimate the mass rates for reactions or transfers that cannot be measured directly in the field. These include the rate at which the contaminant is entering the groundwater from the source (National Research Council 2000).

Example 9.2 shows an example of mass budgeting derived from National Research Council (2000). The example shows how mass budgeting can be used to illustrate that biodegradation is causing contaminant destruction. It is important to recognize that the example illustrates the principles of mass budgeting; however, the method and calculations do not provide biological, physical, and chemical insight into how other processes, in particular geochemical reactions, might affect the calculations.

Example 9.2. Mass Budget Analysis to Determine the LNAPL Depletion Rate (Adapted from National Research Council 2000) Consider a situation in which a leak in an underground storage tank results in a LNAPL source zone at the top of the aquifer. Petroleum hydrocarbons consisting of BTEX and other constituents dissolve into the groundwater and a series of monitoring wells downgradient to the source zone reveal that the plume extends less than 46 m (150 ft) from the LNAPL source. The dissolution rate of the source forming the dissolved phase is unknown from available site records. Field sampling reveals that BTEX concentrations as high as 10 mg/L are detected within the plume. Geochemical measurements in upgradient wells indicate that O_2, NO_3^-, SO_4^{2-}, and CO_2 are available as electron acceptors, while the furthest downgradient well shows that BTEX, O_2, NO_3^-, and SO_4^{2-} are virtually absent, but Fe^{2+} and CH_4 appear. Field geochemical analysis also indicates that alkalinity and pH increase across the plume. Analysis of water table elevations and the available hydraulic conductivity measurements show that the groundwater velocity is 30 m/yr and the porosity is 0.25. Table 9.3 summarizes the upgradient and downgradient values of geochemical constituents of interest.

Stoichiometric relationships among the measured constituents and for the possible reactions can be used to assess whether biodegradation is occurring, causing the destruction of BTEX, and to estimate the NAPL depletion rate. Stoichiometric

Table 9.3. Field Measurements

Constituent	Upgradient	Downgradient	Change
BTEX, mg/L	0	0	0
O_2, mg/L	8	0.2	−7.8
NO_3^-, mg/L	7	0.1	−6.9
SO_4^{2-}, mg/L	9	1	−8.0
Fe^{2+}, mg/L	0	40	40
CH_4, mg/L	0	1	1
Alkalinity, mg/L as $CaCO_3$	10	130	120
pH	4.7	6.1	1.4
Total CO_2, mg/L as C	29	44	15

Table 9.4. Stoichiometric Ratios

Reaction	g C_7H_8/g acceptor	g CO_2-C/g acceptor	g Alkalinity as $CaCO_3$/g acceptor
Aerobic (O_2)	0.319 g C_7H_8/g O_2	−0.29 g C/g O_2	0 g as $CaCO_3$/g O_2
Denitrification (NO_3^- as N)	0.917 g C_7H_8/g N	−0.83 g C/g N	−3.57 g as $CaCO_3$/g N
Sulfate reduction (SO_4^{2-} as S)	0.637 g C_7H_8/g S	−0.53 g C/g S	−3.13 g as $CaCO_3$/g S
Iron reduction (Fe^{2+} generated)	−0.046 g C_7H_8/g Fe^{2+}	0.042 g C/g Fe^{2+}	1.79 g as $CaCO_3$/g Fe^{2+}
CH_4, mg/L	−1.28 g C_7H_8/g CH_4	0.42 g C/g CH_4	0 g as $CaCO_3$/g CH_4

ratios using C_7H_8 to represent BTEX are presented in Table 9.4 (see also Chapter 3, Figure 3.13).

The observed changes in the electron acceptors is combined with the ratios provided in the table of reactions to compute the predicted changes in BTEX, CO_2, and alkalinity based on the changes in the advecting acceptors, as shown in Table 9.5.

A comparison of computed changes in inorganic carbon and alkalinity (Table 9.5) with the changes observed in field measurements (Table 9.3) show that there is good agreement. That is, gains of approximately 15 mg/L in total CO_2, as mg/L as C, and 120 mg/L as $CaCO_3$ are seen. The mass balance calculations support that biodegradation causes BTEX destruction, because the intrinsic supply rates for the acceptors are consistent with the observed footprint measures.

The total BTEX biodegradation computed in Table 9.5 is 17 mg/L of C_7H_8. One can then calculate the BTEX depletion rate using the groundwater velocity of 30 m/yr and plume cross section with an area of 10 m by 2 m deep as

$$17\frac{mg}{L} \times 10\,m \times 2\,m \times 0.25 \times 10^3 \frac{L}{m^3} \times 10^{-3} \frac{g}{mg} \times 30\frac{m}{yr} = 2550\frac{g\ C_7H_8}{yr}$$

It is important to note that the majority of inorganic carbon generation and BTEX degradation results from denitrification and sulfate reduction (see

Table 9.5. Computed Changes in BTEX (as C_7H_8), Inorganic Carbon (as C), and Alkalinity (as $CaCO_3$)

	Computed Changes			
Reaction	Observed Change in Acceptor Concentration (mg/L)	Total CO_2 (mg C/L)	Alkalinity (mg as $CaCO_3$/L)	BTEX (mg as C_7H_8/L)
Aerobic (O_2)	−7.8	2.3	0	2.5
Denitrification (NO_3^- as N)	−6.9	5.7	24.6	6.3
Sulfate reduction (SO_4^{2-} as S)	−8	4.6	25.0	5.1
Iron reduction (Fe^{3+})	+40	1.7	71.6	1.8
Methanogenesis (CH_4)	+1	0.4	0	1.3
Total		14.7	121.2	17.0

Table 9.5). However, the majority of the alkalinity gain results from iron(III) reduction. One must recognize that the natural source of ferric iron (iron oxide solids) could become depleted over time, making these calculations invalid.

A mass balance approach can be useful in obtaining a first-order approximation to whether biodegradation is occurring or not. It may also provide the corresponding rate constants. However, in many field problems advection does not dominate the transport process of geochemical constituents. The transfer of constituents among various phases can add or remove contaminants, which can invalidate the analysis. Some examples include (National Research Council, 2000):

- Transfer of oxygen from the soil air.
- Dissolution of calcareous materials.
- Transfer of volatile compounds to the gas phase, and adsorption of hydrophobic compounds to aquifer solids.

These nonadvective inputs (or outputs), depending on their importance with respect to advection, may complicate the evaluation using this approach. Also, it is important to keep in mind that when the contaminant's input rate exceeds the natural attenuation rate, then the contaminant plume will continue to grow until the source is depleted.

Mass Flux Analysis

Another method to evaluate whether a plume is shrinking, stable, or expanding is to estimate the mass fluxes of contaminants at specified transects. If the total mass flux, that is, the amount of contaminant mass migrating through cross sections of

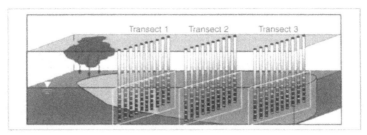

Figure 9.5. Investigation of a contaminant plume using well transects to determine whether a plume is shrinking, stable, or expanding where mass flux is defined according to Equation 9.2 (Adapted from American Petroleum Institute, 2003).

the aquifer perpendicular to groundwater flow, decreases over time, then it is one indication that natural attenuation is taking place (Figure 9.5). EPA (1999) also considers the mass-based approach as a key consideration in the evaluation of natural attenuation.

We summarize below the methodology developed by Einarson and MacKay (2001a,b; see also American Petroleum Institute 2003) to estimate contaminant mass flux using well transects. Contaminant mass flux can be estimated using sampling transects consisting of either traditional monitoring wells or multilevel samplers discussed in Chapter 7. The pros and cons of each type of sampling approach are also discussed in the chapter. An array of monitoring wells or multi-level samplers should be positioned perpendicular to the flow path of the contaminant plume (Figure 9.5), with the transect extending both horizontally and vertically beyond the extent of the contaminant plume. Under natural gradient conditions, the transect will be a straight line but in a converging or divergent flow field, the transect will be a curve (Einarson and MacKay 2001b). Groundwater samples are collected from each monitoring point and analyzed for the contaminant that is under scrutiny. One also calculates the specific discharge using the local hydraulic conductivity and hydraulic gradient (see Chapter 7 on how to calculate this). A cross section of the subsurface is made at the location of the transect and a contour plot of concentrations is made. Polygons are defined based on the cross section and contour plot. The total contaminant flux is then calculated using

$$M_d = \sum_{i=1}^{n} M_{di} = \sum_{i=1}^{n} C_i A_i q_i \qquad [9.2]$$

where M_d is the total mass flux from the source zone $[MT^{-1}]$, M_{di} is the contaminant mass flux through polygon i $[MT^{-1}]$, C_i is the contaminant concentration within the flow area of polygon i $[ML^{-3}]$, A_i is the cross-sectional area of polygon i $[L^2]$, and q_i is the specific discharge through polygon i $[LT^{-1}]$. As discussed in Chapter 4, the specific discharge can be calculated using Eq. [4.14]. Here, we emphasize that groundwater velocity is estimated from the hydraulic gradient and hydraulic conductivity measured for the part of the site where the polygon is located. It is important that one recognizes that in these calculations the retardation of contaminants is not taken into account. One reason for this is that mass flux will be higher if the groundwater velocity is used instead of contaminant velocity subject to retardation. Not taking into account retardation can be regarded as being more conservative in calculating contaminant mass flux. In addition, sorption sites available to contaminants could also be filled after the plume flows over an extended period, causing the contaminants to be transported at the same velocity as the groundwater.

Einarson and MacKay (2001b) present an example in which they applied this technique to estimate contaminant mass flux of cis-1,2-DCE flowing in a dissolved contaminant plume at Site 1, Alameda Point, California. Figure 9.6a shows contours of cis-1,2-DCE concentrations measured along the transect. The concentrations

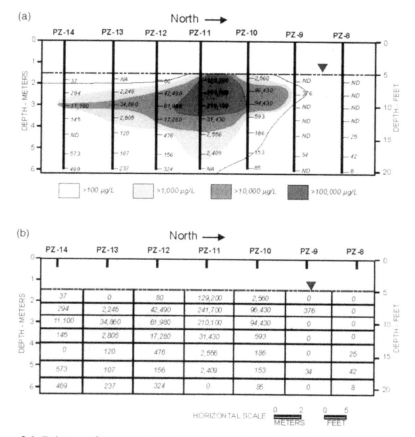

Figure 9.6. Estimates of contaminant M_d can be obtained by sampling a transect of multilevel monitoring points installed perpendicular to the axis of the dissolved plume: (a) contours of cis-1,2-DCE concentrations measured along the transect; (b) discrete concentration measurements are assigned to rectangular cells centered around each monitoring point. The mass discharge within each cell (M_{di}) is calculated by multiplying the concentration value by the flux of groundwater through the cell. Total contaminant M_d is obtained by summing the individual M_{di} values. (Adapted from Einarson and MacKay 2001b.)

were obtained from groundwater samples taken from a seven-zone multilevel sampler installed at the site. Each concentration value obtained from the sampling point is assigned to a polygon (in this case rectangular cells) with the sampling point located at the center of the polygon. The contaminant mass flux in each polygon is then calculated by multiplying the averaged contaminant concentration, with the area of the polygon and the specific discharge. The total contaminant mass flux is obtained by summing the individual M_{di} values. Barker et al. (1999) determined an average hydraulic conductivity value of 6.6×10^{-3} cm/s for the unit under consideration. Together with the measured horizontal hydraulic gradient of 0.0029, the calculated cis-1,2-DCE M_d is approximately 31 g/day (Einarson and MacKay 2001b).

9.2.5. Analytical Solute Transport Modeling of Solute Transport

We have shown in Chapter 5 that analytical models can be very useful tools to evaluate contaminant transport in groundwater and to characterize potential exposure pathways in risk assessment efforts. The key difference between the mass budgeting approach and analytical solute transport models is that an analytical model is based on mass balances that incorporate processes such as advection, dispersion, chemical reactions, and biodegradation of the target contaminants as a function of time. It allows for one to predict in both space and time the concentrations of a particular contaminant being transported in a uniform groundwater flow field under steady-state conditions. Analytical models are useful in predicting how far a contaminant plume will migrate and how long it will take for a plume to reach steady state if site conditions remain constant. The latter is useful in predicting when the plume will become stable, which could mean that natural attenuation is working.

Analytical models can also be used to assess the age and stability of contaminant releases. One common approach is to use the available data on the spatial distribution of the target contaminant(s) with appropriate hydrogeologic parameters to calibrate the model. Using standard data fitting techniques, such as nonlinear regression, groundwater professionals can estimate site-specific migration rates and assess the age of a release. Another use of an analytical model is the estimation of a biodegradation rate. Analytical models can be fit to plume concentrations by trial and error or nonlinear regression by adjusting the degradation rate. It is important to keep in mind that the degradation rate is assumed to be first-order for many analytical models. The assumption of a first-order decay rate implies that the microbes and reactants that are currently driving biodegradation will remain at all necessary locations into the future. A constant first-order rate is almost never accurate because field conditions frequently change in space and time (National Research Council 2000).

The key assumptions that go into analytical solute transport models have been described in Chapter 5. Deviations from model assumptions are commonly due to heterogeneity of the aquifer material, which causes spatial variability in model parameters such as hydraulic conductivity, porosity, sorption capacity, and biodegradation kinetics (due to changes in availability of nutrients or electron acceptors needed for microbial activity). Other deviations from ideal behavior include seasonal changes in hydraulic gradient and associated variability in groundwater flow velocity and direction, artificial hydraulic gradients caused by pumping wells, and variable source flux with undefined shape. Such complexities cannot be incorporated in analytical models, which feature their relative simplicity as both their main advantage and disadvantage. Therefore, analytical model results should be interpreted with caution. These analytical solutions are often used in screening studies (Tiers 1 and 2) and are generally limited to steady, uniform flow and should not be used for groundwater flow or solute transport problems in strongly anisotropic or heterogeneous aquifers. These models additionally should not be applied under nonuniform flow conditions. It should be kept in mind that analytical models are best utilized for order-of-magnitude estimations, since a number of potentially important processes are treated in the models in an approximate manner or sometimes ignored totally. Analytical models are also useful in exploring a variety of worst case scenarios for plume migration, if site conditions are relatively simple and the assessments are preliminary (National Research Council 2000).

When dealing with uncertainty and variability of model parameters, it is good practice to conduct sensitivity analyses such as discussed in Chapter 5, using a reasonable range of all parameters that are not well defined for the site of interest. This type of analysis allows the modeler to demonstrate worst case and best case scenarios, and to provide valuable input for risk assessment and control decisions. The use of analytical fate-and-transport models is likely to increase with the adoption of risk-based corrective action (RBCA) in the environmental protection rules of numerous countries.

9.2.6. Numerical Modeling of Flow and Solute Transport

Numerical flow and solute transport models are needed when the site is complicated and the complexities have to be captured. A discussion of why one would construct a numerical model, the modeling process, data requirements, as well as descriptions of commonly used models and case studies for assessing natural attenuation is provided in Chapter 6.

Numerical models differ from analytical solutions in that the boundary value problem consisting of governing equation(s), initial condition(s), boundary condition(s), and source/sink term(s) comprise a system of algebraic equations. Alternatively, solute transport can be simulated by tracking a large number of particles in a known velocity field. The main advantage of a numerical approach over an analytical solution is that it is flexible and can handle complex geometries. This flexibility allows one to incorporate complex geological features and spatial variability in flow and transport parameters. It can also incorporate spatial variation in the initial condition and both space and time variations in boundary conditions (as well as biogeochemical processes in both space and time). Time and space variations in source/sink terms are readily incorporated in numerical models. If site heterogeneities are large, but well known, it will probably be more appropriate to use a numerical model that discretizes the site into small elements or blocks and assigns appropriate parameters to different domains. Conditions under which a numerical solute transport models should be used include (National Research Council 2000):

- When chemical reactions take place and these need to be incorporated to accurately track the contaminant transport.
- When multiple reactions take place; the products of key reactions participate in other reactions (e.g., precipitation or complexation) that affect aqueous phase concentrations.
- When partitioning of contaminant materials to other phases occurs.
- When loss reactions occur in multiple steps that produce and consume intermediates.
- When the site hydrogeology is complex and/or highly transient in nature, both of which have to be incorporated.

Numerical models are highly useful in modeling natural attenuation processes. When the model is properly calibrated, it can be used as a predictive tool for natural attenuation. However, there are a number of issues that one must consider when using numerical models. Table 9.6 summarizes some of these potential problems that a modeler may encounter. It is important that one spends a significant amount of time developing a comprehensive conceptual model of the site hydrogeology and geochemical and biotransformation processes. The availability of different conceptualizations necessitates a systematic approach to choose the optimal model. Otherwise, the subsequent analysis can be grossly in error. Despite the recognition that the system is open and complex, we note that hydrogeochemical analyses in assessing natural attenuation typically rely on a single conceptual-mathematical model. Yet hydrogeologic as well as biogeochemical environments are open and intricate, rendering them prone to multiple interpretations and mathematical descriptions. Adopting only one conceptual-mathematical model may lead to statistical bias and underestimation of uncertainty. A conceptual model of a site can only be built with the accumulation of data and refinement of that conceptualization (Neuman and Wierenga 2003). It is important to remember that conceptual modeling includes proposing several alternate models, ranking them, and discarding those that are in conflict with available data (Neuman and Wierenga 2003). This means that the mathematical models (analytical or numerical) produced from the conceptual model should be evaluated, and one should strive to document the degree of conservatism in modeling natural attenuation processes, and the level of conservatism presumed by the modeler must be commensurate with the data and conceptual model uncertainty. However, once the processes that are operating at the site are well understood, one can translate the conceptual model to a mathematical or, in this case, numerical model to assess natural attenuation.

Table 9.6. Common Problems with Models (adapted from NRC 2000)

Type of Problem	Examples	Solution
Model Framework: Poor Assumptions or Input		
Applying an inappropriate model or concept to the problem	Using a first-order rate law for biodegradation; simulating reactions that do not occur at the site or assuming a reactant is present in excess	Check that site geochemical data support the model formulations.
Relying on parameter values taken from publications unrelated to the site	Sorption coefficients, biodegradation coefficients, hydraulic conductivities	Use site-specific measurements to obtain reasonable values of parameters.
Failing to meet conditions assumed in the model	Assuming that climatic conditions and anthropogenic effects will remain the same	Evaluate the uncertainty associated with this assumption and its effect on results.
Weighting observations inappropriately in the calibration	Errors associated with inaccuracy and imprecision of the measuring device and process or human error	Weight the observations using the inverse of the measurements that established the value of the observation.
Model Application: Closed Mind During the Modeling Process		
Failing to consider alternate conceptual models	Filling gaps in hydraulic conductivity measurements according to a single conceptual model	Use multiple realizations of conceptual models of a site; combine all available data types to reduce uncertainty.
Forcing the model to predict the expected outcome	Changing the input parameter values to match the data	Evaluate whether processes that control the fate of the plume may have been overlooked; constrain parameter values to reasonable ranges.
Model Use and Presentation		
Extrapolating beyond the model's capability	Using a flow model calibrated to steady-state conditions to predict transient flow fields	Collect new data for calibration of storage coefficient or other uncalibrated features.
Overstating accuracy or reliability	Reporting only a single value for the prediction of interest, with numerous significant figures	Provide a range of possible outcomes, reflecting the range of uncertainty associated with input values.

A Word of Caution on Using Contamination Trends in Assessing Natural Attenuation

As discussed in previous sections, the documented loss of contaminant mass at the field scale is the first and the most direct evidence of whether natural attenuation is taking place. However, we would like to provide a word of caution on solely relying on contaminant trends in assessing natural attenuation for the following reasons:

- *Establishment of Credible Background Concentrations.* In an attempt to establish the documented loss of contaminants over space and time, one must first establish credible background concentrations within the plume as well as away from the plume. The establishment of a credible background is important because if one is to show the regulators and public that bioremediation is actually occurring, one must understand the background conditions. In fact, adequate monitoring is one of the most important aspects of demonstrating that bioremediation is occurring in the subsurface.

- *Seasonal Effects in Groundwater Flow and Contaminant Concentrations.* Transient effects in groundwater flow can affect the direction of plume movement and can also contribute to increased dispersion and dilution. Therefore, thorough understanding of the transient nature of the aquifer mechanics should be a prerequisite for understanding the evolution of contaminant concentrations at a given site. A popular saying that relates to this concept is: "If you do not understand flow, you will not understand transport." Temporal and spatial variability in the flow system can cause the apparent increase or decrease in plume concentrations. There are a number of factors that can cause the reduction of contaminant concentrations in the field; however, this may not necessarily be the case of contaminant destruction or degradation, but instead other mechanisms such as sorption, dispersion, or dilution of contaminants can have an impact on subsurface contaminant concentrations. One such example is dilution, which can be caused by infiltration of rainwater. Therefore, monitoring of contaminants during heavy rainfall events or soon after could cause the contaminant concentrations in the aquifer to decrease, giving a false impression that contaminant mass is being destructed. It is also possible for contaminant concentrations to increase following a precipitation event when the contaminants trapped in the vadose zone are mobilized (through infiltration) and enter the aquifer. Improper groundwater sampling can also have a large effect on contaminant concentrations. To avoid being misled by transient effects in groundwater, contaminant losses should be documented over an area that encompasses the longitudinal axis and fringes of the plume over several years (National Research Council 2000). Also, established sampling protocols should be followed for groundwater sampling to avoid inadvertent contaminant mass increase or loss, which could cloud the investigation.

- *Recognition of Directional Effects in Mechanical Dispersion.* It is important to remember that longitudinal dispersivity is about 10 times larger than transverse dispersion. This could mean that plumes are, in general, narrow and have a prolate spheriodal (think of long cigars) shape, which could easily bypass monitoring networks consisting of a sparse number of wells.

- *Development of Preferential Channels.* The subsurface is heterogeneous and this is the rule rather than the exception. There is no such thing as a homogeneous aquifer. However, it has been a popular choice to conceptualize the subsurface to be geologically simple because this simplifies the monitoring program as well as the subsequent analysis of the data. Thus, we are now also starting to recognize that nature created aquifers that are physically heterogeneous and, at some sites, highly complex, which results in natural geological processes exhibiting complex behavior. The fact that the physical heterogeneity is complex translates into complex flow patterns and transport pathways. Our tendency to simplify the subsurface has caused monitoring networks to miss plumes that are moving through preferential channels developed in buried alluvial channels or in flowing fractures. We are also now finally acknowledging the heterogeneous nature of geochemistry and microbiology as well.

- *Monitoring Scale.* Sampling of groundwater can be very tricky. If a monitoring well with a long screen is used, the sample obtained is an integrated sample of clean and contaminated water. This could potentially lower the concentrations below the regulatory standard.

- *Monitoring Locations.* Groundwater plumes are three-dimensional objects. Despite this recognition, groundwater sampling continues with monitoring wells with long screens, which essentially causes depth averaging of contaminant concentrations. This depth-averaging process essentially leads to the two-dimensional characterization of what is really a three-dimensional plume. In addition, contour plots are prepared for two-dimensional cases but there is a legitimate need to characterize plumes in three dimensions.

- *Transformation into More Harmful Compounds.* Contaminants may also transform through biological or chemical means to products that are more toxic than the parent compound. One such example is the reductive dechlorination of trichloroethene to vinyl chloride. Vinyl chloride is a potent carcinogen that is considered to be more toxic than TCE.

- *Limits to Sorption Sites and Reversible Sorption.* Sorption of contaminants depends on the number of sorption sites. Some models, like the linear model, conceptualize that sorption sites are unlimited; however, this is certainly not the case. When there is a large plume with high concentrations, sorption sites may indeed become filled. Also, sorption kinetics could be such that contaminants could become mobile if the geochemical conditions change in the future, making sorption a reversible process.

For these reasons, regulators and investigators should not rely on simple rules of thumb (e.g., trends in contaminant concentration data over a short period) to evaluate the success of natural attenuation.

Table 9.7. Parameters for Evaluation of Natural/Intrinsic Bioremediation

---

Down Hole Parameters After Well Purging

Dissolved oxygen	Temperature
Redox potential	Specific conductance
pH	

Laboratory Parameters

Target compound volatile organics	Sulfate
Nitrate	Sulfide
Nitrite	Total organic carbon
Nitrogen, ammonia	Chlorides
Manganese (total and dissolved)	Light hydrocarbon scan (ethene, ethane)
Iron (total, dissolved, ferrous)	Permanent gases (carbon dioxide, oxygen, nitrogen, methane, and carbon monoxide)

Optional Groundwater Parameters

Chemical oxygen demand	Biological oxygen demand
Total organic carbon	

Source: Davee and Sanders (2000).

9.2.7. Geochemical Indicators of Natural Attenuation

The second line of evidence relies on hydrogeochemical data to demonstrate indirectly the type(s) of natural attenuation processes that are active at the site, and the rate at which such processes will reduce contaminant concentrations to required levels. For example, characterization data listed in Table 9.7 may be used to quantify contaminant sorption, dilution, or volatilization, or to demonstrate and quantify biodegradation rates occurring at the site. In particular, we focus in this section on the use of geochemical indicators to indirectly determine and/or demonstrate the type of degradation processes that are active at a site. Geochemical parameters that are commonly collected are summarized in Table 9.7.

Some changes in geochemical parameters are known as "footprints" of natural attenuation (National Research Council 2000). Such footprints or indicators can be used as secondary evidence to demonstrate whether natural attenuation is occurring at a site or not. Table 9.8 lists case studies from sites across the country with a variety of contaminants, whether the contaminants were controlled or not, and their corresponding footprints. For example, at Dover Air Force Base in Delaware, USA, contamination of the aquifer with TCE and TCA is of great concern. According to Table 9.8, the contaminants are controlled and the geochemical footprints at the site suggest that reductive dechlorination is most likely taking place. The presence or absence of these compounds alone is not proof of natural attenuation, but their presence and the changes in concentration over time can be indicative of intrinsic remediation.

Table 9.8. Natural Attenuation Footprints from National Research Council (2000) Case Studies

Case Study	Contaminant(s)	Contaminants Controlled?	Footprints
Traverse City	BTEX	Yes	Depletion of O_2; formation of CH_4 and Fe^{2+}
Vandenberg Air Force Base	MTBE	No	Insignificant decrease in O_2 and SO_4^{2-} concentrations; extension of MTBE plume far beyond BTEX plume
Borden Air Force Base	Five chlorinated solvents	Partially	Detection of metabolites of solvent degradation
St. Joseph	TCE	Partially	Formation of CH_4; detection of degradation by-products (vinyl chloride and ethene)
Edwards Air Force Base	TCE	No	Documentation of high NO_3^- and SO_4^{2-} concentrations; demonstration that TCE moves with water
Dover Air Force Base	TCE, TCA	Yes	Formation of degradation by-products (cis-1,2-DCE, 1-1-DCA, vinyl chloride, and ethene); CH_4 and H_2S formation; increase in Cl^- concentration
Hudson River	PCBs	Partially	Detection of breakdown products; detection of unique transient metabolites; observation of microbial metabolic adaptation
South Glens Falls	PAHs	Yes	Depletion of O_2; detection of unique metabolic by-products; detection of genes for degrading PAHs in site microorganisms; rapid PAH degradation in soils taken from site
Pinal Creek Basin	Metals, acid	Yes, but may not be sustainable	Observation of carbonate dissolution, leading to pH increase coincident with metal precipitation; observation of Mn oxide precipitates in stream sediments
Hanford 216-B-5	Radionuclides	Yes	Observation of sorbed radionuclides site samples
Anonymous field site	BTEX	Yes	Loss of O_2, NO_3^-, and SO_4^{2-}; formation of Fe^{2+} and CH_4; increase in inorganic carbon concentration; increase in alkalinity
Bemidji	Petroleum hydrocarbons	Partially	Loss of O_2; formation of Fe^{2+}, Mn^{2+}, and CH_4; formation of intermediate metabolites; observation of selective degradation of petroleum hydrocarbons relative to more stable chemicals

Note: BTEX = benzene, toluene, ethylbenzene, and xylene; MTBE = methyl *tert*-butyl ether; TCE = trichloroethylene; TCA = trichloroethane; PCBs = polychlorinated biphenyls; PAHs = polycyclic aromatic hydrocarbons; DCE = dichloroethene; and DCA = dichloroethane.

Attenuation of Petroleum Hydrocarbons

Intrinsic biodegradation is currently best documented and the processes are best understood for petroleum hydrocarbon sites. As discussed in Chapter 3, BTEX (benzene, toluene, ethylbenzene, and xylene) compounds will biodegrade through microbial activity and ultimately produce nontoxic end products such as water and carbon dioxide. At sites where the source is controlled and where the microbial activity is fast relative to groundwater flow, the dissolved BTEX plume may stabilize. When the source is removed, the dissolved BTEX plume may shrink over time. The National Research Council (2000) lists footprints that are indicative of natural attenuation of petroleum hydrocarbons taking place at a site. These include:

- Loss of electron acceptors (mainly O_2, NO_3^-, Fe(III), and SO_4^{2-}). (In particular one should look for O_2, NO_3^-, and SO_4^{2-} levels below background in the core of the plume; see also Figures 6.10, 6.11, and 6.13).
- Generation of the products of acceptor reduction (such as Fe(II) and CH_4). (Again, Fe(II) and CH_4 levels should be highest in the core of the plume; see also Figures 6.12 and 6.14).
- Presence of organic acids that are known intermediate products of petroleum hydrocarbon degradation.
- An increased concentration of dissolved inorganic carbon (CO_2) and a characteristic change in the alkalinity.

The above are general guidelines that apply to most petroleum spill sites. However, for large sources a short-term criterion may not be justified. That is, in some cases, electron acceptors may be completely depleted when the hydrogeochemical conditions are such that the sustainability of electron acceptors is in question. If this is the case, the biodegradation rate may change over time, causing the natural attenuation of contaminants to fail. In such a case, the slowest sustainable degradation mechanism (which may be methanogenesis) and its corresponding degradation rate have to be compared to the minimum travel time to humans or sensitive ecosystems (National Research Council 2000).

Another factor to consider is the kinetics of the reactions that are taking place. If microbial kinetics are limiting the biodegradation rate, anaerobic electron acceptors (nitrate and sulfate) would be constantly decreasing in concentration as one moved downgradient from the source zone, and anaerobic by-products such as (Fe(II) and methane) would be constantly increasing in concentration. In contrast, if microbial kinetics are considered to be relatively fast, electron acceptors through anaerobic degradation (nitrate and sulfate) would be mostly or totally consumed in the source zone, while the anaerobic by-products (Fe(II)) and methane would be found in the highest concentration in the source zone (Newell et al. 1996) (Figure 9.7).

A good example showing a number of geochemical footprints that indicate intrinsic bioremediation is taking place is a spill site in New York. Figure 9.8 shows BTEX and electron acceptor distributions at a former fire training area at Plattsburgh Air Force Base, New York. Groundwater flow at the site is to the

If microbial kinetics were limiting the rate of biodegradation:	If microbial kinetics were relatively fast (instantaneous):
Anaerobic electron acceptors (nitrate and sulfate) would be constantly **decreasing** in concentration as one moved downgradient from the source zone, and	Anaerobic electron acceptors (nitrate and sulfate) would be mostly or totally **consumed in the source zone**, and
Anaerobic by-products (ferrous iron and methane) would be constantly increasing in concentration as one moved downgradient from the source zone.	Anaerobic by-products (ferrous iron and methane) would be **found in the highest concentrations in the source zone.**

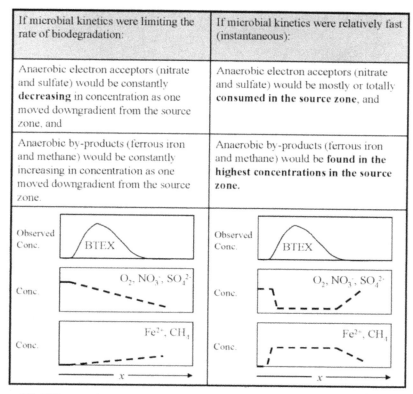

Figure 9.7. Differences in microbial kinetics and their effects on BTEX, electron acceptors, and metabolic by-product concentrations versus distance along the centerline of a plume. (Adapted from Newell et al. 1996.)

southeast and the dissolved BTEX plume extends about 2000 ft downgradient from the source zone and the plume has a maximum width of approximately 500 ft. Within the source zone, total BTEX concentrations as high as 17 mg/L (17,000 µg/L) have been observed. It is evident from Figure 9.8 that the BTEX plume and its electron acceptors are migrating in the direction of groundwater flow. We also see the strong correlation between elevated BTEX concentrations with depleted electron acceptors. The absence of electron acceptors suggests that intrinsic bioremediation is taking place through aerobic respiration as well as anaerobic processes of nitrate and sulfate reduction. Figure 9.9 shows contour plots of the BTEX plume and its metabolic by-products at the same site. It reveals that Fe(II) and methane concentrations are highest in the core of the BTEX plume, suggesting that BTEX degradation is coupled to Fe(III) reduction and methanogenesis.

Another good geochemical indicator for the natural attenuation of petroleum hydrocarbons is H_2 concentrations. Lovley and Goodwin (1988) proposed the use of H_2 concentration as a microbially based, nonequilibrium alternative to p_e analysis for determining which oxidation–reduction reactions are taking place in anoxic

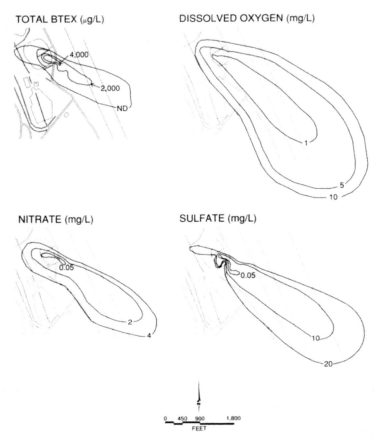

Figure 9.8. Contour plots of BTEX and electron acceptors, measured in 1995, from Plattsburgh Air Force Base, New York. (Adapted from Wiedemeier et al. 1999.)

sedimentary environments. Molecular hydrogen (H_2) is produced through micro-organisms metabolizing natural or anthropogenic organic matter under anaerobic conditions. The produced H_2 is then utilized by microorganisms that most commonly utilize Fe(III), sulfate, or carbon dioxide as terminal electron acceptors (Chapelle et al. 1997). The technique then is based on the fact that each of these terminal electron-accepting processes (TEAPs) has a different affinity for H_2 uptake.

The technique is highly useful because, in some cases, these redox processes are hard to determine just from the geochemical footprints as suggested by the National Research Council (2000). Large differences in geochemical conditions may obscure the operating TEAPs. Despite the large differences in factors such as the pH and organic matter decomposition rates, Lovley and Goodwin (1988; see also Lovley et al. 1994) found that aquatic sediments in which methane produc-tion was the predominant TEAP typically had H_2 concentrations of 7–10 nM;

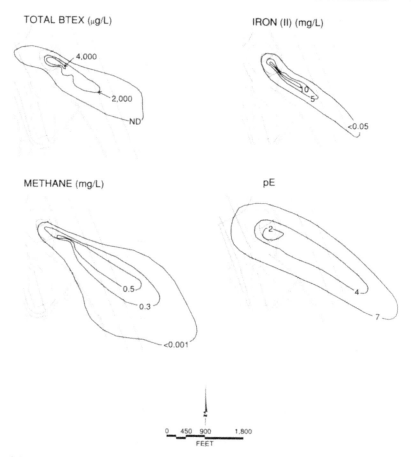

Figure 9.9 Contour plots of BTEX and metabolic by-products, measured in 1995, from Plattsburgh Air Force Base, New York. (Adapted from Wiedemeier et al. 1999.)

sediments with sulfate reduction as the TEAP had H_2 concentrations of 1–1.5 nM; Fe(III)-reducing sediments had H_2 concentrations of 0.2 nM; and sediments in which nitrate or Mn(IV) reduction was the predominant TEAP had H_2 concentrations less than 0.05 nM. Table 9.9 summarizes the dominant redox processes associated with H_2 concentrations.

Attenuation of Chlorinated Solvents

Chlorinated solvents such as TCE, TCA, and PCE represent another class of contaminants that are commonly found in groundwater (Chapter 1). Because of their high density relative to water, they are referred to as DNAPLs (dense nonaqueous phase liquids). DNAPLs are prevalent at a large number of sites throughout the world. The high densities, low interfacial tensions, and low viscosities of these compounds can lead to deep DNAPL penetration (Pankow and Cherry 1996). In

Table 9.9. Dominant Redox Processes Associated
with H_2 Concentrations

Characteristic H_2 Concentration (nM)	Dominant TEAP
0.01–0.05	Denitrification
0.1–0.3	Mn(IV) reduction
0.2–0.8	Fe(III) reduction
1.0–4.0	Sulfate reduction
>5.0–15.0	Methanogenesis

Source: Chapelle (2001).

porous media, much of the DNAPL mass remains in the groundwater as persistent source zones. The variable release history and geologic heterogeneity make the distribution of DNAPLs in the source zone complex, where DNAPLs exist as residuals or as pools of pure phase. Without remediation, these source zones can contribute to long-term groundwater contamination for decades to centuries. Research over the last decade has identified several mechanisms that can intrinsically biodegrade these compounds. Research also suggests that the intrinsic bioremediation of these compounds to the point that is compliant under regulatory standards only takes place under hydrogeologically and biogeochemically favorable conditions that are anticipated to occur at a limited number of sites. For these compounds, reductive dechlorination is thought to be the most widely applicable mechanism (see Chapter 3 for further details).

For reductive dechlorination, indicators for attenuation include many of the same indicators as for attenuation of petroleum hydrocarbons. The presence of electron donors is the most important screening criterion used to determine the potential for reductive dechlorination of chlorinated solvents (National Research Council 2000). That is, one should look for low redox potential (< -100 mV) above background Cl^-, and VFAs. For example, degradation of chlorinated solvents under anaerobic conditions involves total depletion of oxygen, nitrate, and sulfate in the source area and downgradient, as well as the appearance of ferrous iron, methane, acetate, and high levels of organic carbon. For the same chlorinated solvents, under aerobic conditions, depletion of oxygen and nitrate and the appearance of methane, ferrous iron, and acetate in the source area, and the reappearance of oxygen and disappearance of methane downgradient of the plume, illustrate that natural attenuation is occurring. Another indication of natural attenuation can be illustrated by comparing dissolved oxygen (DO) concentrations with redox potentials. As DO decreases, redox potential also decreases. Low DO values typically have negative redox potentials, which are indicative of anaerobic conditions that can lead to reductive dechlorination. Other footprints that one should look for in the reductive dechlorination of chlorinated solvents include:

- Increase in degradation rates or appearance of daughter products. For example, the reductive dechlorination of PCE will yield TCE, which in turn degrades to *cis*-DCE and finally VC in the core of the plume (see Figures 6.15–6.18). Because of the nonsorptive, sequential nature of the reaction kinetics, the plume center for each degradation component will be further downgradient compared to the parent components.

- Generation of the products such as CO_2, ethane, and chloride should be highest downgradient of the parent plume.

Attenuation of Other Contaminants

The footprints of contaminants other than petroleum hydrocarbons and chlorinated solvents, such as inorganic contaminants, are less well established. According to EPA (1999), a number of mechanisms for natural attenuation of inorganic compounds have been postulated but the scientific basis for the postulation is weak. This will likely change over time but, as of 1999, EPA advocates the strategy for evaluating natural attenuation of inorganic contaminants by postulating several geochemical footprints. Natural attenuation processes besides intrinsic bioremediation, such as sorption, decay, and redox reactions, can effectively mitigate the concentrations of these contaminants or transform them into nontoxic forms. Many of the reactions responsible for these are discussed in Chapter 2.

It is of interest to note that many of the inorganic contaminants are persistent in the subsurface, as they are not degraded readily by natural attenuation, with the exception of radionuclides, which undergo radioactive decay. Therefore, natural attenuation is applicable to inorganic compounds that are relatively immobile and undergo radioactive decay (EPA 1999). The key issue is that these processes (in particular sorption) need to be irreversible, such that remobilization of these contaminants does not take place with changes in geochemical conditions at a later time.

9.2.8. Microcosm Studies

The tertiary line of evidence relies on data from field or microcosm studies, which directly demonstrate the occurrence of a particular natural attenuation process at the site and its ability to degrade the contaminants of concern. Currently, microbiological investigations related to bioremediation are conducted through in situ microcosms (see Chapter 7) or by obtaining samples from the field which are incubated and rates of contaminant degradation or immobilization are quantified. Such studies provide estimates of metabolic activity of the microbial community, but provide little insight into the actual mechanisms for bioremediation or how some amendments may or may not stimulate microbial activity (Lovley 2003).

The habitat in which the microorganisms live is very intricate. It involves the continuous cycling of nutrients, water, and gases, as well as chemical constituents. Within such a habitat, there are complex interactions among various species of microorganisms, which are very difficult to decipher and quantify. We also know that the subsurface is highly heterogeneous at a multiplicity of scales (i.e., from micrometers to kilometers) and that the habitat of microorganisms depends on the pore structure. Very little is known about these microhabitats because the act of studying them involves destroying the habit. Unlike other organisms that are commensurate in scale with humans (e.g., plants in landscapes), detailed knowledge of where microorganisms live is very difficult to obtain because of scale- and sampling-related physical characteristics of microhabitats and microorganisms therein (Madsen 1998).

Therefore, field sampling will undoubtedly disturb the habitat of microorganisms. It is also well known that the conditions in the field are very difficult to duplicate in the laboratory; thus, biodegradation rates determined in the laboratory may not be entirely representative of field conditions.

There are many questions that cannot be addressed through field and laboratory microcosm studies that are relevant to the quantification of intrinsic bioremediation (Madsen 1998). Some of these include:

- *What microorganisms are there?* Microbes are responsible for much of the recycling of nutrient and chemical compounds, but determining the presence of these microorganisms and distinctly proving that they are responsible for the chemical changes are not easy. To determine the presence of microorganisms, microbiologists have developed three different assays: viable plate counts of organisms able to grow on selective agar media; extraction and analysis of nucleic acids, phospholipids, or other cellular biomarkers; and microscopic examination of fixed, stained samples (Madsen 1998). All of these techniques have limitations and there is a distinct possibility that some of the microbial communities can be overlooked.

- *What are microorganisms doing?* This question can be subdivided into *What is the general physiological status of the cells?* and *What specific geochemical activities are the cells engaged in?* Similar to procedures investigating the composition of microbial communities, information can be obtained from laboratory measurements of incubated samples obtained from the field, from biomarkers extracted, and by means of microscopic and molecular techniques.

- *When are the microorganisms active?* To answer this question field chamber studies (Conrad 1996) and microelectrode investigations (Glud et al. 1994) have been conducted and applied to the study of biodegradation of contaminants by Wilson and Madsen (1996). However, *when the microorganisms are active* and *when they facilitate key biogeochemical reactions* still remain uncertain and are inferred after the fact (Madsen 1998).

Biodegradation can also be determined using more innovative laboratory analyses on environmental samples. These techniques include documentation of enhanced numbers of protozoan predators inside but not outside the contaminant plume, extracting total RNA to compute relative abundances of each phylogenetic group at different depths, extracting nucleic acids from sediment to identify unique transient intermediary metabolites indicative of biodegradation, assessment of growth using flow cytometry, and cloning, sequencing, and analyzing 16S rRNA genes to identify novel organisms (Madsen 1998). The quest for improved under-standing of biodegradation mechanisms continues. A comprehensive mechanistic understanding of such processes will have to involve the comprehension of processes that regulate the genes that encode the proteins that actually degrade the compounds. More of this exciting topic is discussed in a section on microbial analysis.

9.2.9. Stable Isotope Analysis

Degradation pathways for a number of contaminants are known, and we can analyze footprints of potential pathways to learn which degradation processes (if any) are taking place at a particular site. One common obstacle in identifying biodegradation processes is the difficulty in analyzing for metabolites at trace levels. Metabolites are often quite polar and/or more easily degradable than the parent compound and may also occur as primary contaminants at the site (Schmidt et al. 2004).

The quantification of biodegradation processes especially in a field setting is even more difficult because of the need for a conclusive mass balance, which is often impossible to make due to lack of knowledge on the groundwater flow regime, the limited number of sampling wells, and insufficient observation times (Schmidt et al. 2004). However, in assessing the progress of natural attenuation, it is of para-mount importance to quantify all transport and transformation and attenuation mechanisms at a given site so that contaminant destruction can be quantified. For-tunately, while many of the transport processes discussed in Chapter 4 (advection, dispersion, volatilization, dissolution, and sorption) do not leave chemical sig-natures that can be analyzed, biodegradation often causes a substantial kinetic isotope effect.

The weaker bond strength of light isotopes results in the preferential enrichment of the lighter isotopes in the product of a reaction while the heavy isotopes accu-mulate in the nonreacted material. Therefore, degradation of contaminants increases the number of heavier isotopes as biodegradation proceeds; thus, the ratio between heavy and light isotopes is expected to increase (Day et al. 2002). The idea that fractionation occurs during biodegradation derives from its prevalence in natural geologic systems. For example, Figure 9.10 summarizes the carbon isotope compo-sition of various substances relative to the international standard (i.e., the Vienna Pee Dee Belemnite, VPDB, a limestone fossil of *Belemnitella americana* from a geologic formation Pee Dee in South Carolina). This figure shows that carbon fixed by autotrophic organisms is enriched in ^{12}C, making $\delta^{13}C$ less negative (Madigan et al. 2003).

In Figure 9.10, isotopic ratios are reported as the difference between the inves-tigated sample ratio and a standard ratio, divided by the standard ratio and expressed as δ values in per mil (‰) (Schmidt et al. 2004) according to

$$\delta = \left[\frac{R_{sample} - R_{reference}}{R_{reference}} \right] \times 10^3 \text{ ‰} \qquad [9.3]$$

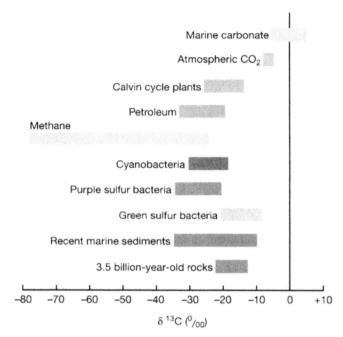

Figure 9.10. Carbon isotopic compositions of various substances. The values are given in parts per thousand (‰) and were calculated using Eq. [9.3]. Note that carbon fixed by autotrophic organisms is enriched in ^{12}C. (From Madigan et al. 2003.)

where R_{sample} is the ratio of the heavy isotope to the light isotope for some compound (e.g., $^{13}C/^{12}C$) and $R_{reference}$ is the corresponding ratio for the international VPDB standard. The reference $^{13}C/^{12}C$ ratio for VPDP is 0.011180 (Schmidt et al. 2004).

A negative δ indicates that the sample being investigated is depleted in the less common isotope with respect to the standard. A positive δ indicates a sample in which the less common isotope is in greater abundance compared to the isotopic standard. It is important to recognize that relative isotope ratios can only be determined with the required precision. For example, a $\delta^{13}C$ value of +10 per mil then corresponds to a sample with an isotope ratio 1% higher than that of the VPDB standard. Reporting the differences in relative ratios in this manner allows for the correction of mass-discriminating effects in a single instrument and facilitates the comparison of published GC/IRMS data (Schmidt et al. 2004).

A significant carbon isotope fractionation accompanies in situ microbial degradation of a variety of common pollutants, including BTEX, chlorinated solvents, methyl-*tert*-butyl ether (MTBE), and *tert*-butyl alcohol (TBA) (Day et al. 2002; Morasch et al. 2002). In addition, isotopic signatures can be used to quantify the relative extent of biodegradation between different zones of the contaminant plume

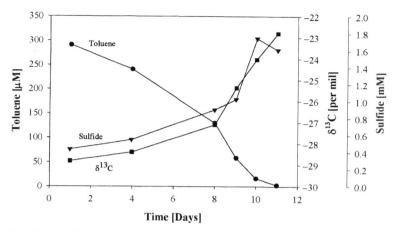

Figure 9.11. Results from a growth experiment with the sulfate-reducing strain TRM1. Toluene and sulfide concentrations are measured over time together with the isotope value δ^{13}C in the residual toluene fraction. (Data are obtained from Meckenstock et al. 1999.)

(Lollar et al. 2001). Figure 9.11 presents an example of isotope fractionation by sulfate-reducing bacteria degrading toluene. In this example, toluene concentrations decrease with time while sulfide concentrations and δ^{13}C increase over time, indicating a kinetic isotope effect. Other examples relying on the stable isotope analysis of carbon to verify the biodegradation of various common priority pollutants include those by Migaud et al. (1996), Richnow et al. (1998), Hanson et al. (1999), Meckenstock et al. (1999), Pelz et al. (1999, 2001), Sherwood-Loller et al. (1999), Fang et al. (2000), Wilks et al. (2000), Rodgers et al. (2000), and Morasch et al. (2001).

The fractionation of an isotope between a substrate and its degradation product is expressed using either the fractionation factor α or the enrichment factor ε according to the following equations:

$$\alpha = \frac{R_{\text{product}}}{R_{\text{reactant}}} = \frac{10^{-3}\delta_p + 1}{10^{-3}\delta_r + 1} \qquad [9.4]$$

and

$$\varepsilon = \left(\frac{R_{\text{product}}}{R_{\text{reactant}}} - 1\right) \times 1000 = (\alpha - 1) \times 10^3 \text{‰} \qquad [9.5]$$

Here, the subscripts p and r refer to product and reactant, respectively, while R_{product} and R_{reactant} are the heavy to light isotope ratios in the degradation product and the substrate, respectively, that appear in an infinitely short period of time

(Clark and Fritz 1997). The enrichment factor ε or the fractionation factor α is determined from the relationship between substrate concentration change and isotope fractionation:

$$\frac{R_t}{R_0} = \left[f \frac{(1 + R_0)}{(1 + R_t)} \right]^{(\alpha - 1)} = \left[f \frac{(1 + R_0)}{(1 + R_t)} \right]^{\varepsilon} \qquad [9.6]$$

Here R_t and R_0 are heavy to light isotope ratios in the reactant r at some initial time $t = 0$ and time t, respectively, while f is the fraction of the reactant at time t that remains after the reaction, which is given as (Hunkeler 2002)

$$f = \frac{L_t + H_t}{L_0 + H_0} = \frac{L_t(1 + R_t)}{L_0(1 + R_0)} \qquad [9.7]$$

where L_0 and H_0 are the light and heavy isotope concentrations, respectively, at some initial time $t = 0$, and L_t and H_t are the light and heavy isotope concentrations, respectively, at time t. Equation [9.6] can be approximated using the classical Rayleigh-type equation [9.8] when the heavy isotopes are in low abundance (i.e., $H + L \approx L$) or when the fractionation of isotopes is very small (i.e., $1 + R_t \approx 1 + R_0$):

$$\frac{R_t}{R_0} = f^{(\alpha - 1)} \qquad [9.8]$$

Upon taking the natural logarithm transformation and using Eq. [9.5], we obtain

$$\ln \left(\frac{R_t}{R_0} \right) = (\alpha - 1) \ln f = \frac{\varepsilon}{1000} \ln f \qquad [9.9]$$

which then results in

$$1000 \ln \left(\frac{10^{-3} \delta_{r,t} + 1}{10^{-3} \delta_{r,0} + 1} \right) = \varepsilon \ln f \qquad [9.10]$$

Here, $\delta_{r,0}$ and $\delta_{r,t}$ are the heavy to light isotope ratios in the reactant r at some initial time $t = 0$ and time t, respectively, expressed in the δ notation.

The quantitative relationship between the extent of biodegradation and isotope fractionation for a given compound is in many cases first studied in pure cultures or enrichments. Experimental results are then interpreted by plotting a graph of $\ln(R_t/R_0)$ versus $\ln f$ (Eq. [9.9]), and the fractionation factor α or the enrichment factor ε is obtained from the slope of the linear regression line (Schmidt et al. 2004). Results from the laboratory can then be used to quantify biodegradation in the field if the assumption is made that the same biogeochemical processes operate in the field and the constants α or ε obtained in the laboratory are also consistent in the field.

Table 9.10 summarizes batch-, column-, and field-scale studies that document the use of compound-specific stable isotope analyses to investigate biodegradation of organic compounds. It is of interest to note that most studies in Table 9.10 report a substantial isotope fractionation during microbial degradation of investigated compounds. This implies that the technique is promising; however, there are a few studies that revealed a lack of stable isotope fractionation despite biodegradation taking place (Drenzek et al. 2001; Mazeas et al. 2002; Morasch et al. 2002), implying that further research is needed to understand such departures from the norm.

Another promising bioremediation forensics approach is the use of microbial lipids as indicators of pollutant degradation and substrate utilization (Abraham et al. 1998; Boschker et al. 1998; Pelz et al. 2001; Pombo et al. 2002). Lipids are primary components of the bacterial cell membrane, accounting for 40–70% of the membrane mass (White et al. 1979). Lipid profiling has been used to detect changes in microbial communities in contaminated aquifers (Fang et al. 1997; Fang and Barcelona 1998; MacNaughton et al. 1999; Rooney-Varga et al. 1999). The analysis of lipids combined with the use of ^{13}C-labeled tracers allows effective tracking of substrate usage and carbon flow during biodegradation of pollutants. However, Fang et al. (2004) showed that differences in fatty acid concentrations between cells grown on natural versus ^{13}C-labeled substrates (i.e., toluene) can affect the interpretation of lipid profiles for microbial community analysis as indicated by principal component analysis of fatty acids. Therefore, caution should be exercised in linking lipid data with microbial population shifts in biodegradation experiments with ^{13}C-labeled tracers.

9.2.10. Chemical Fingerprinting

Similar to human fingerprints, chemical fingerprints can provide unique signatures of different contaminants and demonstrate changes resulting from natural weathering processes, including biodegradation (Stout et al. 1998). The measurement of stable isotopic ratios discussed in the previous section is a form of chemical fingerprinting. However, the most common approach for chemical fingerprinting involves analysis of contaminant mixtures (usually petroleum hydrocarbons) by gas chromatography/mass spectrometry (GC/MS).

Numerous fingerprinting techniques have been used (and misused) in the environmental profession to assess the age and liability of petroleum product releases. Such techniques often involve chemical analysis of groundwater or free product sampled from the subsurface in conjunction with precise knowledge of historical usage and chemical compositions. For example, high-resolution gas chromatography might be useful in liability assessment to distinguish among different types of fuel types that might be present at a single location. The presence of gasoline additives can be particularly revealing, yielding valuable insight for forensic age dating. For example, tetraethyl lead was added to gasoline prior to 1985 (at 400–800 mg/L) to suppress preignition and enhance the octane rating (Watts 1997). Therefore, the presence of organic lead in free product is indicative of a relatively old (pre-1985) release. Similarly, methyl-*tert*-butyl ether (MTBE) has been used as a gasoline

Table 9.10. Biodegradation of Organic Contaminants Investigated by Compound-Specific Stable Isotope Analysis

Compound (Classes)	Scale	Assumed Terminal Electron Acceptors	Isotopes Measured	Isotope Fractionation[a]	Reference
n-Alkanes, phenanthrene	Batch	Aerobic	$^{13}C/^{12}C$	Not significant	Mazeas et al. (2002)
n-Alkanes	Batch	Aerobic	D/H	Max. $\Delta(\delta^2 H) = 26\,‰(C16)$, lower for longer chain alkanes	Pond et al. (2002)
Tetrachloroethene (PCE)	Batch, field	Methanogenic, sulfate-reducing	$^{13}C/^{12}C$	$\varepsilon(C) = -2\,‰$ (batch) Max. $\Delta(\delta^{13}C) = 6\,‰$ (field)	Hunkeler et al. (1999)
Chlorinated ethenes	Batch	Methanogenic	$^{13}C/^{12}C$	$\varepsilon(C) = -2.5/-6.6\,‰$ (TCE) $\varepsilon(C) = -14.1/-16.1\,‰$ (cis-DCE) $\varepsilon(C) = -21.5/-26.6\,‰$ (VC)	Bloom et al. (2000)
PCE, TCE	Field	Anaerobic	$^{13}C/^{12}C$	Max. $\Delta(\delta^{13}C) = 6.4\,‰$ (PCE) Max. $\Delta(\delta^{13}C) = 8\,‰$ (TCE)	Lollar et al. (2001)
PCE	Batch	Methanogenic	$^{13}C/^{12}C$	$\varepsilon(C) = -1.8/-5.5\,‰$	Slater et al. (2001)
Trichloroethylene (TCE)	Batch	Aerobic	$^{13}C/^{12}C$	$\varepsilon(C) = -18.2/-20.7\,‰$	Barth et al. (2002)
trans-Dichloroethene (trans-DCE)	Batch	Aerobic	$^{13}C/^{12}C$	$\varepsilon(C) = -3.5/-6.7\,‰$	Brungard et al. (2003)
Toluene	Batch	Aerobic, sulfate-, nitrate-, and Fe(III)-reducing	$^{13}C/^{12}C$	$\varepsilon(C) = -1.5/-2.6\,‰$	Meckenstock et al. (1999)
Benzene, styrene, other hydrocarbons	Batch, field	Aerobic	$^{13}C/^{12}C$	Max. $\Delta(\delta^{13}C) = 2.2\,‰$ (benzene, batch) Max. $\Delta(\delta^{13}C) = 1.7\,‰$ (styrene, batch)	Meckenstock and Richnow (2002)
Toluene	Batch	Methanogenic, sulfate-reducing	$^{13}C/^{12}C$	$\varepsilon(C) = -0.5/-0.8\,‰$	Ahad et al. (2000)
Toluene	Batch	Methanogenic	D/H	Max. $\Delta(\delta^2 H)60\,‰$	Ward et al. (2000)
Benzene	Batch	Aerobic	$^{13}C/^{12}C$, D/H	$\varepsilon(H) = -11.2/-12.8\,‰$ $\varepsilon(C) = -1.5/-3.5\,‰$	Hunkeler et al. (2001b)

494

Compound	Experiment	Process	Isotope ratio	Enrichment factor / results	Reference
Toluene	Batch	Aerobic, sulfate-, nitrate-, and Fe(III)-reducing	D/H	$\varepsilon(H) = -198$ to $-730‰$ (using deuterium-labeled toluene)	Morasch et al. (2001)
Benzene, ethylbenzene	Field	Methanogenic, partly sulfate- and Fe(III) reducing	$^{13}C/^{12}C$, D/H	Max. $\Delta(\delta^{13}C) = 2‰$; Max. $\Delta(\delta^2 H) = 27‰$ (benzene)	Mancini et al. (2002)
Toluene	Column, field	Sulfate-reducing	$^{13}C/^{12}C$	Max. $\Delta(\delta^{13}C) = 3‰$ (field)	Meckenstock et al. (2002)
Aromatic hydrocarbons	Batch	Aerobic	$^{13}C/^{12}C$, D/H	$\varepsilon(C) = -0.1/-3.3‰$ (depending on enzyme mechanism); $\varepsilon(H) = -16/-905‰$ (using deuterium-labeled substrates)	Morasch et al (2002)
Aromatic hydrocarbons, naphthalene	Batch, field	Sulfate-reducing	$^{13}C/^{12}C$	$\varepsilon(C) = -1.1‰$ (naphthalene, batch); Max. $\Delta(\delta^{13}C) = 3.3‰$ (benzene), 8.1‰ (toluene), 3.7‰ (ethylbenzene), 9.5‰ (o-xylene), 6.8‰ (m-/p-xylene), 3.3‰ (naphthalene), 1.4‰ (1-methylnaphthalene), 2.3‰ (1-methylnaphthalene) (all field)	Griebler et al. (2003)
Benzene	Batch	Nitrate- and sulfate-reducing, methanogenic	$^{13}C/^{12}C$, D/H	$\varepsilon(C) = -1.9/-3.6‰$; $\varepsilon(H) = -29/-79‰$	Mancini et al. (2003)
Aromatic hydrocarbons, naphthalene	Column, field	Sulfate-reducing (column), nitrate reducing, sulfate-reducing	$^{13}C/^{12}C$	$\varepsilon(C) = -1.1‰$ (o-xylene, column); $\varepsilon(C) = -1.5‰$ (toluene, column)	Richnow et al. (2003a,b)

Table 9.10. (*Continued*)

Compound (Classes)	Scale	Assumed Terminal Electron Acceptors	Isotopes Measured	Isotope Fractionation[a]	Reference
Aromatic hydrocarbons	Field	Methanogenic, sulfate- and Fe(III)-reducing	$^{13}C/^{12}C$	$\varepsilon(C) = -1.5‰$ (ethylbenzene) $\varepsilon(C) = -2.1‰$ (m-p-xylene)	Richnow et al. (2003)
PAHs	Field	Aerobic, unsaturated	$^{13}C/^{12}C$	$\Delta(\delta^{13}C) = 2-8‰$	Yanik et al. (2003)
PCBs	Batch	Anaerobic	$^{13}C/^{12}C$	No significant fractionation	Drenzek et al. (2001)
Phenol, benzoate	Batch	Aerobic	$^{13}C/^{12}C$	$\Delta(\delta^{13}C) = 2-8‰$	Hall et al. (1999)
MTBE	Batch	Aerobic	$^{13}C/^{12}C$	$\varepsilon(C) = -1.5/-2.0‰$	Hunkeler et al. (2001a)
MTBE	Batch	Aerobic	$^{13}C/^{12}C$, D/H	$\varepsilon(C) = -1.4/-2.4‰$ $\varepsilon(H) = -29/-66‰$	Gray et al. (2002)
MTBE	Batch, field	Methanogenic	$^{13}C/^{12}C$	$\varepsilon(C) = -9.2‰$ (batch) $\varepsilon(C) = -8.1‰$ (field)	Kolhatkar et al. (2002)

[a] $\Delta(\delta^{13}C)$ and $\Delta(\delta^2 H)$, reported difference in isotopic composition; max. $\Delta(\delta^{13}C)$, maximum difference in isotopic composition.

Source: Adapted from Schmidt et al. (2004).

oxygenate since 1979 (Steffan et al. 1997), and its national use increased rapidly in the early 1980s at 40% per year (Suflita and Mormile 1993). MTBE is not used in all current gasoline formulations, but this additive has been used in amounts up to 15% by volume by many marketers. Thus, high MTBE concentrations in contaminated groundwater suggest a relatively recent (post-1980) gasoline release to the subsurface. It should be kept in mind, however, that atmospheric deposition of MTBE that had been previously volatilized can be a nonpoint source in urban areas (Pankow et al. 1997). Thus, traces of MTBE in groundwater (<10 ppb) do not necessarily indicate recent gasoline contamination. Forensic analysis based on trace constituents is further confounded by the existence of additional unknown sources of contaminant release, which is common in urban areas.

Fingerprinting analysis and examination of spatial and temporal concentration trends of groundwater contaminants are *bona fide* components of liability assessment, but caution should be exercised to avoid stretching forensic inferences beyond scientific constraints. For example, there is a strain of folklore asserting that the age of a petroleum product release can be established in any situation by comparing aqueous concentration ratios such as benzene to xylenes (B/X) or benzene-plus-toluene to ethylbenzene-plus-xylenes ([B + T]/[E + X]). This provocative idea is based on the fact that benzene and toluene are more soluble than ethylbenzene and xylenes and, thus, are preferentially solubilized from the gasoline. Thus, when solubilization is the sole fate and transport mechanism considered, B/X and [B + T]/[E + X] ratios tend to decrease near the source as the time since the spill occurred increases. Such changes in BTEX ratios can be described empirically with an exponential decay function that permits solving for the time required to achieve a given ratio and estimating the age of a release. However, inferring the age of a release based on BTEX ratios alone is a valid approach only in rare, well-constrained situations (Alvarez et al. 1998). Different attenuation mechanisms can affect the relative proportion of BTEX compounds at different rates, and sometimes in opposing directions. For example, biodegradation could increase B/X and [B + T]/[E + X] ratios in groundwater near the source, especially when anaerobic conditions prevail. This is so because benzene degrades very slowly if at all under anaerobic conditions. Downgradient from the source, BTEX ratios are predominantly influenced by adsorption-induced "chromatographic" separation (i.e., retardation), as BTEX compounds migrate at different velocities through the aquifer, and by differential biodegradation. Retardation tends to increase downgradient B/X and [B + T]/[B + X] ratios due to the higher solubility of benzene, whereas biodegradation can either increase or decrease such ratios, depending on the substrate preference of the prevailing phenotypes and the available electron acceptor pools. The aquifer properties that control the rate and extent of different attenuation mechanisms are site-specific. Therefore, one should not use the relative concentrations of individual BTEX compounds in groundwater to determine the age of a petroleum product release reliably. Such attempts bear a burden of proof that is often beyond the limits of scientific constraints and require far more data than can be obtained given common technical and economic constraints.

The above discussion illustrates that liability assessment via fingerprinting techniques requires consideration of how the composition may have weathered since its release, because of physical, chemical, and biological processes. Changes in fingerprints are inevitable but occur to different degrees and at different rates in nearly every situation (Stout et al. 1998).

Another caveat for liability assessment is the presumption that the presence of trace levels of polynuclear aromatic hydrocarbons (PAHs) is due to contamination from anthropogenic sources. PAHs are also by-products of incomplete combustion of organic matter. Thus, natural sources such as forest fires and volcanic activity can contribute to background PAH contamination. However, such (more recent) naturally occurring PAHs are typically alkylated (Mantseva et al. 2002). The anthropogenic contribution of these contaminants (e.g., from coal gasification and petroleum refining operations), which far exceeds natural sources (Blumer 1976), can thus be discerned by the lack of alkyl substituents.

Forensic analysis is further confounded by the existence of additional unknown sources of contaminant release. This is common in urban areas.

9.2.11. Microbial Analysis and Molecular Techniques

As discussed in Chapter 3, communities of microorganisms rather than single strains are most important in bioremediation. Microbial communities usually contain significant metabolic diversity as well as some metabolic redundancy that contributes to process robustness, and their characterization can yield valuable information to assess the efficacy of bioremediation. In particular, establishing that specific microorganisms that can degrade the target pollutants are present and that their concentrations are higher in the treatment zone compared to background samples is an important line of evidence to demonstrate that bioremediation is working. Microbial analysis is also important to identify the main organisms responsible for the reactions of interest, and to evaluate the efficacy of biostimulation or bioaugmentation approaches.

Historically, microbial analyses involved isolation and culture of specific microorganisms by virtue of their ability to grow on the pollutant of interest, followed by their identification based on morphological, physiological, and metabolic traits (e.g., Gram stain, enzyme activity, substrate utilization patterns). However, such traditional techniques are inadequate for microbial characterization of contaminated sites because they (1) are selective and not quantitative (Vestal and White 1989; White et al. 1997), (2) provide little insight into microbial consortium interactions (White et al. 1997), and (3) may introduce disturbance artifacts because these techniques involve subsampling and separation of microorganisms from the environmental matrix (Findlay et al. 1990). Furthermore, most microorganisms in the environment are viable but not cultivable (Xu et al. 1982; McCarthy and Murray 1996). Viable counts of bacteria in environmental samples determined with classical methods represent only a small fraction (0.1–10%) of the active microbial community (White et al. 1997). Furthermore, the most important "players" in the microbial community might not have been yet isolated and characterized. Such limitations have motivated the development of chemical and molecular characterization techniques to determine microbial biomass and community structure without prior isolation and cultivation of microorganisms.

Chemical Analyses

Current approaches used for chemical characterization of microbial populations in natural environments include two techniques that analyze the cell membrane phospholipids. These are (1) phospholipid ester-linked fatty acid (PLFA) analysis by gas chromatography/mass spectrometry (White et al. 1979), and (2) intact phospholipid profiling (IPP) using liquid chromatography/electrospray ionization/mass spectrometry (LC/ESI/MS) analysis of bacterial membrane phospholipids (Fang et al. 2000b). Both techniques rely on the fact that phospholipids are found in the membranes of all living cells, but not in storage lipids, and are rapidly turned over in dead cells. Thus, their quantification provides an estimation of viable biomass (Balkwill et al. 1988).

Identification of microorganisms by either PLFA or IPP is possible, but this approach is subject to potential confounding effects of overlapping phospholipid profiles and potential changes in phospholipid composition due to differences in growth conditions (Haack et al. 1993; White et al. 1997). Nevertheless, both techniques can give valuable insight into microbial community structure, based on the premise that there are a great number of dissimilar fatty acids in bacterial phospholipids. Some bacteria also contain unique fatty acids that can be used as biomarkers for their identification.

Molecular Analyses

DNA is present in every independently living cell in order to translate the genetic information into working enzymes and other proteins (see Chapter 3, Section 3.1.3). The advent of high-throughput methods of nucleic acid synthesis and the analysis of gene expression and function revolutionized the field of environmental microbiology and provided a boon to bioremediation (Lovley 2003). These new molecular-based tools allow us to interrogate the genetic information of community members to learn who is there, what they are doing (including their metabolic state), and how they interact, and how the organisms responsible for the reactions of interest respond to manipulations of environmental conditions. Common targets for microbial molecular analysis and the type of information gained are summarized in Table 9.11.

The most basic molecular target for microbial identification is the ribosomal RNA (rRNA) (Lane et al. 1985). All living organisms contain ribosomal RNA, and the small subunit of prokaryotic rRNA (known as 16S rRNA) has approximately 1500 nucleotide bases that can easily be sequenced. Thus, sequences are already known for thousands of bacteria (Cole et al. 2003). An example of the nucleotide sequence of 16S rRNA is depicted in Figure 9.12. The finding that 16S rRNA sequences are highly conserved allows the use 16S rRNA as an evolutionary clock that reveals phylogenetic relationships among members of complex microbial communities, as well as a measure of phylogenetic identity (Pace et al. 1986).

Table 9.11. Targets for Microbial Interrogation by Molecular Methods

Target	Information Gained	Question Answered
Ribosomal (r)RNA	Phylogenetic identity	Who is there?
Genes for rRNA on DNA	Phylogenetic identity	Who is there?
Catabolic genes on DNA	Phenotypic potential	What can they degrade?
Messenger (m)RNA	Phenotypic activity	What genes are being expressed?
Protein or other products	Phenotypic activity or phylogenetic identity	Who is active and what are they doing?

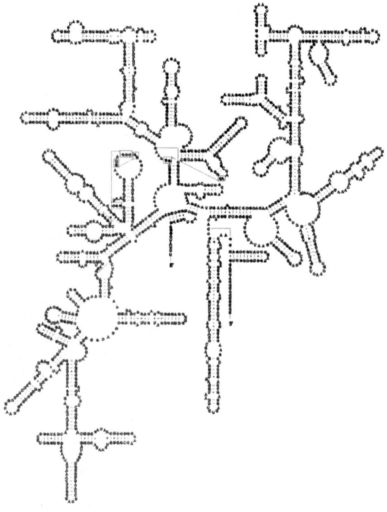

Figure 9.12. Secondary structure of the small subunit (16S) of ribosomal RNA from *Bacillus subtillus*.

Analysis of 16S rRNA is important to identify and quantify the presence of specific strains without having to rely on (biased) isolation and culturing techniques. Thus, depending on the primers used to amplify the 16S rRNA sequence (see PCR procedures below), researchers can now track and quantify specific strains or groups of phylogenetically related microorganisms.

Microorganisms with a known 16S rRNA sequence can also be identified, quantified, and visualized without culturing by using fluorescent in situ hybridization analysis (FISH) (Yang and Zeyer 2003; Thurnheer et al. 2004). Cells are hybridized with a molecular probe that is tagged with a fluorescent molecule, enabling microscopic visualization. Multiple different microorganisms can be visualized and counted by using probes tagged with different fluorescent molecules that vary in color.

Note that the same phyologenetic information inferred by 16S rRNA sequences can be obtained by analyzing the genes that code for 16S rRNA on the bacterial chromosome (Table 9.11). Phylogentic probes commonly used to detect and quantify the presence of specific degraders are listed in Table 9.12. Interestingly, 16S rRNA analyses have revealed that the microorganisms responsible for bioremediation are often phylogenetically related to those that can be cultured (Watanabe and Baker 2000). This has in many cases validated the relevance of studying isolates to learn about their biodegradation reactions and aspects of their physiology that are likely to control their growth and performance in contaminated environments.

Another common molecular target is a catabolic or another functional gene on chromosomal or plasmid DNA, such as those listed in Table 3.4, which code for enzymes in aerobic biodegradation pathways. Recently, several gene sequences associated with anaerobic biodegradation have been identified and targeted (Table 9.13). These include *bssA*, which codes for benzyl succinate synthase (Beller and Edwards 2000)—an enzyme that initiates anaerobic degradation of toluene and xylenes—and *vcr*, which codes for vinyl chloride reductase (Krajmalnik-Brown et al. 2004)—an enzyme associated with dehalorespiration and conversion of vinyl chloride to ethene. In general, there is a positive correlation between the abundance of catabolic genes of interest and the potential for contaminant degradation (Lovley 2003). Thus, determining which genes are present is important to establish what functions can be performed by the microbial community (i.e., phenotypic characteristics). However, since the presence of a gene does not guarantee its expression, this approach can only address the *phenotypic potential*, or "what the microorganisms could do." As mentioned above, the gene that codes for the 16S rRNA can also be targeted instead of the rRNA itself to address *phylogenetic identity*.

The messenger RNA (mRNA) can also be targeted by molecular probes to assess gene expression (see Table 9.10). Recall that mRNA is produced by transcription of the genes to form an RNA product that faithfully carries the genetic code from the DNA to the ribosome, where it is translated into an enzyme or another protein (see Figure 3.5). Thus, detection of specific mRNA that is complementary to a gene of interest means that the microorganisms are expressing or "turning on" that specific genes. Thus, mRNA analysis provides genetic information on what the cells "are

Table 9.12. Common Phylogenetic Probes Used to Detect and Quantify Specific Degraders

Target Group	Primer Name	Sequence	Reference
Bacteria: universal primers to determine total bacteria concentration	Unibac 8F Unibac 1541R	5'-AGAGTTTGATCCTGGCTCAG-3' 5'-AAGGAGGTGATCCAGCCGCA-3'	Löffler et al. (2000)
Archaea: universal primers to determine total *Archaea* concentration (including methanogens).	ARCH1-1369F ARCH2-1369F PROK1541R Probe	5'-CGGTGAATACGTCCCTGC-3' 5'-CGGTGAATATGCCCCTGC-3' 5'-AAGGAGGTGATCCTGCCGCA-3' FAM-5'-CTTGTACACACCGCCCG-TC-3'-BHQ-1	Suzuki et al. (2000)
Pseudomonas (16S rDNA), which is a catabolically versatile genus	Forward Reverse	5'-ACTGCATCCAAAACTGGCAA-3' 5'-TCTCTGCATGTCAAGGCCT-3'	Duteau et al. (1998)
Dehalococcoides (16S rDNA), which dechlorinate PCE or TCE to ethene	FpDHC1 RpDHC1212	5'-GATGAACGCTAGCGGCG-3' 5'-GGATTAGCTGTTCACACT-3'	Hendrickson et al. (2002)
Desulfitobacterium dehalogens (16S rDNA), which exhibits dechlorinating activity	Dd1 Dd2	5'-AATACCGNATAAGTTTATCCC-3' 5'-TAGCGATTCCGACTTCATGTTC-3'	El Fantroussi et al. (1997)
Desulfomonile (16S rDNA), which exhibits dechlorinating activity	Dt1 Dt2	5'-CAAGTCGTACGAGAAACATATC-3' 5'-GAAGAGGATCGTCTTTCCACGA-3'	El Fantroussi et al. (1997)
Geobacter sp. SZ (16S rDNA), an iron-reducing bacterium with dechlorinating activity	*Geo*-F *Geo*-R	5'-GAATATGCTCCTGATTC-3' 5'-ACCCTCTACTTTCATAG-3'	Sung (2004)
Bacteriophage (λ) primers used as internal standard to calculate DNA recovery (%) from environmental samples	Forward Reverse Probe	5'-ACGCCACGCGGGATG-3' 5'-AGAGACACGAAACGCCGTTC-3' FAM-5'-CTTCTGGTTCTTCTGCACCTTGGACACC-3'-TAMRA	Beller et al. (2002)

Table 9.13. Catabolic Gene Probes Used to Detect and Quantify Biodegradation Potential

Target Group	Primer Name	Sequence	Reference
		Aerobic Processes	
Catechol 2,3-dioxygenase, which catalyzes aerobic degradation of BTEX, phenol, naphthalene, and biphenyl	23CAT-F 23CAT-R DEG-F DEG-R QUANT-F	5'-CGACCTGATCTCCATGACCGA-3' 5'-TCAGGTCAGCACGGTCA-3' 5'-CGACCTGATC(AT)(CG)CATGACCGA-3' 5'-T(CT)AGGTCA(GT)(AC)ACGGTCA-3' 5'-CGACCTGATCTCCATGACCGATAACCGCAACGAAGTGTT CTG-3'	Mesarch et al. (2000)
Naphtalene dioxygenase, which initiates the oxidation of napthalene and co-oxidizes other PAHs	NAH-F NAH-R	5'-CAAAA(A/G)CACCTGATT(C/T)ATGG-3' 5'-A(C/T)(A/G)CG(A/G)G(C/G)GACTTCTTTCAA-3'	Baldwin et al. (2003)
Toluene dioxygenase, which initiates aerobic degradation of BTEX and chlorobenzene	TOD-F TOD-R	5'-ACCGATGA(A/G)GA(C/T)CTGTACC-3' 5'-CTTCGGTC(A/C)AGTAGCTGGTG-3'	Baldwin et al. (2003)
Xylene dioxygenase, which initiates aerobic xylene degradation	TOL-F TOL-R	5'-TGAGGCTGAAACTTTACGTAGA-3' 5'-CTCACCTGGAGTTGCGTAC-3'	Baldwin et al. (2003)
Biphenyl dioxygenase, which initiates the aerobic degradation of lightly chlorinated PCBs	BPH1-F BPH1-R BPH2-F BPH2- RBPH3-F BPH4-F BPH3-R	5'-GGACGTGATGCTCGA(C/T)CGC-3' 5'-TGTT(C/G)GG(C/T)ACGTT(A/C)AGGCCCAT-3' 5'-GACGCCCGCCCTATATGGA-3' 5'-AGCCGACGTTGCCAGGAAAAT-3' 5'-CCGGGAGAACGGCAGGATC-3' 5'-AAGGCCGGCGACTTCATGAC-3' 5'-TGCTCCGCTGCGAACTTCC-3'	Baldwin et al. (2003)

503

Table 9.13. (*Continued*)

Target Group	Primer Name	Sequence	Reference
Hydroxyl monooxygenases in the large subunits of the toluene monooxygenase gene	RMO-F RMO-R RDEG-F RDEG-R	5'-TCTC(A/C/G)AGCAT(C/T)CAGAC(A/C/G)GACG-3' 5'-TT(G/T)TCGATGAT(C/G/T)AC(A/G)TCCCA-3' 5'-T(C/T)TC(A/C/G)AGCAT(A/C/T)CA(A/G)AC(A/C/G)GA(C/T)GA-3' 5'-TT(A/G/T)TCG(A/G)T(A/G)AT(C/G/T)AC(A/G)TCCCA-3'	Baldwin et al. (2003)
Phenol hydroxylases, which hydroxylate (halo)phenols and other phenolics	PHE-F PHE-R	5'-GTGCTGAC(C/G)AA(C/T)CTG(C/T)TGTTC-3' 5'-CGCCAGAACCA(C/T)TT(A/G)TC-3'	Baldwin et al. (2003)
Anaerobic Processes			
Benzylsuccinate synthase, which initiates anaerobic degradation of toluene and xylenes	bssA-F bssA- R Probe	5'-ACGACGGYGGCATTTCTC-3' 5'-GCATGATSGGYACCGACA-3' FAM-5'-CTTCTGGTTCTTCTGCACCTTGGACACC-3'-TAMRA	Beller et al. (2002)
TCE reductive dehalogenase system, which mediates the dechlorination of TCE to ethane	tceA-797F tceA-2490R	5'-ACGCCAAAGTGCGAAAAGC-3' 5'-GAGAAAGGATGGAATAGATTA-3'	Magnuson et al. (2000)
Vinyl chloride reductive dehalogenase, which dechlorinates *cis*-DCE and VC to ethane	vcrA-F vcrA-R	5'-CTATGAAGGCCCTCCAGATGC-3' 5'-GTAACAGCCCCAATATGCAAGTA-3'	Spormann et al. (2004)
Oxygen-insensitive, nitro- and chromate-reductase subfamily NsfA, which reduces oxidized nitrogen and metal groups.	NfsA-F NfsA-R NfsB-F NfsB-R	5'-GTAGGATCCACGCCAACCATTGAAC-3' 5'-ACTGAATTCTTAGCGCGTCGCCCAAC-3' 5'-GTAGGATCCGATATCATTTCTGTCGC-3' 5'-ACTGAATTCTTACACTTCGGTTAAGGTG-3'	Kwak et al. (2003)

Description	Primer	Sequence	Reference
NiFe hydrogenase gene, present in dissimilatory metal reducing bacteria	Hyd1F	5'-CGCCCGCCGCGCCCCGCGCCCGTCCCGCGCCGCCCCGCCCG-3'	Wawer et al. (1995)
	Hyd5R	5'-GCAGGGCTTCCAGGTAGTGGGCGGTGGGCGATGAGGT-3'	
Dissimilatory sulfite reductase (DSR) gene, present in all sulfate-reducing bacteria	dsr1F	5'-ACSCACTGGAAGCACG-3'	Wagner et al. (1997)
	dsr4R	5'-GTGTAGCAGTTACCGCA-3'	
Nitrite reductases, present in denitrifying bacteria	nirK1F	GG(A/C)ATGGT(G/T)CC(C/G)TGGCA	Braker et al. (1998)
	nirK2F	GC(C/G)(C/A)T(C/G)ATGGT(C/G)CTGCC	
	nirK3R	GAACTTGCCGGT(A/C/G)G(C/T)CCAGAC	
	nirK4R	GG(A/G)AT(A/G)A(A/G)CCAGGTTTCC	
	nirK5R	GCCTCGATCAG(A/G)TT(A/G)TGG	
	nirS1F	CCTA(C/T)TGGCCGCC(A/G)CA(A/G)T	
	nirS2F	TACCACCC(C/G)GA(A/G)CCGCGCGT	
	nirS3F	TTCCT(C/G/T)CA(C/T)GACGGCGGC	
	nirS4F	TTC(A/G)TCAAGAC(C/G)CA(C/T)CCGAA	
	nirS3R	GCCGCCGTC(A/G)TG(A/C/G)AGGAA	
	nirS5R	CTTGTTG(A/T)ACTCG(C/G)(C/G)CTGCAC	
	nirS6R	CGTTGAACTT(A/G)CCGGT	

doing." Higher biodegradation rates are often associated with higher concentrations of mRNA (Fleming et al. 1993). However, extracting and purifying intact mRNA from environmental samples can be a significant challenge due to the labile nature of mRNA (Sessitsch et al. 2002).

Because transcription of a gene does not guarantee translation, researchers can also target the final product of gene expression, often a catabolic enzyme. A common approach to assay enzyme activities is to measure the transformation rate of a specific substrate by whole cells or cell-free extracts, and to normalize the rate to the protein concentration (Table 9.14). The presence of some enzymes can also be detected using enzyme-linked immunoassay (ELISA) techniques with specific antibodies (Lynch et al. 1996), although this approach is subject to considerable interference by soil matrix constituents such as humic acids.

Table 9.14. Selected Microbial Enzyme Assays

Enzyme	Function	Method Reference
Arylphosphatase	Hydrolyzes pesticides with arylphosphate esters at the O—P bond	Tabatabai and Bremmer (1970)
Catalase	Conversion of hydrogen peroxide, a powerful and potentially harmful oxidizing agent, to water and molecular oxygen	Day et al. (2000)
Dehalogenase	Degradation of chlorinated compounds by hydrolytic cleavage of carbon–halogen bond, replacing halogen with hydroxyl group	Holloway et al. (1998)
Dehydrogenase	Catalyzes the removal of hydrogen from a substrate	Trevors (1984)
Fluorescein diacetate hydrolase	Simple, sensitive, and rapid method for determining microbial activity in soil	Adam and Duncan (2001)
Hydrogenase	Catalyzes the reversible reductive formation of hydrogen from protons and electrons; used by hydrogen-oxidizing bacteria	Watrous et al. (2003)
Nitrogenase	Nitrogen fixation (i.e., reduction of N_2 to NH_3, which is critical for the N cycle)	Staal et al. (2001)
Phosphatase	Hydrolysis of phosphate ester bonds in organic phosphate pesticides	Sannino and Gianfreda (2001)
Toluene dioxygenase	Aerobic BTEX degradation and cometabolism of chlorinated aliphatic compounds such as TCE	Ensley and Gibson (1983)
Urease	Hydrolyzes urea to ammonia and CO_2	Kandeler and Gerber (1988)
Dechlorination activity (nonspecific)	Colorimetric assay that measures dechlorination activity based on the release of chlorine ions	Holloway et al. (1998)

Chemical analyses of the outcome of the activity of proteins, such as fatty acid methyl esters (FAMEs), can also be conducted not only as an alternative to identify microorganisms, as described above, but also to evaluate microbial stress and nutritional status. For example, the ratio of *trans/cis* C16:1Δ9 fatty acids in cell membrane phospholipids has been used as an indicator of stress due to toxicity, contamination, or oxygen tension on microbial communities (Guckert et al. 1985). Apparently, bacteria synthesize more *trans*-monounsaturated acids under stress conditions as a mechanism to maintain their membrane fluidity.

Because the DNA or RNA of interest is often present at very low, undetectable concentrations, assaying for it generally requires that the DNA be amplified first. This is often accomplished by the polymerase chain reaction (PCR). Nevertheless, only a small fraction of microorganisms in the environment have been isolated and studied. Thus, it is often necessary to track and identify strains for which we have no molecular information. In such cases, molecular fingerprinting techniques such as denaturing gradient gel electrophoresis (DGGE) and terminal restriction fragment length polymorphism (T-RFLP) are important analytical tools that researchers often use to complement the study of microbial communities. An overview of PCR and molecular fingerprinting principles is provided below.

Polymerase Chain Reaction. Polymerase chain reaction (PCR) is a method to amplify DNA by producing multiple copies of a gene to enhance molecular analysis. This process is named after the enzyme polymerase, which is capable of replicating DNA. A DNA polymerase that can function at relatively high temperatures (needed to separate two strands of DNA that make up the double helix) is used in PCR. This enzyme was isolated from *Thermus aquaticus* and is called Taq polymerase. PCR requires first a clean sample of the DNA to be replicated, as well as sufficient nucleotides in the reaction medium (i.e., the building blocks of DNA: A (adenine), G (guanine), T (thymine), and C (cytosine)). Other key requirements for PCR include large quantities of primers (short pieces of nucleotide sequences to get the replication started) and the Taq polymerase mentioned above.

The PCR reaction is often carried out in a single vial (Figure 9.13). The first step is to separate the DNA strands, which is accomplished by heating the vial to 95 °C. The solution is then cooled to between 40 and 60 °C to allow primers to anneal to the target DNA sequence. Replication of DNA is subsequently performed by the Taq polymerase at 72 °C. This process is repeated about 30 times, producing millions of copies of the DNA. The amplified DNA can then be viewed on an agarose gel stained with ethidium bromide. This method can be used with primers specific for the 16S rRNA to determine if certain organisms are present, or with primers that target specific catabolic genes to determine if the biodegradation potential is present. Note that PCR by itself is only a qualitative tool since it can only determine the presence or absence of a gene.

Besides the detection of the target gene sequences, quantification is desirable in many cases to determine, for example, changes in the concentration of specific genotypes resulting from specific engineered manipulations (e.g., biostimulation).

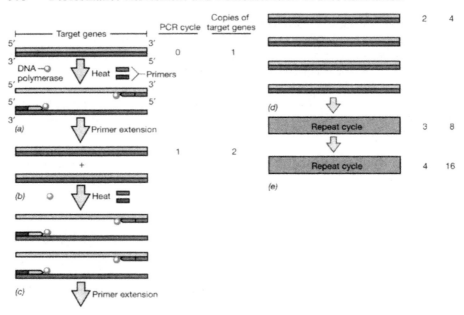

Figure 9.13. The polymerase chain reaction (PCR) for amplifying specific DNA sequences. (a) Target DNA is heated to separate the strands, and a large excess of two oligonucleotide primers, one complementary to the target strand and one to the complementary strand, is added along with DNA polymerase. (b) Following primer annealing, primer extension yield a copy of the original double-stranded DNA. (c) Further heating, primer annealing, and primer extension yield a second double-stranded DNA. (d) The second double-stranded DNA. (e) Two additional PCR cycles yield 8 and 16 copies, respectively, of the original DNA sequence.

Real-time quantitative polymerase chain reaction (RTQ-PCR) was developed for this purpose. This recently developed technology offers several advantages over conventional methods to detect specific genes in environmental samples, including higher sensitivity, accuracy, and the ability to collect quantitative data rapidly in real-time during the amplification process (Livak et al. 1995; Bustin 2000). RTQ-PCR is based on the real-time measurement of the PCR product throughout the amplification reaction using fluorescence labeling (i.e., fluorogenic probe or double-stranded DNA binding dye, such as SYBRgreen or TaqMan) (Heid et al. 1996). In the TaqMan procedure, the $5'$ exonuclease activity of Taq DNA polymerase digests the fluorogenic TaqMan probe that anneals to the target DNA site during primer extension, resulting in the release of a fluorescent molecule. The cycle number threshold (C_T) value is reached when a significant increase in fluorescence emission occurs relative to the background baseline. A larger initial concentration of DNA template results in a lower C_T value. The main goal of RTQ-PCR is thus to quantify the initial target, no longer by postreaction analysis of the PCR product, but by measuring C_T (which is linearly related to the log of the initial target copy) at which the fluorescence signal crosses a fixed threshold during the exponential phase of the PCR. Specifically, RTQ-PCR with SYBRgreen is very similar in principle to traditional PCR. The SYBRgreen procedure is similar. The reaction contains all the same ingredients as traditional PCR, plus SYBRgreen dye. This dye binds to the DNA and fluoresces only when the primers have annealed to the single-stranded DNA and the polymerase begins to replicate the target sequence. Once the DNA is denatured again for the next cycle, the SYBRgreen stops fluorescing; therefore, the RTQ-PCR machine only records the fluorescence at the end of a cycle. Since it is recorded during the exponential phase, the key parameter C_T is free of the fluctuations affecting traditional end-point measurements (Bustin 2000). The RTQ-PCR detection is linear at least over a range of 10^4–10^8 cells/mL, and the detection limit is generally around 10^2 cells/mL (Lendvay et al. 2003).

Fingerprinting Techniques. Molecular fingerprinting is a useful approach to characterize microbial communities and to track the emergence (or disappearance) of important strains or genes, even when the identity of the microorganism of interest is unknown. Fingerprinting relies on the principle that similar base sequences exist for genes that code for similar products in different microorganisms. Thus, a generic primer can be designed for a given class of genes. When such a primer is used in PCR, the amplified (DNA or 16S rRNA) segments are from all members of the microbial community possessing similar genes. Different amplified products can subsequently be separated by electrophoresis to create a spectrum of bands known as the molecular fingerprint. Changes in the molecular fingerprint can be analyzed to track the evolution of the microbial community structure in space and in time.

Denaturing gradient gel electrophoresis (DGGE) is perhaps the most frequently used fingerprinting technique. DGGE involves separating the (selectively amplified) DNA sequences as they migrate under the influence of electrophoresis through a gel. The name of this technique is due to the fact that the gel contains a gradient of a denaturing chemical (usually formamide and urea) that denatures (or separates) the two DNA strands. When separation occurs, the mobility of DNA is retarded due to the branched structure of the single-stranded molecule, and subsequent entangling of the DNA segment in the gel matrix prevents further movement. Different DNA segments are separated and trapped at different positions on the gel. This position depends on the size of the DNA segment and the content of GC base pairs. Smaller DNA segments or those with higher GC content migrate farther through the gel. A high GC content is conducive to resistance to melting (and thus further migration) because GC pairs form triple hydrogen bonds. DNA regions that are rich in double-bonded AT base pairs are easier to melt and do not migrate as far through the gel. The ability of DGGE to separate DNA from different strains is due to the fact that similar DNA molecules that differ by only one nucleotide will have different melting properties that determine the migration distance. The trapped DNA can be visualized as bands in the gel that can be excised and sequenced to obtain

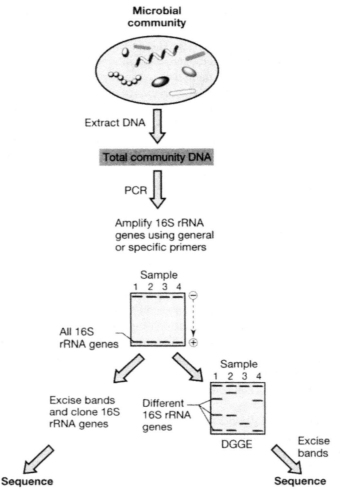

Figure 9.14. Steps in biodiversity analysis of a microbial community using phylogenetic probes. Total community DNA is used with PCR to amplify 16S rRNA genes using universally conserved primers for bacteria or primers that will target only a particular phylum of bacteria. The PCR bands are excised and the different 16S genes are separated by either cloning or DGGE. Note how, in the DGGE gel, samples 1, 2, and 4 share a common band (gene), whereas samples 2 and 3 each contain one unique band.

phylogenetic information (based on 16S rDNA sequences) or knowledge of the function of the gene (Figure 9.14). The obtained sequences can also be used to design oligonucleotide probes to detect and quantify the presence of specific genes in other environmental samples.

Microbial population diversity can also be evaluated by analyzing DGGE bands, with higher numbers of bands corresponding to greater diversity. One approach to quantify diversity is to determine the Shannon–Weaver diversity index (H) (Atlas and Bartha 1998):

$$H = \frac{2.3}{N} \left(N \log_{10} N - \sum n_i \log_{10} n_i \right)$$ [9.11]

where N is the number of species present (taken as the number of DGGE bands) and n_i is the number of individuals of the ith species (estimated by integration of the area of the DGGE bands, using a densitometer).

DNA-microarray technology (also called a "DNA microchip"), which was developed in the medical field to study the genetic basis of many diseases, is also emerging as a valuable approach to expand the capabilities of molecular analysis of microbial communities. This is a powerful tool to simultaneously assess the presence or activity of up to thousands of different genes. Chips could be synthesized with arrays of hundreds to thousand of molecular probes, allowing us to easily track (simultaneously, with one assay) specific strains (by targeting 16S rRNA), catabolic or other functional genes (by targeting amplified DNA sequences), and gene expression (by tracking mRNA) (Wu et al. 2001; Cho and Tiedje 2002; Zhou 2002; Bodorossy et al. 2003; Dennis et al. 2003). However, microarrays can be difficult to use with environmental samples due to currently low detection sensitivity. To be detected by current microarray technology, a target gene sequence may need to comprise at least 5% of the total DNA in the sample (Cho and Tiedje 2002).

Terminal restriction fragment length polymorphism (T-RFLP) is another common molecular fingerprinting technique used for microbial community analysis (Figure 9.15). This technique measures the size polymorphism of terminal restriction fragments from a PCR-amplified marker. Similar to DGGE, T-RFLP can be used for strain identification, for comparative community analysis, and to

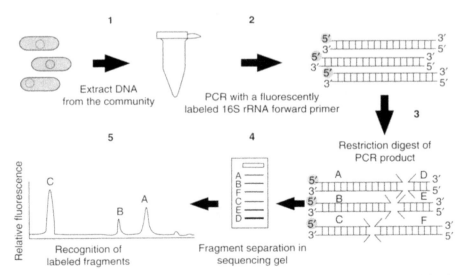

Figure 9.15. Community diversity analysis by T-RFLP (http://rdp.cme.msu.edu/html/t-rflp_jul02.html).

estimate the diversity of a phylogenic group within a community (Marsh 1999). However, T-RFLP may offer greater detection sensitivity to successfully discriminate between common environmental microorganisms (Porteus et al. 2002; Lendvay et al. 2003; Gu et al. 2004). T-RFLP involves performing PCR with a fluorescently labeled primer to amplify 16S rRNA (Figure 9.16). The PCR product

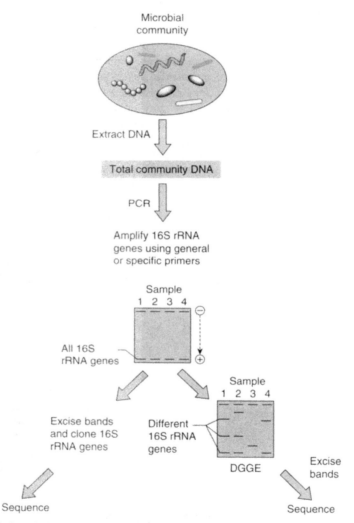

Figure 9.16. Steps in biodiversity analysis of a microbial community using phylogenetic probes. Total community DNA is used with PCR to amplify 16S rRNA genes using universally conserved primers for bacteria or primers that will target only a particular phylum of bacteria. The PCR bands are excised and the different 16S genes are separated by either cloning or DGGE. Note how in the DGGE gel, samples 1, 2, and 4 share a common band (gene), whereas samples 2 and 3 each contain one unique band. (Adapted from Madigan et al. 2003.)

is subsequently digested with several restriction endonucleases that cut the DNA into fragments. Different 16S rRNA fragment sizes are obtained from different microorganisms because each organism has a unique 16S rRNA sequence that is cut by endonucleases at different places. The enzyme digest is then separated in a sequencing gel. Capillary electrophoresis is then used to generate a graph of relative fluorescence versus fragment size in base pairs (bp). The number of labeled fragment peaks is proportional to the number of species in the community.

In summary, similar to stable isotope analysis, an increase in the concentration of catabolic genes could be considered to be very reliable and robust primary evidence of biodegradation, whereas phylogenetic probes, PFLA, and DGGE are useful secondary lines of evidence. We recognize that the presence of a gene does not guarantee its expression. However, catabolic gene copy numbers tend to be temporally quite responsive to changing environmental conditions, and their expression is conducive to bacterial growth on the target pollutant with an associated significant increase in gene copy numbers relative to background levels as a result of bioremediation. Thus, mRNA analysis (which is a much more difficult task due to the instability of mRNA) may not always be required to demonstrate catabolic gene expression.

9.3. CONCLUDING REMARKS

Natural attenuation is the reliance on processes such as biodegradation, dispersion, dilution, sorption, volatilization, chemical or biological stabilization, and transformation or destruction of contaminants to remediate contaminated sites. Bioremediation and natural attenuation are receiving increasing attention due to their potential cost-effectiveness. In particular, natural attenuation showed tremendous growth during the 1990s from a laboratory research project to an approach that is commonly used to clean up contaminated groundwater. Natural attenuation is not to be viewed as a "no action" or "walk-away" approach, but rather as an alternative approach of achieving remediation objectives that may be appropriate for specific, well-documented site circumstances where its use meets the applicable statutory and regulatory requirements. We emphasize that every site where natural attenuation is considered as a remedial option needs to be evaluated very carefully. That is, even if there is a site where natural attenuation may look like it will work, the hydrogeology, geochemistry, and microbiology unique to the site could result in attenuation not occurring or perhaps not completely destroying all the pollutant mass. Because natural attenuation could potentially involve the interaction of a large number of processes, extensive site characterization and understanding of geochemical as well as microbial processes operating in the subsurface become crucial when relying on these approaches for site remediation. To some, natural attenuation may seem like a panacea of remediation technologies, but we have seen in this chapter that it is not universally applicable and effective. Thus, for the wide acceptance of natural attenuation for all interested parties (regulators, responsible parties, public citizens, etc.), there is a critical need to demonstrate whether natural attenuation is actually working in a particular field setting at a rate that is sufficient to protect public health and the environment. Such a demonstration also is referred to as *performance assessment*, which involves integrated performance and risk assessments of natural attenuation in simple to complex hydrogeological settings. In order to demonstrate that natural attenuation is working one must:

1. Demonstrate a clear and meaningful trend of decreasing contaminant concentration utilizing historical groundwater and/or soil data.
2. Provide data of geochemical parameters that demonstrate naturally occurring degradation and provide estimates of attenuation rates.
3. Provide data from field or microcosm studies, or more sophisticated analysis, that demonstrates the occurrence of biological degradation processes.

The demonstration of a clear and meaningful trend of decreasing contaminant concentration is most important to showing that natural attenuation at a given site is occurring, but data on geochemical parameters as well as microcosm studies become important when a trend in contaminant mass destruction cannot be established at sufficient quality and duration. We also discussed innovative approaches, some of which are routinely used and others that are currently under development. These innovative approaches should be used to complement the traditional site characterization techniques. Forensic analysis is used by environmental professionals to determine whether or not natural attenuation is in fact degrading the contaminants at a given site. There are many forms of forensic analysis that include monitoring geochemical parameters for trends indicative of biodegradation, numerical modeling of degrading contaminant concentrations, microcosm studies, chemical fingerprinting, stable isotope analysis, and molecular techniques. Among these, stable isotope analysis and detection of catabolic genes could be considered to be very reliable and robust primary evidence of biodegradation, whereas phylogenetic probes, PFLA, and DGGE are useful secondary lines of evidence.

We have also seen in this chapter that the mechanistic understanding of bioremediation is still in its infancy. Newer technologies will undoubtedly enrich our mechanistic understanding, leading to better assessments and predictions of natural attenuation in the field settings. These methods collectively will allow for the determination of metabolic by-products, the presence of microorganisms, and decreasing contaminant trends that will prove that natural attenuation is occurring, so that the EPA or respective state agency can justify not closing a site.

It is important to recognize that traditional and innovative approaches for characterization including forensic techniques provide disparate information of widely ranging quality and quantity. In order to show that natural attenuation is taking place at a site, one should not rely on only one piece of information. Instead, we advocate the use of multiple lines of evidence, with the traditional approaches described in this chapter taking the central role, while the forensic approaches provide complementary evidence. In some cases, the forensic analysis may provide the key evidence to prove that natural attenuation is working.

We would like to conclude this book by pondering that, since the dawn of socioeconomic development, humankind has exploited natural resources to improve the quality of life. However, in many cases the extraction, processing, utilization, and disposal of natural resources as well as the synthesis of new materials have created enormous pollution problems. Today, natural resources show signs of severe deterioration, and, in particular, our aquifers have been contaminated severely in some areas through the continuous disposal of hazardous chemicals. Pollution prevention, green engineering, and other general efforts to enhance sustainability will be crucial to protect environmental health. However, some extent of environmental pollution seems to be an unavoidable unintended consequence of human activities, and there will always be a (hopefully decreasing) need for pollution control and site remediation. Fortunately, Mother Nature is quite robust and there are signs that natural attenuation and bioremediation will play a crucial role in assimilating and treating many contaminants. Such natural phenomena and ecological forces will play an increasingly important role in hazardous waste treatment.

As robust as bioremediation appears, we should not misuse or overprescribe this technology in situations where our mechanistic understanding is incomplete regarding the degradation of exotic compounds, or the effects of engineered manipulations on microbial activities cannot be reasonably predicted. We have shown in this book that bioremediation is an emerging technology that is based on sound ecological and engineering principles, but further research is needed to understand response variability as a function of site specificity, and to formulate unifying principles that enhance process efficacy and reliability.

We hope that this book has enriched the reader's understanding of the physical, chemical, and biological processes that form the foundation for both bioremediation and natural attenuation, including the use of mathematical models that can enhance performance assessment and risk management. We also genuinely hope to have increased the reader's interest in learning more about this fascinating and exciting topic and contributing to its continuing development. Thus, we conclude with a quote from Sir Winston Churchill:

Now this is not the end. It is not even the beginning of the end.
But it is, perhaps, the end of the beginning.

REFERENCES

Abraham, W.-R., C. Hesse, and O. Pelz (1998). Ratios of carbon isotopes in microbial lipids as an indicator of substrate usage. *Appl. Environ. Microbiol.* **64**:4202–4209.

Adam, G. and H. Duncan (2001). Development of a sensitive and rapid method for the measurement of total microbial activity using fluorescein diacetate (FDA) in a range of soils. *Soil Biol. Biochem.* **33**:943–951.

Ahad, J.M.E., B.S. Lollar, E.A. Edwards, G.F. Slater, and B.E. Sleep (2000). Carbon isotope fractionation during anaerobic biodegradation of toluene: implications for intrinsic bioremediation. *Environ. Sci. Technol.* **34**(5):892–896.

Alvarez, P.J.J., R.C. Heathcote, and S.E. Powers (1998). Caution against interpreting gasoline release dates based on BTEX ratios in ground water. *Ground Water Monitoring & Remediation* **Fall**:69–76.

Alvarez, P.J.J. (2001). *Foundation in Bioremediation Course Pack*. Department of Civil and Environmental Engineering, University of Iowa.

American Petroleum Institute (2003). *Groundwater Remediation Strategies Tool*. Regulatory Analysis & Scientific Affairs Department, Publication Number 4730.

Atlas, R.M. and R. Bartha (1998). *Microbial Ecology*, 4th ed. Benjamin/Cummings, Menlo Park, CA.

Azadpour-Keeley, A., J.W. Keeley, H.H. Russell, and G.W. Sewell (2001). Monitored natural attenuation of contaminants in the subsurface: applications. *Ground Water Monitoring & Remediation* **21**(3):136–143.

Baldwin, B.R., C.H. Nakatsu, and L. Nies (2003). Detection and enumeration of aromatic oxygenase genes by multiplex and real-time PCR. *Appl. Environ. Microbiol.* **6**(69):3350–3358.

Balkwill, D.L., F.R. Leach, J.T. Wilson, J.F. McNabb, and D.C. White (1988). Equivalence of microbial biomass measures based on membrane lipid and cell wall components, adenosine triphosphate and direct counts in subsurface aquifer sediments. *Microbial Ecol.* **16**:73–84.

Barth, J.A.C., G. Slater, C. Schuth, M. Bill, A. Downey, M. Larkin, and R.M. Kalin (2002). Carbon isotope fractionation during aerobic biodegradation of trichloroethene by *Burkholderia cepacia* G4: a tool to map degradation mechanisms. *Appl. Environ. Microbiol.* **68**(4):1728–1734.

Beller, H.R. and E.A. Edwards (2000). Anaerobic toluene activation by benzylsuccinate synthase in a highly enriched methanogenic culture. *Appl. Environ. Microbiol.* **66**:5503–5505.

Berthouex, P.M. and L.C. Brown (1994). *Statistics for Environmental Engineers*. Lewis Publishers, Boca Raton, FL.

Bloom, Y., R. Aravena, D. Hunkeler, E. Edwards, and S.K. Frape (2000). Carbon isotope fractionation during microbial dechlorination of trichloroethene, cis-1,2-dichloroethene, and vinyl chloride: implications for assessment of natural attenuation. *Environ. Sci. Technol.* **34**(13):2768–2772.

Blumer, M. (1976). Polycyclic aromatic compounds in nature. *Sci. Am.* **234**:35–45.

Bodorossy, L., N. Stralis-Pavese, J.C. Murrell, S. Radajewski, A. Weilharter, and A. Sessitsch (2003). A development and validation of a diagnostic microbial microarray for methanotrophs. *Environ. Microbiol.* **5**:1170.

Boschker, H.T.S., S.C. Nold, P. Wellsbury, D. Bos, W. de Graaf, R. Pel, R.J. Parkes, and T.E. Cappenberg (1998). Direct linking of microbial populations to specific biogeochemical processes by ^{13}C-labeling of biomarkers. *Nature* **392**:801–805.

Braker, G., A. Fesefeldt, and K.P. Witzel (1998). Development of PCR primer systems for amplification of nitrite reductase genes (*nirK* and *nirS*) to detect denitrifying bacteria in environmental samples. *Appl. Environ. Microbiol.* **64**(10):3769–3775.

Brungard, K.L., J. Munakata-Marr, C.A. Johnson, and K.W. Mandernack (2003). Stable carbon isotope fractionation of *trans*-1,2-dichloroethylene during co-metabolic degradation by methanotrophic bacteria. *Chem. Geol.* **195**(1-4):59–67.

Buscheck, T.E., K.T. O'Reilly, and G. Hickman (1997). Intrinsic anaerobic biodegradation of chlorinated solvents at a manufacturing plant. In *In Situ and On-Site Bioremediation*, Vol. 3. Battelle Press, Columbus, OH, pp. 149–154.

Bustin, S.A. (2000). Absolute quantification of mRNA using real-time reverse polymerase chain reaction assays. *J. Mol. Endocrinol.* **25**:169–193.

Chapelle, F.H., D.A. Vroblesky, J.C. Woodward, and D.R. Lovley (1997). Practical considerations for measuring hydrogen concentrations in groundwater. *Environ. Sci. Technol.* **31**(10):2873–2877.

Chapelle, F.H. (2001). *Ground-Water Microbiology and Geochemistry*, 2nd ed. John Wiley & Sons, Hoboken, NJ.

Cho, J.C. and J.M. Tiedje (2002). Quantitative detection of microbial genes by using DNA microarays. *Appl. Environ. Microbiol* **68**:1425.

Clark, C. and B.A. Berven (1984). *Results of the Groundwater Monitoring Program Performed at the Former St. Louis Airport Storage Site for the Period of January 1981 through January 1983*. National Technical Information Service, Springfield, VA, ORNL-TM-8879.

Clark, I. and P. Fritz (1997). *Environmental Isotopes in Hydrogeology*, CRC Press, Boca Raton, FL.

Cole, J.R., B. Chai, T.L. Marsh, R.J. Farris, Q. Wamg, S.A. Kulam, S. Chandra, D.M. McGarrell, T.M. Schmidt, G.M. Garrity, and J.M. Tiedje (2003). The Ribosomal Database Project (RDP-II): previewing a new autoaligner that allows regular updates and the new prokaryotic taxonomy. *Nucleic Acids Res.* **31**:442.

Conrad, R. (1996). Soil microorganisms as controllers of atmospheric trace gases (H_2, CO, CH_4, OCS, N_2O, and NO). *Microbiol. Rev.* **60**(4):609–640.

Da Silva, M.L. and P.J.J. Alvarez (2002). Effects of ethanol versus MTBE on BTEX migration and natural attenuation in aquifer columns. *ASCE J. Environ. Eng.* **128**(9):862–867.

Davee, K.W. and D.A. Sanders (2000). Petroleum hydrocarbon monitored natural attenuation: essential framework for remedial managers. *Environ. Geosci.* **7**(4):190–202.

Day, W.A., J.L. Sajecki, T.M. Pitts, and L.A. Joens (2000). Role of catalase in *Campylobacter jejuni* intracellular survival. *Infect. Immunity* **68**(11):6337–6345.

Day, M., R. Aravena, D. Hunkeler, and T. Gulliver (2002). Application of carbon isotopes to document biodegradation of *tert*-butyl alcohol under field conditions. *Contam. Soil Sediment Water*, 88–92.

Dennis, P., E.A. Edwards, S.N. Lisss, and R. Fulthorpe (2003). Monitoring gene expression in mixed microbial communities by using DNA microarrays. *Appl. Environ. Microbiol.* **69**:769.

Drenzek, N.J., T.I. Eglinton, C.O. Wirsen, H.D. May, Q. Wu, K.R. Sowers, and C.M. Reddy (2001). The absence and application of stable carbon isotopic fractionation during the reductive dechlorination of polychlorinated biphenyls. *Environ. Sci. Technol.* **35**:3310–3313.

Dutwau, N.M., J.D. Rogers, C.T. Bartholomay, and K.F. Reardon (1998). Species-specific oligonucleotide for enumeration of *Pseudomonas putida* F1, *Burkholderia* sp. strain JS150, and *Bacillus subtillis* ATCC 7003 in biodegradation experiments. *Appl. Environ. Microbiol.* **64**:4994–4999.

Einarson, M.D. (2001). Flux-Based Corrective Action. Princeton Groundwater Remediation Course, Denver, CO.

Einarson, M.D. and D.M. MacKay (2001a). Predicting the impacts of groundwater contamination. *Environ. Sci. Technol.* **35**(3):66A–73A.

Einarson, M.D. and D.M. MacKay (2001b). Supplementary information to accompany ES&T feature article. Predicting the impacts of groundwater contamination. *Environ. Sci. Technol.* **35**(3):1–13.

El Fantroussi, S. et al. (1997). Introduction and PCR detection of *Desulfomonile tiedjei* in soil slurry microcosms. *Biodegradation* **8**(2):125–133.

EPA (Environmental Protection Agency) (1999). Use of Monitored Natural Attenuation at Superfund, RCRA Corrective Action, and Underground Storage Tank Sites. Office of Solid Waste and Emergency Response, OSWER Directive Number 9200.4-17P.

Fang, J. and M.J. Barcelona (1998). Structural determination and quantitative analysis of phospholipids using liquid chromatography/electrospray ionization/mass spectrometry. *J. Microbiol. Methods* **33**:23–35.

Fang, J., M.J. Barcelona, and C. West (1997). The use of aromatic acids and phospholipid ester-linked fatty acids for delineation of processes affecting an aquifer contaminated with JP-4 fuel. In *Molecular Markers in Environmental Geochemistry*, R.P. Eganhouse (Ed.). American Chemical Society, Washington, DC, pp. 65–76.

Fang, J., M.J. Barcelona, and P.J. Alvarez (2000a). Phospholipid compositional changes of five Pseudomonad archetypes grown with and without toluene. *Appl Microbiol. Biotechnol.* **54**:382–389.

Fang, J., M. Barcelona, and P.J. Alvarez (2000b). A direct comparison between fatty acid analysis and intact phospholipid profiling for microbial identification. *Org. Geochem.* **31**:881–887.

Fang J., N. Lovanh, and P.J.J. Alvarez (2004). The use of isotopic and lipid analysis techniques linking toluene degradation to specific microorganisms: applications and limitations. *Wat. Res.* (in press).

Findlay, R.H., M.B. Trexler, J.B. Guckert, and D.C. White (1990). Response of a benthic microbial community to biotic disturbance. *Mar. Ecol. Prog. Ser.* **62**:135–148.

Fleming, J.T., J. Sanseverino, and G.S. Sayler (1993). Quantitative relationship between naphthalene catabolic gene frequency and expression in predicting PAH degradation in soils at town gas manufacturing sites. *Environ. Sci. Technol.* **27**:1068–1074.

Gibbons, R.D. and D.E. Coleman (2001). *Statistical Methods for Detection and Quantification of Environmental Contamination*. John Wiley & Sons, Hoboken, NJ.

Gilbert, R.O. (1987). *Statistical Methods For Environmental Pollution Monitoring*. Van Nostrand Reinhold, New York.

Glud, R., N.J.K. Gundersen, N.P. Revsbech, and B.B. Jorgensen (1994). Effects on the benthic diffusive boundary layer imposed by microelectrodes. *Limnol. Oceanography* **39**(2):462–467.

Gray, J.R., G. Lacrampe-Couloume, D. Gandhi, K.M. Scow, R.D. Wilson, D.M. Mackay, and B.S. Lollar (2002). Carbon and hydrogen isotopic fractionation during biodegradation of methyl *tert*-butyl ether. *Environ. Sci. Technol.* **36**(9):1931–1938.

Griebler, C., M. Safinowski, A. Vieth, H.H. Richnow, and R.U. Meckenstock (2004). Combined application of stable carbon isotope analysis and specific metabolites determination for assessing in situ degradation of aromatic hydrocarbons in a tar oil-contaminated aquifer. *Environ. Sci. Technol.* **38**(2):617–631.

Grossman, E.L. (1997). In *Manual of Environmental Microbiology*, C.J. Hurst, G.R. Knudsen, M.J. McInerney, L.D. Stetzenbach, and M.V. Walter (Eds.), ASM Press, Washington, DC, pp. 565–576.

Gu, A.Z., B.P. Hedlund, J.T. Staley, and H.D. Stensel (2004). Analysis and comparison of the microbial community structures of two enrichment cultures capable of reductively dechlorinating TCE and cis-DCE. *Environ. Microbiol.* **6**:45.

Guckert, J.B., C.P. Antworth, P.D. Nichols, and D.C. White (1985). Phospholipid ester-linked fatty acid profile as reproducible assays for changes in prokaryotic community structure of estuarine marine sediments. *FEMS Microbiol. Ecol.* **31**:147–158.

Haack, S.K., H. Garchow, D.A. Odelson, L.J. Forney, and M.J. Klug (1993). Accuracy, reproducibility, and interpretation of fatty acid methyl ester profiles of model microbial communities. *Appl. Environ. Microbiol.* **60**:2483–2493.

Hall, J.A., R.M. Kalin, M.J. Larkin, C.C.R. Allen, and D.B. Harper (1999). Variation in stable carbon isotope fractionation during aerobic degradation of phenol and benzoate by contaminant degrading bacteria. *Organic Geochem.* **30**(8A):801–811.

Hanson, J.R., J.L. Macalady, D. Harris, and K.M. Scow (1999). Linking toluene degradation with specific microbial populations in soil. *Appl. Environ. Microbiol.* **65**:5403–5408.

Heid, C.A., J. Stevens, K.J. Livak, and P.M. Williams (1996). Real time quantitative PCR. *Genome Res.* **6**:986.

Hendrickson, E.R. et al. (2002). Molecular analysis of dehalococcoides 16s ribosomal DNA from chloroethene-contaminated sites throughout North America and Europe. *Appl. Environ. Microbiol.* **68**(2):485–495.

Holloway, P., J.T. Trevors, and H. Lee (1998). A colorimetric assay for detecting haloalkane dehalogenase activity. *J. Microbiol. Methods* **32**:31–36.

Hunkeler, D. (2002). Quantification of isotope fractionation in experiments with deuterium-labeled substrate. *Appl. Environ. Microbiol.* **68**(10):5205–5206.

Hunkeler, D., R. Aravena, and B.J. Butler (1999). Monitoring microbial dechlorination of tetrachloroethene (PCE) in groundwater using compound-specific stable carbon isotope ratios: microcosm and field studies. *Environ. Sci. Technol.* **32**(16):2733–2738.

Hunkeler, D., B.J. Butler, R. Aravena, and J.F. Barker (2001a). Monitoring biodegradation of methyl *tert*-butyl ether (MTBE) using compound-specific carbon isotope analysis, *Environ. Sci. Technol.* **35**(4):676–681.

Hunkeler, D., N. Anderson, R. Aravena, S.M. Bernasconi, and B.J. Butler (2001b). Hydrogen and carbon isotope fractionation during aerobic biodegradation of benzene. *Environ. Sci. Technol.* **35**(17):3462–3467.

Kandeler, E. and H. Gerber (1988). Short-term assay of soil urease activity using colorimetric determination of ammonium. *Biol. Fertil. Soils* **6**:68–72.

Keith, L.H. (Ed.) (1986). *Principles of Environmental Sampling*, 2nd ed. ACS Professional Reference Book, Washington, DC.

Kolhatkar, R., T. Kuder, P. Philp, J. Allen, and J.T. Wilson (2002). Use of compound-specific stable carbon isotope analyses to demonstrate anaerobic biodegradation of MTBE in groundwater at a gasoline release site. *Environ. Sci. Technol.* **36**(23):5139–5146.

Krajmalnik-Brown, R., T. Hölscher, I.N. Thomson, F.M. Saunders, K.M. Ritalahti, and F.E. Loeffler (2004). Genetic identification of a putative vinyl chloride reductase in *Dehalococcoides* sp. strain BAV1. *Appl. Environ. Microbiol.* **70**:6347–6351.

Kwak, Y.H., D.S. Lee, and H.B. Kim (2003). *Vibrio harveyi* nitroreductase is also a chromate reductase. *Appl. Environ. Microbiol.* **69**:4390–4395.

Lane, D.J., B. Pace, G.J. Olsen, D.A. Stahl, M.L. Sogin, and N.R. Pace (1985). Rapid determination of 16S ribosomal RNA sequences for phylogenetic analysis. *Proc. Natl. Acad. Sci. USA* **82**:6955.

LeBlanc, D.S., P. Garabedian, K.M. Hess, L.W. Gelhar, R.D. Quadri, K.G. Stollenwerk, and W. Wood (1991). Large-scale natural gradient tracer test in sand and gravel, Cape Cod, Massachussetts. 1. Experimental design and observed tracer movement. *Water Resour. Res.* **27**(5):895–910.

Lendvay, J.M., F.E. Löffler, M. Dillhopft, M.R. Aiello, G. Daniels, B.Z. Frathepure, M. Gebhard, R. Heine, R. Helton, J. Shi, R. Krajmalnik-Brown, C.L. Major, M.J. Barcelone, E. Petrovskis, R. Hickey, J.M Tiedje, and P. Adriaens (2003). Bioreactive barriers: a comparison of bioaugmentation and biostimulation for chlorinated solvent remediation. *Environ. Sci. Technol.* **37**:1422.

Livak, K.J., S.J. Flood, J. Marmaro, W. Giusti, and K. Deetz (1995). Oligonucleotides with fluorescent dyes at opposite ends provide a quenched probe system useful for detecting PCR product and nucleic acid hybridization. *PCR Methods Appl.* **4**:357–362.

Löffler, F.E. et al. (2000). 16S rRNA gene-based detection of tetrachloroethene-dechlorinating *Desulfuromonas* and *Dehalococcoides* species. *Appl. Environ. Microbiol.* **66**(4):1369–1374.

Lollar, B.S.G.F. Slater, B. Sleep, M. Witt, G.M. Klecka, M. Harkness, and J. Spivak (2001). Stable carbon isotope evidence for intrinsic bioremediation of tetrachloroethene and trichloroethene at Area 6, Dover Air Force Base. *Environ. Sci. Technol.* **35**(2):261–269.

Lovley, D.R. (2003). Cleaning up with genomics: applying molecular biology to bioremediation. *Nature Rev. Microbiol.* **1**:35–44.

Lovley, D.R. and S. Goodwin (1988). Hydrogen concentrations as an indicator of the predominant terminal electron-accepting reactions in aquatic sediments. *Geochim. Cosmochim. Acta* **52**:2993–3003.

Lovley, D.R., F.H. Chapelle, and J.C. Woodward (1994). Use of dissolved H_2 concentrations to determine distribution of microbially catalyzed redox reactions in anoxic groundwater. *Environ. Sci. Technol.* **28**(7):1205–1210.

Lynch, N.A., H. Jiang, and D.T. Gibson (1996). Rapid purification of the oxygenase component of toluene dioxygenase from a polyol-responsive monoclonal antibody. *Appl. Environ. Microbiol.* **62**:2133–2137.

MacDonald, J.A. (2000). Evaluating natural attenuation for groundwater cleanup. *Environ. Sci. Technol.* **34**(15):346A–353A.

MacNaughton, S.J., J.R. Stephen, A.D. Venosa, G.A. Davis, Y.-J. Chang, and D.C. White (1999). Microbial population changes during bioremediation of an experimental oil spill. *Appl. Environ. Microbiol.* **65**:3566–3574.

Madigan, J.T., J.M. Martinko, and J. Parker (2003). *Brock Biology of Microorganisms*, 10th ed. Prentice Hall, Upper Saddle River, NJ.

Madsen, E.L. (1998). Epistemology of environmental microbiology. *Environ. Sci. Technol.* **32**(4):429–439.

Magnuson, J.K. et al. (2000). Trichloroethene reductive dehalogenase from *Dehalococcoides ethenogenes*: sequence of *tceA* and substrate range characterization. *Appl. Environ. Microbiol.* **66**(12):5141–5147.

Mancini, S.A., A.C. Ulrich, G. Lacrampe-Couloume, B. Sleep, E.A. Edwards, and B.S. Lollar (2003). Carbon and hydrogen isotopic fractionation during anaerobic biodegradation of benzene. *Appl. Environ. Microbiol.* **69**(1):191–198.

Mancini, S.A., G. Lacrampe-Couloume, H. Jonker, B.M. Van Breukelen, J. Groen, F. Volkering, and B.S. Lollar (2002). Hydrogen isotopic enrichment: an indicator of biodegradation at a petroleum hydrocarbon contaminated field site. *Environ. Sci. Technol.* **36**(11):2464–2470.

Mantseva, E., A. Malanichev, and N. Vulykh, (2002). Meteorological Synthesizing Centre (EAST) Technical Note 9/2002—Polyaromatic Hydrocarbons in the Environment. Cooperative Programme for Monitoring and Evaluation of the Long-Range Transmission of Air Pollutants in Europe (EMEP) (http://www.msceast.org/reps/TN9-2002.pdf).

Marsh, T.L. (1999). Terminal restriction fragment length polymorphism (T-RFLP): an emerging method for characterizing diversity among homologous populations of amplification products. *Curr. Opinion Microbiol.* **2**:323.

Matthew, B., A.B. Mesarch, C.H. Nakatsu, and L. Nies (2000). Development of catechol 2,3-dioxygenase-specific primers for monitoring bioremediation by competitive quantitative PCR. *Appl. Environ. Microbiol.* **66**(2):678–683.

Mazeas, L., H. Budzinski, and N. Raymond (2002). Absence of stable carbon isotope fractionation of saturated and polycyclic aromatic hydrocarbons during aerobic bacterial biodegradation. *Org. Geochem.* **33**(11):1259–1272.

McAllister, P.M. and C.Y. Chiang (1994). A practical approach to evaluating natural attenuation of contaminants in groundwater. *Ground Water Monitoring & Remediation* **14**(2):161–173.

McBean, E.A. and F.A. Rovers (1998). *Statistical Procedures for Analysis of Environmental Monitoring Data and Risk Assessment.* Prentice-Hall, Upper Saddle River, NJ.

McCarthy, C.M. and L. Murray (1996). Viability and metabolic features of bacteria indigenous to a contaminated deep aquifer. *Microbial Ecol.* **32**:305–321.

McNab, W.W. Jr. and B.P. Dooher (1998). A critique of a steady-state analytical method for estimating contaminant degradation rates. *Ground Water* **36**(6):983–987.

Meckenstock, R. and H.H. Richnow (2002). Quantification of isotope fractionation in experiments with deuterium-labeled substrate—authors' reply. *Appl. Environ. Microbiol.* **68**(10):5206–5207.

Meckenstock, R.U., B. Morasch, R. Warthmann, B. Schink, E. Annweiler, W. Michaelis, and H.H. Richnow (1999). C-13/C-12 isotope fractionation of aromatic hydrocarbons during microbial degradation. *Environ. Microbiol.* **1**(5):409–414.

Meckenstock, R.U., B. Morasch, M. Kästner, A. Vieth, and H.H. Richnow (2002). Assessment of bacterial degradation of aromatic hydrocarbons in the environment by analysis of stable carbon isotope fractionation. *Water Air Soil Pollut. Focus* **2**(3):141–152.

Migaud, M.E., J.C. Chee-Sanford, J.M. Tiedje, and J.W. Frost (1996). Benzylfumaric, benzylmaleic, and Z- and E-phenyltaconic acids: characterization and correlation with a metabolite generated by *Azoarcus tolulyticus* Tol-4 during anaerobic toluene degradation. *Appl. Environ. Microbiol.* **62**:974–978.

Morasch, B., H.H. Richnow, B. Schink, and R.U. Meckenstock (2001). Stable hydrogen and carbon isotope fractionation during microbial toluene degradation: mechanistic and environmental aspects. *Appl. Environ. Microbiol.*, **67**(10):4842–4849.

Morasch, B., H.H. Richnow, B. Schink, A. Vieth, and R.U. Meckenstock. (2002). Carbon and hydrogen stable isotope fractionation during aerobic bacterial degradation of aromatic hydrocarbons. *Appl. Environ. Microbiol.*, **68**(10):5191–5194.

Müller, J.A. et al. (2004). Molecular identification of the catabolic vinyl chloride reductase from *Dehalococcoises* sp. strain VS and its environmental distribution. *Appl. Environ. Microbiol.* **70**(8):4880–4888.

National Research Council (2000). *Natural Attenuation for Groundwater Remediation.* National Academy Press, Washington, DC.

National Research Council (2001). *Conceptual Models of Flow and Transport in the Fractured Vadose Zone.* National Academy Press, Washington, DC.

Neuman, S.P. and P.J. Wierenga (2003). *A Comprehensive Strategy of Hydrogeologic Modeling and Uncertainty Analysis for Nuclear Facilities and Sites,* NUREG/CR-6805. U.S. Nuclear Regulatory Commission, Office of Nuclear Regulatory Research, Washington, DC.

Newell, C.J., R.K. McLeod, and J.R. Gonzales (1996). *BIOSCREEN Natural Attenuation Decision Support System User's Manual,* Version 1.3, EPA/600/R-96/087, August, Robert S. Kerr. Environmental Research Center, Ada, OK.

Pace, N.R., D.A. Stahl, D.J. Lane, and G.J. Olsen (1986). The analysis of natural populations by ribosomal RNA sequences. *Adv. Gen. Microbiol. Ecol.* **9**:1–55.

Pankow, J.F. and J.A. Cherry (1996). *Dense Chlorinated Solvents and Other DNAPLs in Groundwater: History, Behavior, and Remediation.* Waterloo Press.

Pankow, J.F., N.R. Thompson, R.L. Johnson, A.L. Baehr, and J.S. Zogorski (1997). The urban atmosphere as a non-point source of the transport of MTBE and other volatile organic compounds (VOCs) to shallow ground water. *Environ. Sci. Technol.* **31**:2821–2828.

Pelz, O., M. Tesar, R.-M. Wittich, E.R.B. Moore, K.N. Timmis, and W.-R. Abraham (1999). Towards elucidation of microbial community metabolic pathways: unraveling the network of carbon sharing in a pollutant-degrading bacterial consortium by immunocapture and isotopic ratio mass spectrometry. *Environ. Microbiol.* **1**:167–174.

Pelz, O., A. Chatzinotas, N. Andersen, S.M. Bernasconi, C. Hesse, W.R. Abraham, and J. Zeyer (2001). Use of isotopic and molecular techniques to link toluene degradation in denitrifying aquifer microorganisms to specific microbial populations. *Arch. Microbiol.* **175**:270–281.

Pombo, S.A., O. Pelz, M.H. Schroth, and J. Zeyer (2002). Field-scale ^{13}C-labeling of phospholipid fatty acids (PLFA) and dissolved inorganic carbon: tracing acetate assimilation and mineralization in a petroleum hydrocarbon-contaminated aquifer. *FEMS Microbiol. Ecol.* **41**:259–267.

Pond, K.L., Y.S. Huang, Y. Wang, and C.F. Kulpa (2002). Hydrogen isotopic composition of individual *n*-alkanes as an intrinsic tracer for bioremediation and source identification of petroleum contamination. *Environ. Sci. Technol.* **36**(4):724–728.

Porteus, L.A., F. Widmer, and R.J. Seidler (2002). Multiple enzyme restriction fragment length polymorphism analysis for high resolution distinction of *Pseudomonmas* (sensu stricto) 16S rRNA genes. *J. Microbiol. Methods* **51**:337.

Richnow, H.H., A. Eschenbach, R. Seifert, P. Wehrung, P. Albrecht, and W. Michaelis (1998). The use of ^{13}C-labeled polycyclic aromatic hydrocarbons for the analysis of their transformation in soils. *Chemosphere* **36**:2211–2224.

Richnow, H.H., E. Annweiler, W. Michaelis, and R.U. Meckenstock (2003a). Microbial in situ degradation of aromatic hydrocarbons in a contaminated aquifer monitored by carbon isotope fractionation. *J. Contam. Hydrol.* **65**(1–2):101–120.

Richnow, H.H., R.U. Meckenstock, L.A. Reitzel, A. Baun, A. Ledin, and T.H. Christensen (2003b). In situ biodegradation determined by carbon isotope fractionation of aromatic hydrocarbons in an anaerobic landfill leachate plume (Vejen, Denmark), *J. Contam. Hydrol.* **64**(1–2):59–72.

Rodgers, R.P., E.N. Blumer, M.R. Emmett, and A.G. Marshall (2000). Efficacy of bacterial bioremediation: demonstration of complete incorporation of hydrocarbons into membrane phospholipids into *Rhodococcus* hydrocarbon degrading bacteria by electrospray ionization Fourier transform ion cyclotron resonance mass spectrometry. *Environ. Sci. Technol.* **34**:535–540.

Rooney-Varga, J.N., R.T. Anderson, J.L. Fraga, D. Ringelberg, and D.R. Lovley (1999). Microbial communities associated with anaerobic benzene degradation in a petroleum-contaminated aquifer. *Appl. Environ. Microbiol.* **65**:3056–3063.

Sannino, F. and L. Gianfreda (2001). Pesticide influence of soil enzymatic activities. *Chemosphere* **22**:1–9.

Schlotelburg, C. et al. (2002). Microbial structure of an anaerobic bioreactor population that continuously dechlorinates 1,2-dichloropropane. *FEMS Microbiol. Ecol.* **39**(3): 229–237.

Schmidt, T.C., L. Zwank, M. Elsner, M. Berg, R.U. Meckenstock, and S.B. Haderlein (2004). Compound-specific stable isotope analysis of organic contaminants in natural environments: a critical review of the state of the art, prospects, and future challenges. *Anal. Bioanal. Chem.* **378**:283–300.

Sessitsch, A., G. Gyamfi, N. Stralis-Pavese, A. Weilharter, and U. Pfeider (2002). RNA isolation from soil for bacteria. Community and functional analysis: evaluation of different extraction and soil conservation protocols. *J. Microbiol. Methods* **51**:171.

Sherwood-Loller, B., G.F. Slater, J. Ahad, B. Sleep, J. Spivack, M. Brennan, and P. MacKenzie (1999). Contrasting carbon isotope fractionation during biodegradation of trichloroethylene and toluene: implications for intrinsic bioremediation. *Org. Geochem.* **30**:813–820.

Slater, G.F., B.S. Lollar, B.E. Sleep, and E.A. Edwards (2001). Variability in carbon isotopic fractionation during biodegradation of chlorinated ethenes: implications for field applications. *Environ. Sci. Technol.* **35**(5):901–907.

Staal, M., S. Lintel-Hekkert, F. Harren, and L. Stal (2001). Nitrogenase activity in cyanobacteria measured by the acetylene reduction assay: a comparison between batch incubation and on-line monitoring. *Environ. Microbiol.* **3**(5):343–351.

Stapleton, R.D., G.S. Sayler, J.M. Boggs, E.L. Libelo, T. Stauffer, and W.G. Macintyre (2000). Changes in subsurface catabolic gene frequencies during natural attenuation of petroleum hydrocarbons. *Environ. Sci. Technol.* **34**:1991–1999.

Steffan, R.J., K. McClay, S. Vainberg, C.W. Condee, and D. Zhang (1997). Biodegradation of the gasoline oxygenates methyl *tert*-butyl ether, ethyl *tert*-butyl ether, and *tert*-amyl methyl ether by propane oxidizing bacteria. *Appl. Environ. Microbiol.* **63**:4216–4222.

Stout, S.A., A.D. Uhler, T.G. Naymik, and K.J. McCarthy (1998). Environmental forensics unraveling site liability. *Environ. Sci. Technol.* 260A–263A.

Suflita, J.M. and M.M. Mormile (1993). Anaerobic biodegradation of known and potential gasoline oxygenates in the terrestrial subsurface. *Environ. Sci. Technol.* **27**:976–978.

Sung, Y. (2004). Primer sequence for *Geobacter* sp. strain SZ. Personal communiction, Atlanta, GA.

Suzuki, M.T., L.T. Taylor, and E.F. DeLong (2000). Quantitative analysis of small-subunit rRNA genes in mixed microbial populations via 5'-nuclease assays. *Appl. Environ. Microbiol.* **66**:4605–4614.

Tabatabai, M.A. and J.M. Bremmer (1970). Arulsulphatase activity of soils. *Soil Sci. Soc. Am. Proc.* **34**:427–429.

Thurnheer T., G. Gmur, and B. Guggenheim (2004). Muliplex FISH analysis of a six-species bacterial biofilm. *J. Microbiol. Methods* **56**:37.

Trevors, J.T. (1984). Dehydrogenase in soil: a comparison between the INT and TTC assay. *Soil. Biol. Biochem.* **16**:673–674.

Tulis, D., P.F. Prevost, and P. Kostecki (1998). Study points to new trends in use of alternative technologies at LUST sites. *Soil Groundwater Cleanup*, 12–17.

United States Environmental Protection Agency (2000). *Guidance for Data Quality Assessment, Practical Methods for Data Analysis*, Report EPA/600/R-96/084. U.S. Environmental Protection Agency, Washington, DC.

Vestal, J.R. and D.C. White (1989). Lipid analysis in microbial ecology. *BioScience* **39**:535–541.

Wagner, M., A.J. Roger, J.L. Flax, G.A. Brusseau, and D.A. Stahl (1998). Phylogeny of dissimilatory sulfite reductase supports and early origin of sulfate respiration. *J. Bacteriol.* **180**:2975–2982.

Ward, J.A.M., J.M.E. Ahad, G. Lacrampe-Couloume, G.F. Slater, E.A. Edwards, and B.S. Lollar (2000). Hydrogen isotope fractionation during methanogenic degradation of toluene: potential for direct verification of bioremediation. *Environ. Sci. Technol.* **34**(21):4577–4581.

Watanabe, K. and P.W. Baker (2000). Environmentally relevant microorganisms. *J. Biosci. Bioeng.* **89**:1–11.

Watrous, M.M., S. Clark, R. Kutty, S. Huang, F.B. Rudolph, J.B. Hughes, and G.N. Bennett (2003). 2,4,6-Trinitrotoluene reduction by an Fe-only hydrogenase in *Clostridium acetobutylicum*. *Appl. Environ. Microbiol.* **69**(3):1542–1547.

Watts, R.J. (1997). *Hazardous Wastes: Sources, Pathways, Receptors*. John Wiley & Sons, Hoboken, NJ.

Wawer, C., M.M. Jetten, and G. Muyzer (1997). Genetic diversity and expression of the [NiFe] hydrogenase large-subunit gene of *Desulfovibrio* spp. in environmental samples. *Appl. Environ. Microbiol.* **63**(11):4360–4369.

White, D.C., R.J. Bobbie, J.D. King, J.S. Nickels, and P. Amoe (1979). Lipid analysis of sediments for microbial biomass and community structure. In *Methodology for Biomass Determination and Microbial Activities in Sediments*, C.D. Litchfield and P.L. Seyfried (Eds.), ASTM STP 673. American Society for Testing and Materials, Philadelphia, PA, pp. 87–103.

White, D.C., H.C. Pinkart, and A.B. Ringelberg (1997). Biomass measurements: biochemical approaches. In *Manual of Environmental Microbiology*, C.J. Hurst, G.R. Knudson, M.J. McInerney, L.D. Stetzenbach, and M.V. Walter (Eds.). ASM Press, Washington, DC, pp. 91–101.

Wiedemeier, T.H., H.S. Rifai, C.J. Newell, and J.T. Wilson (1999). *Natural Attenuation of Fuels and Chlorinated Solvents in the Subsurface*. John Wiley & Sons, Hoboken, NJ.

Wilks, H., C. Boreham, G. Harms, K. Zengler, and R. Rabus (2000). Anaerobic degradation and carbon isotopic fractionation of alkylbenzenes in crude oil by sulphate-reducing bacteria. *Org. Geochem.* **31**:101–115.

Wilson, M.S. and E.L. Madsen (1996). Field extraction of a transient intermediary metabolite indicative of real time in situ naphthalene biodegradation. *Environ. Sci. Technol.* **30**:2099–2103.

Wu, L., D.K. Thompson, G. Li, R.A. Hurt, and J.M. Tiedje (2001). Development and evaluation of functional gene arrays for detection of selected genes in the environment. *Appl. Environ. Microbiol.* **67**:5780.

Xu, H.-S., N. Roberts, F.L. Singleton, R.W. Atwell, D.J. Grimes, and R.R. Colwell (1982). Survival and viability of non-culturable *Escherichia coli* and *Vibrio cholerae* in the estuarine and marine environment. *Microbial Ecol.* **8**:313–323.

Yang, Y. and J. Zeyer (2003). Specific detection of *Dehalococcoides* species by fluorescence in situ hybridization with 16S-rRNA-targeted oligonucleotide probes. *Appl. Environ. Microbiol.* **64**:112.

Yanik, P.J., T.H. O'Donnell, S.A. Macko, Y.R. Qian, and M.C. Kennicutt (2003). The isotopic compositions of selected crude oil PAHs during biodegradation. *Org. Geochem.* **34**(2):291–304.

Young, H.K., D.S. Lee, and H.B. Kim (2003). *Vibrio harveyi* nitroreductase is also a chromate reductase. *Appl. Environ. Microbiol.* **69**(8):4390–4395.

Zhou, J. (2002). Microarrays for bacterial detection and microbial community analysis. *Curr. Opin. Microbiol.* **13**:204.

APPENDIX A

CHEMICAL PROPERTIES OF VARIOUS ORGANIC COMPOUNDS

Chemical Name	Formula Weight (g/mol)	Water Solubility (mg/L)	Vapor Pressure (mm Hg)	Henry's Constant (atm-m^3/mol)	K_{oc} (mL/g)	log K_{ow}
Acenaphthene	154	3.42E + 00	1.55E − 03	9.20E − 05	4600	4.00
Acenaphthylene	152	3.93E + 00	2.90E − 02	1.48E − 03	2500	3.70
Acetone	58	1.00E + 06	2.70E + 02	2.06E − 05	2.2	−0.24
Acetonitrile	41	1.00E + 06	7.40E + 01	4.00E − 06	2.2	−0.34
2-Acetylaminofluorene	223	6.50E + 00			1600	3.28
Acrylic acid	72	1.00E + 06	4.00E + 00			0.13
Acraylonitrile	53	7.90E + 04	1.00E + 02	8.84E − 05	0.85	0.25
Aflatoxin	312					
Aldrin	365	1.80E − 01	6.00E − 06	1.60E − 05	96000	5.30
Allyl alcohol	58	5.10E + 05	2.46E + 01	3.69E − 06	3.2	−0.22
Aluminum phosphide	58					
4-Aminobiphenyl	169	8.42E + 02	6.00E − 05	1.59E − 08	107	2.78
Amitrole	84	2.80E + 05			4.4	−2.08
Ammonia	17	5.30E + 05	7.60E − 03	3.21E − 04	3.1	0.00
Anthracene	178	4.50E − 02	1.95E − 04	1.02E − 03	14000	4.45
Antimony and compounds	122		1.00E + 00			
Arsenic and compounds	75		0.00E + 00			
Asbestos						
Auramine	267	2.10E + 00			2900	4.16
Azaserine	173	1.36E + 05			6.6	−1.08
Aziridine	43	2.66E + 06	2.55E − 02	5.43E − 06	1.3	−1.01
Barium and compounds	137					
Benzene	78	1.75E + 03	9.52E + 01	5.59E − 03	83	2.12
Benzidine	184	4.00E + 02	5.00E − 04	3.03E − 07	10.5	1.30
Benz[a]anthracene	228	5.70E − 03	2.20E − 08	1.16E − 06	1380000	5.60
Benz[c]acridine	229	1.40E + 01			1000	4.56

Bioremediation and Natural Attenuation: Process Fundamentals and Mathematical Models
By Pedro J. J. Alvarez and Walter A. Illman   Copyright © 2006 John Wiley & Sons, Inc.

Chemical Name	Formula Weight (g/mol)	Water Solubility (mg/L)	Vapor Pressure (mm Hg)	Henry's Constant (atm-m^3/mol)	K_{oc} (mL/g)	log K_{ow}
Benzo[a]pyrene	252	1.2–3	5.60E − 09	1.55E − 06	5500000	6.06
Benzo[b]fluoranthene	252	1.40E − 02	5.00E − 07	1.19E − 05	550000	6.06
Benzo[ghi]perylene	276	7.00E − 04	1.03E − 10	5.34E − 08	1600000	6.51
Benzo[k]fluoranthene	252	4.30E − 03	5.10E − 07	3.94E − 05	550000	6.06
Benzly chloride	127	3.30E + 03	1.00E + 00	5.06E − 05	50	2.63
Beryllium and compounds	9		0.00E + 00			
Bis(2-chloroethyl)ether	143	1.02E + 04	7.10E − 01	1.31E − 05	13.9	1.50
Bis(2-chloroisopropyl)ether	171	1.70E + 03	8.50E − 01	1.13E − 04	61	2.10
Bis(chloromethyl)ether	115	2.20E + 04	3.00E + 01	2.06E − 04	1.2	0.38
1,3-Butadiene	54	7.35E + 02	1.84E + 03	1.78E − 01	120	1.99
Cacodylic acid	138	8.30E + 05			2.4	0.00
Cadmium and compounds	112		0.00E + 00			
Captan	301	5.00E − 01	6.00E − 05	4.75E − 05	6400	2.35
Carbaryl	201	4.00E + 01	5.00E − 03			2.36
Carbon disulfide	76	2.94E + 03	3.60E + 02	1.23E − 02	54	2.00
Carbon tetrachloride	154	7.57E + 02	9.00E + 01	2.41E − 02	110	2.64
Cholodane	410	5.60E − 01	1.00E − 05	9.63E − 06	140000	3.32
Chlorobenzene	113	4.66E + 02	1.17E + 01	3.72E − 03	330	2.84
Chlorobenzilate	325	2.19E + 01	1.20E − 06	2.34E − 08	800	4.51
Chlorodibromomethane	208		1.50E + 01			2.09
Chloroform	119	8.20E + 03	1.51E + 02	2.87E − 03	31	1.97
Chloromethyl methyl ether	81					0.00
4-Chloro-o-toluidine HCl	142					
Chromium(III) and compounds	52		0.00E + 00			
Chromium(VI) and compounds	52		0.00E + 00			
Chrysene	228	1.80E − 03	6.30E − 09	1.05E − 06	200000	5.61
Copper and compounds	64		0.00E + 00			
Creosote						
Cresol	108	3.10E + 04	2.40E − 01	1.10E − 06	500	1.97
Barium cyanide	189					
Calcium cyanide	92					
Copper cyanide	90					
Cyanogen	52	2.50E + 05				
Cyanogen chloride	61	2.50E + 03				0.00
Hydrogen cyanide	27	1.00E + 06				−0.25
Nickel cyanide	182					
Potassium cyanide	65	5.00E + 05				
Potassium silver cyanide	199					
Silver cyanide	134					
Sodium cyanide	49	8.20E + 05				
Zinc cyanide	117					
Cyclophosphamide	261	1.31E + 09			0.042	−3.22
DDD	320	1.00E − 01	1.89E − 06	7.96E − 06	770000	6.20
DDE	318	4.00E − 02	6.50E − 06	6.80E − 05	4400000	7.00
DDT	355	5.00E − 03	5.50E − 06	5.13E − 04	243000	6.19

Chemical Name	Formula Weight (g/mol)	Water Solubility (mg/L)	Vapor Pressure (mm Hg)	Henry's Constant (atm-m^3/mol)	K_{oc} (mL/g)	log K_{ow}
Diallate	274	1,4e1	6.40E − 03	1.65E − 04	1000	0.73
2,4-Diaminatoluene	122	4.77E + 04	3.80E − 05	1.28E − 10	12	0.35
1,2,7,8-Dibenzopyrene	305	1.10E − 01			1200	6.62
Dibenzo[a, h]anthracene	278	5.00E − 04	1.00E − 10	7.33E − 08	3300000	6.80
1,2-Dibromo-3-chloropropane	236	1.00E + 03	1.00E + 00	3.11E − 04	98	2.29
Dibutylnitrosoamine	152					
Dibutyl phthalate	278	1.30E + 01	1.00E − 05	2.82E − 07	170000	5.60
1,2-Dichlorobenzene	147	1.00E + 02	1.00E + 00	1.93E − 03	1700	3.60
1,3-Dichlorobenzene	147	1.23E + 02	2.28E + 00	3.59E − 03	1700	3.60
1,4-Dichlorobenzene	147	7.90E + 01	1.18E + 00	2.89E − 03	1700	3.60
3,3'-Dichlororobenzidine	253	4.00E + 00	1.00E − 05	8.33E − 07	1553	3.50
Dichlorodifluoramethane	121	2.80E + 02	4.87E + 03		58	2.16
1,1-Dichloroethane	99	5.50E + 03	1.82E + 02	4.31E − 03	30	1.79
1,2-Dichloroethane (EDC)	99	8.52E + 03	6.40E + 01	9.78E − 04	14	1.48
1,1-Dichloroethylene	97	2.25E + 03	6.00E + 02	3.40E − 02	65	1.84
1,2-Dichloroethylene (trans)	97	6.30E + 03	3.24E + 02	6.56E − 03	59	0.48
1,2-Dichloroethylene (cis)	97	3.50E + 03	2.08E + 02	7.58E − 03	49	0.70
Dichloromethane	85	2.00E + 04	3.62E + 02	2.03E − 03	8.8	1.30
2,4-Dichlorophenol	163	4.60E + 03	5.90E − 02	2.75E − 06	380	2.90
2,4-Dichlorophenoxyacetic Acid (2,4-D)	221	6.20E + 02	4.00E − 01	1.88E − 04	20	2.81
4-(2,4-Dichlorophenoxy)-butyric Acid (2,4-DB)						
Dichlorophenylarsine	223					
1,2-Dichloropropane	113	2.70E + 03	4.20E + 01	2.31E − 03	51	2.00
1,3-Dichloropropane	111	2.80E + 03	2.50E + 01	1.30E − 03	48	2.00
Dieldrin	381	1.95E − 01	1.78E − 07	4.58E − 07	1700	3.50
Diepoxybutane	86					
Diethanolnitrosamine	134					
Diethyl arsine	134	4.17E + 02	3.50E + 01	1.48E − 02	160	2.97
1,2-Diethylhydrazine	88	2.88E + 07			0.3	−1.68
Diethylnitrosamine	102		5.00E + 00			0.48
Diethyl phthalate	222	8.96E + 02	3.50E − 03	1.14E − 06	142	2.50
Diethylstilbestrol (DES)	268	9.60E − 03			28	5.46
Dihydrosafrole	164	1.50E + 03			78	2.56
Dimethoate	229	2.50E + 04	2.50E − 02			2.71
3,3'-Dimethoxybenzidine	244					
Dimethylamine	45	1.00E + 06	1.52E + 03	9.02E − 05	2.2	−0.38
Dimethyl sulfate	126	3.24E + 05	6.80E − 01	3.48E − 07	4.1	−1.24
Dimethylaminoazobenzene	225	1.36E + 01	3.30E − 07	7.19E − 09	1000	3.72
7,12-Dimetylbenz[a]anthracene	256	4.40E − 03			476000	6.94
Dimethylcarbamoyl chloride	108	1.44E + 07	1.95E + 00	1.92E − 08	0.5	−1.32
1,1-Dimethylhydrazine	60	1.24E + 08	1.57E + 02	1.00E − 07	0.2	−2.42
1,2-Dimethylhydrazine	60					
Dimethylnitrosoamine	74	1.00E + 06	8.10E + 00	7.90E − 07	0.1	−0.68
1,3-Dinitrobenzene	168	4.70E + 02			150	1.62

Chemical Name	Formula Weight (g/mol)	Water Solubility (mg/L)	Vapor Pressure (mm Hg)	Henry's Constant (atm-m^3/mol)	K_{oc} (mL/g)	log K_{ow}
4,6-Dinitro-o-cresol	198	2.90E + 02	5.00E − 02	4.49E − 05	240	2.70
2,4-Dinitrophenol	184	5.60E + 03	1.49E − 05	6.45E − 10	16,6	1.50
2,3-Dinitrotoluene	182	3.10E + 03			53	2.29
2,4-Dinitrotoluene	182	2.40E + 02	5.10E − 03	5.09E − 06	45	2.00
2,5-Dinitrotoluene	182	1.32E + 03			84	2.28
2,6-Dinitrotoluene	182	1.32E + 03	1.80E − 02	3.27E − 06	92	2.00
3,4-Dinitrotoluene	182	1.08E + 03			94	2.29
Dinoseb	240	5.00E + 01				
1,4-Dioxane	88	4.31E + 05	3.99E + 01	1.07E − 05	3.5	0.01
N,N-Diphenylamine	169	5.76E + 01	3.80E − 05	1.47E − 07	470	3.60
1,2-Diphenylhydrazine	184	1.84E + 03	2.60E − 05	3.42E − 09	418	2.90
Dipropylnitrosoamine	130	9.90E + 03	4.00E − 01	6.92E − 06	15	1.50
Epichlorohydrin	93	6.00E + 04	1.57E + 01	3.19E − 05	10	0.15
Ethanol	46	1.00E + 06	7.40E + 02	4.48E − 05	2,2	−0.32
Ethyl methanesulfonate	124	3.69E + 05	2.06E − 01	9.12E − 08	3.8	0.21
Ethyl benzene	106	1.52E + 02	7.00E + 00	6.43E − 03	1100	3.15
Ethyl dibromide (EDB)	188	4.30E + 03	1.17E + 01	6.73E − 04	44	1.76
Ethylene oxide	44	1.00E + 06	1.31E + 03	7.56E − 05	2,2	−0.22
Ethylenethiourea	102	2.00E + 03			67	−0.66
1-Ethyl-nitroseurea	117	3.31E + 08			0.1	
Ferric dextran	7500					
Fluoranthene	202	2.06E − 01	5.00E − 06	6.46E − 06	38000	4.90
Fluorene	116	1.69E + 00	7.10E − 04	6.42E − 05	7300	4.20
Fluorides						
Formaldehyde	30	4.00E + 05	1.00E + 01	9.87E − 07	3.6	0.00
Formic acid	46	1.00E + 06	4.00E + 01			−0.54
Glycidaldehyde	72	1.70E + 08	1.97E + 01	1.10E − 08	0.1	−1.55
Diethylene glycol, monoethyl ether	134					
2-Ethoxyethanol	90	1.00E + 06				0.00
Ethylene glycol, monobutyl ether	118	1.00E + 06				
2-Methoxyethanol	76	1.00E + 06				0.00
Propylene glycol, monoethyl ether						
Propylene glycol, monomethyl ether						
Heptachlor	374	1.80E − 01	3.00E − 04	8.19E − 04	12000	4.40
Heptachlor epoxide	389	3.50E − 01	3.00E − 04	4.39E − 04	220	2.70
Hexachlorobenzene	285	6.00E − 03	1.09E − 05	6.81E − 04	3900	5.23
Hexachlorobutadiene	261	1.50E − 01	2.00E + 00	4.57E + 00	29000	4.78
Hexachlorocyclopentadiene	273	2.10E + 00	8.00E − 02	1.37E − 02	4800	5.04
α-Hexachlorocyclohexane (HCCH)	291	1.63E + 00	2.50E − 05	5.87E − 06	3800	3.90
β-HCCH	291	2.40E − 01	2.80E − 07	4.47E − 07	3800	3.90

Chemical Name	Formula Weight (g/mol)	Water Solubility (mg/L)	Vapor Pressure (mm Hg)	Henry's Constant (atm-m^3/mol)	K_{oc} (mL/g)	log K_{ow}
γ-HCCH	291	7.80E + 00	1.60E − 04	7.85E − 06	1080	3.90
δ-HCCH	291	3.14E + 01	1.70E − 05	2.07E − 07	6600	4.10
Hexachloroethane	237	5.00E + 01	4.00E − 01	2.49E − 03	20000	4.60
Hexachlorophene	407	4.00E − 03			91000	7.54
Hydrazine	32	3.41E + 08	1.40E + 01	1.73E − 09	0.1	−3.08
Hydrogen sulfide	34	4.13E + 03				
Indeno[1,2,3-cd]pyrene	276	5.30E − 04	1.00E − 10	6.86E − 08	1600000	6.50
Iodomethane	142	1.40E + 04	4.00E + 02	5.34E − 03	23	1.69
Iron and compounds	56					
Isoprene	68		4.00E + 02			
Isosafrole	168	1.09E + 03	1.60E − 08	3.25E − 12	93	2.66
Kepone	491	9.90E − 03			55000	2.00
Lasiocarpine	412	1.60E + 03			76	0.99
Lead and compounds (inorganic)	207		0.00E + 00			
Linuron						
Malathion	330	1.45E + 02	4.00E − 05			2.89
Manganese and compounds	55					
Mercury and compounds (alkyl)						
Mercury and compounds (inorganic)	201		2.00E − 03			
Mercury fulminate						
Methyl chloride	50	6.50E + 03	4.31E + 03	4.40E − 02	35	0.95
Methyl ethyl ketone	72	2.68E + 03	7.75E + 01	2.74E − 05	4.5	0.26
Methyl ethyl ketone peroxide						
Methyl methacrylate	100	2.00E + 01	3.70E + 01	2.43E − 01	840	0.79
Methyl parathion	263	6.00E + 01	9.70E − 06	5.59E − 08	460	1.91
2-Methyl-4-chloro-phenoxyacetic acid	201					
3-Methylchloranthrene	268					
4-4′-Methylene-bis-2-chloroaniline	267					
Methylnitrosourea	103	6.89E + 08			0.1	−3.81
Methylthiouracil	142					
Methylvinylnitrosoamine	86	7.60E + 05	1.23E + 01	1.83E − 06	2.5	−0.23
N-Methyl-N′-nitro-N-nitrosoguanadin						
Mitomycin C	334					
Mustard gas	159	8.00E + 02	1.70E − 01	4.45E − 05	110	1.37
1-Napthylamine	143	2.35E + 03	6.50E − 05	5.21E − 09	61	2.07
2-Napthylamine	143	5.86E + 02	2.56E − 04	8.23E − 08	130	2.07
Nickel and compounds	59		0.00E + 00			
Nitric oxide	30					
Nitrobenzene	123	1.90E + 03	1.50E − 01		36	1.85

Chemical Name	Formula Weight (g/mol)	Water Solubility (mg/L)	Vapor Pressure (mm Hg)	Henry's Constant (atm-m³/mol)	K_{oc} (mL/g)	log K_{ow}
Nitrogen dioxide	46					
Nitrosomethylurethane						
N-Nitrosopoperidine	114	1.90E + 06	1.40E − 01	1.11E − 08	1.5	−0.49
N-Nitrosopyrrolidine	100	7.00E + 06	1.10E − 01	2.07E + 09	0.8	−1.06
5-Nitro-o-toluidine						
Pentachlorobenzene	250	1.35E − 01			13000	5.19
Pentachloronitrobenzene	295	7.11E − 02	1.13E − 04	6.18E − 04	19000	5.45
Pentachlorophenol	266	1.40E + 01	1.10E − 04	2.75E − 06	53000	5.00
Phenacetin	179					
Phenanthrene	178	1.00E + 00	6.80E − 04	1.59E − 04	14000	4.46
Phenobarbital	232	1.00E + 03			98	−0.19
Phenol	94	9.30E + 04	3.41E − 01	4.54E − 07	14.2	1.46
Phenyl mecuric acetate	337	1.67E + 03				
Phosphine	34					
Phenylalanine mustard	305					
Polychlorinated biphenyls (PCBs)	328	3.10E − 02	7.70E − 05	1.07E − 03	530000	6.04
Propane sultone						
Propylenimine	57	9.44E + 05	1.41E + 02	1.12E − 05	2.3	−0.48
Pyrene	202	1.32E − 01	2.50E − 06	5.04E − 06	38000	4.88
Pyridine	79	1.00E + 06	2.00E + 01			0.66
Saccharin	183					2.53
Safrole	162	1.50E + 03	9.10E − 04	1.29E − 07	78	
Selenium and compounds	79		0.00E + 00			
Selenious acid	129					
Selenourea						
Thallium selenite	488					
Silver and compounds	108		0.00E + 00			
Streptozocin	457					
Strychnine	334	1.56E + 02				
1,2,4,5-Tetrachlorobenzene	216	6.00E + 00			1600	4.67
2,3,7,8-ICCD (dioxin)	322	2.00E − 04	1.70E − 06	3.60E − 03	330000	6.72
1,1,1,2-Tetrachloroethane	168	2.90E + 03	5.00E + 00	3.81E − 04	54	
1,1,2,2-Tetachloroethane	168	2.90E + 03	5.00E + 00	3.81E − 04	118	2.39
Tetrachloroethylene	166	1.50E + 02	1.78E + 01	2.59E − 02	364	2.60
2,3,4,6-Tetrachlorophenol	142	1.00E + 03			98	4.10
Tetraethyl lead	323	8.00E − 01	1.50E − 01	7.97E − 02	4900	
Thallium and compounds	204		0.00E + 00			
Thallium acetate	263					
Thallium carbonate	469					
Thallium chloride	240	2.90E + 03	0.00E + 00			
Thallium nitrate						
Thallium oxide						
Thallium sulfate	505	2.00E + 02	0.00E + 00			
Thioacetamide	75					−0.46

Chemical Name	Formula Weight (g/mol)	Water Solubility (mg/L)	Vapor Pressure (mm Hg)	Henry's Constant (atm-m^3/mol)	K_{oc} (mL/g)	log K_{ow}
Thiourea	76	1.72E + 06			1.6	−2.05
o-Lolidine	212	7.35E + 01			410	2.88
Touene	92	5.35E + 02	2.81E + 01	6.37E − 03	300	2.73
o-Toluidine	107	1.50E + 04	1.00E − 01	9.39E − 07	22	1.29
Toxaphene	414	5.00E − 01	4.00E − 01	4.36E − 01	964	3.30
Tribromomethane (bromoform)	253	3.01E + 03	5.00E + 00	5.52E − 04	116	2.40
1,2,4-Trichlorobenzene	181	3.00E + 01	2.90E − 01	2.31E − 03	9200	4.30
1,1,1-Trichloroethane	133	1.50E + 03	1.23E + 02	1.44E − 02	152	2.50
1,1,2-Trichloroethane	133	4.50E + 03	3.00E + 01	1.17E − 03	56	2.47
Trichloroethylene	131	1.10E + 03	5.79E + 01	9.10E − 03	126	2.38
Trichlorofon	257	1.54E + 05	7.80E − 06	1.71E − 11	6.1	2.29
Trichloromonofluoromethane	137	1.10E + 03	6.67E + 02		159	2.53
2,4,5-Trichlorophenol	197	1.19E + 03	1.00E + 00	2.18E − 04	89	3.72
2,4,6-Trichlorophenol	198	8.00E + 02	1.20E − 02	3.90E − 06	2000	3.87
1,1,2-Trichloro-1,2,2-trifluoroethane	187	1.00E + 01	2.70E + 02			2.00
Tris(2,3-dibromopropyl) phosphate	698	1.20E + 02			310	4.12
Trypan blue	961					
Uracil mustard	252	6.41E + 02			120	−1.09
Uranium and compounds	238					
Urethane	89					
Vanadium and compounds	51					
Vinyl chloride	63	2.67E + 03	2.66E + 03	8.19E − 02	57	1.38
o-Xylene	106	1.75E + 02	1.00E + 01			2.95
m-Xylene	106	1.30E + 02	1.00E + 01			3.26
p-Xylene	106	1.98E + 02	1.00E + 01			3.15
Xylenes	106	1.98E + 00	1.00E + 01	7.04E − 03	240	3.26
Zinc and compounds	65		0.00E + 00			

APPENDIX B

FREE ENERGY AND THERMODYNAMIC FEASIBILITY OF CHEMICAL AND BIOCHEMICAL REACTIONS

B.1. DEFINITIONS

$\Delta G°$ Free energy of a chemical reaction at standard temperature and pressure (STP) conditions: that is, 1 molar concentration of products and reactants. (Thus, $[H^+] = 1$ M and pH $= 0$, partial pressure $= 1$ atm for gases, and $T = 298$ K (25 °C))

$\Delta G°$ Free energy at STP and physiological conditions: that is, 1 molar concentration of products and reactants except $[H^+] = 10^{-7}$ M (pH $= 7$)

ΔG Free energy under specific conditions: concentrations of products or reactants other than 1 molar (or 1 atm for gases), T other than 25 °C, and so on

Note:

$\Delta G = 0 \Rightarrow$ equilibrium (no reaction)

$\Delta G < 0 \Rightarrow$ favorable, reaction *could* proceed as written

$\Delta G > 0 \Rightarrow$ unfavorable, reaction will not proceed as written

Whether reaction proceeds depends on the *activation energy* (E_a) required to break bonds and bring molecules to reactive state (Figure B.1).

Another useful concept to investigate the feasibility of oxidation–reduction reactions is

$\Delta E°$ The difference between reduction potentials of the electron donor and electron acceptor at standard conditions (positive when feasible)

Bioremediation and Natural Attenuation: Process Fundamentals and Mathematical Models
By Pedro J. J. Alvarez and Walter A. Illman   Copyright © 2006 John Wiley & Sons, Inc.

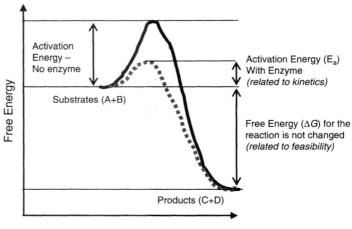

Figure B.1. Progress of a hypothetical reaction $(A + B \rightarrow C + D)$ and the concept of activation energy. Some reactions do not proceed spontaneously even if the reaction is thermodynamically feasible $(\Delta G < 0)$ unless the reactants are first activated. This activation energy represents a barrier that reacting molecules must overcome. Catalysts such as enzymes lower this barrier and enhance the rate of reaction without affecting ΔG.

$\Delta G°$ and $\Delta E°$ are related by Eq. [B.1]:

$$\Delta G° = -nF \, \Delta E° \qquad [B.1]$$

where n = moles of electrons transferred (given by half-reaction of e^- donor)
        F = Faraday's constant = 96,630 J/V-mol (23.06 kcal/V-mol)

Values of $E°$ for different compounds are given in Table B.1.

B.2. BALANCING REDOX REACTIONS

Chemical reactions must be balanced correctly to compute the correct value of $\Delta G°$. Most metabolic reactions involve the transfer of electrons (i.e., oxidation and reduction reactions). Balancing such reactions is best accomplished by adding the appropriate half-reactions for the electron donor and the electron acceptor.

   To construct such half-reactions:

1. Balance the element to be oxidized (or reduced).
2. Add H_2O on appropriate side to balance oxygen.
3. Add H^+ on appropriate side to balance hydrogen.
4. Add e^- on appropriate side to balance charge.

Table B.1. Oxidation–Reduction Potential for Compounds that Are Important from a Biochemical and Microbiological Perspective

Redox Pair	E° (volts)
SO_4^{2-}/HSO_3^-	−0.52
CO_2/formate	−0.42
$2H^+/H_2$	−0.41
$S_2O_3^{2-}/HS^- + HSO_3^-$	−0.40
Ferrodoxin ox/red	−0.39
Flavodoxin ox/red	−0.37
NAD^+/NADH	−0.32
Cytochrome c_3 ox/red	−0.29
CO_2/acetate$^-$	−0.29
S^0 / HS^-	−0.27
CO_2/CH_4	−0.24
FAD/FADH	−0.22
SO_4^{2-} / HS^-	−0.217
Acetaldehyde/ethanol	−0.197
Pyruvate$^-$/lactate$^-$	−0.19
FMN/FMNH	−0.19
Dihydroxyacetone phosphate/glycerolphosphate	−0.19
$HSO_3^-/S_3O_6^{2-}$	−0.17
Flavodoxine ox/red	−0.12
HSO_3^-/HS^-	−0.116
Menaquinone ox/red	−0.075
APS/AMP + HSO_3^-	−0.075
Rubredoxine ox/red	−0.057
Acylyl-CoA/propionyl-CoA	−0.015
Glycine/acetate$^-$ + NH_4^+	−0.010
$S_4O_6^{2-}/S_2O_3^{2-}$	+0.024
Fumarate^{2-}/succinate^{2-}	+0.033
Cytochrome b ox/red	+0.035
Ubiquinone ox/red	+0.113
AsO_4^{3-}/AsO_3^{3-}	+0.139
Dimethyl sulfoxide (DMSO)/dimethylsulide (DMS)	+0.16
$Fe(OH)_3 + HCO_3^-/FeCO_3$	+0.20
$S_3O_6^{2-}/S_2O_3^{2-} + HSO_3^-$	+0.225
Cytochrome c_1 ox/red	+0.23
NO_2^-/NO	+0.36
Cytochrome a_3 ox/red	+0.385
Chlorobenzoate$^-$/benzoate$^-$ + HCl	+0.297
NO_3^-/NO_2^-	+0.43
SeO_4^{2-}/SeO_3^{2-}	+0.475
Fe^{3+}/Fe^{2+}	+0.77
Mn^{4+}/Mn^{2+}	+0.798
O_2/H_2O	+0.82
ClO_3^-/Cl^-	+0.13
NO/N_2O	+1.18
N_2O/N_2	+1.36

Source: Madigan et al. (2000).

These steps are illustrated below for the oxidation of glucose ($C_6H_{12}O_6$, an e^- donor) to carbon dioxide (CO_2):

1. $C_6H_{12}O_6 \rightarrow 6CO_2$
2. $C_6H_{12}O_6 + 6H_2O \rightarrow 6CO_2$
3. $C_6H_{12}O_6 + 6H_2O \rightarrow 6CO_2 + 24H^+$
4. $C_6H_{12}O_6 + 6H_2O \rightarrow 6CO_2 + 24H^+ + 24e^-$

$$(1)$$

Similarly, we can use the same approach to obtain a half-reaction for the reduction of oxygen (O_2, an e^- acceptor) to water (H_2O):

1. $O_2 \rightarrow H_2O$
2. $O_2 \rightarrow 2H_2O$
3. $O_2 + 4H^+ \rightarrow 2H_2O$
4. $O_2 + 4H^+ + 4e^- \rightarrow 2H_2O$

$$(2)$$

These two half-reactions can be combined on an electron equivalent basis (i.e., electrons donated by glucose must be balanced with electrons accepted by O_2). Multiplying (2) by 6 and adding from (1), the overall equation then becomes

$$C_6H_{12}O_6 + 6O_2 \rightarrow 6CO_2 + 6H_2O$$

This reaction involves the transfer of 24 electrons per mole of glucose.

B.3. CALCULATION OF $\Delta G°$

$\Delta G°$ can be calculated from the free energy of formation ($\Delta G_f°$) of products and reactants as

$$\Delta G° = \Sigma \Delta G_f°(\text{products}) - \Sigma \Delta G_f°(\text{reactants}) \qquad [\text{B.2}]$$

Free energy of formation data for different compounds are presented in Table B.2.

Example   Consider the aerobic degradation of glucose:

$$C_6H_{12}O_6 + 6O_2 \rightarrow 6CO_2 + 6H_2O$$

From Table B.2,

$$
\begin{aligned}
\Delta G° \text{ for this reaction} &= \Sigma \Delta G_f°(\text{products}) - \Sigma \Delta G_f°(\text{reactants}) \\
&= [(6)(-394.9) + (6)(-237.57)] - [(1)(-917.61) - 0.0] \\
&= -2877.21 \text{ kJ/mol} = 687.6 \text{ kcal/mol}
\end{aligned}
$$

Table B.2. Free Energy of Formation $(G_f°)$ for Selected Substances (kJ/mol)

Carbon Compounds	Metal	Inorganic, Nonmetal	Nitrogen Compounds
CO, −137.34	Cu^+, +50.28	H_2, O	N_2, O
CO_2, −394.4	Cu^{2+}, +64.94	H^+, 0 at pH 7	NO_2, +51.95
H_2CO_3, −623.16	Fe^{2+}, −78.87	(−5.69 per pH unit)	NO_2^-, −37.2
HCO_3, −586.85	Fe^{3+}, −4.6	O_2, 0	NO_3^-, −111.34
CO_3^{2-}, −527.90	$FeCO_3$, −673.23	OH^-, 157.3 at pH 14;	NH_3, −26.57
Acetate, −369.41	FeS_2, 150.84	−198.76 at pH 7;	NH_4^+, −79.37
Alanine, −371.54	$FeSO_4$, −828.62	−237.57 at pH 0	N_2O, +104.18
Aspartate, −700.4	PbS, −92.59	H_2O, −237.17	
Benzoic acid, 245.6	Mn^{2+}, −227.93	H_2O_2, −134.1	
Butyrate, −352.63	Mn^{3+}, −82.12	PO_4^{3-}, −1026.55	
Caproate, −335.96	MnO_4^{2-}, −506.57	Se^0, 0	
Citrate, −1168.34	MnO_2, −456.71	H_2Se, −77.09	
Crotonate, −277.4	$MnSO_4$, −955.32	SeO_4^{2-}, −439.95	
Cysteine, −339.8	Hgs, −49.02	S^0, 0	
Ethanol, −181.75	MoS_2, −225.42	SO_3^{2-}, −486.6	
Formaldehyde, −130.54	ZnS, −198.60	SO_4^{2-}, −744.6	
Formate, −351.04		$S_2O_3^{2-}$, −513.4	
Fructose, −915.38		H_2S, −27.87	
Fumarate, −604.21		HS^-, +12.05	
Gluconate, −1128.3		S^{2-}, +85.8	
Glucose, −917.22			
Glutamate, −699.6			
Glycerate, −658.1			
Glycerol, −488.52			
Glycine, −314.96			
Glycolate, −530.95			
Guanine, +46.99			
Lactate, −517.81			
Lactose, −1515.24			
Malate, −845.08			
Mannitol, −942.61			
Methanol, −175.39			
Oxalate, −674.04			
Phenol, −47.6			
n-Propanol, −175.81			
Propianate, −361.08			
Pyuvate, −474.63			
Ribose, −757.3			
Succinate, −690.23			
Sucrose, −370.90			
Urea, −203.76			
Valerate, −344.34			

Source: Madigan et al. (2000).

Example   Compare the thermodynamic feasibility of acetate biodegradation under aerobic versus anaerobic conditions. Using a similar approach, we can write balanced reactions for acetate biodegradation and calculate the corresponding $\Delta G°$ values:

Aerobic $CH_3COO^- + H^+ + 2O_2 \rightarrow 2CO_2 + 2H_2O,$ $\Delta G° = -895\,kJ$
Anaerobic $CH_3COO^- + H^+ \rightarrow CO_2 + CH_4,$ $\Delta G° = -76\,kJ$

Since protons are involved in these reactions, the free-energy change will be a function of pH. In most cells, the pH of the cytoplasm is close to 7.0, so $\Delta G°'$ is more appropriate than $\Delta G°$. To calculate $\Delta G°'$, use the following formula:

$$\Delta G°' = \Delta G° + m\,\Delta G_f°(H^+) \qquad [B.3]$$

where m = number of moles of H^+ in the reaction (negative when consumed, positive when evolved)
$\Delta G_f°(H^+)$ = free energy of formation of a proton at pH $7 = -40.01\,kJ$

In the previous example, one proton is consumed for each acetate molecule degraded. Therefore, $\Delta G°'$ for aerobic and anaerobic acetate utilization is obtained by adding $40\,kJ$ to the previous $\Delta G°$ values:

$$\Delta G°' = -855\,kJ \text{ (aerobic)} \quad \text{and} \quad -37.67\,kJ \text{ (anaerobic)}$$

The $\Delta G°'$ calculated can be verified using Eq. [B.1]:

$$\Delta G°' = -n\,F\,\Delta E°'$$

Here, $n = 8$ (based on the half-reaction of acetate: $CH_3COO^- + 2H_2O \rightarrow 2CO_2 + 7H^+ + 8e^-$) and

$$\Delta E°' = E°'(e^-\text{acceptor}) - E°'(e^-\text{donor}) \qquad [B.4]$$

From Table B.1,

$$\Delta E°' = 0.82 - (-0.29) = 1.11\,V \quad \text{(for the aerobic case)}$$

Thus, according to Eq. [B.1],

$$\Delta G°' = -8(96.6)(1.11) = -855\,kJ, \quad \text{as before}$$

B.4. DETERMINATION OF ΔG

ΔG is a function of the concentrations of products and reactants present. As a reaction proceeds toward equilibrium, the ΔG of the reaction approaches 0. In cells, the concentrations of reactants or products is rarely 1 molar; therefore, the free energy available to the cell is not representative of calculated $\Delta G^{\circ\prime}$ values. To calculate ΔG, consider the general stoichiometry

$$aA + bB + \cdots \rightarrow cC + dD + \cdots$$

To calculate ΔG, we need to know ΔG° and the activity of individual species participating in the reaction:

$$\Delta G = \Delta G^{\circ} + RT \ln \left(\frac{[C]^c [D]^d \cdots}{[A]^a [B]^b \cdots} \right) \qquad \text{[B.5]}$$

where R = ideal gas constant 0.00829 kJ/mol-K
     T = absolute temperature (K)
     $[\ ]$ = activity of reacting species
        $\approx$ molar concentrations for dissolved species
        $\approx$ partial pressure for gaseous species
        $\approx$ 1 for water and solid species

Example Estimate the amount of energy that is potentially available from five different substrates (hydrogen, acetate, methane, NADH, and FADH) under four different electron acceptor conditions (aerobic, denitrifying, sulfidogenic, and methanogenic). Assume STP, pH 7, and $\Delta G' = \Delta G^{\circ\prime}$ for simplicity. Which substrate yields the most energy/mole? Which oxidant is most favorable?

    Calculate $\Delta E^{\circ\prime}$ using Table B.1 with Eq. [B.4] (i.e., $\Delta E^{\circ\prime} = E^{\circ\prime}$ (e^- acceptor) $-$ $E^{\circ\prime}$ (substrate)):

$\Delta E^{\circ\prime}$ (volts)

| | Electrons Acceptors | | | |
Substrates	O_2	NO_3^-	SO_4^{2-}	CO_2
Hydrogen	1.23	0.84	0.193	0.17
Acetate	1.11	0.72	0.073	0.05
Methane	1.06	0.67	0.023	0
NADH	1.14	0.75	0.103	0.08
FADH	1.037	0.647	0.003	-0.02

To convert to $\Delta G^{\circ\prime}$, use Eq. [B.1] (i.e., $\Delta G^{\circ\prime} = -nF \Delta E^{\circ\prime}$), where $n = 2$ for hydrogen, 8 for acetate, 8 for methane, 2 for NADH, and 2 for FADH and $F = 96.63$ kJ/V.

$$\Delta G^{\circ\prime} \text{ (kJ/mol)}$$

Substrate	Electrons Acceptors			
	O_2	NO_3^-	SO_4^{2-}	CO_2
Hydrogen	−237.7	−162.3	−37.3	−32.8
Acetate	−858.1	−556.6	−56.4	−38.7
Methane	−819.4	−517.9	−17.8	0
NADH	−220.3	−144.9	−19.9	−15.5
FADH	−200.4	−125.0	−0.6	+3.9

This example illustrates that the biological oxidation of a substrate is thermodynamically more feasible using stronger electron acceptors, Thus, the potentially available energy would decrease according to the following respiratory regimes: aerobic (O_2) > denitrifying (NO_3^-) > sulfate reducing (SO_4^{2-}) > methanogenic (CO_2) (Figure B.2).

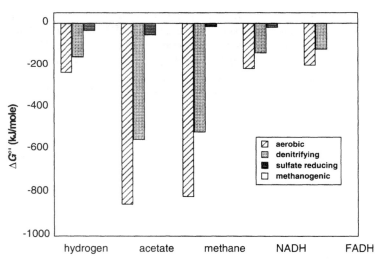

Figure B.2. Thermodynamic feasibility for the oxidation of different substrates under different electron-accepting regimes.

APPENDIX C

COMMONLY USED NUMERICAL GROUNDWATER FLOW AND SOLUTE TRANSPORT CODES*

Name of Code	Description	Distributor
BIO1D®	One-dimensional model for simulation of biodegradation and sorption of hydrocarbons. Transport of substrates and electron acceptors is considered, assuming a uniform flow field. Several reaction options are available for biodegradation and sorption. Has a preprocessor and display graphics. A proprietary code developed by GeoTrans, Inc.	GeoTrans, Inc; IGWMC
BIOPLUME III	Two-dimensional model for reactive transport of multiple hydrocarbons under the influence of advection, dispersion, sorption, first-order decay, and reactant-limited biodegradation. Development commissioned by the Air Force Center for Environmental Excellence (AFCEE). Has interactive, graphical pre- and postprocessing capabilities. Based on the United States Geological Survey (USGS) two-dimensional Method of Characteristics (MOC) model (including a finite difference flow model) by Konikow and Bredehoeft (1978). Oxygen-limited biodegradation is a reactive transport process. A public domain code with a menu-driven preprocessor and limited postprocessing capabilities.	AFCEE

*Modified from Wiedemeier et al. (1999).

Bioremediation and Natural Attenuation: Process Fundamentals and Mathematical Models
By Pedro J. J. Alvarez and Walter A. Illman   Copyright © 2006 John Wiley & Sons, Inc.

Name of Code	Description	Distributor
FEHM	Numerical simulation code for three-dimensional, time-dependent, multiphase, multicomponent, nonisothermal, reactive flow through porous and fractured media. It can accurately represent complex three-dimensional geologic media and structures and their effects on subsurface flow and transport.	Los Alamos National Laboratory
FEFLOW	Three-dimensional flow and transport code for saturated and unsaturated zones with capabilities for heat transport as well. Sorption is modeled through Henry, Freundlich, or Langmuir isotherms. It also considers variable fluid density due to temperature or (salt) concentration effects and additional capabilities include one-dimensional/two-dimensional finite elements for flow and transport in fractures, channels, or tubes.	Waterloo Hydrogeologic, Inc.
FEMWATER, FEMWASTE	Finite element flow (FEMWATER) and transport (FEMWASTE) models. FEMWATER can simulate variably saturated conditions in two and three dimensions. FEMWASTE can simulate transport in one, two, and three dimensions. The system may be heterogeneous and anisotropic, and the code can account for advection, dispersion, first-order decay, and three types of sorption. Public domain codes developed by researchers at Oak Ridge National Laboratories. Some proprietary versions of FEMWATER are available: they are based on the Department of Defense Groundwater Modeling System (GMS), a modeling and data management package.	Oak Ridge National Laboratories, NTIS, distributors of proprietary GMS program
FLONET[R], FLOTRANS[R]	Two-dimensional steady-state groundwater flow (FLONET) and transient solute transport (FLOTRAS) models for cross-sectional problems. FLOTRANS is an extension of FLONET that can simulate transport under the influence of advection, dispersion, linear sorption, and first-order decay. A proprietary program with an interactive graphical user interface and extensive pre- and postprocessing capabilities. Developed by Waterloo Hydrogeologic Software, Inc.	IGWMC; Waterloo Hydrogeologic Software, Inc.
FRAC3DVS	FRAC3DVS is a three-dimensional, finite element model for simulating steady-state or transient, variably saturated groundwater flow and advective–dispersive solute transport in porous or discretely fractured porous media.	Waterloo Hydrogeologic, Inc.

Name of Code	Description	Distributor
FRACTRAN	The code was developed by E.A. Sudicky, at the Waterloo Centre for Groundwater Research (WCGR), and R. Therien, at Laval University. FRACTRAN is a two-dimensional (cross-section and areal) finite element model for simulating steady-state groundwater flow and transient contaminant transport in a discretely fractured, porous medium. The porous medium is represented by block elements while fractures (optional) are represented by line elements. Unlike other fracture flow models, FRACTRAN accounts for flow and transport mechanisms through both the discrete fractures and the matrix block. The powerful solving algorithm makes use of the LTG scheme, which solves a 20 year or a 2000 year simulation in almost the same run-time. FRACTRAN also has the option of solving for flow and transport in nonfractured porous media.	Waterloo Hydrogeologic, Inc.
FTWORK	Block centered finite difference model for one-, two-, and three-dimensional flow and transport. The transport model includes advection, dispersion, first-order decay, and two types of sorption (linear and nonlinear equilibrium). A public domain code that may be acquired with a proprietary (IGWMC) textual and menu-driven preprocessor and postprocessor. Originally developed by Faust et al. (1990) at GeoTrans, Inc.	IGWMC; GeoTrans, Inc.
Groundwater Modeling System (GMS)	Graphical user interface (GUI)-based software that combines applications in three-dimensional groundwater flow and contaminant transport modeling. It support both finite difference and finite element models in two and three dimensions including MODFLOW 2000, MODPATH, MT3DMS/RT3D, SEAM3D, ART3D, UTCHEM, FEMWATER, PEST, UCODE, MODAEM, and SEEP2D. Model calibration can be done with a parameter estimation package PEST implemented in the code. Preprocessing of input data and postprocessing of results can be done within the software.	Environmental Modeling Systems, Inc.
HST3D	Program for simulating groundwater flow and associated heat and solute transport in three dimensions. Solute transport is for a single solute with advection, dispersion, linear sorption, and first-order decay. A public domain code with no pre- and postprocessors. Prepared by K.L. Kipp of the USGS.	IGWMC

Name of Code	Description	Distributor
MOC, USGS2D-MOC	Two-dimensional model for simulation of groundwater flow and nonconservative solute transport. Derived from the original model developed by Konikow and Bredehoeft (1978). The latest version (March 1995) simulates transport under the influence of advection, dispersion, first-order decay, reversible equilibrium-controlled sorption, and reversible equilibrium-controlled ion exchange. The flow model is a finite difference model, while transport is simulated using MOC methods. A public domain code with an interactive preprocessor.	IGWMC
MODFLOW	Block centered finite difference code for steady-state and transient simulation of groundwater flow in two and three dimensions. Consists of main program and a large number of subroutines (modules) that are used to simulate a wide variety of boundaries and stresses on the hydrogeologic system. Originally coded by McDonald and Harbaugh (1988) of the USGS. Possibly the most widely used flow model in the United States and Canada, MODFLOW can be used to generate flow fields that may be coupled with a wide variety of transport models (e.g., MT3D, BioTrans®, or RAND3D).	USGS; IGWMC; in addition, many companies have developed pre- and postprocessing programs with a wide variety of capabilities and features
MODFLOWP	Extension of MODFLOW that includes a package that uses nonlinear regression techniques to estimate model parameters under constraints given by the modeler. Model input includes statistics for analyzing the parameter estimates and the model to quantify the reliability of the resulting model, to suggest changes in model construction, and to compare results of models constructed in different ways. Prepared by M. C. Hill of the USGS. Requires a user with advanced skills.	USGS; IGWMC
MODFLOW-SURFACT	The code is based on the USGS MODFLOW code with flow and transport modeling capabilities for both the saturated and unsaturated zones. The code has an axisymmetric option for quick analyses of slug tests or tracer tests. It can accommodate up to five contaminant species in a single simulation including first-order chain reactions, which may occur due to radioactive decay (species dependent) or biochemical transformation (species and soil/location dependent). Other highlights include options for linear or nonlinear (Freundlich)	Scientific Software Group

Name of Code	Description	Distributor
	equilibrium adsorption and two-, three-, and four-component dispersivity options for various anisotropic conditions. Under isotropic conditions, the three- and four-dispersivity formulations collapse to Scheidegger's two-component (longitudinal and transverse) dispersivity equation.	
MODPATH	A particle tracking code that is used in conjunction with MODFLOW. Particles are tracked through time, assuming that advection is the only transport mechanism. User can define locations of particles. The code is useful for defining capture zones, well head protection zones, and potential locations for contaminant source zones.	USGS
MT3D	Three-dimensional transport model for simulation of advection, dispersion, linear or nonlinear sorption, and first-order decay of single species. Uses a modular structure similar to that of MODFLOW. Intended for use with any block centered finite difference model, such as MODFLOW, on the assumption that concentration changes will not affect the flow field. MT3D uses one of three methods (all based on MOC) for solution of the transport equation. Prepared by C. Zheng (for S. S. Papadopulos & Associates, Inc.), MT3D is available in public-domain and proprietary versions. Proprietary versions are typically the most advanced in terms of pre- and postprocessing capabilities.	S. S. Papadopulos & Associates, Inc.; IGWMC; many versions available from many companies with pre- and postprocessing programs with a wide variety of capabilities and features. Often coupled with MODFLOW in such codes. Public-domain version may be acquired from USEPA.
MT3DMS	A new version of MT3D that has a comprehensive set of options and capabilities for simulating advection, dispersion/diffusion, and chemical reactions of contaminants in groundwater flow systems under general hydrogeological conditions. The key features of MT3DMS are summarized below. Includes the standard finite difference method; the particle-tracking-based Eulerian–Lagrangian methods; and the higher-order finite volume TVD method. The code also supports an optional, dual-domain formulation for modeling mass transport. It can be used to simulate changes in concentrations of miscible contaminants in groundwater considering	C. Zheng, University of Alabama

Name of Code	Description	Distributor
	advection, dispersion, diffusion, and some basic chemical reactions, with various types of boundary conditions and external sources or sinks. The chemical reactions included in the model are equilibrium-controlled or rate-limited linear or nonlinear sorption, and first-order irreversible or reversible kinetic reactions.	
RT3D	Modification of MT3DMS, developed by Battelle Pacific Northwest National Laboratory. RT3D (Reactive Transport in Three Dimensions) is designed to describe multispecies transport and reactions, including attenuation of chlorinated compounds and their daughter products, and fate of solid phase species. Also included are reaction packages for aerobic, instantaneous BTEX reactions (similar to BIOPLUME II) and multiple-electron-acceptor, kinetically limited BTEX reactions (similar to BIOPLUME III). The code is well suited for simulating natural attenuation and bioremediation.	Battelle Pacific Northwest National Laboratory
SEAM3D	A reactive transport model used to simulate complex biodegradation problems involving multiple substrates and multiple electron acceptors. The code is based on MT3DMS. In addition to the regular MT3DMS modules, SEAM3D includes a biodegradation package and NAPL dissolution package. Developed by Mark Widdowson at Virginia Tech University.	Mark Widdowson at Virginia Tech University
SUTRA	Code for simulating two-dimensional fluid movement and transport of energy or dissolved substances. May be used for saturated systems or variably saturated systems in profile view. Can simulate advection, dispersion, sorption, and first-order decay. A public domain code originally prepared by C.I. Voss of the USGS, IGWMC version has a graphical processor.	IGWMC; USGS
SWMS_2D	Two-dimensional model for simulating water and solute movement in variably saturated media. Includes dispersion, linear sorption, zero-order production, and first-order decay. A public domain code prepared by researchers at the U.S. Salinity Lab.	IGWMC
TMVOC	A numerical simulator for three-phase nonisothermal flow of water, soil gas, and a multicomponent mixture of volatile organic chemicals (VOCs) in multidimensional heterogeneous porous media. It is an extension of the TOUGH2 general-purpose simulation	Lawrence Berkeley National Laboratory

Name of Code	Description	Distributor
	program developed at the Lawrence Berkeley National Laboratory. TMVOC is designed for applications to contamination problems that involve hydrocarbon fuel or organic solvent spills in saturated and unsaturated zones. It can model contaminant behavior under "natural" environmental conditions, as well as for engineered systems, such as soil vapor extraction, groundwater pumping, or steam-assisted source remediation. TMVOC is upward compatible with T2VOC and can be initialized from T2VOC-style initial conditions.	
UTCHEM	A program for modeling transient and steady-state three-dimensional flow and mass transport in the groundwater (saturated) and vadose (unsaturated) zones of aquifers. Physical, chemical, and biological process models important in describing the fate and transport of NAPLs in contaminated aquifers are incorporated into the simulator. These include multiple organic NAPL phases; the dissolution and/or mobilization of NAPLs by nondilute remedial fluids; chemical and microbiological transformations; and changes in fluid properties as a site is remediated. The model allows for nonequilibrium interphase mass transfer; sorption; geochemical reactions; and the temperature dependence of pertinent chemical and physical properties. It can simulate the flow and transport of remedial fluids whose density, temperature, and viscosity are variable, including surfactants, cosolvents, and other enhancement agents. The biodegradation model includes inhibition, sequential use of electron acceptors, and cometabolism and can be used to model a very general class of bioremediation processes.	Gary Pope, University of Texas at Austin
Visual Modflow Pro 4.0	Graphical user interface (GUI)-based software that combines applications in three-dimensional groundwater flow and contaminant transport modeling. Flow and transport codes implemented include MODFLOW2000, MODPATH, MT3D, and RT3D. Model calibration can be done with a parameter estimation package PEST implemented in the code. Preprocessing of input data and post-processing of results can be done within the software.	Waterloo Hydrogeologic, Inc.

Name of Code	Description	Distributor
VSAFT2	A finite element program for solving flow and reactive solute transport in variably saturated porous media (two-dimensional horizontal plane, vertical plane, or axisymmetrical plane). A graphical user interface has been developed for easy generation of irregular solution domain with homogeneous properties, zonal properties, or randomly distributed hydraulic properties, and easy definition of boundary/initial conditions, and observation wells. In addition, a geostatistics package (GSLIB) is being implemented in the program to allow variogram analysis and estimation using kriging and cokriging for parameter estimations. Developed by Yeh et al. (1993) at the University of Arizona.	T.-C. Jim Yeh: Department of Hydrology and Water Resources, The University of Arizona

APPENDIX D

NONPARAMETRIC STATISTICAL TESTS FOR DETERMINING THE EFFECTIVENESS OF NATURAL ATTENUATION[*]

Two nonparametric statistical tests are described here: the Mann–Kendall (S) and Mann–Whitney (U) statistical tests. These tests can be used to show whether groundwater contaminant concentrations in a monitoring well are increasing, stable, or decreasing. However, neither test is able to determine the rate at which the concentrations are changing over time. The Mann–Kendall test can be used with a minimum of four rounds of sampling results; however, the Mann–Kendall test is not valid for data that exhibit seasonal behavior. The Mann–Whitney U test is applicable to data that may or may not exhibit seasonal behavior, but the test requires eight consecutive rounds of quarterly or semiannual sampling results. To demonstrate that natural attenuation is effective, the chosen statistical test must show decreasing contaminant concentrations at an appropriate confidence level, given in the test methodologies that follow.

D.1. MANN–KENDALL TEST

1. Assemble well data for at least four sampling events for each contaminant in the order in which the data was collected. Include all contaminants that have exceeded the ES at one or more monitoring wells. Include data from:

   (a) one or more contaminated monitoring wells near the downgradient plume margin, which may include piezometers;

   (b) a monitoring well near the source zone; and

---

[*]Adapted from Wisconsin Department of Natural Resources.

Bioremediation and Natural Attenuation: Process Fundamentals and Mathematical Models
By Pedro J. J. Alvarez and Walter A. Illman   Copyright © 2006 John Wiley & Sons, Inc.

(c) at least one monitoring well along a flow line between the source zone well and plume margin well.

2. For purposes of the Mann–Kendall test, all nondetect data values should be assigned a single value that is less than the detection limit, even if the detection limit varies over time.

3. *Tests for Seasonality in Data.* For seasonally affected data, either remove the seasonality in the data (e.g., by only testing data from the seasons with the highest contaminant concentrations) or use a statistical test that is unaffected by seasonality, such as the Mann–Whitney U test. To test for data seasonality:

(a) Determine if groundwater flow direction changes with season by comparing a water table map from each season for which the contaminant concentrations are measured. If the flow direction changes from one sampling period to another and shifts the plume away from the wells being used in the statistical test, then data from those season(s) that are shifted away from the centerline monitoring wells cannot be used in the Mann–Kendall test.

(b) Determine if groundwater elevation and contaminant concentration change seasonally. Plot contaminant concentration versus groundwater level for each well to be assessed by the Mann–Kendall test. If groundwater concentrations change as water level changes, then the data is seasonally affected. The seasons with the highest contaminant concentrations should be included in the Mann–Kendall test.

4. Calculate the Mann–Kendall statistic (S) using a manual method or a DNR supplied spreadsheet. Assess all contaminants in the plume for the selected wells being assessed with the Mann–Kendall test. Enter data for each contaminant in the order it was collected.

(a) *Manual Method to Calculate Mann–Kendall Statistic.* Compare data sequentially, comparing sampling event 1 to sampling events 2 through n, then sampling event 2 to sampling events 3 through n, and so on. Each row is filled in with a 1, 0, or −1, as follows:

Along row 2, if:

- Concentration of event x_i > event 1: enter +1.
- Concentration of event x_i = event 1: enter 0.
- Concentration of event x_i < event 1: enter −1.

where n = total number of sampling events and x_i = value of given sample event, with i = 2 to n.

Continue for the remaining rows. Sum each row and enter result at the end of the row. Add the sum of each row down to obtain the Mann–Kendall statistic (S). See Table D.1 as an example.

Table D.1. Mann–Kendall Statistic

	Sampling Event 1	Sampling Event 1	Sampling Event 1	Sampling Event 1	Sampling Event 1	
Contaminant concentration >	100	50	85	75	50	Sum Rows
Compare to event 1	>	−1	−1	−1	−1	−4
Compare to event 2		>	+1	+1	0	+2
Compare to event 3			>	−1	−1	−2
Compare to event 4				>	−1	−1
				Mann–Kendall statistic (total)		−5

(b) *Manual Mann–Kendall Statistic Lookup Table.* Table D.2 gives the maximum S statistic (S_{max}) to accept a declining trend alternative at an α level of significance. If the computed S is greater than S_{max} (or S is a smaller negative number than S_{max}), then there is either a no-trend or an increasing trend in the data.

5. *Test for a Declining Trend.* Evaluate data trends for each contaminant identified in the plume. Evaluate the null hypothesis of no trend against the alternative of a decreasing trend. The null hypothesis can be rejected in favor of a decreasing trend if both of the following conditions are met:

(a) S is a large negative number (see Table D.2 for magnitude of S).

(b) The probability value, given n (number of data) and the absolute value of S, is *less* than the a priori significance level, α, of the test. An $\alpha \leq 0.2$ is acceptable.

Table D.2. Mann–Kendall Statistic Lookup Table

N	Range of S	S_{max} $\alpha = 0.2$[a]
4	−6 to +6	−4
5	−10 to +10	−5
6	−15 to +15	−6
7	−21 to +21	−7
8	−28 to +28	−8
9	−36 to +36	−10
10	−45 to +45	−11

[a]The probability that the computed Mann–Kendall statistic $S \leq S_{max}$ is at most α.

6. *Test for an Increasing Trend.* An increasing trend alternative (i.e., an advancing plume) is shown if both of the following conditions are met:

(a) S is positive.

(b) $S \geq |S_{max}|$ at a given a level of significance (see Table D.2). If the computed S is equal to or greater than the absolute value of S_{max}, then it can be concluded the plume is advancing at an α level of significance.

7. *Test for Plume Stability.* If the Mann–Kendall test indicates no trend is present, perform the coefficient of variation test. As a nonparametric test, the Mann–Kendall test does not take into account the magnitude of scatter in the data. A data set with a great deal of scatter may return a Mann–Kendall test indicating there is no trend, when, in fact, no conclusion can be drawnregarding trend because of data variability. In this case, additional data collection may be necessary to determine that the plume is stable, declining, or advancing. As a simple test, the coefficient of variation can assess the scatter in the data:

$$CV = \sigma/\bar{x}$$

where CV = coefficient of variation, s = standard deviation, and $\bar{x}$ = arithmetic mean. CV should be ≤ 1 to say that the no-trend hypothesis also indicates a stable plume configuration.

D.2. MANN–WHITNEY U TEST

This test is equivalent to the Wilcoxon rank sum test.

1. Assemble well data for the most recent eight consecutive quarterly or semiannual sampling events for each contaminant that has exceeded the ES at one or more monitoring wells. Include data from:

(a) one or more contaminated monitoring wells near the downgradient plume margin, which may include piezometers;

(b) a monitoring well near the source zone; and

(c) at least one monitoring well along a flow line between the source zone well and plume margin well.

2. Enter the data into a DNR supplied spreadsheet or manually assemble the data into a table (e.g., Table D.3) in the order the data was collected. Assign a rank to each sample value, with the smallest value ranked #1 and the largest value ranked #8.

Table D.3. Example Data Set for the Mann–Whitney *U* Statistical Test

Year/Date	Benzene Concentration (µg/L)	Rank	Rank Sum of 1st Year (*Wrs*)
1st year, 1st quarter	160	8	
1st year, 2nd quarter	130	7	
1st year, 3rd quarter	80	4	
1st year, 4th quarter	100	6	$8 + 7 + 4 + 6 = 25$
2nd year, 1st quarter	89	5	
2nd year, 2nd quarter	0	1	
2nd year, 3rd quarter	53	3	
2nd year, 4th quarter	24	2	
			$U = 26 - Wrs = 1$

3. For purposes of the Mann–Whitney U test, all nondetect values should be assigned a data value of zero (0).

4. Sum the ranks for the data in the 1st year. Denote this sum as Wrs (or the Wilcoxon rank sum).

5. Calculate the U statistic. $U = 26 - Wrs$.

6. *Interpreting U Statistic.* For two groups of four samples, at $U \leq 3$, the probability that year 2 data show a decrease relative to year 1 data is at least 90%; and so $U \leq 3$ shows that the contaminant concentration is declining.

7. If there are ties in sample data, calculate an average rank value for the tied data and assign this average rank to the tied sample data. See example in Table D.4.

8. *Probability and the U Statistic.* Table D.5 shows the α value and the confidence interval for values of U calculated for two groups of four samples each.

9. If more than eight consecutive rounds of data are available, a Mann–Whitney U statistic can be calculated similar to the method presented here. Each set of data

Table D.4. Example of Rank Sum Value for Tied Data

Year/Date	Benzene Concentration (µg/L)	Check for Ties	Rank	Rank Sum of 1st Year (*Wrs*)
1st year, 1st quarter	300		8	
1st year, 2nd quarter	280		7	
1st year, 3rd quarter	105		4	
1st year, 4th quarter	110	*	5.5	$8 + 7 + 4 + 5.5 = 24.5$
2nd year, 1st quarter	83		3	
2nd year, 2nd quarter	50	√	1.5	
2nd year, 3rd quarter	110	*	5.5	
2nd year, 4th quarter	50	√	1.5	
				$U = 26 - Wrs = 1.5$

Table D.5. Probability and U Statistic (for Two Groups of Four Samples Each)

U Statistic	Level of Significance (α)	Confidence Level (%)
0	0.014	98.6
1	0.029	97.1
2	0.057	94.3
3	0.100	90.0

to be compared should represent the same span of time (e.g., 1 year) and the same time interval between samples (e.g., quarterly). The test must be conducted at a level of significance $\alpha \leq 0.10$

REFERENCES

Conover, W.J. (1971). *Practical Nonparametric Statistics*, 2nd ed. John Wiley & Sons, Hoboken, NJ, pp. 216–223.

Gilbert, R.O. (1987). *Statistical Methods for Environmental Pollution Monitoring*. Van Nostrand Reinhold, New York, pp. 204–240, 272.

APPENDIX E

CRITICAL VALUES OF THE STUDENT t-DISTRIBUTION

Degrees of Freedom, ν	Significance Level, α				
	0.10	0.05	0.025	0.01	0.005
1	3.078	6.314	12.706	31.821	63.657
2	1.886	2.920	4.303	6.965	9.925
3	1.638	2.353	3.182	4.541	5.841
4	1.533	2.132	2.776	3.747	4.604
5	1.476	2.015	2.571	3.365	4.032
6	1.440	1.943	2.447	3.143	3.707
7	1.415	1.895	2.365	2.998	3.499
8	1.397	1.860	2.306	2.896	3.355
9	1.383	1.833	2.262	2.821	3.250
10	1.372	1.812	2.228	2.764	3.169
11	1.363	1.796	2.201	2.718	3.106
12	1.356	1.782	2.179	2.681	3.055
13	1.350	1.771	2.160	2.650	3.012
14	1.345	1.761	2.145	2.624	2.977
15	1.341	1.753	2.131	2.602	2.947
16	1.337	1.746	2.120	2.583	2.921
17	1.333	1.740	2.110	2.567	2.898
18	1.330	1.734	2.101	2.552	2.878
19	1.328	1.729	2.093	2.539	2.861
20	1.325	1.725	2.086	2.528	2.845
21	1.323	1.721	2.080	2.518	2.831
22	1.321	1.717	2.074	2.508	2.819

Bioremediation and Natural Attenuation: Process Fundamentals and Mathematical Models
By Pedro J. J. Alvarez and Walter A. Illman   Copyright © 2006 John Wiley & Sons, Inc.

Degrees of Freedom, ν	Significance Level, α				
	0.10	0.05	0.025	0.01	0.005
23	1.319	1.714	2.069	2.500	2.807
24	1.318	1.711	2.064	2.492	2.797
25	1.316	1.708	2.060	2.485	2.787
26	1.315	1.706	2.056	2.479	2.779
27	1.314	1.703	2.052	2.473	2.771
28	1.313	1.701	2.048	2.467	2.763
29	1.311	1.699	2.045	2.462	2.756
∞	1.282	1.645	1.960	2.326	2.576

GLOSSARY

Abiotic   Referring to a system without living organisms.

Absorption   The process that occurs when the solute diffuses into the soil particle and sorbs onto interior surfaces.

Acidic   A compound that releases hydrogen ions (H^+) when dissolved in water or yields positive ions upon dissolution; a solution with a pH value lower than 7.0.

Acidophiles   Microorganisms that prefer to grow at low pH (e.g., pH 2).

Activated carbon   A porous material used in water conditioning as an adsorbent for organic matter and certain dissolved gases.

Activation energy   Energy needed to increase the reactivity of substrate molecules; enzymes catalyze biotransformations by lowering the activation energy.

Active site   The part of the enzyme that is directly involved in binding substrate(s).

Adsorption   The interface (surface) accumulation of solid, liquid, or gaseous molecules.

Advection   The process that causes solute transport through flowing groundwater.

Aerobes   Microorganisms that require the presence of air or molecular oxygen to grow.

Aerobic   Presence of air or molecular oxygen; respiring molecular oxygen.

Air sparging   a remediation approach that relies on mass transfer of contaminants through volatilization of groundwater contaminants to air injected into aquifers.

Air stripping   Transfer of mass from the solution to the gas phase, usually air.

Alkaline   Solutions with a basic pH (pH >7); a condition in which hydroxyl (OH^-) ions are in abundance.

Amensalism   An interactive association between two populations that is detrimental to one while the other is not adversely affected.

Anabolism   The biochemical processes involved in the synthesis of new cell constituents from simpler molecules, which usually requires energy.

Anaerobes   Organisms that grow in the absence of air or molecular oxygen; organisms that do not respire molecular oxygen.

Anaerobic   The absence of molecular oxygen.

Anaerobic respiration   Use of an electron acceptor other than O_2 to obtain energy in an electron transport-based oxidation that generates a proton motive force.

Bioremediation and Natural Attenuation: Process Fundamentals and Mathematical Models
By Pedro J. J. Alvarez and Walter A. Illman   Copyright © 2006 John Wiley & Sons, Inc.

Anisotropic   Directional dependence of some physical property such as hydraulic conductivity.

Antagonism   The inhibition, injury, or killing of one species by another; a microbial interpopulation relationship in which one population has a deleterious effect on another.

Archaea   A phylogenetic domain of prokaryotes consisting of the methanogens, extreme halophiles, and hyperthermophiles.

Aquiclude   A geological unit through which virtually no water moves.

Aquifer   A water-bearing geological formation, such as subsurface water bodies that supply water for wells and springs; a permeable layer of rock or soil capable of holding and transmitting water.

Aquitard   A saturated geological unit that does not yield water freely to a well or a spring but may transmit appreciable water to or from adjacent aquifers.

Artesian well   A well tapping a confined aquifer in which the piezometric surface is higher than the land surface, causing free flowing of groundwater out of the well.

Artificial recharge   Recharge of water back into the aquifer such as irrigation, or induced infiltration from streams or wells.

Assimilation   The uptake of nutrients into the biomass of an organism.

ATP   Adenosine triphosphate, which is the principal energy carrier of the cell that releases energy upon hydrolysis.

Autotrophs   Organisms that utilize inorganic carbon (e.g., CO_2) as carbon source for cell synthesis.

Bacteria   All prokaryotes that are not members of the phylum Archaea.

Bacteriophage   A virus that infects and replicates within bacterial cells.

Bacteriostatic   An agent that inhibits bacterial growth and reproduction without necessarily killing them.

Baseline monitoring   A designed surface or groundwater monitoring program for continuous or periodic measurements that will be compared with future observations.

Bedrock   A general term for solid rock that underlies soil or other unconsolidated material.

Bentonite   A material composed of clay minerals commonly used in drilling mud.

Beta-oxidation (β-oxidation)   Central metabolic pathway for the oxidation of fatty acids resulting in the formation of acetate and a new fatty acid that is two carbon atoms shorter than the original fatty acid.

Bioaugmentation   The addition of microorganisms with specific biodegradation capabilities to contaminated sites or biological treatment reactors.

Biocatalysis   The use of microorganisms or their enzymes to synthesize a product or carry out a specific biochemical transformation.

Biocide   An agent that kills microorganisms.

Biodegradable   A substance that can be transformed and broken down into smaller molecules by microorganisms.

Biofilm   Microbial colonies encased in an adhesive material, usually polysaccharides, and attached to a surface.

Biodegradation   The breakdown of a substance into smaller products caused by microorganisms or their enzymes.

Biodeterioration The alteration of a product by microorganisms that decreases the usefulness of that product for its intended purpose.

Biofilter A device used for the biodegradation of air pollutants consisting of an immobilized microbial community as a biofilm through which contaminated air is passed.

Biogeochemical cycling Biologically mediated transformations of elements that result in their global cycling, including transfer between air, water, and soil compartments.

Biological oxygen demand (BOD) The amount of dissolved oxygen required to stabilize organic matter in sewage or water by aerobic biodegradation.

Bioluminescence The generation of light by certain microorganisms generally possessing proteins called luciferins that are converted in the presence of oxygen to oxyluciferins by luciferase, producing light.

Biomagnification An increase in the concentration of a substance in animal tissue as the substance is passed to higher members of a food chain.

Biomass The dry weight, volume, or other quantitative estimation of organisms in a given ecosystem.

Bioremediation A spontaneous or managed biological treatment process involving the use of microorganisms or their enzymes to remove pollutants from air, soils, and waters.

Biosensor An immunological or genetic method of detecting chemicals or microbial activities (e.g., based on the emission of light or electrical detection).

Biostimulation Addition of nutrients, electron acceptors (or electron donors), and sometimes also auxiliary substrates to stimulate the growth and activity of specific indigenous microbial populations.

Biotic Of or relating to living organisms.

Bioventing An in situ remediation technology that uses indigenous microorganisms to biodegrade organic constituents adsorbed to soils in the unsaturated zone.

Breakthrough curve A graph of relative concentration versus time, where relative concentration is defined as the ratio between solute concentration at some point in space and time and the intial concentration.

Brownian motion The random motion of atoms and molecules in fluids due to fluid temperature.

Capillary fringe The zone above the water table in which the pores are fully saturated but the water is under tension.

Capillary pressure The difference in pressure between the nonwetting and wetting phase fluids caused by interfacial tension between the two phases.

Carcinogen A substance that initiates tumor formation, frequently also causing DNA mutation.

Catabolism The biochemical process involved in the degradation of organic or inorganic substrates, usually leading to the production of building blocks and energy for anabolism.

Catalysis Increase in the rate of a chemical reaction by lowering its activation energy.

Catalyst A substance that promotes a chemical reaction without itself being consumed.

Catabolic pathways A metabolic pathway in which molecules are broken down into smaller ones.

Cations Positively charged ions.

Cell The fundamental unit of life.

Chemolithotroph   An organism that obtains its energy from the oxidation of inorganic compounds.

Chemoorganotroph   An organism that obtains its energy from the oxidation of organic compounds.

Chelator   A substance that binds metallic ions.

Chemostat   A continuous culture reactor controlled by the concentration of the limiting nutrient and the dilution rate.

Chemotaxis   Motion of bacteria (toward or away) in response to a chemical concentration gradient.

Chromosome   A genetic element harboring genes essential to cellular function. Prokaryotes typically have a single circular chromosome whereas eukaryotes tend to have multiple chromosomes, each containing a linear DNA molecule.

Coliforms   Gram-negative, lactose-fermenting, enteric rod-shaped bacteria (e.g. *Escherichia coli*).

Cometabolism   The fortuitous degradation of a pollutant, often in the presence of a primary substrate; the pollutant that is cometabolized serves no metabolic purpose and the bacteria that degrade it do not benefit from this reaction.

Comensalism   A unidirectional interaction between two different microbial populations where one discards compound or degradation by-products that benefit the other. The organism generating the "table scraps" is not affected. Comensalism often results from cometabolism.

Competition   An interaction between two species, both of which need some limited environmental factor for growth and thus grow at suboptimal rates because they must share the growth-limiting resource.

Competitive exclusion principle   The notion that competitive interactions tend to bring about the ecological separation of closely related populations; precludes two populations from occupying the same ecological niche in a given habitat.

Competitive inhibition   The inhibition of enzyme activity caused by an inhibitor compound that binds to the substrate-binding site on the enzyme.

Compressibility   The ratio of the percent change in volume to the change in pressure applied to a fluid or rock.

Concentration gradient   The change in solute concentration per unit distance driving Fickian diffusion (spreading) of solutes from regions of highest to lowest concentrations.

Confined aquifer   A water-bearing geological unit that is confined by low-permeability formations.

Consortium   A bacterial culture (or natural assemblage) consisting of multiple populations in which each organism benefits from the others.

Contamination   The degradation of natural water quality due to anthropogenic activities.

Contour   A curve that includes points of equal value and separates points of higher value from points of lower value.

Contour interval   The separation value between two adjacent contours.

Contour map   A map displaying curves that include points of equal value and separate points of higher value from points of lower value.

Covalent bond   A nonionic bond between two atoms formed by the sharing of electrons.

Cross section A diagram of a vertical section through a volume, as opposed to the plan view of a map.

Culture A particular strain or kind of microorganism growing in a laboratory medium.

Curve fitting The graphical comparison of theoretically derived type curves with experimental data by means of manual or automated procedures.

Darcian velocity See *Specific discharge.*

Darcy's Law An empirical law obtained by Henri Darcy in 1856, which states that the flow velocity through the porous medium is directly proportional to the hydraulic gradient assuming that the flow is laminar and inertia can be neglected.

Datum An agreed on and known value of elevation for a benchmark to which other measurements are corrected.

Dehalorespiration Microbial utilization of chlorinated compounds such as perchloroethene and trichloroethylene as electron acceptors in energy-yielding metabolic reactions.

Denaturation Irreversible destruction of a macromolecule, such as the deformation of a protein by heat.

Denaturing gradient gel electrophoresis (DGGE) An electrophoretic technique that separates nucleic acid fragments of the same size but that differ in sequence.

Dendrograms graphic representations of taxonomic analyses that show the phylogenetic relationships between the organisms examined.

Denitrification Reduction of nitrate or nitrite to nitrogen gases under anoxic conditions.

Density Mass per unit of volume.

Deoxyribonucleic acid (DNA) A polymer of nucleotides connected via a phosphate–deoxyribose sugar backbone; the genetic blueprint of the cell.

Desorption The reverse process of sorption.

Deterministic methods Methods that use equations or algorithms that have been previously developed and do not involve stochastic or statistical approaches.

Diffusion The transport of ions or molecules from regions of high concentration to low concentration within a solution.

Diffusion coefficient A characteristic constant appearing in the diffusion equation.

Diffusion equation A partial differential equation that describes the space–time variation of a physical quantity such as temperature, hydraulic head, and solute concentration governed by the diffusion process.

Dilution Reduction of solute concentrations in contaminant plumes through the addition of fresh water through groundwater recharge.

Dispersion The spreading and mixing of solutes in groundwater caused by diffusion and mixing due to microscopic variations in velocities within and between pores.

Dispersion coefficient The sum of the mechanical dispersion and molecular diffusion coefficients in a porous or fractured medium.

Dispersion, longitudinal Spreading and mixing of solutes in the direction of bulk groundwater flow.

Dispersion, mechanical See *Mechanical dispersion.*

Dispersion, transverse Spreading and mixing of solutes perpendicular to the direction of bulk groundwater flow.

Dispersivity A characteristic property of a porous medium which determines the dispersion characteristics of the medium by relating the components of pore velocity to the dispersion coefficient.

Distribution coefficient The quantity of chemical constituents that sorbs onto the solid per unit weight of solid divided by the quantity dissolved in the water per unit volume of water.

Diversity The heterogeneity of a system; the variety of different types of organisms present.

Diversity index A mathematical measure describing the species richness and apportionment of species within a community.

DNA fingerprinting Use of molecular techniques to determine the origin of DNA in a sample tissue.

DNAPLs (dense nonaqueous phase liquids) Compounds that are denser than water, thus sinking in aquifers.

Doubling time The time needed for a population to double.

Drawdown The distance between the static water level prior to pumping and the surface of the cone of depression caused through pumping.

Ecological niche The functional role of an organism in an ecosystem.

Ecology Study of the interrelationships between organisms and their environment.

Ecosystem A community of organisms and their natural environment.

Effective porosity The interconnected pore volume or void space that excludes isolated pore space in a rock that contributes to fluid flow and solute transport in geologic media.

Electron acceptor A substance that accepts electrons during an oxidation–reduction reaction, usually associated with respiration.

Electron donor A compound that donates electrons in an oxidation–reduction reaction, usually serving as a fuel molecule.

Electron transport phosphorylation Synthesis of ATP involving a membrane-bound electron transport chain and the creation of a proton motive force. Also called respiration and oxidative phosphorylation.

Electrophoresis Separation of charged molecules in an electric field.

ELISA Enzyme-linked immunosorbent assay. An analytical immunoassay tool that uses specific antibodies to detect specific proteins or other antigens or antibodies in environmental or other samples. The antibody-containing complexes are visualized through enzyme coupled to the antibody. Addition of substrate to the enzyme–antibody–antigen complex results in a colored product.

Enrichment culture A microbial culture in a liquid medium that results in an increase in a given type of organism while minimizing the growth of any other organism present.

Enzyme A catalyst, usually a protein that promotes specific reactions or groups of reactions.

Equipotential line Line along which the hydraulic head or potential is constant.

Eubacteria Former name for Bacteria; prokaryotes other than archaebacteria.

Eukaryotes Organisms with a membrane-bound nucleus within which the genome of the cell is stored as chromosomes; eukaryotic organisms include algae, fungi, protozoa, plants, and animals.

Eutrophic Environment with high nutrient concentrations, such as a lake with high phosphate concentration that will support excessive algal blooms.

Eutrophication The enrichment of natural waters with inorganic materials, especially nitrogen and phosphorus, that stimulates excessive growth of photosynthetic organisms.

Evolution The process of change of organisms by which descendants become distinct in form and/or function from their ancestors.

Exponential growth Microbial growth where the number of cells doubles in a fixed time period.

Extreme environments Environments characterized by extremes in temperature, salinity, pH, and water availability, among other growth conditions.

Facultative Denotes metabolic versatility, often referring to an organism capable of growing in either the presence or absence of an environmental factor (e.g., "facultative aerobe").

Fermentation Catabolic reactions involving substrate-level phosphorylation for producing ATP; organic compounds serve as both primary electron donor and ultimate electron acceptor.

Fickian diffusion Solute transport from regions of higher to regions of lower concentrations caused by the concentration gradient.

FISH Fluorescent in situ hybridization, a process of labeling microbial cells with a specific nucleic acid probe that contains an attached fluorescent dye.

Flow line A general path in which water particles flow under laminar flow conditions.

Flow net A graph of flow and equipotential lines for two-dimensional, steady-state groundwater flow.

Flow path A pathway that a water molecule or solute would follow in a given groundwater velocity field.

Flow, steady A flow system in which the magnitude and direction of specific discharge are constant in time at any point throughout the flow domain.

Flow, unsteady A flow system in which the magnitude and/or direction of the specific discharge changes with time.

Flow velocity See *Specific discharge*.

Fluid potential The mechanical energy per unit mass of a fluid at any given point in space and time with respect to an arbitrary datum.

Fluorescent Having the ability to emit light of a specific wavelength when activated by light of another wavelength.

Food web A trophic interrelationship among organisms in which energy is transferred from one organism to another.

Formation A body of strata of predominantly one type or combination of types that is sufficiently continuous that it can be mapped.

Fracture A crack or breakage within consolidated or unconsolidated deposits as well as rocks.

Free energy Energy available to do useful work.

Fresh water Water that contains less than 1000 milligrams per liter (mg/L) of dissolved solids or chemical constituents.

Fungi Nonphototrophic eukaryotic microorganisms with rigid cell walls.

Gel   An inert polymer, usually made of agarose or polyacrylamide, used for electrophoretic separation of macromolecules such as DNA and proteins.

Gene   A unit of heredity consisting of a DNA segment that codes for a particular protein or polypeptide chain, a tRNA or an rRNA molecule.

Genetic engineering   The manipulation of the genetic properties of an organism by the application of recombinant DNA technology.

Genetically modified organism (GMO)   An organism whose genome has been altered by genetic engineering.

Genetics   The study of heredity and variation of organisms.

Genome   The complete set of genes present in an organism.

Genotype   The genetic constitution of an organism.

Genus   A taxonomic group of related species.

Geochemistry   The study of the Earth's chemistry within solid bodies including the distribution, circulation, and abundance of elements, molecules, minerals, rocks, and fluids.

Geologic map   A topographic map showing the spatial distribution of rocks and deposits, as well as structural features at the surface of the Earth.

Geology   The study of the history, structure, geochemical composition, life forms, and processes that continue to change the Earth.

Geophysics   The study of the electrical, gravitational, and magnetic fields and propagation of seismic waves within the Earth.

Geostatistics   The study of statistics using geospatial data to conduct interpolation and uncertainty analysis.

Glycocalyx   The capsule outside the bacterial cell wall consisting mainly of polysaccharides.

Glycolysis   The conversion of glucose to pyruvate.

Gram-negative cell   Bacteria with cell wall containing relatively little peptidoglycan and possessing an outer membrane composed of lipopolysaccharide, lipoprotein, and other complex macromolecules.

Gram-positive cell   Bacteria with cell wall containing mainly peptidoglycan and lacking the outer membrane of gram-negative cells.

Gravitational head   The component of total hydraulic head expressed in units of length related to the position of a given mass of water relative to an arbitrary datum.

Groundwater   Subsurface water in terrestrial environments.

Groundwater basin   The subsurface volume with defined boundaries from which groundwater drains.

Groundwater discharge   Water flow from the zone of saturation usually into surface water bodies.

Groundwater divide   A ridge in the water table or potentiometric surface from which groundwater moves away perpendicularly from the ridge line in both directions.

Groundwater mound   A raised area in a water table or potentiometric surface created due to recharging groundwater.

Groundwater recharge   Addition of water to the subsurface that does not undergo evapotranspiration.

Groundwater travel time   The time required for a unit volume of groundwater to travel between two locations along a flow path.

Growth   An increase in the number of microbial cells.

Growth rate   The rate at which growth occurs.

Habitat   The location where an organism resides.

Head, static   The height of water column above a standard datum (or another liquid) that is supported by the static pressure at a given point. A quantitative definition is the sum of pressure and gravitational head under static conditions.

Head, total hydraulic   The total hydraulic head of a liquid given in length units at a given point is the sum of the elevation head, which is equal to the elevation of the point above a datum, the pressure head, which is the height of a column of static water that can be supported by the static pressure at the point, and the velocity head, which is the height to which the kinetic energy of the liquid is capable of lifting the liquid. The velocity head is generally neglected in most groundwater applications.

Heterogeneity   A characteristic of the geological medium in which material properties vary spatially.

Heterogeneous medium   A geological medium with spatially varying rock properties.

Heterotroph   Organisms utilizing organic compounds as carbon source.

Homoacetogens   Bacteria that produce acetate as the sole product of sugar fermentation, or from $H_2 + CO_2$.

Homogeneity   A characteristic of a geological medium in which material properties are identical spatially.

Homogeneous medium   A geological medium with rock properties that do not change spatially.

Horizontal (lateral) gene transfer   A gene in an organism that was transferred from another organism.

Humic acids   Soil organic matter composed primarily of high molecular weight irregular organic polymers with acidic character.

Humus   The organic portion of the soil remaining after microbial decomposition.

Hydraulic conductivity   A proportionality constant relating specific discharge to hydraulic gradient.

Hydraulic diffusivity   The ratio between hydraulic conductivity and the specific storage.

Hydraulic fracturing   Fracturing of geological formations by applying pressure in boreholes to increase the hydraulic conductivity.

Hydraulic gradient   The change in hydraulic head per unit of distance in a given direction.

Hydrodynamic dispersion   The macroscopic spreading and mixing of solutes during transport resulting from both mechanical dispersion and molecular diffusion.

Hydrogen bond   A weak chemical bond between a hydrogen atom and a second, more electronegative atom, usually oxygen or nitrogen.

Hydrogeology   The study of the interrelationships of groundwater with geologic materials and processes.

Hydrolase   Enzyme that catalyzes hydrolysis.

Hydrology   The study of the occurrence, distribution, and chemistry of all forms of water of the Earth.

Hydrolysis   Breakdown of a bond (usually ester, glycosidic, or peptide bonds) by the addition of water.

Hydrophilic   Attraction for water by some chemical constituent or material surface.

Hydrophobic   Repellency for water by some chemical constituent or material surface.

Immiscible   Liquids or other phases that cannot mix to form a homogeneous mixture. For example, oil and water are immiscible liquids.

Immobilization   The binding of a substance so that it is no longer available to react or circulate freely.

Impermeable   A rock that is incapable of transmitting fluids because of extremely low hydraulic conductivity.

Infiltration   Water flow from the land surface into the subsurface.

Injection well   A well constructed for the purpose of injecting water and chemical constituents directly into the ground.

Inverse problem   The problem of determining the value or spatial variation of a physical property by comparing measurements of state variables to model predictions.

In vitro   In glass, away from the living organism.

In vivo   In the body, in a living organism.

In situ   In site; in the natural location or environment (usually belowground).

Induction   The process by which a gene that codes for an enzyme is turned on and the enzyme is synthesized in response to the presence of an inducer.

Indicator organism   An organism used to identify a particular condition, such as *E. coli* or other coliforms as indicators of fecal contamination.

Indigenous   Native or autochthonous to a particular habitat.

Interspecies hydrogen transfer   Synthrophic relation by which organic compounds are degraded, involving the interaction of several groups of microorganisms in which H_2 production and H_2 consumption are closely coupled to enhance the thermodynamic feasibility of the process.

Isotopes   Different forms of a given element containing the same number of protons and electrons but differing in the number of neutrons.

Isotropic   Directional independence of some physical property such as hydraulic conductivity.

Land farming (land treatment)   The mixing of toxic organic wastes with surface soils for the purpose of biodegradation.

Leachate   Liquids that carry chemical constituents or other substances in solution or suspension through the geological formation.

Leakage   Water flow from one geological unit to another.

Leaky aquifer   Aquifer, whether confined or unconfined, that loses or gains water through adjacent less permeable geological units.

LNAPLs (light nonaqueous phase liquids)   Compounds that are lighter than water, thus floating on the water table.

Lysis   Rupture of a cell that results in the loss of cell contents.

Matrix   The fine grain, interstitial particles that lie between larger particles or in which larger particles are embedded in sedimentary rocks. It also refers to the unfractured portion of a fractured rock or formation.

MCL (maximum contaminant level) The largest contaminant concentration allowed in drinking water without causing risk to human health. The MCL is a designation given by the U.S. Environmental Protection Agency (EPA) to drinking water standards promulgated under the Safe Drinking Water Act.

Mechanical dispersion Mass transport flux of solutes caused by velocity variations at the microscopic level.

Messenger RNA (mRNA) An RNA molecule transcribed from DNA that contains the genetic information necessary to synthesize a particular protein.

Metabolism All biochemical reactions in a cell, both anabolic and catabolic.

Methanogen A methane-producing anaerobic prokaryote of the Archaea phylum.

Methanogenesis Biological production of methane (CH_4) by methanogens.

Methanotroph An organism capable of oxidizing methane, usually aerobically.

Methylation The process of substituting a hydrogen atom with a methyl group.

Methylotroph An organism capable of metabolizing organic compounds containing a single carbon atom; if able to oxidize CH_4, also a methanotroph.

Microaerophilic Requiring O_2 at a lower levels than atmospheric (e.g., 1 mg/L or less).

Microbial ecology The discipline that examines the interactions of microorganisms with their biotic and abiotic surroundings.

Microbiology The study of microorganisms and their activities.

Mineral An inorganic, naturally occurring, crystalline substance with definite atomic structure that has a unique or limited range of chemical compositions.

Mineralization The microbial breakdown of organic materials into inorganic materials such as CO_2, water, and mineral salts.

Miscible Liquids or other phases that can mix to form a homogeneous mixture. For example, hydrocarbon gases and liquids are commonly miscible.

Mixotroph An organism that can assimilate organic compounds as carbon sources while using inorganic compounds (usually H_2) as electron donors for energy metabolism.

Model A mathematical, physical, or conceptual representation of an entity or process with assumptions, which is used to compare against observations and/or make predictions.

Moisture content, gravimetric The ratio of the weight of water and the weight of solid particles expressed as moisture weight percentage.

Moisture content, volumetric The ratio of the volume of water to the volume of solid particles expressed as moisture volume percentage in a given volume of porous medium.

Monitored natural attenuation A passive remediation approach that relies on the combination of natural biological, chemical, and physical processes that act without human intervention to reduce the mass, toxicity, mobility, volume, or concentration of the contaminants (e.g., intrinsic bioremediation, dispersion, dilution, sorption, and volatilization).

Monitoring well A well that is generally of small diameter used to measure the hydraulic head or to obtain samples of groundwater.

Monte Carlo simulation An approach to performing risk analysis on a given situation with uncertain input data. Generally, random fields are generated from representative input data and then used in iterative calculations to find the most likely outcome and the range of probable outcomes useful in uncertainty analysis.

Most probable number (MPN) A method to determine the concentration of viable organisms using statistical analyses and successive dilution of the sample to reach a point of extinction.

Multiphase flow The simultaneous flow of more than one fluid phase through a geologic medium.

Mutagen An agent that induces a mutation, such as UV radiation or certain chemicals.

Mutualism A microbial interaction in which two organisms of different species live in close physical association, both deriving benefit from the symbiotic association.

Mutant A strain differing from its predecessor because of a mutation.

Mutation An inheritable change in the base sequence of the DNA of an organism.

Neutralism The lack of any recognizable interaction between two different microbial populations.

Niche The functional role of a microorganism in an ecosystem.

Nonpolar Hydrophobic (water-repelling) substance not easily dissolved in water.

Nucleic acid probe A nucleic acid strand that can be labeled and used to hybridize to a complementary molecule from a mixture of other nucleic acids. In microbial ecology, short oligonucleotides of unique sequences are used as hybridization probes for identifying specific degraders or other organisms of interest.

Numerical methods Mathematical methods that require iterative processing of data using algorithms implemented on a calculator or a computer.

Numerical model A rendering of a model of the subsurface in entirely numerical formats.

Numerical simulation The simulation of groundwater flow and solute transport to analyze and predict fluid and solute transport behavior over space and time.

Nutrient A substance taken by a cell from its environment to support catabolic or anabolic reactions.

Obligate A qualifying adjective referring to an environmental characteristic that is always required for growth (e.g., "obligate anaerobes" cannot tolerate O_2).

Oligonucleotide A short nucleic acid molecule that is obtained from an organism or synthesized chemically.

Oligotrophic A habitat in which nutrients are scarce.

Open reading frame (ORF) A DNA sequence which, if transcribed, could yield a protein of known length and composition. A functional ORF is one that encodes a cell protein.

Operator A specific region of the DNA at the beginning of a gene, where the repressor protein binds and blocks RNA polymerase from synthesizing mRNA.

Operon A cluster of prokaryotic genes whose expression is controlled by a single operator.

Osmosis Diffusion of water through a membrane from a region of lower to higher solute concentration.

Osmotic pressure The pressure resulting from differences in salt or other solute concentrations on opposite sides of a semipermeable membrane.

Oxic Microbial habitat containing molecular oxygen; aerobic.

Oxidation Process by which a compound donates electrons (or H atoms) and becomes oxidized.

Oxidation–reduction (redox) reaction A pair of reactions in which one compound becomes oxidized while another becomes reduced and accepts the electrons released in the oxidation reaction.

Oxidative (electron transport) phosphorylation The nonphototrophic production of ATP when a proton motive force (formed by electron transport) is dissipated.

Oxygenase Aerobic enzyme mediating the oxidative biotransformation of a molecule by inserting one (monooxygenase) or two (dioxygenase) atoms from molecular oxygen.

Packer A device that is inserted into a borehole with a smaller initial outside diameter that then expands to seal the borehole.

Partitioning coefficient A coefficient describing the distribution of a reactive solute between solution and other phases.

Peclet number A dimensionless number expressed as the ratio of the product of the average interstitial velocity, times the characteristic length, divided by the coefficient of molecular diffusion. Small Peclet numbers indicate dominance of diffusive transport, while large values indicate dominance of advective transport.

Partial penetration An incompletely drilled portion of the aquifer.

pH A measure of acidity, calculated as the logarithm to the base 10 of the reciprocal of the hydrogen ion concentration.

Perched aquifer An aquifer that is situated above the regional water table created by an underlying layer of low-permeability material.

Permeability The intrinsic ability of a geologic medium to transmit fluids.

Permeable reactive barriers Barriers made with granular material installed perpendicular to the flow direction of the plume to intercept the contaminant transport from the source and reduce contaminant flux from the source or treat the dissolved or aqueous phase contamination in the plume.

Phenotype The observable traits of an organism. Compare with *Genotype*.

Piezometer A devise used to measure groundwater pressure head at a point in the subsurface.

Photoautotroph An organism that can use light as its sole source of energy and CO_2 as its sole carbon source.

Photoheterotroph An organism that can use light as a source of energy and organic materials as carbon source.

Phototrophs Organisms that use light as the sole or principal source of energy.

Photosynthesis The harvesting of radiant (light) energy by specialized pigments of a cell and the subsequent conversion to chemical energy; the ATP formed in the light reactions is used to fix carbon dioxide, with the production of organic matter.

Phylogeny The ordering of species into higher taxa and the construction of evolutionary trees based on evolutionary (natural) relationships.

Phytoremediation A biological treatment process that utilizes natural processes harbored in (or stimulated by) plants to enhance degradation and removal of contaminants in contaminated soil or groundwater.

Piezometer A type of monitoring well, usually narrow in diameter, that has a short screen allowing for discrete sampling.

Plasmid   An extrachromosomal genetic element that is not essential for growth and codes for functions that may enhance the ability of a cell to cope with its environment (e.g., antibiotic or heavy metal resistance).

Plume   An underground pattern of contaminant concentrations created by the movement of groundwater beneath a contaminant source.

Point mutation   A mutation that involves one or very few base pairs.

Pollutants   Naturally occurring or anthropogenic substances that contaminate air, soil, or water.

Polymerase chain reaction (PCR)   A method to amplify a specific DNA sequence in vitro by repeated synthesis cycles using specific primers and DNA polymerase.

Pore   A void within a geologic formation, which may contain air, water, contaminants, or other fluids.

Porosity   The ratio of pore volume or void space to the volume of the geologic medium.

Porous medium   A rock or soil with interconnected pores that allow for fluid flow and solute transport through the medium.

Potentiometric surface   The level to which water will rise above the water level in a confined aquifer. If the potential level is higher than the land surface, the well will overflow and it is called an artesian well.

Pressure   Force distributed over a surface.

Pressure head   Hydrostatic pressure expressed as the height of a water column given in dimensions of length that the pressure can support at the point of measurement.

Pressure, hydrostatic   The pressure exerted by the weight of water at any given point in a body of water under static conditions.

Primer   A molecule (usually a polynucleotide) to which DNA polymerase attaches the first deoxyribonucleotide during DNA replication.

Prokaryote   A unicellular organism lacking a nucleus and other membrane-enclosed organelles, usually having its DNA in a single circular chromosome.

Promoter   The site on DNA where the RNA polymerase binds to begin transcription.

Protein   A biopolymer consisting of one or more polypeptides.

Proteomics   The large-scale or genome-wide study of the structure, function, and regulation of proteins.

Prototroph   The parent from which an auxotrophic mutant originates.

Protozoa   Unicellular eukaryotic microorganisms that lack a cell wall.

Pure culture   A culture that contains cells of one kind (a.k.a. axenic).

Pump-and-treat systems   Removal of contaminants in the aqueous phase through pumping of groundwater followed by treatment in an aboveground facility.

Pyrite   A common mineral containing iron disulfide (FeS_2).

Recalcitrant   A chemical that is resistant to microbial attack.

Recharge   Natural or artificial addition of water to an aquifer.

Recharge area   An area in a groundwater basin where water reaches the zone of saturation by surface infiltration.

Radioisotope   An isotope of an element that is unstable, causing radioactive decay or spontaneous disintegration emitting radiation.

Radionuclide   Synonymous to radioisotope.

Redox See *Oxidation–reduction reaction.*

Reduction Process by which a compound accepts electrons to become reduced.

Reduction potential ($E^{\circ\prime}$) The tendency, measured in volts, of the oxidized compound of a redox pair to become reduced.

Reductive dechlorination Removal of chlorine atoms from an organic compound as chlorides, usually by reducing the carbon atom from C—Cl to C—H.

Regulation Process that controls the rates of protein synthesis, such as induction and repression.

Remediation Removal, treatment, or containment of contaminated groundwater.

Reporter gene A gene incorporated into a vector because its product is easy to detect.

Repression The inhibition of the synthesis of an enzyme by the presence of an external substance, the repressor.

Repressor protein A regulatory protein that binds to a specific site on DNA and blocks transcription; involved in negative control of gene expression.

Respiration Catabolic reactions producing ATP in which either organic or inorganic compounds serve as electron donors and organic or inorganic compounds are the terminal electron acceptors.

Retardation factor The ratio of the average linear velocity of groundwater to the velocity of the retarded constituent at $C/C_0 = 0.5$ point on the concentration profile of the retarded solute.

Reverse transcription The process of copying information from RNA into DNA.

Rhizosphere The region in the immediate vicinity of plant roots.

Rhizosphere effect The influence of plant roots on bacteria, resulting in microbial populations that are higher within the rhizosphere (the region directly influenced by plant roots) than in unplanted soil.

Ribonucleic acid (RNA) A nucleotide polymer connected via a phosphate–ribose backbone; involved in protein synthesis or as genetic material of some viruses.

Ribosomal RNA (rRNA) Type of RNA serving as structural component of the ribosome; some rRNAs catalyze protein synthesis.

Ribosome A cytoplasmic particle composed of ribosomal RNA and proteins where mRNA is translated into proteins (i.e., the main protein synthesis machine).

Rock An aggregate of minerals, organic matter, or volcanic glass.

Saturated zone Portions of the subsurface in which all voids are filled with water under pressure greater than atmospheric.

Saturation The relative amount of gas, nonaqueous phase liquid, or water in the pores of geologic media, usually as a percentage of volume.

Seepage velocity See *Specific discharge.*

Selection Conditions that favor the growth of particular genotypes.

Self-purification Inherent capability of natural systems to cleanse themselves of pollutants based on biogeochemical cycling.

Semiconfined aquifer See *Leaky aquifer.*

16S rRNA A large polynucleotide ($\sim$1500 bases) that is a part of the small subunit of the prokaryotic ribosome (Bacteria and Archaea) and from whose sequence evolutionary relationships can be obtained; the eukaryotic counterpart is 18S rRNA.

Soil bulk density   The mass of dry soil per unit bulk soil.

Soil moisture   Subsurface liquid water in the unsaturated zone expressed as a fraction of the total porous medium volume occupied by water. It is less than or equal to the porosity, n.

Soil water   See *Soil moisture*.

Solubility   The total amount of solute species that will remain indefinitely in a solution maintained at constant temperature and pressure in contact with the solid crystals from which the solutes were derived.

Solute transport   The net flux of solute through a geologic medium controlled by the flow of subsurface water and transport mechanisms.

Sorption   Processes that remove solutes from the fluid and concentrate them on the solid particles.

Species   Of prokaryotes, a collection of closely related (>97% 16S rRNA sequence homology and >70% genomic hybridization) strains sufficiently different from all other strains so that they are recognized as a distinct unit.

Specific discharge   The rate of volumetric discharge of groundwater per unit area of a porous medium.

Specific storage   The water volume released from or taken into storage per unit volume of the porous medium per unit change in head.

Specific yield   The ratio of the water volume released from saturated medium by gravity to the volume of the porous medium.

Stochastic analysis   An analysis related to a process involving a randomly determined sequence of observations, each of which is considered as a sample of one element from a probability distribution.

Storage coefficient   The water volume that an aquifer releases from or takes into storage per unit surface area of the aquifer per unit change in head.

Storativity   See *Storage coefficient*.

Substrate   A molecule that undergoes a specific reaction with an enzyme.

Substrate-level phosphorylation   Synthesis of high-energy phosphate bonds involving the reaction of inorganic phosphate with an activated organic substrate.

Superoxide anion (O^{2-})   A reactive derivative of O_2 capable of oxidative destruction of cell components.

Symbiosis   An obligatory relationship between two organisms.

Synergism   An interactive but nonobligatory association between two populations in which both populations benefit.

Syntrophy   A nutritional situation in which two or more organisms share their metabolic capabilities to degrade a substance not capable of being catabolized by either one alone.

Taxonomy   The study of scientific classification and nomenclature.

Three phase flow   The simultaneous flow of three fluids such as nonaqueous phase liquids, gas, and water.

Tolerance range   The range of a parameter, such as temperature or pH, over which microorganisms survive.

Total dissolved solids   The total concentration of dissolved constituents in solution, usually expressed in milligrams per liter.

Total hydraulic head   See *Head, total hydraulic*.

Total porosity The total pore volume per unit volume of rock.

Trophic level The position of an organism or population within a food web, such as primary producer, grazer, and predator.

Transcription Synthesis of an RNA molecule complementary to one of a gene.

Transduction Transfer of genes from one cell to another via a virus.

Transfection Transformation of a prokaryotic cell by DNA or RNA from a virus.

Transfer RNA (tRNA) A type of RNA that carries amino acids to the ribosome during protein synthesis; contains the anticodon.

Transformation Acquisition of genetic information by uptake of free DNA.

Transgenic organisms Organisms (usually plants or animals) that stably pass on cloned DNA that has been inserted into them.

Transmissivity The rate at which water is transmitted through a unit width of the aquifer.

Turbulent flow Flow condition in which inertial forces are dominant over viscous forces and in which head loss is not linearly related to velocity.

Two phase flow The simultaneous flow of two fluids in the subsurface.

Type curves Families of paired drawdown (and in some cases its derivative) computed from an analytical or numerical model.

Type curve analysis A method for quantifying aquifer parameters such as hydraulic conductivity and specific storage by comparing the drawdown (and in some cases its derivative) of the acquired data to type curves.

Uncertainty The degree to which data or prediction may be in error quantified in terms of variance or standard deviation.

Unconfined aquifer An aquifer in which the water table is at atmosphere pressure and is the upper boundary of the aquifer.

Unconsolidated rock Geologic formation composed of sands and gravel that are loosely bound.

Unsaturated zone The zone below the soil surface and the regional water table.

Unsaturated flow Water flow in a porous medium in which the pore spaces are not filled to capacity with water.

Vadose zone See *Unsaturated zone.*

Variogram A two-point function that describes the correlation between sample values as separation between them increases.

Vector An agent, usually a plasmid or virus, able to carry DNA from one host to another.

Velocity, average interstitial The average rate of groundwater flow in connected pores expressed as the product of hydraulic conductivity and hydraulic gradient divided by the effective porosity.

Virus A genetic element containing either DNA or RNA and protein that replicates in cells and can have an extracellular state.

Viscosity A property of fluids and slurries that indicates their resistance to flow, defined as the ratio of shear stress to shear rate.

Void ratio The ratio of the volume of void space to the volume of solid particles in a given soil mass.

Volatiles   Substances with relatively large vapor pressures. Many organic substances are almost insoluble in water so that they occur primarily in a gas phase in contact with water, even though their vapor pressure may be very small.

Water activity (aw)   An measure of the relative availability of water in a substance. Pure water has an aw of 1.000.

Water content   See *Moisture content.*

Water table   The upper surface of a zone of saturation where pressure is equal to atmospheric except where that surface is formed by a confining unit.

Water table aquifer   See *Unconfined aquifer.*

Windrow method   A slow composting process that requires turning and covering with soil or compost.

Xenobiotic   A synthetic chemical compound foreign to life (i.e., not naturally occurring on Earth).

INDEX

Abiotic
  reduction, metal biotransformation, 99
  weathering processes, 7
Aboveground treatment, 15, 380
Absolute temperature, 43
Absorption, 25, 38
Acclimation, 74–76, 162
Acenaphthene, 5
Acetate, 69, 87, 97–98, 362
Acetic acid, 26, 90
Acetoclastic methanogens, 97
Acetogenesis, interspecies hydrogen transfer,
  97–98
Acetogenic bacteria, 97
Acetogens, interspecies hydrogen transfer, 97
Acetone, 14, 138, 152
Acetyl-CoA, 79–80
Acid, *see specific types of acids*
  leaching, 357
  mine drainage, 404–405
  organic, 12, 482
Acinetobacter, 61
Acrolein, 325
Activated carbon
  adsorption, 353
  bioaugmentation systems, 412
  characteristics of, 37, 40, 385
Activated sludge
  characteristics of, 356, 418
  treatment process, 14–16
Activation, defined, 59
Active bioremediation, 358
Adaptation, biodegradation process, 74–75
Adenine, 54
Adenosine diphosphate (ADP), 53
Adenosine monophosphate (AMP), 53
Adenosine triphosphate (ATP), 51–52, 64, 88
Adhesion-deficient bacteria, 415

Adsorption
  analytical fate-and-transport models,
    170, 176
  characterized, 353
  groundwater flow, 154–156
  mechanisms, 38–39
Advection
  biodegradation rate determination, 341
  contaminant transport and reaction, 175, 334
  groundwater flow, 142–144, 158
  numerical fate-and-transport prediction models,
    226–227, 233
  in sensitivity analysis, 190
Advection-dispersion equation (ADE)
  groundwater flow, 149–151
  numerical fate-and-transport prediction model,
    204, 209–211
  solute transport in saturated media, 149–151
Advection-dispersion reaction, 100
Advective-dispersive processes, groundwater flow,
  147
Advective movement, 136
Advective transport, 149
Advective velocity
  benzene, 184
  groundwater flow, 127–128, 146
Aeration, 15
Aerobes, 53
Aerobic/aerobic conditions
  bacteria, treatment trains and, 430
  biodegradation, 59, 65, 67–69, 82, 92, 95, 376,
    430
  bioremediation, 359–360
  biotransformation of metals, 100
  cometabolism, 406
  degradation
    instantaneous, 230, 233–237
    natural attenuation model case study, 270

Bioremediation and Natural Attenuation: Process Fundamentals and Mathematical Models
By Pedro J. J. Alvarez and Walter A. Illman   Copyright © 2006 John Wiley & Sons, Inc.

Aerobic/aerobic conditions (*continued*)
   hydrocarbon degradation kinetics, 64
   hydrolysis, 88
   metabolism, in biodegradation, 64, 73
   oxidation, 406
   processes, 16
   respiration
      biodegradation rates, 341
      BTEX degradation, 237–238
   ring cleavage, 95
Agricultural contaminants, mass flux calculation,
      143–144
Air
   in bioremediation processes, generally, 360, 362
   diffusion coefficients for organic compounds,
      138
   organic concentrations in, 43
Air curtain biobarrier, 403
Air Force Center for Environmental Excellence
   (AFCEE), 543
Air pumps, 363
Air sparging
   air spargers, functions of, 362
   biostimulation system, 380, 387, 392–397
   in situ, 397–402
Air strippers/air stripping, 353, 367, 385
Aklyl chains, 81
Alachlor, 325
Alcaligenes, 75
Alcohol(s)
   biotransformations, 88
   characteristics of, 26, 59
   monitoring natural attenuation (MNA),
      439–440
   primary, 78, 88
   secondary, 78
Aldehydes, 59, 79, 88
Aldrin, 6, 9, 59, 82, 138, 352
Algae, 13
Aliphatic
   compounds, nonvolatile, 440
   hydrocarbons, 326
Alkaline earth metals, 39
Alkanes, 26, 78–80, 494
Alkenes, 61, 78
Alkylbenzenes, degradation of, 82
Alkylbenzene sulfonates (ABS), 60, 92
Alkyl halides, 26–27
Alkylsulfides, 61
Alluvial aquifer, spills in, 8
Alluvium, hydraulic connectivity, 329
Alpha radiation, 31
Alternative site models, numerical fate-
      and-transport models, 224–225

Aluminum, 42, 369
Amidase enzymes, 88–90
Amides, 26–28, 77, 89
Amino acids
   biodegradation process, 69, 91
   characteristics of, 78
   complexation process, 44
   in gene expression, 55
   in hydrolysis, 90
Amitrole, acclimation phase of, 75
Ammonia monooxygenase, 61
Ammonia ntirogen, production of, 390
Ammonium chloride, 390
Amorphous ferric oxide, 40
Anabolism, 51, 91
Anaerobes/anaerobic
   biobarriers, 403
   biodegradation process, 59, 68, 73, 92, 94–95
   biosparging, 397
   biostimulation, 362
   biotransformation of metals, 100
   characteristics of, 53, 98, 359–360
   conditions, 16
   corrosion, 378
   degradation
      BTEX, 266
      hydrocarbons, 227
      implications of, 69
   electron acceptors, 64
   environments, 52
   gasoline-contaminated sites, 66
   hydrolysis, 88
   mineralization, 99
   oxidation, 406
   reductive dechlorination, 272, 406
Anaerobiosis, 362
Analytical fate-and-transport model
   application in contaminant hydrogeology,
      183–189
   boundary conditions, 170–175
   characteristics of, 169–170
   for contaminant transport and reaction
      processes, 175–183
   initial boundary conditions, 170–175
   limitations of, 193–197
   number solutions compared with, 174–175
   reliability of, 192–193
   sensitivity analysis, 189–192
   solutions, analytical vs. numerical, 174–175
   source and sink terms, 174
Analytical solute transport modeling, 473–476
Analytical solutions, analytical fate-and-transport
   models, 174–175
Aniline, 40, 88

Anion(s)
exchange, 40–41, 43–44
groundwater flow, 137–138
Anisotropic aquifers, 174
Anisotropy, 126, 134, 274, 316
Anthracene, 5, 138
Antibiotics, 12
Applied biotechnology, 12
AQTESOLV, 306
Aquifer(s), *see specific types of aquifers*
anaerobic, 85
characteristics of, 116–117
common contaminants of, 7
confined, 303–304
groundwater contamination, 5–6
homogenous isotropic, 182
hydrocarbon-contaminated, 358
matrix, 314, 322–323
mechanics of, 478
solids, 35, 39
unconfined, 303, 315–318
AquiferTest, 306
Aquitards, 116–117, 303
Archaea, 502
Arginine, 57
Aristotle, 13, 15
Aroclor 1254/Aroclor 1260, 352
Aromatic compounds
acclimation phases, 75
aerobic biotransformations, 85
chlorinated, 78, 86
microbial degradation of, 82–84
rhizoremediation process, 368
Aromatic hydrocarbons
characteristics of, 4, 26, 40
microbial degradation, 83
stable isotope analysis, 494–496
Aromatic rings, 82
Aromatics
chlorinated, 59
correlations estimate, 326
partitioning coefficients, 325
solubilities, 325
Arrhenius relationship, 71–72
Arsenate, 369
Arsenic, 4, 41, 78, 100, 369, 371, 441
Arsenite, 41
Artesian well, 117
Aryl alkyl sulfides, 61
Aryl nitro groups, 421
Asian cholera epidemic, 15
Assimilative capacity, 356
Atmospheric CO_2, 1
Atrazine, 62, 81, 325

Atrichous bacteria, 50
Autotrophs, 53
Average linear velocity, groundwater flow, 127
Aviation fuel spill, water circulation system case
study, 388–392

BAAP soil, 421–422
Bacillus, 61–62
Back-barrier deposits, 326
Backward difference approximation, 208
Bacteria
biodegradation, generally, 58–60, 73
classification by metabolic traits, 53
gene expression, 53–58
iron-reducing, 36, 99
metabolism, 51–53
methane-degrading, 73
phylogenetic probes in degradation detectin,
502
physiology, 50–51
sulfate-reducing, 36
Bacterial DNA, 50
Bacterial heterotrophic metabolism, 52
Baigon, 6, 9
Bail tests, defined, 297. *See also* Slug test analysis
Barriers
biobarriers, 381, 402–405
groundwater flow and, 136–137
numerical fate-and-transport prediction models,
216
Basalt, porosity and density values, 314
Base-catalyzed hydrolysis, 27
Batch reactors, 363, 419
Batch studies, 41, 494–496
Becquerel (Bq), 31
Bedrock, boundary conditions illustration,
173–174
Benz(*a*)anthracene, 5
Benzene(s), *see* BTEX
analytical fate-and-transport model, 184–185,
189, 193–197
biodegradation
kinetics, 109
models, 195–197
process, 68–69, 73
rate, 338
characteristics of, 5, 7, 62, 138, 352
chlorinated, 75
contaminant transport, 324, 335
contamination scenarios, 183–184
in drinking water, 342
fuel spills, 380
monitoring natural attenuation (MNA), 440
numerical fate-and-transport model, 221, 238

Benzene(s), *see* BTEX (*continued*)
  partitioning coefficients, 325
  RBCA Tier 3 modeling LUST case study, 252,
      257–259
  rings, 92
  stable isotope analysis, 494–496
  treatment train case study, 432–433
Benzidine, 352
Benzo(*a*)pyrene, 5, 7, 325, 352
Benzo(*b*)fluoranthene, 5, 352
Benzo(*g,h,i*)perylene, 5
Benzo(*k*)fluoranthene, 5
Benzoate, 96, 496
Benzoic acid, 138
Benzoyl-CoA, 85–86
Benzylsuccinate
  characteristics of, 85
  synthase, 84, 501, 505
Bernoulli equation, 120
Best equation, 104–108
Beta radiation, 31, 34
Beta-oxidation, 77–79, 85, 92
Bicarbonate, in biodegradation process, 64
Bichromate, 100
Bioaccumulation, 9–10, 78, 91
Bioaugmentation
  biostimulation distinguished from, 360–361
  case study, 415–418
  characteristics of, 16, 63, 356–357,
      410–415
  pilot study, 416–417
Bioavailability
  best equation, 105–108
  implications of, 16, 38, 63, 74, 92
  number, 106–107
  significance of, 104–105
Biobarriers, biostimulation system, 381,
      402–405
Biocassettes, 415
Biochemical applications, 359
Biodegradability/biodegradation
  analytical fate-and-transport models, 170
  defined, 58–60
  kinetics, 192, 194
  principles of, *see* Biodegradation principles
  process, 16–17, 26–28, 38, 91–93
  rate
    coefficient, 183, 332–342
    contaminant fate and transport model, 283,
        332–342
    determination using common field methods,
        333
    determination using in situ microcosms
        (ISM), 339

resistance to, 7
  in sensitivity analysis, 190
  toxic products, 20
  in treatment train, 433
Biodegradation principles
  acclimation, 74–76
  bacterial engine
    bacterial metabolism, 51–53
    bacterial physiology, 50–51
    biodegradation terms, 58–60
    gene expression, 53–58
  bioavailability, 104–108
  common biotransformation mechanisms
    effect of organic contaminant structure
        on biodegradability, 91–93
    hydrolytic and other biotransformations
        not involving redox processes, 88–91
    overview of, 76–78
    oxidative transformation, 78–85
    reductive transformations, 85–88
    synthetic reactions, 91
  enhanced biodgradation, cooperation between
      different microbial species
    commensalism, 93–95
    interspecies hydrogen transfer, 97–99
    syntrophism, 95–96
  kinetics applications, 100–104
    to fate-and- transport modeling, 108–109
  metals, biotransformation of, 99–100
  overview of, 49
  recalcitrance, implications of, 72–74
  requirements for biodegradation
    absence of toxic or inhibitory substances, 72
    accessibility of target pollutants to the
        microorganisms, 63
    adequate pH and buffering capacity, 70
    adequate temperature, 70–72
    availability of appropriate electron acceptors
        and/or electron donors, 64–69
    availability of nutrients, 69–70
    existence of organism with required
        degradation potential, 60–62
    induction of appropriate degradative
        enzymes, 64
    presence of specific degrader in the
        contamination zone, 62–63
Biodegradation rate coefficient, contamination
    scenario parameters, 183
Biodiversity analysis, 1, 510–513
Biofilm reactors, 418
Biogenesis, 10
Biogeochemical environment, 478
Biogeochemistry, 447
BIOID, 543

Biological contactors, rotating, 356
Biological remediation treatments, 356–357
Biomethylated lead, 370
BIOPLUME II, 233
BIOPLUME III, 227–228, 543
Bioreactors, 356, 361. *See also specific types of reactors*
Bioremediation
  applications, 69, 96
  costs of, 3, 15
  defined, 3
  environmental contamination by hazardous substances
    common groundwater pollutants, 3–10
    emerging pollutants, 10–12
    magnitude of, 2
  historical perspectives, 12–19
  in situ, 63
  limitations of, 19–22
  merits of, 19–22
  research, 18
  technologies, overview of
    aerobic vs. anaerobic, 359–360
    applications, 352–353
    bioaugmentation, 360–361, 410–418
    biostimulation, 360–361, 379–410
    epistemology of, 445–448
    ex situ (aboveground), 361–363, 418–429
    heavy metals, 368–371
    historical perspectives, 351–352
    information resources, 358
    in situ, 361–363, 371–379
    microbial vs. plant-based, 364–368
    monitored natural attenuation, 433–444
    political challenges for, 444–448
    popularity of, 358
    scientific challenges for, 444–448
    site remediation, 353–357
    treatment trains, 429–433
Bioscreens, *see* Biobarriers
Biosparge treatment, treatment trains, 429–430
Biostimulation
  augmentation distinguished from, 360–361
  case studies, 388–392, 397–402, 407–410
  chlorinated solvent spill treatment
    case study, 407–410
    overview of, 405–407
  fuel spills
    air sparging, 380, 392–397
    biobarriers, 381, 402–405
    bioventing, 380–384
    case studies, 388–392, 397–402
    overview of, 379–381
    water recirculation systems, 380, 385–388

oil spill treatment, 379–388, 392–397, 402–405
  techniques, 356
Biotechnology
  landmarks in, 13–14
  products, 12
Biotransformation
  characteristics of, 16
  defined, 58
  natural attenuation and, 440
Bioventing, biostimulation system, 380–384
Biowalls, *see* Biobarriers
Biphenyl
  biodegradation, 503
  characteristics of, generally, 59
  chlorinated, 59
  dioxygenase, 61
    biodegradation potential, 503
Bismuth, 30
"Black box" computational codes, 218
Block centered finite difference grid, numerical fate-and-transport prediction model, 205–206, 211–212
Bound residues, in phytodegradation, 366
Boundary conditions
  in biodegradation process, 170–174
  groundwater flow, 134, 139, 162
  hydraulic connectivity calculation, 305
  numerical fate-and-transport prediction models, 214, 217, 219, 222–223
Bouwer and Rice method, 298
Bromide
  biodegradation rate determination, 341
  characteristics of, 234, 322
  contour maps, 465–466
  porosity of, 321–322
Bromoform, 138, 325
BTEX
  biodegradation process, 7, 43, 62, 64, 67, 162, 170, 323, 337, 470–471, 503
  bioremediation systems
    air sparging, 395
    fuel spills, 380–381
    water recirculation case study, 389–392
  biotransformation pathways, 406
  concentration contour map, 342–343
  continuous spill, 235
  land farming case study, 426
  mass budget analysis, 469
  monitored natural attenuation (MNA), 435, 439–440
  natural attenuation, 483, 485–486, 491
    implications of, 483, 485–486, 491
    model case study, 262–268

BTEX (*continued*)
  numerical fate-and-transport prediction models,
      205, 224, 230–243, 266
  phytovolatilization process, 367
  stable isotope analysis, 497
  treatment train case study, 432–433
Bubble curtains, *see* Biobarriers
Buffering/buffer zone
  analytical fate-and-transport models, 185
  in biodegradation process, 70
Bulk density
  contaminant transport modeling, 323
  contamination scenario parameters, 183
  parameter determination for contaminant
      fate and transport models, 283
Bulk diffusion
  coefficient, groundwater flow, 137
  groundwater flow, 144, 147
Burkholderia, 61, 83, 411, 413, 421
Buscheck and Alcantar method, biodegradation
      rate coefficient determination, 333–337,
      342
Butane, 360, 384, 406
Butanol, 14
Butyrate, interspecies hydrogen transfer, 97
By-products, wastewater chlorination, 12
BZ-2, biodegradation process, 96

Cadmium, 4, 34, 42, 43, 364, 369, 440
Calcium
  characteristics of, 69
  phosphate, 41
  urinate, 41
Calibration, numerical fate-and-transport models,
      220–221, 223–224, 254–255, 266–267,
      276, 280
cAMP-CAP complex, in gene expression, 57
Camphor, 76
Cancer, 31
Capillary fringe, 115–116
Carbamates, 6, 9, 26, 28–29, 41, 90
Carbaryl, 6, 9, 28–29
Carbazole, 61
Carbon
  activated, *see* Activated carbon
  characteristics of, 136
  dechlorination process, 272
  ex situ biodegradation, 421
  isotopes, international standard, 489
  natural organic values, 326
  organic, 283, 326
Carbon dioxide
  biodegradation process, 64, 69, 227
  BTEX degradation process, 238

hydroxylation process, 80
  generation of, 487
  implications of, 77, 96–97, 469, 487
  treatment train, 430
Carbon disulfide, 138
Carbon tetrachloride (CT)
  bioaugmentation systems, 411
  biodegradation process, 69
  characteristics of, 5, 8, 26, 43, 138,
      325, 405
  first-order rate coefficients, 245
  phytovolatilization process, 367
Carbon-carbon bonds, 27, 77, 92
Carbon-nitrogen bonds, 77, 90
Carbon-oxygen bonds, 77
Carbonate, 41, 44
Carboxylic acid
  characteristics of, 22, 59, 88–89
  acid esters, 26–27, 29
Carboxylic groups, 40
Carcinogens, 8, 12, 479
Cartesian systems, 169
Casing radius, hydraulic connectivity calculation,
      299
Catabolic capacity, 19
Catabolic enzymes, 64, 76
Catabolic genes, biodegradation process, 75
Catabolic plasmids, 76
Catabolism, defined, 52
Catabolite
  activator protein (CAP), 57
  repression, 57–59
Catalysis, 99
Cataracts, 31
Catechol 2,3-dioxygenase, 503
Cation exchange(s)
  in bioremediation processes, 370
  in complexation reaction, 43–44
  implications of, 40
  sorbent processes, 41
Cations
  exchange, *see* Cation exchange(s)
  groundwater flow, 137–138
  sorption of, 40
Cauchy condition, 172, 174
Cell membrane, in bacterial cell, 51
Cell wall, in bacterial cell, 51
Celsius scale, 43
CERCLA Priority List, 352
Cesium, 30, 32, 441
Chemical data, numerical fate-and-transport
      prediction models, 222
Chemical degradation, 91
Chemical fingerprinting, 458, 493, 497–498

Chemical redox reactions, analytical
    fate-and-transport models, 170
Chemical spill, sensitivity analysis,
    189
Chemical transformation processes
    chemical reduction and oxidation reactions,
        34–37
    hydrolysis
        alkyl halides, 26–27
        amides, 27–28
        carbamates, 28–29
        carboxylic acid esters, 27
        characterized, 26
        epoxides, 28
        nitriles, 28
        organophosphate ester, 29–30
        sulfonylureas, 29
    radioactive decay, 30–34
Chemotrophs, 53
Chlordane, 6, 9, 138, 352
Chloride, 42, 278, 390, 486
Chloride ions, 272
Chlorinated aliphatic compounds, 7, 36, 384, 397,
    406
Chlorinated aromatic compounds, 7
Chlorinated compounds, 5, 7–9, 78, 352
Chlorinated ethane, natural attentuation model
    case study, 268–280
Chlorinated ethenes, stable isotope analysis,
    494
Chlorinated organics, 37
Chlorinated solvents, 4, 38, 42, 52, 59, 93,
    485–487, 490
Chlorine, 63, 77
Chloroalkanes, 61
Chlorobenzene, 59, 61, 77, 96, 138, 325
Chlorobenzoate, 96, 412
3-Chlorobenzoate, 76
4-Chlorobenzoate, 69
Chloroethane, 325, 405–406
Chloroethene, 26
Chloroethylene, 7
Chloroform, 8, 61, 138, 245, 325, 352, 405
Chloromethane, 325, 405–406
Chloroneb, 81
Chlorophenol, 59, 325
2-Chlorophenol, 138
Chlorpropham, 28–29
Chromate, 40, 100
Chromium, 4, 36, 43, 100, 358, 369, 428,
    441
Chrysene, 5
Civil engineers, functions of, 201
C-labeled substrates/tracers, 493

Clays
    bioaugmentation
        case study, 416
        systems, 412
    biodegradation process, 73
    characteristics of, 136
    hydraulic conductivity, 366
    porosity and density values, 314
    sorption processes, 40
Clean Air Act (CAA) of 1970, 16
Cleanup process, *see* Site
    cometabolism for, 60
    data collection, 214–216, 284, 473
    time frame for, 381
Clean Water Act (CWA) of 1972, 16
Cleavage, 29, 63, 81–85
Closed-form solutions, analytical fate-and-
    transport models, 174
CoA-esters, 83
CoA-ligase, 78
Coal, 370
Coal tar creosote, 352
Cobalt, 99, 441
Colloidal matter, biodegradation process, 73
Colloid transport, 158
Column studies, 493
Comanmonas, 61
Cometabolic
    biodegradation, 70, 360, 424
    commensalism, 60, 93
Cometabolism
    aerobic, 406
    biodegradation process, 73
    bioremediation systems, 60, 360, 406
Commensalism, 93–95
Commercially available cultures, bioaugmentation
    systems, 412
Committee on Intrinsic Remediation, 442
Community diversity analysis, 512
Compartmentation, in phytodegradation,
    366
Complementary error function, 140, 178, 180
Complexation reaction, 40, 43–44
Composting
    benefits of, 419
    characteristics of, 423
    ex situ case study, 423–425
Compounds, chemical properties of,
    527–533
Comprehensive Environmental Response,
    Compensation, and Liability Act of 1980
    (CERCLA), 17–19, 352
Computation code selection, numerical fate-
    and-transport model, 218–219

Computer software
  analytical models, 184
  commercially available, 465
  numerical fate-and-transport prediction model,
    202–203
  spreadsheets, *see* Spreadsheet applications
Confidence interval, 468
Confined, defined, 117
Congeners, 61
Conjugate
  gradient method, 209, 226
  reactions, phytodegradation, 366
Conservatism, numerical fate-and-transport
    model, 218
Constant-rate biodegradatoin, 161
Constitutive enzymes, 55, 57, 64
Consumer products, as pollutants, 11
Contaminant concentration
  analytical fate-and-transport models, 176–182,
    196
  calculation methods, 175
  implications of, 465, 478
Contaminant detection, volatilization process, 42
Contaminant hydrogeology, analytical models
    applications
  continuous planar source with constant source
    contamination, 184–186
  continuous radioactive source, 187–188
  continuous source with decaying source
    concentration, 186–189
  gasoline spill as an instantaneous point source,
    189
  overview of, 183–184
  site-specific parameters, 183, 192
Contaminant migration, 118, 155, 161–163
Contaminant plume trend, 344
Contaminant spills, 174
Contaminant transport, groundwater flow
  advection, 142–144
  advection-dispersion equation (ADE), solute
    transport in saturated media, 149–151
  common parameters, 119
  diffusion, 136–142
  hydrodynamic dispersion, 144–145, 162
  mechanical dispersion, 145–149
  modeling
    biodegradation rate, 333
    biodegradation rate coefficient, 332–342
    dispersivity, 327–331
    hydrodynamic dispersion coefficient,
      327–331
    retardation coefficient, 322–327
Contaminant travel time, estimation of, 158–159
Contamination, dimensions of, 196

Continental drift, 447
Contour maps, 465–466
Convection-dispersion equation (CDE), 331
Conventional bioreactors, 356
Conversion reactions, phytodegradation, 366
Copper, 42, 369, 441
Corn, in phytoextraction process, 364
Cornyebacterium, 363
Corrective action, numerical fate-and-transport
    prediction models, 214
Cortisone, 14
Cosmetics, as pollutant, 12
Cosolvent washing, 357
Covalent bonds, 78
Cramer's method, 209
Credible backgroud, establishment of, 478
Creeks, numerical fate-and-transport prediction
    case study, 255
Creosote, 440
Crick, Francis, 14
Cross-fertilization, 13
Crucifers, in phytoextraction process, 364
Cultures, in bioaugmentation, 41
Curie (Ci), 31
Curie, Marie, 31
Curve matching, hydraulic connectivity
    calculation, 306–307
CXTFIT, 331
Cyanide, 28, 90
Cyclic AMP (cAMP) synthesis, 57–58
Cyclodextrin, 406
Cyclohexane, 61, 93
Cytoplasm, in bacterial cell, 51
Cytosine, 54

Dakota Formation, 258
Damkohler numbers, 104–105
Dandelion, in phytoextraction process, 364
Darcy's Law, 124–126, 134–135, 149, 305
Darwin, Charles, 13
Darwinian evolution theory, 76
Data
  acquisition, groundwater flow modeling, 284
  collection techniques
    bioremediation systems, 472
    numerical fate-and-transport models,
      214–216
Daughter products, 30
DDD, 138, 352
DDE, dehydrohalogenation, 90–91
DDT
  biodegradation process, 73
  dehydrohalogenation, 90–91
  implications of, 6, 9, 43, 138, 325, 352

Dealkylation, 81
Decarboxylation, 77
Decay
  coefficient, 196–197, 334
  natural attenuation, 487
  rates, analytical fate-and-transport models, 192
Dechlorination
  anaerobic reductive, 406
  characteristics of, 36, 353
  reductive, 69, 86–87, 91–92, 95, 378, 407–410,
    415–418, 486–487
  using water recirculation system, 386
Degradation
  hydrolytic, 26
  potential, 60–62
  process, groundwater flow, 159–161
  rate
    implications of, 101–105
    numerical fate-and-transport prediction
      models, 215, 222
Degrader detection, phylogenetic probes used,
    502–505
Dehalobacter restrictus, 87
Dehalococcoides, 87, 406, 412, 502
Dehalogenases, 77, 90
Dehalogenation, reductive, 85–87, 90
Dehalorespiration, 360, 406
Dehydrogenases, 77
Dehydrohalogenation, 90
Deinococcus radiodurans, 359
Denaturing gradient gel electrophoresis (DGGE),
    507, 509–513, 514
Denitrification
  BTEX degradation, 238
  implications of, 270
  natural attenuation model case study, 270
Dense nonaqueous phase liquid (DNAPL)
  analytical models for contaminant transport
    and reaction, 175
  characteristics of, 8, 141, 486
  groundwater flow, 117–118
  hydraulic connectivity calculation, 304
  natural attenuation modeling, 228,
    244–252
  numerical fate-and-transport models, 204, 223
  physicochemical treatment, 354, 357
  well installation and, 375
Density, groundwater flow, 125, 133, 135, 158,
    204. *See also specific types of
    contamination*
Deoxyribonucleic acid (DNA)
  bioremediation system, 507–509
  characterized, 50
  double-stranded, 14

hydrolysis, 88
  -microarray technology, 509–511
  sequence, 55
Depletion, numerical fate-and-transport prediction
    models, 223
Deprotonation, 36
Derepression, gene, 56
Desorption
  in air sparging system, 395
  biodegradation kinetics, 106
  characteristics of, 377
  numerical fate-and-transport prediction
    model, 227
Desulfitobacterium, 87, 502
Desulfomonile, 87, 96, 412–413, 502
Desulfovibrio, 100, 359
Detoxification
  components of, 59
  using physicochemical treatments, 354
  phytodegradation, 366
  reaction, 78
Diauxy, 58–59
Diazinon, 6, 9
Dibenz(*a,h*)anthracene, 5
Dibenzofurans, 61
Dibenzo[*a,h*]anthracene, 352
Dichlorfop-methyl, 90
Dichlorobenzene
  biaugmentation system, 415–416, 418
  biotransformation pathways, 406
  characteristics of, 59
  hydrogen sparging, 409–410
  intrinsic bioremediation, 473
  natural attenuation, 486
  partitioning coefficient, 325
  retardation coefficient, 323
1,2-Dichlorobenzene, 138
1,4-Dichlorobenzene, 138
Dichlorodifluoromethane, 325
Dichloroethane
  bioaugmentation systems, 415
  characteristics of, 405
  natural attenuation modeling case study,
    268–280
  plume mass, 278
  treatment train case study, 430, 433
1,1-Dichloroethane (DCA), 138
1,2-Dichloroethane (DCA), 59, 138–139,
    247–248, 416
1,1-Dichloroethanol (1,1-DCE), 26–27
cis-Dichloroethene, 69, 86, 272
Dichloroethylene
  biodegradation, 7, 61
  bioventing system, 384

Dichloroethylene (*continued*)
 characteristics of, 405
 monitoring natural attenuation (MNA), 440
 numerical fate-and-transport model, 227,
  272–273, 276
 treatment train case study, 432
1,1,-Dichloroethylene (1,1-DCE), 26, 138
cis-1,2-Dichloroethylene, 138
Dichloromethane
 aerobic metabolism, 406
 characteristics of, 61, 325, 405
 first-order rate coefficients, 245
2,4-Dichlorophenol, 245
2,4-Dichlorophenoxy acetic acid, 76
Dichromate, 100
Dieldrin, 6, 9, 59, 73, 82, 138, 325, 352
Differentiation, product rule of, 135
Diffusion
 biodegradation
  kinetics, 106, 108
  rate determination, 341
 coefficient, *see* Diffusion coefficient
 groundwater flow, 136–142
Diffusion coefficient
 importance of diffusion as a transport process,
  142
 solute concentration prediction in clay landfill
  liner example, 140–142
Diffusivity, determination using Jacob–Cooper
 method, 309, 313–320
Dihaloelimination, 86
Dilution, biodegradation process, 7, 73,
 75, 151
Dimerization, 91
2,3-Dimethylbutane, 4
Dimethylether, 61
Dimethylmercury, 370
Dinitroaniline herbicides, 326
Dintitro-*ortho*-cresol (DNOC), 75
4,6-Dinitro-*o*-cresol, 87
2,4-Dinitrophenol, 91
Dinitrotoluene (DNT), 421–422
2,4-Dinitrotoluene, 59
2,6-Dinitrotoluene, 59
1,4-Dioxane, 11–12, 62–63, 359
Dioxins, 9, 61, 95
2-4 Diphenoxyacetic acid (2,4-D), 81
Directional dependence, groundwater flow,
 125–126
Directional effects, 479
Dirichlet
 boundary, 173
 condition, 171

Discharge, defined, 116
Dispersion
 biodegradation rate determination, 341
 bulk diffusion and, 144, 147
 contaminant transport and reaction, 175
 groundwater flow, 136
 hydrodynamic, 144–145, 149
 mechanical, 145–149
 numerical fate-and-transport prediction models,
  226–227
 in sensitivity analysis, 190
 transport, groundwater flow, 149
Dispersivity
 characteristics of, generally, 327–331
 contaminant fate and transport model
  parameters, 283, 327–331
 defined, 146
 numerical fate-and-transport prediction models,
  215, 220, 240
Dissimilatory sulfite reductase (DSR),
 biodegradation potential, 505
Dissolution, 152–154, 377
Dissolved oxygen (DO), 65, 391, 400–401, 488
Distribution coefficient, groundwater flow,
 155–156
DNT, 421–422
Documentation requirements, 463
Dodecyltrimethyl-ammonium chloride,
 acclimation phase, 75
Downgradient wells, 469
Drains
 buried, 174
 numerical fate-and-transport prediction
  models, 225
Drawdown
 defined, 303
 pumping test applications, 303–309, 313, 315
Dredging, 357
Drilling, hydraulic connectivity, 291, 321
Drinking water, 100
 analytical fate-and-transport models, 185–186
 benzene concentration, 342–343
 contaminant transport modeling, 328
 contaminated, 12, 100
Dual-phase extraction system, 376
Dupuit assumptions, 316
Dynamic viscosity, groundwater flow, 125

Economic feasibility
 bioremediation systems, generally, 447, 514
 bioventing, 380
 monitored natural attenuation (MNA), 434, 442
 treatment costs and, 15
 water recirculation systems, 386

Eddy, Harrison P., 16
Effective porosity
  contamination scenario parameters, 183
  groundwater flow modeling, 321–322
  hydraulic connectivity, determination of
    aquifer materials and, 322
  implications of, 321–322
  values of, 314
Elastic energy, 120
Electric resistive heating, 357
Electrodeposition, 353
Electron acceptor
  anaerobic, 84, 406
  analytical fate-and-transport models, 176
  biodegradation kinetics, 109
  biodegradation process, 53, 64–69, 73
  bioremediation processes and, 359
  BTEX degradation process, 237–243
  contaminant transport modeling, 332
  fuel spills, 380
  functions of, generally, 35–36, 542
  hydrocarbon biodegradation, 227
  in situ aerobic biostimulation system, 362
  in situ bioremediation system, 378
  mass budgeting, 469
  metals as, 99
  natural attention, 482
  numerical fate-and-transport prediction models,
    237–238
  plume, numerical fate-and-transport prediction
    models, 228–229
  reductive dechlorination, 87
Electron donor
  biodegradation process, 52, 93
  bioremediation processes, 60, 64–69, 100
  heavy metals and, 371
  hydrogen biosparging, 407
  in situ bioremediation, 362, 378
  numerical fate-and-transport models, 235
  oxidation-reduction reactions, 160
Electron tower, in biodegradation process, 66–68
Electron transfer models, 37
Electrostatic attraction, 38–40, 44
Elevation head, groundwater flow, 121–122
Elliptical plumes, contaminant concentration
    analysis, 181–182
Emergency response teams, underground storage
    tank spills, 189
Emerging pollutants, classification of, 10–12
Emerging technology, 515
Encapsulation, bioaugmentation system, 412, 415
Endocrine disrupting compounds (EDCs), 10–11
Endosulfan, 139
Endrin, 6, 9, 139

Energy transduction, in bacterial cell, 51
Engineered bioremediation, 358
Enhanced biodegradation
  commensalism, 93–95
  interspecies hydrogen transfer, 97–99
  syntrophism, 95–96
Enrichment cultures in bioaugmentation, 412
Enthalpy, biodegradation process, 71
Environmental contamination by hazardous
    substances
  common groundwater pollutants, 3–10
  emerging pollutants, 10–12
  magnitude of, 2
Environmental engineers, functions of, 192, 217
Environmental forensic analysis, 457, 459–460
Environmental pollutants, biodegradation of, 63
Environmental stress, response to, 57
Enzyme synthesis, 20
Enzymes
  biodegradation process, 12, 69–71, 73–74, 77
  catabolic, 76
  constitutive, 55, 57, 64
  co-oxidation of organic pollutants, 61
  degradative, 64
  dehalogenase, 90
  exocellular, 51, 74
  hydroxylation, 79–80
  inducible, 55
  microbial degradation, 82–83
  synthesis, 20
Eolian deposits, 326
EPN, 87
Epoxidation, 59, 81–82
Epoxides, 26, 28, 77
Equilibrium
  biodegradation process, 71
  secular, 33–34
  sorption processes, 39
Ereky, Karl, 12
Error estimate, numerical fate-and-transport
    models, 221
Error function, 140, 178
Escherichia coli, 57–59
Esterases, 77, 88
Ester(s)
  characteristics of, 77
  monitoring natural attenuation (MNA), 44, 439
Ethane(s) (ETHs)
  calibrated degradation constants, 278
  chlorinated, 93–94
  contamination from, 405
  generation of, 487
  numerical fate-and-transport model, 268, 276,
    278

Ethanol
  characteristics of, 59, 64, 69
  fermentation of, 97
  interspecies hydrogen transfer, 97–98
Ethene(s)
  bioaugmentation systems, 418
  chlorinated, 69, 93–94, 96
  contamination from, 405
  plume mass and, 278
  respiration, chlorinated, 98–99
Ethers, 11, 26
Ethyl groups, 81
Ethylbenzene
  characteristics of, 5, 7, 61–62, 139, 325
  fuel spills, 380
  stable isotope analysis, 495
  treatment train case study, 432–433
Eukaryotes, 50
Evapotranspiration
  analytical fate-and-transport models, 174
  numerical fate-and-transport prediction models,
    222, 225
Excavation, 353, 357, 362–363, 386
Exchange coefficient, biodegradation kinetics, 106
Exocellular enzymes, 74
Explosives
  monitoring natural attenuation (MNA), 440
  waste, treatment of, 363
Exponential decay, 173
Exponential integral formula, 305
Ex situ (aboveground) bioremediation
  case studies, 421–429
  characteristics of, 418–423, 425–426
  in situ vs., 361–363
Extracellular matrix, in bacterial cell, 51
Extraction
  of groundwater, 353, 357
  techniques, 356
  thermal, 357
  wells, 354

Falling head test, 297
Faraday's constant, 67
Fate-and-transport modeling
  analytical, see Analytical fate-and-transport
    model
  groundwater flow, 155
  implications of, 108–109, 147, 151–152
  numerical, see Numerical fate-and-transport
    prediction models
Fate processes, groundwater flow, 136
Fatty acid(s)
  beta-oxidation of, 85
  in biodegradation proces, 93

concentration of, 78, 493
hydroxylation, 79–80
methyl esters (FAMEs), 506
protonated, 73
volatile, 70
Federally funded programs, 17–18
Federal research grants, 203
FEFLOW, 544
FEHM, 544
FEMWASTE, 544
FEMWATER, 544
Fenton's reagent, 354
Fermentation process, 13, 15, 97, 406
Fermentative bacteria, 96–97, 99
Ferrous iron
  BTEX degradation, 266–267, 271
  characteristics of, 36–37, 42, 160
  natural attenuation model case study, 270
  oxides, 64, 66
  physicochemical treatments, 354
Fertilizers, 358
Fickian diffusion, 103, 108
Fickian dispersion, 176
Fick's First Law, 137, 139, 153
Field-scale studies, 492–493
Field testing, see Contaminant transport
    modeling
Film penetration model, groundwater flow, 154
Fingerprinting
  analysis, 497
  techniques, 509–513
Finite difference grid, numerical fate-and-
    transport prediction model, 205–207, 209
Finite element, numerical fate-and-transport
    prediction model, 205–206, 212
First-order
  biodegradation, 194, 334
  decay
    coefficients, 196–197
    generally, 108, 169, 227, 332
    radioactive, 226
  degradation, 101
  kinetics, 31, 108, 159–160, 176, 194–195
  reaction rate, anaerobic, 279
Fish, pollutants detected in, 11, 172
Fixed coordinate systems, 210
Flagellum, in bacterial cell, 50
Flammable waste, 363
Flamprop-methyl, 90
FLONET, 544
FLOTRANS, 544
Flow and solute transport, numerical modeling,
    475–479
Flow-and-transport model, development of, 269

Flowing artesian well, 117
Flow-through aquifer, 321
Fluid flow-and-transport model, 224
Fluoranthene, 5
Fluorene, 5
Fluorescent in situ hybridization (FISH), 501
Fluoride, 44
Fluvial-deltaic deposits, 326
Fluvial deposits, hydraulic connectivity, 329
Foam, biaugmentation systems, 412
Footprints, geochemical, 480, 482–485
Forensic analysis, 442, 514–515
Formamide, 501
Formate, 61, 69, 87
Forward difference approximation, 208
Forward problems, numerical fate-and-transport models, 212–213
Fowler, Gilbert, Dr., 16
FRAC3DVS, 544–545
Fractionation factor, 491
FRACTRAN, 545
Fractured aquifers, 175
Fractured rock
    aquifers, 322
    groundwater flow and, 137
Fragrances, as pollutants, 12
Free energy, 67–68, 535, 539
Free-product recovery, 357
Free radicals, 73, 84
Freundlich isotherms, 157, 159
FTWORK, 545
Fuel additives, 59
Fuel-contaminated sites, 66–67
Fulvic acid, 40
Fungi/fungicides, 9, 50, 363, 411
Funnel-and-gate system, 431–432
Furans, 9

β-Galactosidase, 57
Gamma radiation, 31, 34
Garbage-in garbage-out (GIGO) system, 203
Gas chromatography/mass spectrometry
    (GC/MS), 493, 499
Gasoline
    -contaminated sites, 66, 117
    leaks, 2, 117, 184, 193–197
    monitoring natural attenuation, 440
    oxygenates, 47
    removal by air sparging, 393
    spill, as instantaneous point source, 189
    streams, contaminated, 419
    underground storage tanks, 2–4
Gaussian distribution, 140, 182
Gaussian elimination, 209

Gene expression, 53–58, 509
Genetically modified organisms (GMOs), 412
Genetic markers, 14
Genetic rearrangements, 75
Geobacter, 99, 100, 359, 502
Geochemical analysis, 17, 21, 442, 514
Geochemical attenuation mechanisms
    abiotic processes, 25
    chemical transformation processes, 26–37
    immobilization and phase change processes, 37–44
Geochemical changes, sensitivity to, 41
Geochemical conditions, numerical
    fate-and-transport prediction models, 216–217, 219
Geochemical data
    implications of, 480–487
    numerical fate-and-transport models, 224
Geologic
    deposits 372
    materials, groundwater flow, 125–26
    media, characterized, 128
Geological units, numerical fate-and-transport prediction models, 215, 217
Geostatistical analyses, applications of, 130
Geosynclines, 447
Glacial deposits, 326, 329, 416
Glacial sediments, porosity and density values, 314
Glaciofluvial deposits, 326
Glass production, 370
Global warming, 223–224
Glucose, 57–58, 538, 540
Glutathione, biodegradation process, 91
Glutathione S-transferases, 78
Glycerides, degradation process, 160
Glycerol, 14
Glycocalyx, in bacterial cell, 51
Glycols, 26
Glycolysis, 58, 77
GOAL SEEK, 184–185
Goethite, 36, 40
Grain size analysis, hydraulic connectivity, 292–296
Granite, porosity and density values, 314
Granular iron, 355
Graphical analysis, 458, 463–467
Graphical user interface (GUI), 219
Gravels
    organic carbon content, 326
    porosity and density values, 314
Gravity, contaminant transport models, 315–316
Green liver concept, 366
Green Pages, 411

Green rust, 36
Green sulfur bacterium, 51
Groundwater, defined, 115
Groundwater flow
  adsorption, 154–155
  attenuation of contaminant migration, effect of
    processes on, 161–162
  Darcy's law, 124–126, 134–135, 149
  contaminant transport
    advection, 142–144
    advection-dispersion equation, solute
      transport in saturated media, 149–151
    diffusion, 136–142
    hydrodynamic dispersion, 144–145
    mechanical dispersion, 145–149
  degradation process, 159–161
  dilution, 151
  dissolution, 152–154
  fate-and-transport model, numerical modeling,
    201–208
  hydraulic conductivity, 125–131
  hydraulic head, 119–124
  modeling parameters
    effective porosity, 321–322
    hydraulic conductivity, 290–321
    hydraulic gradient, 284–290
    total porosity, 321–322
  rate, 181
  retardation, 154–159
  saturated groundwater flow equation, 131–135
  volatilization, 151–152
Groundwater modeling systems (GMS), 219, 231,
  545
Groundwater pollutants
  chlorinated compounds, 5, 7–9
  organic, 4
  overview of, 3–4
  pesticides, 9–10
  petroleum hydrocarbons, 4–7
  sources of, 3
  statistical research, 2
  25 most frequently detected, 4
Groundwater recirculation rate, 417
Groundwater systems, abiotic transformation
  processes, 25
Groundwater temperature, significance of, 26
Growth rate, biodegradation kinetics, 101
Guanine, 54

Half-life
  analytical fate-and-transport models, 187
  benzene biodegration, 196
  contaminant concentration analysis, 181
  of radioactive elements, 31–33

Halobenzoates, acclimation phase of, 75
Halogenated aliphatics, 325
Halogenated aromatics, 440
Halogenated compounds, reductive
  biotransformations, 85
Halorespiring bacteria, 96, 99
Hard-to-withdraw hydrophobic pollutants,
  20
Hayes, William, 14
Hazardous wastes, organic, 20, 93, 352
Hazen, Allen, 16
Hazen method
  grain size analysis, 294–295
  hydraulic connectivity determination using,
    295–296
γ-HCH (Lindane), 139
Health risks radiation, 31
Heating techniques, 357
Heavy metals, *see specific types of metals*
  biodegradation process, 75
  bioremediation technologies, 358–359,
    362, 368–372
  biotransformation of, 99
  characteristics of, 20, 368
  effective bioremediation treatments, 362
  methylation of, 78
  numerical fate-and-transport prediction model,
    227
  physicochemical treatment, 354, 357
  phytoextraction process, 364–365
  radioactive, 30
  redox reactions, 37
  redox-sensitive, 99
  sorption processes, 40
  toxic, 4, 43, 72
Helium, biosparging system, 408
Hematite, 40
Hemicellulose, 366
Henry's constant, 43, 151–152, 367, 384
Henry's Law, 42–43, 151–152
Heptachlor, 6, 9, 139
Herbicides, 9, 90, 326
Heredity laws, 14
Heterocyclic hydrocarbons, 26
Heterocyclic pollutants, 7
Heterogeneous aquifers, 174
Heterogeneous hydraulic conductivity, 126
Heterotrophic microorganisms, 66
Heterotrophs, 53
Hexachlorethane (HCE), 68–69
Hexachlorobenzene (HCB), 360, 325
Hexachlorobutadiene, 352
Hexachlorocyclohexane, 104
Hexachloroethane (HCA), 86, 139, 325

Hexahydro-1,3,5-trinitro-1,3,5-triazine (RDX) 11, 72, 87, 423–425, 440
High molecular weight (HMW), 368
High recharge rate zone, 151
Hippocrates, 13
Holism, 445–446
Homogeneous hydraulic conductivity, 126
Horizontal dispersion coefficient, 184
Horizontal dispersivity
  contamination scenario parameters, 183
  numerical fate-and-transport prediction models, 215
  transverse, 177–178
Horizontal extraction wells, 381–382
Horizontal hydraulic
  connectivity, 319, 321
  gradients
    groundwater flow, 124
    numerical fate-and-transport prediction models, 215
Hormones, as pollutant, 9, 12
Hospitals, radioactive iodine spill, 32–33
HST3D, 545–546
Human blood, pollutants detected in, 11
Humic acids, 40, 44
Humic materials, 40–41
Humic substances
  biodegradation process, 73
  biotransformations, 88
Humification, 25
Hybrid method of characteristics (HMOC), 226
Hydratase enzymes, 78, 80
Hydraulic conductivity
  air sparging system, 395–396
  analytic fate-and-transport model, 125–131, 192
  contamination scenario parameters, 183
  for geological formations, 329
  groundwater flow
    components of, 124–125
    heterogeneity in, 128–129
    modeling, 290–321
    scale effect, 130
  intrinsic bioremediation, 473
  numerical fate-and-transport prediction models, 204, 215, 220, 222, 254, 256, 268, 271
  one-dimensional column experiments, 329–331
  parameter determination for contaminant fate and transport models, 283
  site investigation, 372–373
Hydraulic connectivity, mass budget analysis, 469
Hydraulic control, 365–367, 378–379
Hydraulic gradient(s)
  analytical fate-and-transport models, 174, 192, 215

contamination scenario parameters, 183
groundwater flow modeling
  implications of, 123–124, 143
  pressure measurements, 284–287
  trigonometric approach, 287–290
parameter determination for contaminant fate and transport models, 283
Hydraulic head
  analytic fate-and-transport model, 119–124, 134–135
  numerical fate-and-transport prediction models, 209, 211, 213, 215, 219–220, 224–225, 232, 266
Hydrocarbon(s)
  aromatic, 4
  biodegradation, 59, 227
  -contaminated sites, 64
  -contaminated soil, 363
  degradation, 360–361
    in soil, optimum environmental conditions for, 363
  halogenated, 326
  hydroxylation, 80
  monoaromatic, 5, 7, 379–380
  natural attenuation modeling, 228–252
  oxygenated, 440
  petroleum, 4–7
Hydrodehalogenation, 85
Hydrodynamic dispersion
  analytical fate-and-transport models, 170, 176
  coefficient, contaminant fate and transport model, 327–331
  groundwater flow, 144–145
  implications of, 149, 158, 162
Hydrogen
  in biodegradation process, 69, 97, 441
  biosparging case study, 407–410
  bonding, 38
  methane fermentation process, 98
  partial pressures, 98–99
  transfer, interspecies, 97–99
Hydrogenolysis, 85–86
Hydrogenophaga paleronii, 421
Hydrogen peroxide, 354, 360, 362, 389–390
Hydrogen Releasing Compound (HRC®), 378, 406
Hydrogen sulfide, 72
Hydrogeologic data, numerical fate-and-transport prediction models, 215, 217, 219, 222
Hydrogeologic environment, 17, 478
Hydrogeological heterogeneities, 67
Hydrogeologist, functions of, 192, 201, 217
Hydrogeology, 513
Hydrolases, 77

Hydrolysis
    adenosine triphosphate (ATP), 52–53
    analytical fate-and-transport models, 170
    biodegradation process, 38, 77
    biotransformation without redox process, 88–90
    microbial degradation, 85
Hydrolytic dehalogenation, 90
Hydrolytic reactions, degradation process, 160
Hydrophobic
    contaminants, 323, 354
    expulsion, 37–38
    interactions, 37
    pollutants, 354, 361–363, 366
Hydrophobicity, 154–155
Hydroquinones, 36–37
Hydrostatic forces, 116
Hydrous titanium oxide, 40
Hydroxo complexes, complexation process, 44
Hydroxyapatite solubility, 42
Hydroxylase, 78
Hydroxylation, 75–80
Hydroxyl groups, 77
Hydroxyl ions, 44
Hydroxyl monooxygenases, 504
Hydroxyl radicals, 367
2-Hydroxypyridine, 76
Hydroxypyromorphite, 42
Hyperaccumulators, 364
Hyperbolic equations, biodegradation kinetics,
    101–102

Immobilization
    in bioaugmentation systems, 412
    bioremediation technologies and, 358
    complexation and, 43–44
    heavy metals and, 371
    implications of, 25, 37–38, 332, 352
    intrinsic bioremediation, 488
    natural attenuation and, 440
    precipitation, 41–42
    rate-limited sorption reaction, 249
    sorption, 37–41
    volatilization, 42–43
Impermeable zones, groundwater flow, 124
Incineration, treatment costs, 15
Increasing-rate biodegradation, 161
Indan, 61
Indeno(1,2,3-cd)pyrene, 5
Indian mustard, in phytoextraction process, 364
Inducible enzymes, 55
Induction, in gene expression, 56–57
Industrial chemicals, historical perspectives, 14
Industrial ecology, 351
Industrial solvents, 117

Industrial wastewater, 356
Infiltration
    field, groundwater flow, 124
    water recirculation systems, 386
Inflow rate, determination of, 299
Inhalation hazards, 30–31
Inhibitory substances, biodegradation processes,
    72
Initial conditions
    in biodegradation process, 170–174
    groundwater flow, 139, 162
    hydraulic connectivity calculation, 305
    numerical fate-and-transport prediction models,
        214, 222–223
Injection systems, 174
Injection wells, 362, 416
Inner-sphere adsorption, 39
Inorganic
    contaminants, physicochemical treatment, 354
    mercury, 370
    nutrients, in biodegradation process, 72
    pollutants, 4
Insecticides, 9, 12, 26, 28, 326
In situ aquifer sparging (IAS), 396
In situ biodegradation, 91–92, 360
In situ bioremediation
    applications, 16–17, 19, 63, 356
    distinguished from ex situ, 361–363
    implementation approach
        nutrient delivery system selection, 377–379
        overview, 371–372
        site investigation, 372–374
        spreading prevention measures, 374–377
        stimulatory material delivery system
            selection, 377–379
        water recirculation system case study,
            388–392
In situ chemical treatment system, 355
In situ microcosms (ISM), 339
In situ reductive dechlorination using
    bioaugmentation, 415–418
Instantaneous reaction, 227, 230
Instantaneous spill event, contaminant
    concentration analysis, 181–182
Intact phospholipid profiling (IPP), 499
Intellectual property, 12
Interdiction well field, installation of, 374–375
Interspecies hydrogen transfer, 97–99
Intrinsic biodegradation, 195–196
Intrinsic bioremediation
    applications, 358, 486
    direct evidence of contaminant mass loss,
        460–463
    documentation of contaminant mass loss, 463

graphical analyses, 463–467
statistical analyses, 463–467
Inverse problem, numerical fate-and-transport
prediction model, 212–213
In-well aeration (IWA), 396
Iodine, 30, 32–33
Ion exchange, 25, 39–42, 227
Ion exclusion effects, 158
Ionization, 31
Iron
-bearing minerals, 36
biobarrier systems, 404
biodegradation of, 70, 227–228, 369, 469
BTEX degradation process, 238, 240
carbonate, 36
ferrous, *see* Ferrous iron
-manganese oxides, 136
nanoparticles, 355
natural attenuation model case study,
270
oxides, 36, 136
-reducing bacteria, 99
treatment train case study, 429–430
Irrigation, 174
Isomerases, 78
Isopentane, 4
Isopropyl groups, 81
Isotherm adsorption, 155, 157
Isotropic hydraulic connectivity, 126

Jacob–Cooper method, semilogarithmic, 309,
313–320
Jet fuel, 4, 393
JM1, 413
Johannsen, Wilhelm, 14

Karst aquifers, 175
Kelvin scale, 43
Ketones, 26, 59, 439–440
Kinetics
aerobic hydrocarbon degradation, 64
analytical fate-and-transport models,
192
in biodegradation process, 100–104,
108–109, 120
dechlorination compounds, 272
first-order, 31, 108, 159–160, 176, 194–195
numerical fate-and-transport prediction models,
223
zero-order, 195
Klebsiella pneumonaie, 71
Koch, Robert, 14–15
Krebs cycle, 58, 77, 80–82
Kühne, Wilhem Friedrich, 77

Laboratory tests, hydraulic connectivity, 292
Lac operon, 56, 58
Lactate, in bioremediation processes, 360, 362,
386, 415, 417
Lactic acid bacteria, 13
Lacustine deposits, 326
Lagoons, 3, 174
Lakes
analytical fate-and-transport models, 174
numerical fate-and-transport prediction models,
215, 219
Land farming, 363, 419, 425–429
Landfilling, treatment costs, 15
Landfill(s)
leachate, analytical fate-and-transport models,
187
municipal solids and hazardous waste, 3
Langmuir isotherms, 157, 159
Laplace equation, 134, 207–208
Lawrence Experimental Station (Massachusetts),
15–16
Leachate
heavy metals and, 368
landfills, 187
Leaching, 369
Lead
characteristics of, 4, 30, 370, 440
phosphates, insoluble, 42
removal, precipitation process, 42
Leaf compost, biobarrier systems, 404
Leaking underground storage tank (LUST)
analytical fate-and-transport models
characteristics of, 175, 177
benzene simulation, 193–197
BTEX degradation, 240
incidence of, 352
numerical fate-and-transport models
benzene, 257–259
BTEX degradation, 240
characteristics of, 221, 231
RBCA Tier 3 modeling case study,
252–261
xylene, 257–258, 260–261
removal of, 354
Lederberg, Joshua, 14
Leeuwenhoek, Anton van, 13
Legislation, 16, 18–19
Lepidocrocite, 36
Liebig, Justis, 13
Ligamentum, 44
Ligand exchange, 38–39, 44
Ligases, 78
Lighter than water nonaqueous phase liquids
(LNAPL)

Lighter than water nonaqueous phase liquids (LNAPL) (*continued*)
  air sparging system, 395
  characteristics of, 7–8, 117
  mass budget analysis, 469
  physicochemical treatment, 354, 357
  well installation and, 375
Lignin, 366
Limestone
  formations, 41
  hydraulic connectivity, 329
  porosity and density values, 314
Lindane, 73, 90–92, 325
Linear adsorption isotherm, 157
Linear partitioning, 39
Linear regression, applications of, 338
Lines of evidence
  environmental forensic analysis, 459
  intrinsic bioremediation, 460
Lipids, 493
Liquid chromatography/electrospray/ionization/ mass spectrometry (LC/ESI/MS) analysis, 499
Liquid injection, in bioaugmentation systems, 412
Lithium, contour maps, 465–466
Lithotrophs, 53
Litigation, benzene fate-and-transport analysis, 193–197
Loading, boundary conditions and, 172–173
Locations, monitoring, 479
Locomotion, in bacterial cell, 51–52
Loess
  organic carbon content, 326
  porosity and density values, 314
Longitudinal dispersion, 158, 176, 184, 188, 334
Longitudinal dispersivity
  contaminant transport modeling, 327–328
  contamination scenario parameters, 183
  hydrodynamic, 331
  implications of, 44, 146–148, 151, 177, 327, 329
  numerical fate-and-transport prediction models, 215, 231, 242, 271
Long-term remediation technology, 381
Loyala, San Ignacio de, 447
Lyases, 77

Macronutrients, in biodegradation process, 69
Magnesium, 69
Magnetite, 36, 40
Malathion, 6, 9, 90
Manganese, 64, 66
Manganese salts,100
Mann–Kendall statistical test, 344
Mann–Whitney statistical test, 344

Marine deposits, 326
Marine wildlife, pollutants detected in, 11
Mass balance technique, biodegradation rate, 332–333
Mass budgeting, 468–471
Mass conservation law, 132, 134, 150
Mass flux analysis, 471–473
Mass flux calculation
  bioremediation systems, 472
  groundwater flow, 143–144
Mass inflow rate, 132
Mass loss, contaminant
  documentation of, 463
  graphical analysis, 463–467
  statistical analysis, 463–467
Mass outflow rate, 132
Mass transfer coefficient, 152–154
Mass transfer model, rate-limited, 249
Mass transfer rate, 154, 222
Mathematical modeling, 359
Maximum contaminant levels (MCLs), 18, 193, 418, 429
Maximum initial benzene concentration, 83
Maximum likelihood Bayesian model averaging (MLBMA), 218
Measurement(s)
  of radiation units, 31
  site-specific, 214
Mechanical dispersion, 145–149, 478–479
Mecoprop, acclimation phase of, 75
Medical application tracers, 30
Medical therapeutic agents, 370
Mendel, Gregor, 14
Mercury, 78, 370, 441
Mesh centered finite difference grid, 205–206
Messenger RNA (mRNA)
  functions of, 501
  in transcription process, 54–55
Meta, oxidative ring cleavage, 82, 85
Metabisulfide, 355
Metabolism, *see specific types of metabolism*
  anaerobic, 85–86
  in biodegradation process, 53
  defined, 52
Metal(s), *see specific types of metals*
  alloys production, 370
  biotransformation of, 99–100
  complexation, 38–39
  ions, 40
  natural attenuation and, 441, 481
  organic, 36
  phosphates, low-solubility, 42
  -reducing bacteria, 64, 67

Methane, 61, 360, 482–483
  BTEX degradation, 266–267, 271
  in bioventing system, 384
  characteristics of, 61, 360, 483
  chlorinated, 8, 93–94
  contamination from, 405
  halogenated, 36
  monooxygenase, 61, 81
  natural attenuation, 486
Methanogenesis, 66, 238, 240, 266, 270
Methanogenic capacity (MC), 238
Methanogenic conditions, 68–69
Methanogens, 64, 98
Methanol, 27, 59, 61, 69, 362
Methanospirillium, 96
Method of characteristics (MOC)
  numerical fate-and-transport prediction model,
    204, 209–212, 228, 257
  numerical groundwater flow, 546
Methylation, 78
Methyl bromate, 26–27, 139
Methylene chloride, 59, 139, 440
Methyl fluoride, 61
Methyl group, 82
Methylmicrobium, 61
Methylococcus, 61
Methylosinus trichosporium, 413
Methylosynus, 61
Methyl parathion, 6, 9
Methylphosphonate, biodegradation process, 72
Methyl-*tert*-butyl ether (MTBE)
  bioaugmentation, 359, 361, 411
  bioremediation systems
    air sparging case study, 397–402
    phytovolatilization of, 367
  characteristics of, 11–12, 59, 61–63
  natural attenuation, 440, 490
  stable isotope analysis, 496
Michaelis–Merton enzyme kinetics, 102
Microarray technology, 511
Microbeads, bioaugmentation systems, 412
Microbial
  analysis
    characteristics of, 442, 498–499
    chemical analysis, 499
    molecular analysis, 499–501, 506–513
  bioremediation, 364–368
  degradation, 159
  ecology, 17, 358, 447
  fences, *see* Biobarriers
  infallibility hypothesis, 16–17, 62
  interrogation, 500
  kinetics, 482
  metabolism, 70, 359

  population diversity, 511
  processes, 21
  transport, bioaugmentation systems, 414, 415
Microbiological
  analysis, 442
  applications, 17, 359, 513
  data collection, numerical fate-and-transport
    prediction models, 217, 219, 222
Microcosm studies, 458, 487–489
Microfiltration, 353
Micronutrients, in biodegradation process, 70
Microorganisms, in biotechnology, 12–13
Microtox toxicity, land farming, 428
Migration, *see specific contaminants*
  risk assessment, 175
  site-specific, 192
Milk, pollutants detected in, 11
Military waste products, 30
Mill tailings piles, analytical fate-and-transport
    models, 174
Mine excavation, 30
Mineralization, in biodegradation process, 36,
    59–60, 93–97
Mineral spirits, 393
Mirex, 6, 9
Mitochondrial bioenergetics, 10
Mixed boundary condition, 172
MODFLOW, 206, 219, 225–227, 229, 232, 251,
    253, 274, 546
MODFLOWP, 546
MODFLOW-SURFACT, 546–547
Modified methods of characteristics (MMOC),
    226, 233
MODPATH, 547
Molasses, in bioremediation processes, 360, 362,
    386, 415
Molecular analysis
  fingerprinting techniques, 499–513
  implications of, 499–501, 458, 506–507
  polymerase chain reaction (PCR), 507–509
Molecular oxygen, 64, 77–78, 160
Molybdate, 465–466
Molybdenate, 40
Molybdenum, biotransformation of, 99
Monitored natural attenuation (MNA)
  characteristics of, 353, 433–437
  protocols, 442–443
  selection factors, 437–439, 442–444
  success factors, 440–441
Monitoring
  chlorinated ethane compounds, 272
  contaminant concentration, 180
  frequency, 179, 186
  significance of, 20

Monitoring well(s)
  bioaugmentation systems, 418
  contaminant concentration in, 344
  contaminant transport modeling, 327
  hydrogen biosparging, 407
  importance of, 461
  network, 462
Monoaromatic hydrocarbons, 5, 7, 42, 59, 85, 379–380
Monod constants, 240
Monod kinetics, 105, 107, 195, 227
Monod's equation, 100–102, 104, 107, 195
Monod's half-saturation coefficient, 108
Monogenases, microbial degradation, 85
Monohaloalkanes, 77
Monooxygenase enzymes, 78, 82
Monte Carlo simulations, 218, 224
MT3D, 204–205, 225–232, 547
MT3DMS, 226–228, 257, 547–548
MTBE, see Methyl-tert-butyl ether (MTBE)
Multipump bioslurping extraction well, 377
Municipal landfill leachates, 70
Municipal wastewater, 14, 356
Musk ketone, 11
Mutations, 31, 75–76
m-Xylene, 5, 139
Mycobacterium, 61, 93, 363

NADH, 78, 85
NADPH, 78, 85
n-Alkanes, 61
Nanofiltration, 353
Naphthalene
  biodegradation, 503
  characteristics of, 5, 76, 139
  dioxygenase, 61, 77
  partitioning coefficients, 325
  stable isotope analysis, 495
Natural attenuation
  capacity (NAC), 336–337
  contamination trends and, 478–479
  defined, 434
  effectiveness, nonparametric statistical tests for determination of
    Mann–Kendall test, 551–556
    Mann–Whitney U test, 554
  geochemical indicators of
    chlorinated solvents, 485–487
    miscellaneous, 487
    overview of, 480
    petroleum hydrocarbons, 482–485
  hydrogen biosparging, 407
  recommended levels of, 464

Natural bioremediation, 358
Natural degradation, 49
Natural exponential decay, degradation process, 159–160
Natural gradient conditions, 473
Natural gradient pulse injection tracer test, 145
Natural Research Council, 444, 463
Natural selection, 13
n-Butane, 4
NDMA (n-nitrisodimethylamine), 11–12
Nernst's equation, 67
Nettles, in phytoextraction process, 364
Neumann condition, 171, 174
Neuman type curve method, hydraulic connectivity determination
  aquifer parameters using, 320–321
  characteristics of, 318–320, 329
New biodegradation, 62
Nichols, William R., 16
Nickel, 42, 70, 99, 364, 441
Nickel-cadmium batteries, 369
Nicotine/nicotinate, 76
Nitrate
  biodegradation of, 4, 42, 64, 66, 227
  bioremediation processes, 359
  BTEX degradation, 271
  cometabolism, 406
  depletion of, 486
  ex situ bioremediation system, 421
  in situ bioremediation system, 378
  natural attenuation, 482
Nitrilases, 88s, 90
Nitriles, 26, 28
Nitrite, 64
Nitrite reductases, biodegradation potential, 506
Nitroacetic acid (NTA), acclimation phase of, 75
Nitroaromatic compounds, 40, 59, 78, 85
Nitrobenzene(s), 36, 59, 61, 77, 88, 139, 325
Nitrogen, biodegradation process, 21, 69, 72, 427
Nitro-group reduction, 87–88
Nitrophenol, 325
4-Nitrophenol, 75
Nitrosomonas, 61
Nitrotoluene, 59
Nitrotoluene isomers, 61
Nitrous oxide, 64, 469
No Action Required (NAR), 214, 358
n-Octane, 76
No-flow boundary, numerical fate-and-transport prediction models, 219
No-flux boundary condition, 171
Nonaqueous phase liquids (NAPLs)
  in air sparging system, 395
  bioavailability, 104, 108

biodegradation process, 63, 74
contaminant concentration, analytical models, 181
depletion rate, 469
emulsion of, 420
groundwater flow, 117, 152–153
physicochemical treatment, 354
spills, 8
water recirculation systems, 386
Nonlinear regression techniques, 192, 340
Nonreactive contaminants, numerical fate-and-transport models, 232–234
Normal distribution, 140
n-Pentane, 4
n-Propylbenzene, 4
Nuclear region, in bacterial cell, 50
Nucleic acids, biodegradation process, 70
Nucleophilic attack, 29–30
Numerical dispersion, numerical fate-and-transport prediction model, 210
Numerical fate-and-transport prediction models
boundary value problem, 206–207, 223
case studies
natural attenuation model of BTEX compounds, 262–268
natural attenuation model of chlorinated ethane compounds, 268–280
RBCA Tier 3 modeling of LUST storage site, 252–261
description of
calibration process, 220–221, 223–224, 254–255, 266–267, 276, 280
postaudit, 224
prediction process, 223–224
purpose of modeling, 213–214
redesigning the model, 224–225
running the numerical model, 219–220
sensitivity analysis, 222, 257, 279
setting up numerical model, 219–220
site conceptual model development, 216–218
site data, collection and assessment of, 214–216
suitable computation code, selection of, 218–219
validation process, 222–223
fundamentals of
finite difference approach, 205, 207–209
finite element approach, 205, 212
forward vs. inverse problems, 212–213
method of characteristics (MOC), 204, 209–212, 228
overview, 205–207

natural attenuation evaluation
BIOPLUME III, 227–228
of hydrocarbons and DNAPLs, 228–252
MODFLOW, 206, 219, 225–227, 229, 232, 251, 253, 274
MT3D, 204–205, 225–232
MT3DMS, 226–228, 257
RT3D, 204–205, 226–231, 240, 247, 249
purpose of, 201–205
Numerical groundwater flow, 543–550
Numerical solutions, analytical fate-and-transport models, 175
Nutrient(s)
-amended groundwater, recirculation of, 14
analytical fate-and-transport models, 192
bioaugmentation system, 411, 416
biodegradation process, 69–70, 73
bioremediation technologies, 356, 363, 447
delivery system selection, 377–379
in situ bioremediation, 362
Nutritional stress, response to, 57

Observation wells, hydraulic connectivity, 316–317
Occupational Safety and Health Act (OSHA) of 1970, 16
Oceanic trenches, 447
Oceans, analytical fate-and-transport models, 174
Octahydro-1,3,5,7-tetranitro-1,3,5,7-tetrazocine (HMX), 87, 424
Octanol-water coefficient, 39
-OH groups, 77–78, 82
Oil exploration, 4
Olefins, aliphatic, 61
Oligotrophs, 107
Operator, in gene expression, 56
Operator-split (OS) scheme, 230
Operons, in gene expression, 56
OPTIMIZER, 184
Order-of-magnitude
analytical fate-and-transport models, 175, 192
benzene biodegradation, 196
estimation, 175
numerical fate-and-transport models, 202, 222
Organic chemicals, 16
Organic compounds
biodegradation process, 38, 69, 76–77
estimation correlations, 326
hazardous, 93, 352
partitioning coefficients, 325
solubilities, 325
volatilization, 42–43

Organic contaminants, physicochemical
    treatment, 354
Organic pollutants
    acclimation phases, 74–75
    biodegradation process, 15, 36, 67, 82, 356, 363
    biotransformations, 88
    common, 324
    co-oxidation of, 61
    oxidative ring cleavage, 82
    toxic, 352
Organization for Economic Cooperation
    and Development (OECD), 21
Organochlorides, 6, 9
Organophosphate(s)
    characteristics of, 6, 9, 90
    ester, 29–30
    insecticides, 26
Organophosphorodithionates, 29
Organophosphorothionates, 29
Organotrophs, 53
Ortho, oxidative ring cleavage, 82, 85
Osmosis, reverse, 353
Outer sphere adsorption, 39
Oxidants, 353
Oxidase enzymes, 78
Oxidation, *see* Oxidation-reduction
    aerobic, 406
    anaerobic, 406
    biodegradation process, 69
    capacity (OXC), 35–36
    of contaminants, 355
    reaction to, 359
Oxidation-reduction, *see* Redox
    biodegradation process, 65
    degradation processes, 160
    potential (ORP), 64, 537–538
    reactions, 4, 20, 25, 34–37
Oxidative ring cleavage, 82–85
Oxidative transformation, biodegradation process
    dealkylation, 81
    epoxidation, 81–82
    hydroxylation, 75–80
    oxidative ring cleavage, 82–85
Oxidized rocket propellants/energetics, 11
Oxidizers, functions of, 354
Oxidizing agents, physicochemical treatment, 357
Oxidoreductases, 77
Oxyanions, 100, 440
Oxygen
    alternative sources of, 387
    biodegradation process
        BTEX, 266–268, 271
        generally, 227
    bioremediation processes, 363

dechlorination process, 272
depletion of, 486
sparging, 432
treatment trains, 430
water recirculation system, 387–388
Oxygen Releasing Compound (ORC^R), 378,
    387–388
Oxygenase enzymes, 77, 82–83
Oxygenation, air sparging system, 395
o-Xylene, 5, 139

Parathion, 87
Parent-daughter half-lives, 34
Parsimony, 218
Partial differential equation (PDE)
    groundwater flow, 134
    hydraulic connectivity calculation, 304
    numerical fate-and-transport models, 205, 212
Partitioning
    analytical fate-and-transport model, 194
    coefficient
        contaminant transport modeling, 324–325
        contamination scenario parameters, 183
    sorption processes, 39
Pasteur, Louis, 13–15, 77
Pasteurization, 13
Patents, 14
p-Chlorobiphenyl, 76
p-Cresol, 82
Peat
    biobarrier systems, 404
    characteristics of, 40
    organic carbon content, 326
    porosity and density values, 314
Peclet number, 147, 330–331
Penicillin, 14
Pentachlorphenol (PCP), 43, 75, 139, 325, 440
2,3,4,5,6-Pentachlorocyclohexene (PCA), 90–91
Perchlorate, 11, 69
Perchloroethylene (PCE)
    bioaugmentation system, 416
    biodegradation process, 5, 69
    bioremediation technologies, 360
    biotransformation pathways, 406
    calibrated degradation constants, 278
    dehydrohalogenation, 90
    first-order rate coefficients, 245
    groundwater flow, 159
    hydrogen biosparging, 407
    natural attenuation, 485–487
    numerical fate-and-transport models, 204, 227,
        244–248, 268, 277–278
    phytovolatilization process, 367
    plume mass, 278

reductive dechlorination, 86–87
retardation coefficient, 323
sequential anaerobic degradation of,
244–248
source mass release rates, 278
treatment train case study, 430
Percolation, heavy metals and, 368
Perfluorinated octanes (PFOs), 10–11
Performance assessment, 514
Performance monitoring, long-term, 462
Permeability
air sparging system, 395–396
biodegradation kinetics, 106
biodegradation process and, 73
bioventing systems, 383
estimation of, *see* Grain size analysis
numerical fate-and-transport models, 204
Permeable reactive barriers (PRBs), 355–356
PEST, 220
Pesticide(s), *see specific types of pesticides*
applications to, 3–4, 9–10, 29
biodegradation process and, 59, 72–73
chlorinated, 60
correlation estimates, 326
detoxification strategies, 81, 90
epoxidation of, 82
hydrolysis, 89
mineralization of, 93
nitro-group reduction, 87
partitioning coefficients, 325
solubilities, 325
synthetic reactions, 91
Petrachloroethylene, (PCE), 7
Petroleum
abandoned wells, 3
exploration, 352
hydrocarbons, *see* Petroleum hydrocarbons
numerical fate-and-transport prediction models,
213–214
product releases, 337
production, 370
Petroleum hydrocarbons
characteristics of, 4–7
ex situ land farming case study, 426–429
natural attenuation, 482–487
stable isotope analysis, 496
numerical fate-and-transport prediction model,
227
pH
adjustment of, 356, 363
bioaugmentation systems, 411
in biodegradation process, 70
ex situ composting, 424
in situ bioremediation system, 379

significance of, 26, 359, 370
treatment trains, 432
Phaenerochaeta, 423
Pharmaceuticals, as pollutant, 9, 12
Phase change processes, 37–44
Phenanthrene, 5, 61, 367, 494
Phenol(s)
biodegradation, 26, 43, 59, 61, 139, 360, 503
cometabolism, 406
partitioning coefficient, 325
stable isotope analysis, 496
Phenolic groups, 40
Phenolic pollutants, 7
Phenotypic potential, 506
Phenyl hydroxylases, biodegradation potential,
505
Phenylureas, 326
Phosphatases, 88
Phosphate(s)
bioremediation technologies and, 43, 370
esters, 77
mass flux calculation, 144
Phosphoanhydride bonds, 52
Phosphoester bonds, 88
Phospholipid(s)
bioremediation system, 507
ester-linked fatty acid (PLFA) analysis, 499
Phosphorous
bacterial storage, 51
biodegradation process, 21, 41, 69, 72
land farming, 427
production, 390
Photochemical reactions, degradation process, 160
Photographic emulsions, 370
Photolysis, 25, 160, 170
Phototrophs, 53
Phreatophytes, hydraulic control process, 366
Physicochemical treatment, 354–357
Phytodegradation, 365–366
Phytoextraction, 364–365
Phytoremediation
advantages of, 364
applications, 353, 364
hydraulic control, 366–367
phytodegradation, 365–366
phytoextraction, 364–365
phytostabilization, 368
phytovolatilization, 367
rhizofiltration, 365
rhizoremediation, 367–368
Phytostabilization, 365, 368
Phytotoxins, 90
Phytovolatilization, 365, 367
Piezometer, applications of, 122, 268, 284

Pigment production, 370
Pilli, in bacterial cell, 51
Pilot land treatment unit (PLTU), 426
Pipelines, damaged
    leaking, 4
    spills, 14
Pits, unlined, 3–4
Plant-based bioremediation, 364–368
Plant decay, 37
Plasmid(s)
    DNA, 501
    functions of, 50, 76
Plastics, breakdown of, 160
Plug-flow behavior, 162
Plume
    bioremediation process, 357
    capture, 367
    contaminant, 117–118
    contaminant transport and reactions, 175
    defined, 117
    delineation, 372
    groundwater flow, 125
    instantaneous point source, 182
    management with monitored natural attenuation
        (MNA), 435, 438
    migration
        analytical models, 179–180
        biobarrier systems and, 403–405
        hydraulic control of, 374–375
    radioactive, 187
    stable, 466–467
    steady-state, 194–195
    trends, 466–467
    transient behavior, 179
Plume length
    dispersivity and, 328
    hydraulic connectivity and, 329
    sensitivity analysis
        change in, influential factors, 191–192
        implications of, 322
        increase in, 190–191
Plutonium, 30, 32, 43, 441
PM-1, mineralization, 96
Point of compliance (POC)
    analytical fate-and-transport models, 175, 179
    numerical fate-and-transport models, 201–202
Polishing step, in physicochemical treatment,
        354
Pollutants, resistance to hydrolysis, 26.
    See also specific types of pollutants
Pollution prevention strategies, 351
Polonium, 30
Polyacetate esters, 406
    in bioremediation processes, 360

Poly-β-hydroxybutyrate (PHB), 51
Polybrominated diphenyl ethers (PBDEs), 11
Polychlorinated biphenyls (PCBs)
    biodegradation potential, 503
    biodegradation process, 8, 60–61, 63, 352, 362,
        442, 481
    mineralization of, 93
    monitoring natural attenuation, 440
    rhizoremediation process, 367–368
    stable isotope analysis, 496
Polychlorinated compounds, biodegradation of,
        93
Polycyclic aromatic hydrocarbons (PAHs)
    biodegradation potential, 503
    characteristics of, 63, 92, 352, 440, 442
    land farming case study, 426–429
    natural attenuation, 482
    phytodegradation process, 366
    rhizoremediation process, 367–368
    stable isotope analysis, 496, 498
Polycyclic hydrocarbons, 26, 59
Polyhalogenated alkanes, 27
Polymerase chain reaction (PCR), 507–509
Polymers, 12
Polynuclear aromatics, 5, 7–8, 326, 340
Polysaccharides, 90, 160
Polytrichous bacteria, 50
Polyvinyl chlorides (PVCs), 9
Ponds, 3, 219
Porosity
    analytical fate-and-transport models, 192
    contaminant transport modeling, 324
    groundwater flow modeling
        effective, 314, 321–322
        total, 314, 321–322
    hydraulic conductivity, 127
    numerical fate-and-transport prediction models,
        204, 215, 240
    parameter determination for contaminant
        fate and transport models, 283
    site investigation, 372
Postaudit, numerical fate-and-transport models,
        224
Potassium
    diphosphate, 390
    land farming, 427
Potential energy, 120
Precipitation
    analytical fate-and-transport models,
        174
    biodegradation process, 73, 75
    heavy metals, 353
    implications of, 37–38, 41–42
    reactions, 25

Prediction process, numerical fate-and-transport models, 223–224
Predictive simulations, 223
Preferential channels, development of, 479
Pressure head, groundwater flow, 121–122
Pressure measurements, groundwater flow modeling, 286
Primary lines of evidence, 459
Priority pollutants, 36–37, 59, 64
Probabilities, numerical fate-and-transport prediction models, 225
Promoters, in gene expression, 55–56
Propane
  bioventing system, 384
  cometabolism, 406
  implications of, 360
  monooxygenase, 61
Propionate, interspecies hydrogen transfer, 97–98
Protein(s), *see* Amino acids
  biodegradation process, 70, 160
  in hydrolysis, 90
Pseudomonas, 61, 75, 82, 93, 359, 363, 413, 502
Public drinking water systems, 2
Public health, 15
Pulse-type loading, 173
Pump-and-treat (P&T) systems
  analytical fate-and-transport models, 174
  cost factor, 358
  fuel spills, 380
  implementation of, 17, 19–20, 353, 361, 376, 386, 419
Pumping test techniques
  hydraulic connectivity
    aquifer parameters, determination using the Neuman method, 320–321
    characteristics of, generally, 292, 303
    flow to well in confined aquifer, 303–304
    semilogarithmic Jacob–Cooper method, 309, 313–320
    Theis solution, 304–309
    Theis type curve model, 306–307, 310–312, 315
  numerical fate-and-transport prediction models, 220, 223
Pumping well(s)
  analytical fate-and-transport model, 124, 192
  hydraulic connectivity, 316–317
Pure cultures, bioaugmentation systems, 412
Pure oxygen, in bioremediation processes, 360
Push-pull tests, biodegradation rates, 333, 340–342, 458
p-Xylene, 4–5, 139
Pyrene, 5, 139, 428

Pyrite, 36–37, 86
Pyruvate, 87

Quadrilateral finite elements, numerical fate-and-transport prediction model, 212

Rabies vaccine, 14
Racemases, 78
Rad (rad), 31
Radial groundwater flow, 298
Radiation units, 31
Radioactive decay
  analytical fate-and-transport models, 187–188
  common problems, 30
  detrimental effects, 31
  exponential, 32
  measurement units, 31
  numerical fate-and-transport prediction models, 226
  rates, 31
  secular equilibrium, 33–34
  types of, 31
Radioactive waste, 186, 363
Radiofrequency heating, 357
Radionuclides, 4, 31, 354, 358, 364, 440, 481, 487
Radium, 30, 34
Radon, 30, 32, 34
Rainfall, 115–116
Rank sum value, 555
Raoult's Law, 152
Rate-limited sorption reaction, 248–252
Raymond, Richard, 14, 385
Raymond process, 385
RCRA (1976), 17
RDX, *see* Hexahydro-1,3,5-trinitro-1,3,5-triazine (RDX)
Reaction kinetics, numerical fate-and-transport prediction models, 223, 226
Reaction rates, 274
Reactive barrier, 357
Reactor design, 359. *See also* Bioreactors
Real-time quantitative polymerase chain reaction (RTQ-PCR), 508–509
Recalcitrance
  bioaugmentation systems, 411
  in biodegradation process, 62–63, 72–74
  impact of, 356, 359
Recharge
  defined, 116
  rates, numerical fate-and-transport prediction models, 215, 222–223, 225, 254–256, 271
Recycling, of environmental pollutants, 49

Redesign process, numerical fate-and-transport
    models, 224–225
Redi, Francesco, 13
Redox
    active reagents, 357
    conditions, 17
    manipulation, 357
    potential, 356
    processes associated with hydrogen
        concentrations, 486
    reactions
        balancing, 536
        biodegradation, 228
        heavy metals and, 370
        natural attenuation, 486
Redox-sensitive pollutants, 36
Reducing/reduction
    agents, 353
    of contaminants, 353, 355
Reductionism, 445
Reductive biotransformation
    characterized, 85, 88
    nitro-group reduction, 87–88
    reductive dehalogenation, 85–87
Reductive transformation, biodegradation process,
    85–88
Redundancy, metabolic, 497
Refinery wastes, 4
Regression statistics, contaminant transport
    models, 344–345
Regulator, in gene expression, 56
Regulatory bodies, development of, 17
Reinjection wells, 386
Rem (rem), 31
Remediation, 17, 162. *See also* Bioremediation
Remobilization, 487
Representative elementary volume (REV), 149
Repression, in gene expression, 56–57
Research philosophies, 445–448
Resistance, biodegradation kinetics, 102
Respiration
    in bacterial cell, 51
    in biodegradation process, 64
    processes, 360
Retardation
    analytical fate-and-transport models, 169
    coefficient
        contaminant fate and transport model,
            322–327
        parameter determination for contaminant
            fate and transport models, 283
    factor
        benzene, 184
        contaminant transport modeling, 331

    numerical fate-and-transport prediction
        models, 215, 271, 275
    groundwater flow, 136, 154–159
    in sensitivity analysis, 191
    stable isotope analysis, 497
Reverse osmosis, 353
Rhizofiltration, 365
Rhizoremediation, 367–368
Rhodococcus, 61
Riboflavin, 14
Ribonucleic acid (RNA)
    in bacterial cell, 50
    extraction of, 488
    polymerase, in gene expression, 55–57
    ribosomal (rRNA), 55, 488, 499, 501,
        511, 512
Ribosomes, in bacterial cell, 50
Rising head test, 297
Risk assessment
    analytical fate-and-transport models, 192
    groundwater contamination, 162
    numerical fate-and-transport prediction model,
        202
    significance of, 185
Risk-based corrective action (RBCA)
    analytical fate-and-transport model, 193
    importance of, 17–19, 439, 442
    intrinsic bioremediation, 461, 475
    numerical fate-and-transport prediction model,
        202
Risk protection, 358
River(s)
    boundary conditions illustration, 172–174
    finite element approach, 212
    groundwater flow, 124
    numerical fate-and-transport prediction case
        study, 215, 225–226, 255
    sediment deposits, 326
RNA, *see* Ribonucleic acid (RNA)
Robustness, 497, 514
Rock(s), *see specific types of rocks*
    fractured, 137, 322
    hydraulic conductivity, 366
Rodenticides, 9
Roentgen (R), 31
Roman Empire, 13, 15
RT3D code, 204–205, 226–231, 240, 247, 249,
    269, 548
RT3E (reactive transport in 3 directions),
    biodegradation process, 162
Rubber production, 370

Safe Drinking Water Act (SWA) of 1974, 16
Salicylate, 76

Saltwater intrusion, 3
Sand(s)
    bioaugmentation case study, 416
    organic carbon content, 326
    porosity and density values, 314
Sandstone
    aquifer, groundwater flow, 128–129
    hydraulic connectivity, 329
    porosity and density values, 314
Saturated
    aliphatic compounds, 92
    groundwater flow equation, 131–135
    hydrocarbons, 59
    media, solute transport in, 149–151
    zones, 115–116
Sawdust, biobarrier systems, 404
Scale
    effect, groundwater flow, 130
    monitoring, 479
Scotchgard$^{\circledR}$, 10
SEAM3D, 548
Seasonal effects, 478
Secondary lines of evidence, 459
Sediment contamination, physicochemical
        treatment, 357
Seed(s)
    bacteria, 363
    biobarrier systems, 404
Seepage
    numerical fate-and-transport prediction models,
        232
    velocity, 127, 334–335
Selenium, 4, 370, 441
Semiquinones, 37
Semi-volatile compounds, 11
Sensitivity analysis
    analytical fate-and-transport models, 189–192,
        197
    BTEX degradation, 268, 271
    hydraulic connectivity, 322
    numerical fate-and-transport model, 222, 247,
        256–257, 268, 279
Septic systems, 3
Sequential degradation, 94, 269
Sequential hydrolysis, 27
Sequestration reactions, 40
Serpentines, in phytoextraction process, 364
Sewage lagoons, 174
Sewer system leaks, 32
Shale
    hydraulic conductivity, 366
    porosity and density values, 314
Shepherd method, hydraulic connectivity
        determination, 295–296

Shewanella, 99–100, 359
Sigma factor-DNA complex, 56
Silica minerals, 44
Silt
    bioaugmentation case study, 416
    land farming, 427
    organic carbon content, 326
    porosity and density values, 314
Siltstone, porosity and density values, 314
Simulation, *see specific types of simulations*
    analytical fate-and-transport models, 189–192
    numerical fate-and-transport models, 204,
        274–275
Single-well "push-pull" test, 340, 342
Sink(s)
    analytical fate-and-transport models, 174
    biodegradation process, 333
    numerical fate-and-transport prediction models,
        214, 219, 226, 230
Site(s)
    characterization, 461
    cleanup, cometabolism for, 60
    conceptual model development, numerical
        fate-and-transport models, 216–218
    data, numerical fate-and-transport prediction
        models, 217
    fuel-contaminated, 64, 66, 117
    investigation, 372–374
    measurement, 214
    migration and, 192
    remediation
        biological treatment, 356
        implementation of, 26
        physicochemical treatment, 353–357
    risk assessment of, *see* Risk assessment
    -scale models, data collection and assessment,
        214–216
    -specific risk, 175
    -specific standards, 18
    Superfund, 2, 352, 435
Skin penetration, of radiation, 31
Slug test analysis, hydraulic connectivity
        determination
    Bouwer and Rice method, 298–303
    characteristics of, 292, 296–298
Slurry (batch) reactors, 419
Slurry bioreactors, 363, 420–421
Smith, Angus, Dr., 15
Snow, John, Dr., 15
Sodium monophosphate, 390
Soil
    bioremediation technologies, 363
    contaminated
        characteristics of, 353–354, 363

Soil (*continued*)
    physicochemical treatment, 357
    sources of, 5
    gas vapors, 375–376
    hydrocarbon-contaminated, 363
    materials, sorbent processes, 40–41
    monitoring by volatilization, 42
    subsoils, 69
    subsurface,192
    surfaces, complexation process, 44
    vapor extraction (SVE), 383–384, 392, 419
    vapor flow, 381
    washing, 15, 40, 353, 357, 421
Solar batteries, 369
Solar radiation, 160
Solid phase bioremediation, 363
Solid-water interface, 38
Solubility, 37
Solute transport in saturated media, 149–151
SOLVE FOR, 184
Solvents
    chlorinated, 117, 152
    types of, 12
SOLVER, 184
Sorbents, natural, 40–41
Sorption
    analytical fate-and-transport model, 192, 194
    biodegradation process, 73–75, 333
    characteristics of, 7, 12, 25–26, 37–41
    contaminant transport, 323, 334
    contaminated migration, 162
    natural attenuation, 487
    numerical fate-and-transport model, 220, 223,
        226–227, 248–252
    site investigation, 373
Source
    analytical fate-and-transport models, 174
    bioremediation processes, 357
    contaminant transport, 334, 336
    depth, contamination scenario parameters,
        183
    removal, 374
    terms, numerical fate-and-transport prediction
        models, 214, 219, 226
    width, contamination scenario parameters,
        183
    zone
        air sparging case study, 400
        concentration, multidimensional transport
            analytical models, 177
        in situ bioremediation, 362
        numerical fate-and-transport prediction
            models, 214, 216, 230, 272
        site remediation, 353

Sparge curtains, *see* Biobarriers
Spatial distribution, numerical fate-and-transport
    prediction models, 216, 219, 243
Spatial variability
    analytical fate-and-transport models, 192
    groundwater flow, 125–126, 128
Specific adsorption, 39
Specific discharge, groundwater flow, 126
Specific storage, groundwater flow, 135
Sphingomonas, 61
Spills, impact of, 3. *See also specific types of
    spills*
Spodic soil materials, 41
Spontaneous generation, 13, 15
Sprayers, bioremediation processes, 363
Spreading prevention measures, 374–377
Spreadsheet applications
    analytical fate-and-transport models, 177, 184,
        186–187
    biodegradation rate determination, 342
    contaminant concentrations, 175, 344
    diffusion coefficients, 140
    type curve matching, 310
Sprinkler systems, 363
Square grid, numerical fate-and-transport
    prediction models, 209
Stable isotope analysis, 458, 489–493
Stagnant film model, groundwater flow, 153
Starved bacteria, 415
Statistical analysis, 458, 463–467
Steady flow, analytical fate-and-transport model,
    194
Steady state
    analytical fate-and-transport model, 185,
        194–195
    biodegradation kinetics, 106
    in biodegradation process, 170
    concentration, 179–180
    contaminant transport, 334
    groundwater flow, 134
    models, application of, 179
    numerical fate-and-transport prediction models,
        219–220, 232, 274
    plume, determination techniques,
        342–345
    in sensitivity analysis, 189–190
Steam injection, 357
Sterility, 31
Steroids, 12
Stimulatory material delivery system, 377–379
Stochastic(s)
    analytical fate-and-transport models, 174
    numerical fate-and-transport prediction models,
        219, 225

Stoichiometric
  applications, 541
  ratios, 469–470
Storage material, in bacterial cell, 51
Storativity
  determination using the Theis method,
    307–309
  numerical fate-and-transport prediction models,
    215, 222
  parameter determination for contaminant fate
    and transport models, 283
Strain-specific threshold, in biodegradation, 64
Stratified soils, 362
Streams, 215, 219, 226
Streptomyces, 95
Stripping
  air, 353, 367, 385
  groundwater, 395
  volatile organic compounds (VOCs), 379
Strontium, 4, 32, 78, 187–188, 441
Structure-activity correlations, in biodegradation,
  91
Student *t*-distribution, 468, 557–558
Styrene, 139, 494
Subarctic climate, biodegradation process, 70
Subsoils, biodegradation process, 69
Subsurface
  environments, sorbent processes, 40
  remediation, 385
  soil, plume length and, 192
Successive overrelaxation, 209
Sugars
  biodegradation process, 91
  in hydrolysis, 90
Sulfate
  biobarrier systems, 403
  biodegradation of, 41–42, 64, 66–67, 72, 227
  BTEX degradation, 238, 240, 266–268, 271
  cometabolism, 406
  depletion of, 486
  in situ bioremediation system, 378
  natural attenuation, 270, 483
  -reducing bacteria, 371
Sulfides, 36–37, 88
Sulfonic acids, 26
Sulfonylureas, 29
Sulfoxides, biotransformations, 88
Sulfur
  biosparging system, 408
  land farming, 427
Sunflower, in phytoextraction process, 364
Sun Oil pipeline spill, 14
Superfund Amendment and Reauthorization
    Act (SARA), 19

Superfund program (1980), 17–19
Surface
  adsorption, precipitation process, 42
  complexation, 39
  coordination reactions, 38–39
  hydroxyl groups, 36
  soil, bioaugmentation systems, 411
  spills, 3
  water
    finite element approach, 212
    numerical fate-and-transport prediction
      models, 215, 222
Surfactant(s)
  bioaugmentation systems, 415
  flooding, 357
Surfer, computer software program, 465
Suspended-growth reactors, 418
SUTRA, 548
SWMS-2D, 548
Synthetic musk fragrance, 11
Synthetic organic chemicals, 2, 88–89, 160
Synthetic reaction, in biodegradation process, 91
Syntrophism, 95–96
Systems analysis, 359

Table scraps, mineralization of, 93, 95
Taq polymerase, 508
Target pollutants, in biodegradation process, 63
Taylor's expansion, 208
TBU pathway, microbial degradation, 82, 84
Technetium, 4, 441
Tecplot, computer software program, 465
Temperature
  adjustment of, 363
  bioaugmentation systems, 411
  in biodegradation process, 70–73
  bioremediation systems, 447
  ex situ composting, 424
  significance of, 20–21, 359
Temporal distribution, numerical fate-
    and-transport prediction models, 216
Temporary recalcitrance, 73
Terminal electron-accepting processes (TEAPs),
    484–485
Terminal restriction fragment length
    polymorphism (T-RFLP), 507, 511–512
tert-butyl alcohol (TBA), 490
Tertiary lines of evidence, 459
Tetrachloride, monitoring natural attenuation, 440
Tetrachlorodibenzofurane, 440
1,1,2,2-Tetrachloroethane, 139
Tetrachloroethene, 405, 412, 493
Tetrachloroethylene, 139, 325
Tetraethyl lead, 370

Tetramethyl lead, 370
Textile production, 370
T4MO pathway, microbial degradation, 82, 84
Theim equation, 298
Theis solution, 304–306
Theis type curve model, 306–307, 310–312, 315
Thermal desorption, 15, 353, 355
Thermal extraction, 357
Thermodynamics
    in biodegradation process, 65–67, 69, 93
    bioremediation processes, 360
    feasible reactions, 35, 98, 542
    physicochemical treatment, 354–355
    unfeasible, 16
Thermus aquaticus, 508
Thioethers, 61
Thiol groups, 78
Thiomargarita namibiensis, 51
Three-dimensional continuity equation, 134
Thymine, 54–55
Tier 1 risk-based corrective action (RBCA) investigation, 18–19
Tier 2 risk-based corrective action (RBCA) investigation, 18–19, 202–203
Tier 3 risk-based corrective action (RBCA) investigation, 18–19, 202–203
Till Hydrology Research Site (THRS), 288–289
Tilling, in land farming, 427
Tin, 42
TMVOC, 548–549
TOD pathway, microbial degradation, 82, 84
TOL pathway, microbial degradation, 82, 84
Toluene
    biodegradation potential, 503
    catabolism of, 84
    characteristics of, 4–5, 7, 62, 67, 139, 360
    cometabolism, 406
    dioxygenase, 61, 77, 82
    fuel spills, 380
    monooxygenase, 61
    natural attenuation, 491
    partitioning coefficient, 325
    stable isotope analysis, 494
    treatment train case study, 432–433
Toluene/m-p-xylene, 76
TOM pathway, microbial degradation, 82, 84
Total organic carbon (TOC), 427
Total petroleum hydrocarbons (TPHs), 427
Total porosity
    contaminant transport modeling, 324
    contamination scenario parameters, 183
    groundwater flow modeling, 321–322

hydraulic connectivity determination
    implications of, 321–322
    sensitivity analysis, 322
    values of, 314
Total reduction capacity (TRC), 35–36
Total variation diminishing (TVD), 226, 228
Toxaphene, 6, 9
Toxic organic pollutants, 352
Toxic Substance Control Act (TSCA) of 1976, 16
Toxic substances, biodegradation processes, 72
Trace elements, land farming, 427
Trace metals, 21
Trametes hirsuta, 423
trans-1,2-Dichloroethylene, 138
Transcription, 54–55, 506
Transferases, 78
Transfer RNA (tRNA), 55
Transient advection-dispersion equation, numerical fate-and-transport prediction model, 209–210
Transient flow, groundwater flow, 135
Transition metal cations, 40
Transition metals, 39
Translation, 55
Transmissivity
    contaminant transport models, 319
    determination using Neuman type curve method, 320–321
    determination using Theis method, 307–309, 313
Transpiration
    hydraulic control and, 366
    rate, 366
Transport
    advective, in groundwater flow, 143
    analytical models, *see* Analytical fate-and-transport models
    modeling, in biodegradation process, 108–109
    multidimensional transport from finite, planar continuous source of contamination
        under steady-state conditions, 179–180
        under transient conditions, 176–179
    multidimensional transport from finite, planar, decaying source of contamination, under transient conditions, 180–181
    multidimensional transport from an instantaneous point source of contamination under transient conditions, 181–182
    processes, 21, 142
Transposons, biodegradation process, 75
Transverse dispersion, 176
Transverse dispersivity, 145, 147–148, 231, 329

Travel time, groundwater flow, 161
Treatment trains, 429–433
Trench biosparge, see Biobarriers
Triaminotoluene (TAT), 88–89
Triangular finite elements, numerical
    fate-and-transport prediction model,
    206–207, 212
Triazines, 326
Trichloracetic acid, biodegradation process, 95
1,2,4-Trichlorobenzene, 139
1,1,1,-Trichloroethane (1,1,1,-TCA), 5, 8, 26–27,
    139, 245
1,1,2-Trichloroethane, 139
Trichloroethene (TCE)
    biaugmentation system, 411, 415–418
    biodegradation process, 5, 43, 7–8, 63, 69,
        72–73, 77, 139, 325, 352, 405, 440
    biotransformation pathways, 406
    bioventing system, 384
    calibrated degradation constants, 278
    cometabolism, 61
    epoxidation, 81–82
    first-order rate coefficients, 245
    geochemical analysis, 481
    hydrogen biosparging, 407, 409–410
    intrinsic bioremediation, 464
    natural attenuation
        data analysis, 464–465, 485
        monitoring, 438
        performance assessment, 480–482, 286
    numerical fate-and-transport models, 204, 227,
        244, 246–247, 268, 272, 277–280
    oil spill in alluvial aquifer, diffusion coefficient
        example, 142
    phytovolatilization process, 367
    plume mass, 278–280
    reductive dechlorination of, 59, 86–87
    spill in fractured bedrock, diffusion coefficient
        example, 142
    stable isotope analysis, 494
    treatment train case study, 430, 433
Trichlorofluoromethane, 245
Trickling filters, 356, 418
Trimethylbenzene (TMB), biodegradation rate
    characteristics of, 332, 337–338
    determination by normalization method,
        338–342
Trinitrotoluene (TNT)
    biodegradation process, 69, 423, 440
    biotransformations, 87–88
    composting case study, 424–425
    nitro-group reduction, 87–88
2,4,5-Triphenoxyacetic acid (2,4,5-T), acclimation
    phase of, 75

Tritium, 32
TSCA (1976), 17
Type B curves, 318

U(IV), 359
U(VI) reduction, 358–359
UCODE, 220
Ultramicrobacteria, 415
Ultraviolet range, 160
Uncertainty principle, 446–448
Underground storage tank
    leaking, see Leaking underground storage ta█
        (LUST)
    overfilling, 189
Uniform flow, analytical fate-and-transport
    model, 194
Unit thickness, numerical fate-and-transport
    prediction models, 215
United Nations Environment Programme (UN█
    "dirty dozen," 9
U.S. Environmental Protection Agency (USE█
    2, 7, 12, 344, 406, 418, 443–444, 463, ◂
    514
U.S. Geological Survey, 225, 227
U.S. Navy National Environment Test Site,
    Strategic Environmental Research and
    Development Program (SERDP), 397–3█
University of Minnesota Biocatalysis/
    Biodegradation Database, 76–77
Unsaturated aliphatic compounds, 92
Unsaturated hydrocarbons, 59
Unsaturated soil zone, biodegradation proces█
Unsaturated zones, 115–116
Unstable biological molecules, 31
Upgradient wells, 469
Uracil, 55
Uranium
    biotransformation of, 100
    characteristics of, generally, 4, 30, 36, 41█
        441, 467
    complexation process, 43–44
    -contaminated groundwater, 358–359
    half-life of, 32
    waste, 359
Uranyl complexes, 43
Urea, 509
U statistic, 556
UTCHEM, 549

VAAP soil, 421–422
Vadose zone, 115–116, 215, 217, 376
Validation
    monitoring, 461
    numerical fate-and-transport model, 22█

380

of, 393

84
w, 152
57
oremediation processes, 360,

of
58
26–128, 141, 144–145,
472
323
ransport prediction model,
225, 232
fficient, 184

io parameters, 183

insport prediction models,

381–382
ctivity, 319, 321
nts

sport prediction models,

int transport and

, 70

e (VPDB), 490–491

415–416
, 7, 69
ays, 406
nstants, 278
ly, 5, 59, 61, 139, 352

358,

s, 245
7, 409–410
1d, 487
tion, 436, 440
37
rt models, 227,
3, 276, 278
5
367

30, 432–433

–223

Violets, in phytoextraction process, 364
Visual Modflow, 549
Vitamin(s)
  B12, 14, 95
  production in commensalism example, 95
Volatile organic chemicals (VOCs), 42–43, 367,
    377, 379, 392
Volatile pollutants, 367
Volatilization
  analytical fate-and-transport models, 170
  applications, generally, 7, 38, 42–43, 353–354
  biodegradation process, 73, 75
  groundwater flow equation, 151–152
  in treatment train, 433
Volume released, contamination scenario
    parameters, 183
Volumetric flow rate, 299
VSAF2, 550

Wastewater
  bioaugmentation treatment, 411
  organic contaminants, 10
Water
  depth, bioventing system, 383
  diffusion coefficients in, for ions, 138
  filtration, 381
  global distribution of, 2
  partitioning, 39
  pressure, 116
  quality, numerical fate-and-transport prediction
      models, 216
  recirculation systems
    biostimulation system, 380, 385–388
    monitoring recommendations, 388
  saturation, 116
  table
    contaminant concentration and, 180
    contaminant transport models, 316
    groundwater flow, 124, 136
    hydraulic connectivity, 299, 316
    implications of, 116–117
    mass budget analysis and, 469
    wells, 2, 117
Watson, James, 14
Weathering processes, impact of, 75, 181
Wells
  finite element approach, 212
  groundwater flow modeling, 174, 284–290
  hydraulic connectivity calculation, 298–299
  numerical fate-and-transport prediction models,
      214, 225
  pumping tests
    confined aquifer, 303–309, 315
    unconfined aquifer, 315–318, 320–321

Wetlands, analytical fate-and-transport models, 174
Wet oxidation, 357
Whipple, George C., 16
Willows, hydraulic control process, 366
Windrow composting, 424–425
Wisconsin Department of Natural Resources (WDNR), 344
Wood chips, bioaugmentation systems, 412
Wood preservatives, 369

Xenobiotics, 74, 91, 366
Xenophores, 62, 83, 90
Xylene, *see* BTEX
    characteristics of, 7, 62

dioxygenase, biodegradation potential, 503
fuel spills, 380
RBCA Tier 3 modeling LUST case study, 252, 257–258
treatment train case study, 432–433
Xylene/toluene, 76

Yeast, 13

Zeolites, sorption processes, 40
Zero-order kinetics, 195
Zinc, 42, 70, 99, 370–371, 440
Zone of incorporation (ZOI), land farming case study, 427

ENVIRONMENTAL SCIENCE AND TECHNOLOGY

A Wiley-Interscience Series of Texts and Monographs

Edited by JERALD L. SCHNOOR, *University of Iowa*
    ALEXANDER ZEHNDER, *Swiss Federal Institute for Water Resources and Water Pollution Control*

PHYSIOCHEMICAL PROCESSES FOR WATER QUALITY CONTROL
    Walter J. Weber, Jr., Editor
pH AND pION CONTROL IN PROCESS AND WASTE STREAMS
    F. G. Shinskey
AQUATIC POLLUTION: An Introductory Text
    Edward A. Laws
INDOOR AIR POLLUTION: Characterization, Prediction, and Control
    Richard A. Wadden and Peter A. Scheff
PRINCIPLES OF ANIMAL EXTRAPOLATION
    Edward J. Calabrese
SYSTEMS ECOLOGY: An Introduction
    Howard T. Odum
INTEGRATED MANAGEMENT OF INSECT PESTS OF POME AND STONE FRUITS
    B. A. Croft and S. C. Hoyt, Editors
WATER RESOURCES: Distribution, Use and Management
    John R. Mather
ECOGENETICS: Genetic Variation in Susceptibility to Environmental Agents
    Edward J. Calabrese
GROUNDWATER POLLUTION MICROBIOLOGY
    Gabriel Bitton and Charles P. Gerba, Editors
CHEMISTRY AND ECOTOXICOLOGY OF POLLUTION
    Des W. Connell and Gregory J. Miller
SALINITY TOLERANCE IN PLANTS: Strategies for Crop Improvement
    Richard C. Staples and Gary H. Toenniessen, Editors
ECOLOGY, IMPACT ASSESSMENT, AND ENVIRONMENTAL PLANNING
    Walter E. Westman
CHEMICAL PROCESSES IN LAKES
    Werner Stumm, Editor
INTEGRATED PEST MANAGEMENT IN PINE-BARK BEETLE ECOSYSTEMS
    William E. Waters, Ronald W. Stark, and David L. Wood, Editors
PALEOCLIMATE ANALYSIS AND MODELING
    Alan D. Hecht, Editor
BLACK CARBON IN THE ENVIRONMENT: Properties and Distribution
    E. D. Goldberg
GROUND WATER QUALITY
    C. H. Ward, W. Giger, and P. L. McCarty, Editors
TOXIC SUSCEPTIBILITY: Male/Female Differences
    Edward J. Calabrese
ENERGY AND RESOURCE QUALITY: The Ecology of the Economic Process
    Charles A. S. Hall, Cutler J. Cleveland, and Robert Kaufmann
AGE AND SUSCEPTIBILITY TO TOXIC SUBSTANCES
    Edward J. Calabrese

ECOLOGICAL THEORY AND INTEGRATED PEST MANAGEMENT PRACTICE
Marcos Kogan, Editor
AQUATIC SURFACE CHEMISTRY: Chemical Processes at the Particle Water Interface
Werner Stumm, Editor
RADON AND ITS DECAY PRODUCTS IN INDOOR AIR
William W. Nazaroff and Anthony V. Nero, Jr., Editors
PLANT STRESS–INSECT INTERACTIONS
E. A. Heinrichs, Editor
INTEGRATED PEST MANAGEMENT SYSTEMS AND COTTON PRODUCTION
Ray Frisbie, Kamal El-Zik, and L. Ted Wilson, Editors
ECOLOGICAL ENGINEERING: An Introduction to Ecotechnology
William J. Mitsch and Sven Erik Jorgensen, Editors
ARTHROPOD BIOLOGICAL CONTROL AGENTS AND PESTICIDES
Brian A. Croft
AQUATIC CHEMICAL KINETICS: Reaction Rates of Processes in Natural Waters
Werner Stumm, Editor
GENERAL ENERGETICS: Energy in the Biosphere and Civilization
Vaclav Smil
FATE OF PESTICIDES AND CHEMICALS IN THE ENVIRONMENT
J. L. Schnoor, Editor
ENVIRONMENTAL ENGINEERING AND SANITATION, Fourth Edition
Joseph A. Salvato
TOXIC SUBSTANCES IN THE ENVIRONMENT
B. Magnus Francis
CLIMATE-BIOSPHERE INTERACTIONS
Richard G. Zepp, Editor
AQUATIC CHEMISTRY: Chemical Equilibria and Rates in Natural Waters, Third Edition
Werner Stumm and James J. Morgan
PROCESS DYNAMICS IN ENVIRONMENTAL SYSTEMS
Walter J. Weber, Jr., and Francis A. DiGiano
ENVIRONMENTAL CHEMODYNAMICS: Movement of Chemicals in Air, Water, and
Soil, Second Edition
Louis J. Thibodeaux
ENVIRONMENTAL MODELING: Fate and Transport of Pollutants in Water, Air, and Soil
Jerald L. Schnoor
TRANSPORT MODELING FOR ENVIRONMENTAL ENGINEERS AND SCIENTISTS
Mark M. Clark
FORMULA HANDBOOK FOR ENVIRONMENTAL ENGINEERS AND SCIENTISTS
Gabriel Bitton
BIOREMEDIATION AND NATURAL ATTENUATION: Process Fundamentals and
Mathematical Models
Pedro J. J. Alvarez and Walter A. Illman

CPSIA information can be obtained
at www.ICGtesting.com
Printed in the USA
BVHW042017161019
561322BV00010B/74/P